University Texts in the Mathematical Sciences

Textbooks in this series cover a wide variety of courses in mathematics, statistics and computational methods. Ranging across undergraduate and graduate levels, books may focus on theoretical or applied aspects. All texts include frequent examples and exercises of varying complexity. Illustrations, projects, historical remarks, program code and real-world examples may offer additional opportunities for engagement. Texts may be used as a primary or supplemental resource for coursework and are often suitable for independent study.

Dharmaraja Selvamuthu

Introduction to Stochastic Processes

Queues, Finance, and Credit Risk

Springer

Dharmaraja Selvamuthu
Department of Mathematics
Indian Institute of Technology Delhi
New Delhi, Delhi, India

ISSN 2731-9318 ISSN 2731-9326 (electronic)
University Texts in the Mathematical Sciences
ISBN 978-981-97-6151-7 ISBN 978-981-97-6152-4 (eBook)
https://doi.org/10.1007/978-981-97-6152-4

Mathematics Subject Classification: 60G05, 60Jxx, 60K15, 60K25, 60J80, 60J60, 60G55, 60H05

This Springer imprint is published by the registered company Springer Nature Singapore Pte Ltd.
The registered company address is: 152 Beach Road, #21-01/04 Gateway East, Singapore 189721, Singapore

Dedicated to my parents
K. Selvamuthu
and
S. Vellaiammal

Preface

கற்றது கை மண் அளவு கல்லாதது உலகளவு

Quote: ‘“Katrathu Kai Mann Alavu, Kallathathu Ulagalavu”

Meaning: What one knows is only a handful, whereas the unknown is size of the world.

— Avvaiyar (Tamil poet of Sangam Era)

The concept of stochastic processes plays a crucial role in numerous scientific domains and real-life applications. The journey of teaching stochastic processes started for me at IIT Delhi in the year 2003. This book consolidates my 30 years (1995–2024) of experience in learning, understanding and familiarity with stochastic processes.

Learning stochastic processes started for me by solving the problems from the book “A First Course in stochastic processes” by Karlin and Taylor along with my friend Dr. R. B. Lenin (Dillard’s Inc., USA). This was further strengthened by my doctoral and post doctoral research of stochastic processes and their applications in Chemical Kinetics, Fault-Tolerant systems, Cellular Networks (ranging from 2G to 6G) and Finance. This text book draws a notable inspiration from the book stochastic processes by Prof. J. Medhi in understanding the basic concepts. Prof. S. R. S. Varadhan’s Lecture Notes too gave an insight into understanding the theoritical aspects.

This text book serves as a first introduction to stochastic processes for both undergraduate and postgraduate students with a good background in Real Analysis, Linear Algebra and Differential Equations. Also, this book serves as an introduction to stochastic of finance for postgraduate and researchers who are familiar with stochastic processes.

The examples provided in this text book are drawn from real life applications in Natural Sciences, Engineering and Finance. This helps the students to understand the concepts better and also suggest a way of approach to real life problems. In my opinion, these real life examples set apart this book from the already existing ones on stochastic processes.

The book consists of fifteen chapters in total that is divided into two parts. The first part has nine chapters and gives an introduction to stochastic processes whereas the second part with six chapters dives into stochastic of finance.

The book is primarily intended as a freestanding textbook for a course on stochastic processes or a reference guide for any course in a similar fashion. It covers all the basic notations of probability theory and stochastic processes that are important for students to navigate the initial challenges during the undergraduate or postgraduate studies. This book illustrates more than 250 examples and 250 exercises. Appendices A.1–A.9, involving some basic concepts of probability, standard distributions, orthogonal polynomials, Bessel functions, Laplace transforms, continued fractions and determinant of tri-diagonal matrices are given for review purposes.

Chapter 1 deals with the basic concepts on stochastic processes. Some of the standard stochastic processes and its properties are discussed. The classification of stochastic processes are provided along with the fundamental properties such as independent increments, memoryless and stationarity. In Sect. 1.4, some of the important processes such as Poisson process, Gaussian process, and Brownian process are discussed.

The importance of Markov chains lies in the fact that a number of natural, physical, biological and economic phenomena can be described by them. Chapters 2 and 3 are devoted to an elaboration of the basic concepts and methods of discrete time Markov chains (DTMCs). The further sections contain limiting distribution, stationary distribution and steady-state analysis of the DTMCs. In Chap. 3, the reducible Markov chains and time-reversed Markov chains are discussed. At the end, discrete time branching processes are presented.

Chapters 4 and 5 give the basic concept of continuous time Markov chains (CTMCs). Section 4.1 gives the definition and characteristics of the CTMCs. The limiting and stationary distributions of the CTMC are presented in Sects. 4.2 and 4. 3 respectively. Finally, a section is devoted to continuous time branching processes and in particular, Bellman Harris branching process.

The fundamentals of birth and death processes (BDPs) have been analysed in Chap. 5. Various methods of finding time-dependent probabilities for finite state BDP are given. In Sects. 5.3 and 5.4, pure birth processes and pure death processes are discussed respectively. The basic concepts and properties of Poisson processes of time homogeneous and non- time homogeneous have been provided in Sect. 5.5.

Chapter 6 discusses various simple Markov queueing models. In Sect. 6.1, the fundamentals of queueing systems has been given. Further sections provide an insight into various queueing models with different number of servers, different service discipline, different size of queue in the system, etc (for example $M/M/1$, $M/M/1/N$, $M/M/c$, $M/M/c/N$, $M/M/c/c$ and $M/M/\infty$ queuing models are discussed). Both steady-state and time-dependent probabilities are derived for these queuing models and many real- life applications are provided along with their performance measures.

Chapter 7 provides a detailed analysis of the advanced queueing models. Section 7.1 describes Quasi-Birth-Death Processes (QBDs) which are the simplest and well- studied and most well-studied class of multi-dimensional continuous time Markov chains. The solution methodology called the Matrix Geometric Method has been provided to handle these queueing models. Section 7.2 discusses the retrial queuing models. Section 7.3 deals with balking and reneging queuing models. Section 7.4 describes bulk arrival and bulk service queueing models. Section 7. 5 discusses various vacation policies in queueing systems. In Section 7.6, time-dependent probabilities for the fluid queueing models are derived by two methods and also the steady-state probabilities are obtained in closed form.

Chapter 8 discusses certain processes that are non-Markov processes in general. Section 8.1 describes the point processes in terms of arrival times, inter-arrival times and counting process. Section 8.2 discusses the renewal process which is the generalization of Poisson process. Hawkes process that is a non-Markov extension of the Poisson process is discussed in Sect. 8.3. Sections 8.4 and 8.5 discuss respectively the basic concepts of semi-Markov process and Markov regenerative process. The steady-state and transient analysis for various stochastic models have been provided.

Chapter 9 provides a detailed analysis of non-Markov or general queueing models. Both transient and steady-state distributions are derived for various non-Markov queueing models such as $M/G/1$, $GI/M/1$, $GI/G/1$, $M/G/1/N$, $G/M/1/N$, $GI/G/1/N$, $M/G/c$, $GI/M/c$, $G/G/c$, $M/G/c/N$, $M/G/c/c$, and $GI/G/\infty$. Further, a good number of examples of the above non-Markov queueing models are included with retrial and vacation policies. $Hawkes/sdHawkes/\infty$ system is also presented at the end.

Chapter 10 presents in detail the Martingale property, one of the important properties of stochastic processes. This chapter starts with the definition and properties of conditional expectation. A few results related to the martingale such as inequalities, convergence theorem and optimal stopping theorem are given in detail.

Chapter 11 deals with the diffusion and jump-diffusion processes. It starts with the standard Brownian motion and its properties and later introduces the multivariate Wiener process. Then the multivariate Wiener process is introduced. Section 11.4 uncovers the Lévy process, the most generic process with stationary and independent increments whose paths are right continuous with left limits. Section 11.4 also introduces jump processes, point processes and compound Poisson processes. In Sect. 11.6, Hawkes processes that are non-Markov extension of homogeneous Poisson processes is discussed. The chapter concludes with the discussion of self exciting and mutually exciting models.

Chapter 12 describes the stochastic calculus for diffusion processes and jump-processes. In this chapter, the integrals and differentials of stochastic processes with respect to Wiener process are introduced. It starts with the quadratic variation of the stochastic processes followed by Itô process and Itô integral. It also discusses the Itô's formula for jump processes as well as multivariate processes. Itô's formula, also known as the change of variables formula, is an important one in the theory of stochastic integration.

Chapter 13 begins with stochastic differential equations and discusses the weak and the strong solutions. At the end, this chapter introduces various methods for solving the stochastic differential equations. In this section, the conditions under which a solution exists, existence of unique solution and the method to determine the exact solution are addressed.

In Chap. 14, the problem of valuation of financial derivatives is discussed. In this chapter, the underlying securities with various stochastic models such as diffusion models, jump-diffusion models are analysed. This chapter begins with various bond pricing such as Hull and White, Vasicek and CRR models. Then, the Black-Scholes option pricing formula is derived. Finally, option pricing greeks are presented.

Chapter 15 deals with applications of stochastic processes in modeling the credit risk, an important category of risk faced by financial institutions. Section 15.2 gives a brief overview of the objective in credit risk modeling, followed by a brief description of the two broad categories of stochastic models of credit risk, namely firm-value models and reduced-form models. In particular, applications of Markov chains, Markov regenerative processes, jump-diffusion models in credit risk are presented.

Topics such as hidden Markov model, Markov reward models, Markov decision processes, discrete queue, Inventory models, network of queues, simulation and Stochastic Petri nets are not discussed in this version and may be included in the later version.

I look forward to suggestions and comments towards the improvement of this book from students, teachers and researchers.

New Delhi, India Dharmaraja Selvamuthu
June 2025

Acknowledgments

This book is an outcome of the lecture notes given at IIT Delhi for a one semester course each on stochastic processes and Stochastic of finance.

I wish to express my deepest gratitude to my teachers Prof. P. R. Parthasarathy (IIT Madras, India), Prof. Kishor S. Trivedi (Duke Univeristy, USA) and Prof. Attahiru Sule Alfa (University of Manitoba, Canada) who taught me the fundamentals of stochastic processes.

I wish to acknowledge my research collaborators Dr. Viswanathan Arunachalam (Universidad Nacional de Colombia, Bogota, Colombia), Dr. Liliana Blanco (Universidad Nacional de Colombia, Bogota, Colombia), Dr. Paola Tardelli (Università degli Studi dell'Aquila, Italy), Dr. Vidyottama Jain (Central University of Rajasthan), Dr. R. Arumuganathan (PSG Tech., Coimbatore), Dr. R. Sudhesh (Anna University, Tiruchirappalli), Dr. N. Selvaraju (IIT Guwahati) and Dr. Guglielmo D'Amico (Università degli studi G. D'Annunzio Chieti Pescara, Italy) for their helpful suggestions on the first draft of this book. I thank Dr. Yogesh Mani Tripathi (IIT Patna) and Dr. Arivudainambi (Anna University, Chennai) for their comments.

I am grateful to Prof. A. Rangan (IIT Madras, India) for his encouragement, careful reading of the draft and his invaluable advice. Many thanks to my phd students Dr. Vaneeta Jindal, Dr. Vandana Khaitan, Dr. Shruti Kapoor, Dr. Puneet Pasricha and Dr. Anisha Aggarwal whose research works are included as real life applications in this book.

I gratefully acknowledge my current Post Doc student Dr. Raina Raj for her contribution in preparing the LATEX version of this book. I thank my previous Post Doc students Dr. Priyanka Kalita (Bhattadev University, India), Dr. Anubha Goel, Dr. Ankita Sen and Dr. Rasmi K, previous Ph.D. students Dr. Arti Singh, Dr. Nitu Sharma, Dr. Shakti Singh and Dr. Mohammad Sarfraz, current Ph.D. students Ms. Priya Mittal, Ms. Shamiksha Pandey and Mr. Utpal Jyoti Deba Sarma for carefully reading this version and giving useful suggestions.

Many thanks are due to Mr. Shamim Ahmad and the staff at Springer India for their highly professional yet sensitive management of this project. Also, I want to thank the IIT Delhi for providing an excellent working environment.

Finally, I would like to thank my wife K. Uma, my sons D. Kathiravan and D. Madhuvanth for their moral support throughout this long project.

Dharmaraja Selvamuthu
Department of Mathematics, Indian
Institute of Technology Delhi, New
Delhi, Delhi, India

Contents

Acronyms

a.a., a.e., a.s.	Almost all, almost everywhere, almost surely
BDP	Birth-death process
CDF	Cumulative distribution function
C-K	Chapman–Kolmogorov
CLT	Central limit theorem
CTMC	Continuous-time Markov chain
DTMC	Discrete-time Markov chain
EMC	Embedded Markov chain
FIFO	First in first out
IID	Independent identically distributed
LDP	Large deviation principle
LIFO	Last in first out
LT	Laplace transform
LST	Laplace–Stieltjes transform
MAP	Markovian arrival process
MC	Markov chain
MRGP	Markov regenerative process
MTTF	Mean time to failure
NPP	Non-homogeneous Poisson process
ODE	Ordinary differential equation
PDF	Probability density function
PDE	Partial differential equation
PGF	Probability generating function
PH distribution	Phase-type distribution
PP	Poisson process
QBD	Quasi-birth-death process
RV	Random variable/random vector
RW	Random walk
RNPM	Risk-neutral probability measure
SMP	Semi-Markov process
s.t.	Such that

SDE	Stochastic differential equation
SLLN	Strong law of large numbers
SPDE	Stochastic partial differential equation
STD	State transition diagram
TPM	Transition probability matrix
URW	Unrestricted random walk
w.r.t.	With respect to
WLLN	Weak law of large numbers

Mathematical Notations

$\mathcal{B}$	Borel σ—field
D	Deterministic distribution
$\mathcal{E}$	Event
E_k	Erlang type-k distribution
$\text{Exp}(\lambda)$	Exponential distribution with parameter λ
$\mathcal{F}$	σ-Algebra
G	General distribution
H_k	Hyper exponential distribution type-k
I	Indicator function
$N(\mu, \sigma^2)$	Normal distribution with parameters μ and σ^2—field
P	Probability function
$\mathcal{P}$	Power set of Ω
p_{ij}	One-step transition probabilities
$\mathbb{N}$	The natural numbers
Q	Infinitesimal generator matrix
r	Risk-free rate
$\mathbb{R}^n$	The n-dimensional Euclidean space
$\mathbb{R}^+$	The non-negative real numbers
$\mathbb{R}^{n\times m}$	The $n \times m$ matrices (real entries)
S	State space
S_t	Stock price value at time t
S_0	Initial stock price
v	Stationary probability vector
$\{W(t),\ t \geq 0\}$	Wiener process aka Brownian motion
$\{X_n, n = 0, 1, \ldots\}$	Discrete time stochastic process
$\{X(t),\ t \geq 0\}$	Continuous time stochastic process
$\mathbb{Z}$	The integers
$\pi(t)$	Time-dependent probability vector
π	Steady-state probability vector
$\lambda(0)$	Initial intensity at time $t = 0$
Ω	Sample space
ρ	Traffic intensity
η	Equilibrium level of the intensity process
δ	Constant mean-reversion speed of intensity process
σ	Volatility of the diffusion component
φ	Convex function
$(\Omega, \mathcal{F})$	Measurable space
$(\Omega, \mathcal{F}, P)$	Probability space
$(\Omega, \mathcal{F}, \{\mathcal{F}_t\}_{t\in T}, P)$	Filtered probability space
$\square$	Denoting end of proof
$\stackrel{d}{=}$	Equality in distribution

$\xrightarrow{d}$	Convergence in distribution
$\xrightarrow{a.s.}$	Convergence almost surely
$\xrightarrow{p}$	Convergence in probability
$\xrightarrow{r}$	Convergence in *r*-order moments

Chapter 1
Introduction to Stochastic Processes

A stochastic process is a process describing the evolution in time of a random phenomenon. It means that there is a system where observations are made at certain times, and the outcome, which is, the observed value at each time is a random variable (RV). Stochastic process is also called as random process or chance process. From a mathematical point of view, the theory of stochastic processes were started around 1950. Since then, stochastic processes have become a common tool for mathematicians, physicists, and engineers, and the fields of applications ranges from the modeling of communication systems, to stock pricing, and rational option pricing theory.

In this chapter, the basic concepts of stochastic processes are discussed. Further, the properties and some standard stochastic processes are also presented. The classification of stochastic processes are provided along with the fundamental properties such as independent increments, stationarity, memorylessness, and martingale. Additionally, some of the very important processes such as Poisson process, Gaussian process, and Brownian process are discussed.

1.1 Introduction

The field of stochastic processes is essentially a branch of probability theory, treating probabilistic models which mostly evolve with time. Stochastic process is considered as a branch of applied mathematics, starting with the axioms of probability and containing a rich and fascinating collection of results derived from those axioms.

Applying axiomatic probability concepts to a real-world application requires creating a probability model for the given application. The precise mathematical results are obtained for the model and translated back to the real-world applications. If the model fits the application sufficiently well, real-world problems can be solved by analysing the model. However, since models are always simplified approximations

D. Selvamuthu, *Introduction to Stochastic Processes*, University Texts in the Mathematical Sciences, https://doi.org/10.1007/978-981-97-6152-4_1

of reality, precise results for the model become approximations in the real-world problems.

Finding an appropriate probability model is an essential part of this execution. Sometimes an area of real-world application will have customary choices of models, or at least structured ways of selecting them. For example, there is a well-developed taxonomy of queueing models. A very good knowledge of the area of real-world application, combined with a very good knowledge of the behavior of these queueing models, often lets one choose a suitable model for a given issue within the area of real-world application. In other cases, one can start with a particularly simple probabilistic model, use the behavior of that model to gain insights about the real-world application, and use it to iteratively guide the selection of more general probabilistic models.

An important aspect of choosing a probability model for real-world applications is that a prospective choice depends mostly on prior understanding, at both an intuitive and mathematical level of results from the range of probability models that might be involved. This partly explains the title of the text "Introduction to Stochastic Processes: Queues, Finance and Credit Risk". The goal of this book is to guide the reader in both the mathematical and intuitive understanding which is required to create probability models using stochastic processes for studying real-world applications.

Therefore, this book neither uses nor develops measure theory. Measure theory is undoubtedly important in understanding probability at a deep level, but most of the topics useful in many real-world applications can be understood without measure theory. Moreover, the required concepts in probability theory are provided in Appendix as review. Some exposure to linear algebra and analysis (especially, concrete topics like vectors, matrices, and limits) is helpful. The most important prerequisite is having the mathematical maturity and patience to couple precise reasoning with intuition.

1.2 Definition, Classification, and Characterization

In this section, we deal with the fundamental concepts, categorization, and attributes of probability theory and stochastic processes.

1.2.1 Basics of Probability

Within this section, we explore the fundamental principles for defining probability.

Definition 1.1 (**Random Experiment**) An experiment is said to be random if its result cannot be determined beforehand. It is assumed that the set of all possible results of a random experiment is known.

Definition 1.2 **(Sample Space)** The set Ω of all possible results of a random experiment is called a sample space. An element or a result of the random experiment, i.e., $\omega \in \Omega$ is called an outcome or a sample point.

Definition 1.3 **(σ-Field or σ-Algebra and Event)** Let $\Omega \neq \emptyset$. A collection of subsets of Ω, denoted by $\mathcal{F}$, is called a σ-field or a σ-algebra on Ω if it satisfies the following axioms

1. $\Omega \in \mathcal{F}$.
2. If $A \in \mathcal{F}$, then the complement set of A, $A^c \in \mathcal{F}$.
3. If $A_i \in \mathcal{F}, i = 1, 2, \ldots,$ then $\bigcup_{i=1}^{\infty} A_i \in \mathcal{F}$.

The pair $(\Omega, \mathcal{F})$ is called a measurable space, and any element of $\mathcal{F}$ is called an event.

The collection of all subsets of Ω is called the power set of Ω, denoted by $\mathcal{P}(\Omega)$. It is obvious that this collection is a σ-algebra on Ω.

Remark 1.1 The intersection of any collection of σ-fields on Ω is again a σ-field on Ω.

Definition 1.4 **(Generated $\sigma-$Field)** Let $\mathcal{A}$ be any collection of subsets of Ω. The intersection of all σ-fields containing $\mathcal{A}$, is called the σ-field generated by $\mathcal{A}$.

Definition 1.5 **(Borel σ-Field or Borel σ-Algebra and Borel Set)** The σ-field generated by open subsets of $\mathbb{R}$ is called the Borel σ-field or Borel σ-algebra on $\mathbb{R}$, denoted by $\mathcal{B}(\mathbb{R})$. An element of the Borel σ-field is called a Borel set.

Remark 1.2 1. Let $\mathcal{P}(\mathbb{R})$ be the power set of $\mathbb{R}$. Then, $\mathcal{P}(\mathbb{R})$ is a σ-field on $\mathbb{R}$, but not a Borel σ-field on $\mathbb{R}$.
2. The Borel $\sigma-$field on $\mathbb{R}$ can also be generated by the following sets,

$$\begin{aligned}
\mathcal{E}_1 &= \{(a, b) : a, b \in \mathbb{R}, a < b\} \\
\mathcal{E}_2 &= \{[a, b) : a, b \in \mathbb{R}, a < b\} \\
\mathcal{E}_3 &= \{(a, b] : a, b \in \mathbb{R}, a < b\} \\
\mathcal{E}_4 &= \{[a, b] : a, b \in \mathbb{R}, a < b\} \\
\mathcal{E}_5 &= \{(a, \infty) : a \in \mathbb{R}\} \\
\mathcal{E}_6 &= \{(-\infty, a) : a \in \mathbb{R}\} \\
\mathcal{E}_7 &= \{[a, \infty) : a \in \mathbb{R}\}
\end{aligned}$$

and

$$\mathcal{E}_8 = \{(-\infty, a] : a \in \mathbb{R}\}.$$

Definition 1.6 **(Probability Measure)** Let $(\Omega, \mathcal{F})$ be a measurable space. A real-valued function P defined on $\mathcal{F}$ is called a probability measure on the measurable space $(\Omega, \mathcal{F})$ if it satisfies the following conditions

1. $P(A) \geq 0, \ \forall A \in \mathcal{F}$.
2. $P(\Omega) = 1$.
3. Let, $A_1, A_2, \ldots$ be mutually exclusive events in $\mathcal{F}$, i.e., $A_i \bigcap A_j = \emptyset, \ \forall i \neq j$, then

$$P\left(\bigcup_{i=1}^{\infty} A_i\right) = \sum_{i=1}^{\infty} P(A_i).$$

Here, the triplet $(\Omega, \mathcal{F}, P)$ is called a probability space.

Remark 1.3 The different $\mathcal{F}$ and the different P will make more than one probability space for a given Ω.

Definition 1.7 (Random Variable) Let $(\Omega, \mathcal{F}, P)$ be a probability space. A random variable (RV) is a mapping $X : \Omega \to \mathbb{R}$ such that for all $A \in \mathcal{B}(\mathbb{R})$, $X^{-1}(A) \in \mathcal{F}$, where $\mathcal{B}$ $(\mathbb{R})$ is the Borel σ-field on $\mathbb{R}$.

Remark 1.4 A RV is a real-valued function X defined on Ω with the property that for every Borel set B of Borel σ-field $\mathcal{B}(\mathbb{R})$, i.e., $B \in \mathcal{B}$ $(\mathbb{R})$, the subset of Ω, denoted by $\{X \in B\}$, satisfies $\{X \in B\} = \{\omega \in \Omega : X(\omega) \in B\} \in \mathcal{F}$. Thus, the RV X is referred as $\mathcal{F}-$ measurable.

Definition 1.8 (Cumulative Distribution Function) Let $(\Omega, \mathcal{F}, P)$ be a probability space and $X : \Omega \to \mathbb{R}$ be a random variable. The function $F_X : \mathbb{R} \to [0, 1]$ defined by $F_X(x) = P(X \leq x)$ is called the cumulative distribution function (CDF) of the RV X.

1.2.2 Stochastic Processes

In this subsection, the foundational concepts of stochastic processes are defined.

Definition 1.9 (Stochastic Processes) Let $(\Omega, \mathcal{F}, P)$ be a probability space. A stochastic process is a collection of RVs $\{X_t, t \in T\}$, indexed by the parameter t, defined on the probability space $(\Omega, \mathcal{F}, P)$. A stochastic process is also called a random process.

The set of all possible values, denoted by S, of the RVs X_t, for all $t \in T$, is known as the *state space* of the stochastic process and the set of values of the parameter t, denoted by T, is known as the *parameter space* of the stochastic process. The state space and parameter space are subsets of $\mathbb{R}$ and they can be finite, countably infinite, or uncountable.

In case of discrete time, i.e., the parameter space is finite or countably infinite, the parameter generally used is n. Then, the collection of the RVs is represented by $\{X_n, n = 0, 1, 2, \ldots\}$. In case of continuous-time, i.e., when the parameter space is uncountable, both the representations $\{X_t, t \in T\}$ and $\{X(t), t \in T\}$ are used. In general, $\{X_t, t \in T\}$ will be used to denote any general stochastic process.

Definition 1.10 (d-**Dimensional Stochastic Process**) A d-dimensional stochastic process is a stochastic process $\{X_t, t \in T\}$ in which for every $t \in T$, $X_t = (X_1(t), X_2(t), \ldots, X_d(t))$ is a $d-$dimensional random vector where $X_i(t)$, $i = 1, 2, \ldots, d$ is a RV.

Example 1.1 Consider the experiment of throwing a fair die infinitely many times. For $n = 1, 2, \ldots$, if X_n is the outcome of the nth throw, then $\{X_n, n = 1, 2, \ldots\}$ is a sequence of RVs such that for distinct values of n, one gets a distinct RV X_n. Hence, $\{X_n, n = 1, 2, \ldots\}$ is a stochastic process. If Y_n denotes the number of 6's in the first n throws, then $\{Y_n, n = 1, 2, \ldots\}$ is a stochastic process as well.

Example 1.2 Consider the number of students using the school bus service each working day. Let X_n be the RV representing the number of students using the school bus service in the nth working day. Then, the sequence of RVs $\{X_n, n = 1, 2, \ldots\}$ constitutes a stochastic process.

Example 1.3 Suppose X_t is a RV which represents the number of incoming calls at a telephone switchboard in a duration t units from time 0. Then, the collection of the RVs $\{X_t, t \in T\}$ constitutes a stochastic process, where $T = [0, \infty)$.

Example 1.4 Consider an insurance company that sells car insurance policies. Let X_t be the number of claims received from the insurers till time t. The collection of the RVs $\{X_t, t \geq 0\}$ is a stochastic process.

1.2.3 Classification of Stochastic Processes

Stochastic processes can be classified based on the nature of their parameter and state spaces as follows,

1. Discrete parameter, discrete state stochastic processes.
2. Discrete parameter, continuous state stochastic processes.
3. Continuous parameter, discrete state stochastic processes.
4. Continuous parameter, continuous state stochastic processes.

Definition 1.11 (Sample Path) Let $X = \{X_t, t \in T\}$ be a stochastic process and $\omega \in \Omega$ be fixed. The function $t \to X_t(\omega)$, $t \in T$, is called the sample path (or the realization or the ω-trajectory) of the stochastic process X.

The various sample paths of stochastic processes based on the above four classifications are shown in Fig. 1.1.

The list of examples for the above four classifications is as follows

(a) Let X_n denote the number of vehicles entering a city per hour. The set of possible values of X_n is $\{0, 1, \ldots\}$ and the set of values of the parameter space is $\{1, 2, \ldots\}$. Then, $\{X_n, n = 1, 2, \ldots\}$ is a discrete parameter, discrete state stochastic process.

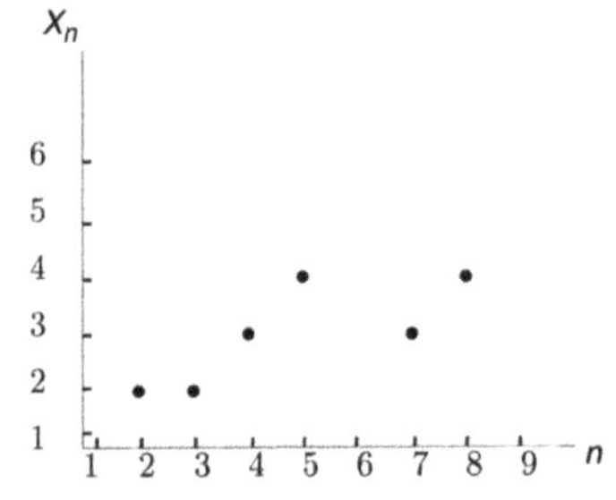

(a) Discrete parameter discrete-state stochastic processes

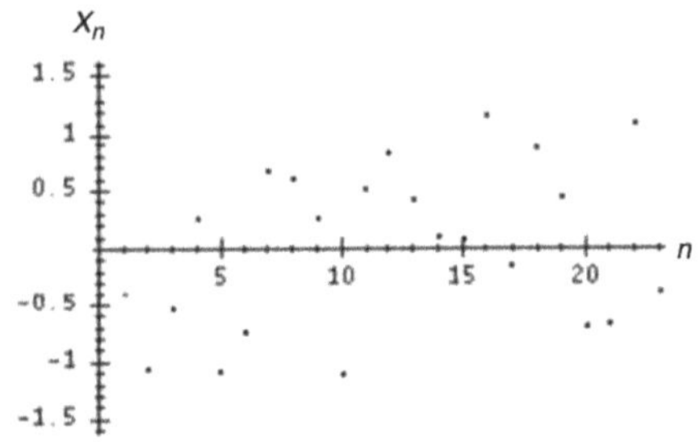

(b) Discrete parameter continuous state stochastic processes

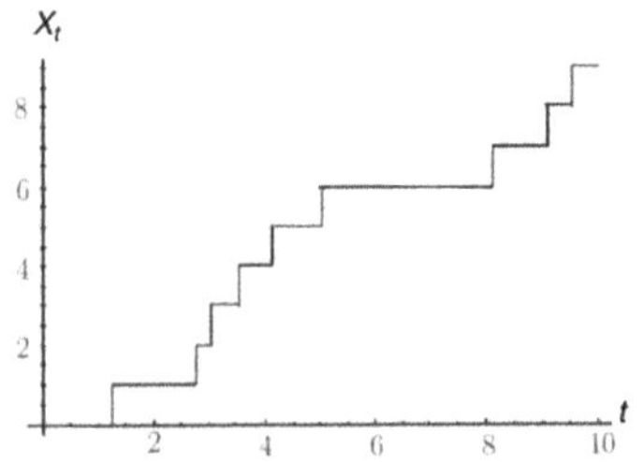

(c) Continuous parameter discrete-state stochastic processes

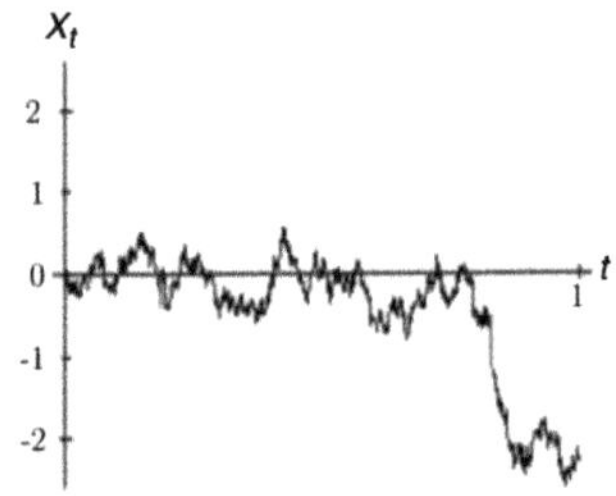

(d) Continuous parameter continuous state stochastic processes

Fig. 1.1 Sample path for different types of stochastic processes

(b) Let X_n denote the GDP growth of a country for the nth year. The set of possible values of X_n is $[0, \infty)$ and the set of values of the parameter space is $\{1, 2, \ldots\}$. Then, $\{X_n, n = 1, 2, \ldots\}$ is a discrete parameter, continuous state stochastic process.

(c) Let X_t denote the number of vehicles entering a city up to time t. The set of possible values of X_t are $0, 1, \ldots$ and the values of the parameter space is $\{0 \leq t < \infty\}$. Then, $\{X_t, t \geq 0\}$ is continuous parameter, discrete state stochastic process.

(d) Let X_t denote the return of a stock at time t. The set of possible values of X_t is $[0, \infty)$ and the set of values of the parameter space is $\{0 \leq t < \infty\}$. Then, $\{X_t, t \geq 0\}$ is a continuous parameter, continuous state stochastic process.

Often, the discrete-state stochastic processes are referred as stochastic chains or simply chains.

1.2.4 *Characterization of Stochastic Processes*

The probabilistic concepts of the stochastic processes are useful to get insights on the relation between the RVs within the stochastic processes.

Definition 1.12 (**Expectation of a RV**) Let $\{X_t,\ t \in T\}$ be a stochastic process defined on a probability space $(\Omega, \mathcal{F}, P)$. For a fixed t, a RV X_t is said to be integrable if $\displaystyle\int_{\Omega} |X_t| dP < \infty$. Then $E[X_t] = \int_{\Omega} X_t dP$ exists and is called the expectation of X_t.

The collection of integrable RVs is denoted by $L^1(\Omega, \mathcal{F}, P)$.

Definition 1.13 (**Mean Function**) Let $\{X_t, t \in T\}$ be a given stochastic process defined on a probability space $(\Omega, \mathcal{F}, P)$. The mean function is defined as $m(t) = E[X_t]$ provided $E[|X_t|] < \infty$.

A RV X_t is called square integrable if $\int_{\Omega} |X_t|^2 dP < \infty$. Then, the variance of X_t can be defined by $\text{var}(X_t) = \int_{\Omega}(X_t - E[X_t])^2 dP$. The collection of square integrable RVs will be denoted by $L^2(\Omega, \mathcal{F}, P)$.

Definition 1.14 (**Second-Order Process**) A stochastic process $\{X_t,\ t \in T\}$ is called a second-order process if the RVs X_t are square integrable for all t, i.e., $E[X_t^2] < \infty$ for all t.

Example 1.5 Suppose that, for $n = 1, 2, \ldots$, X_n is a RV with mean 0 and variance 1. Then, $\{X_n, n = 1, 2, \ldots\}$ is a second-order process. In fact, if one chooses any finite value to mean and any finite positive value to variance of the RVs, then also, $\{X_n, n = 1, 2, \ldots\}$ is a second-order process.

Definition 1.15 (**Covariance Function**) Let $X = \{X_t, t \in T\}$ be a second-order process. The function

$$C(s, t) = \text{cov}(X_s, X_t),\ s, t \in T$$

is called the covariance function of the stochastic process X.

Remark 1.5 The covariance function satisfies the following properties

1. $C(s, t) = C(t, s)$, for all $s, t \in T$.
2. $C(s, t) \leq \sqrt{C(s, s)C(t, t)}$.
3. $C(s, t)$ is non-negative definite.

Definition 1.16 (**Auto-Correlation Function**) Let $X = \{X_t, t \in T\}$ be a second-order process. The function

$$R(s, t) = \frac{C(s, t)}{\sqrt{\text{var}(X_s)}\sqrt{\text{var}(X_t)}},\ s, t \in T$$

is called the auto-correlation function of the stochastic process X.

Example 1.6 Let X and Y be two independent and identically distributed (IID) RVs each with mean 1 and variance 4. Define $Z_t = X + tY$ for $t \in [0, \infty)$. Find the mean function, the covariance function and the auto correlation function of the stochastic process $\{Z_t, t \geq 0\}$.

Solution: Mean function, $m(t)$, is given by

$$\begin{aligned} m(t) &= E[Z_t] = E[X + tY] \\ &= E[X] + tE[Y] \\ &= 1 + t. \end{aligned}$$

Covariance function is given by

$$\begin{aligned} C(s, t) &= \text{cov}(Z_s, Z_t) \\ &= E[Z_s Z_t] - E[Z_s]E[Z_t] \\ &= E[(X + sY)(X + tY)] - (1 + s)(1 + t) \\ &= 5 + (s + t) + 5st - (1 + s)(1 + t) \\ &= 4 + 4st. \end{aligned}$$

Auto-correlation function is given by

$$R(s, t) = \frac{E[Z_s Z_t] - E[Z_s]E[Z_t]}{\sqrt{\text{var}(Z_s)}\sqrt{\text{var}(Z_t)}} = \frac{(1 + st)}{\sqrt{(1 + s^2)(1 + t^2)}}.$$

1.3 Some Important Properties of Stochastic Processes

The RVs in the stochastic processes that we come across, may have dependency among themselves. Based on these dependencies, various types of stochastic processes can be studied, such as processes with independent increments, stationary increments, martingale, etc.

In this section, stochastic processes having some important properties, such as independent increment property and stationary property are discussed. Martingale property will be discussed in detail in Chap. 10.

The various stochastic processes to be discussed in this book based on the different properties are shown in Fig. 1.2.

1.3.1 Stochastic Process with Independent Increments

Independent increment is an important characteristic of the stochastic processes and is defined as follows.

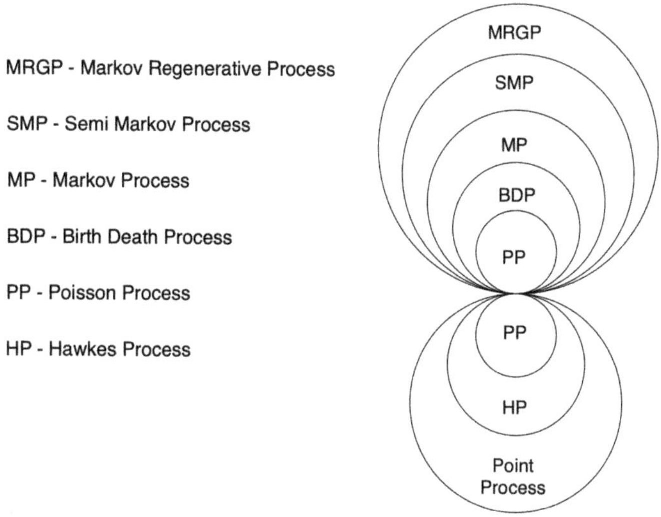

Fig. 1.2 Various stochastic processes

Definition 1.17 **(Independent Increments)** Let $\{X_t, t \in T\}$ be a stochastic process. If, for all $t_0, t_1, t_2, \cdots, t_n \in T$ with $t_0 < t_1 < t_2 < \cdots < t_n$, for all n, the RVs

$$X_{t_1} - X_{t_0}, X_{t_2} - X_{t_1}, X_{t_3} - X_{t_2}, \ldots, X_{t_n} - X_{t_{n-1}}$$

are mutually independent, then the stochastic process $\{X_t, t \in T\}$ is said to be a process with independent increments.

The well known examples for the stochastic processes with independent increments are the Poisson process and the Wiener process. These stochastic processes will be discussed in Sect. 1.4.

Example 1.7 Let $X_1, X_2, \ldots$ be a sequence of IID RVs each with $P(X_i = 0) = p, P(X_i = 1) = 1 - p, 0 < p < 1$. Define, for $n = 1, 2, \ldots$, $S_n = \sum_{i=1}^{n} X_i$. Then, the stochastic process $\{S_n, n = 1, 2, \ldots\}$ has independent increments with state space $S = \{0, 1, \ldots\}$. $\{X_n, n = 1, 2, \ldots\}$ is known as the Bernoulli process and $\{S_n, n = 1, 2, \ldots\}$ is known as the Binomial process.

1.3.2 Markov Processes

In this subsection, we describe the Markov property, in which the future path of a stochastic process depends only on its present state, independent of its past trajectory.

Definition 1.18 (Markov Property) Let $\{X_t, t \in T\}$ be a stochastic process with state space S. For any $t_0 < t_1 < t_2 < \cdots < t_n < t$, any state $x_0, x_1, \ldots, x_n$ and $x \in \mathbb{R}$, if

$$P\{X_t \le x | X_{t_0} = x_0, X_{t_1} = x_1, \cdots, X_{t_n} = x_n\} = P\{X_t \le x | X_{t_n} = x_n\} \ \forall\, n,$$

then the stochastic process $\{X_t, t \in T\}$ is said to have the Markov property. This property is also known as no aging property. A stochastic process with the Markov property is called a Markov process.

Remark 1.6 1. A Markov process is a stochastic process $\{X_t, t \in T\}$ such that, given the value of X_s, the distribution of $X_t,\ t > s$, does not depend on the values of $X_u,\ u < s$.
2. A Markov process with discrete state space is known as a Markov chain.
3. A stochastic process $\{X_t, t \in T\}$ with independent increments satisfies the Markov property. On the other hand, Markov property does not imply independent increments.

Poisson process and Wiener process are the important examples of Markov processes.

Example 1.8 Consider Example 1.7. Since $\{S_n, n = 1, 2, \ldots\}$ has independent increments, Markov property is satisfied. Thus, $\{S_n, n = 1, 2, \ldots\}$ is a Markov chain, as the state space is discrete.

1.3.3 Stationary Processes

In order to derive certain useful and general properties about stochastic processes, some constraints, and additional assumptions are needed to be imposed on the processes. In this regard, Markov property and stationary property are the two main assumptions that are applied to control the dependence. When a process is described as stationary, it means that it maintains its position over time. For a stochastic process to be considered stationary, all its statistical properties must remain constant over time. However, this is a stringent constraint, and though any stochastic process can also be considered stationary with weaker conditions, such as the mean and covariance values remain constant.

Definition 1.19 (Strict-Sense Stationarity) The stochastic process $\{X_t, t \in T\}$ is said to be strict-sense stationary, if for every $n = 1, 2, \ldots$ and arbitrary $t_1, t_2, \ldots, t_n$, such that $t_1 < t_2 < \cdots < t_n$ the joint distribution of the n-dimensional random vectors $\left(X_{t_1}, X_{t_2}, \ldots, X_{t_n}\right)$and $\left(X_{t_1+h}, X_{t_2+h}, \ldots, X_{t_n+h}\right)$ are same for every time lag $h > 0$ such that $t_i + h \in T$.

When $\{X_t, t \in T\}$ is a strict-sense stationary stochastic process, the marginal distribution of X_t doesn't depend on t. Further, the joint distribution of the two-dimensional random vector (X_{t_1}, X_{t_2}) is independent of the location of t_1 and t_2, while only the distance between t_1 and t_2 matters. As a consequence, the mean function $m(t)$ is a constant, and the covariance function $C(s,t)$ is a function of $|t-s|$ only. Also, the higher-order moments, remains unchanged if one adds a constant time shift to s and t.

Definition 1.20 **(Wide-Sense Stationarity)** A second-order stochastic process $\{X_t, t \in T\}$ is said to be wide-sense stationary, or covariance stationary, or weak stationary if it satisfies the following conditions.

1. Its mean function $m(t) = E[X_t]$ is independent of t.
2. Its covariance function $C(s,t)$ is a function of $|t-s|$, for all $t, s \in T$, i.e., $C(s,t) = f(|t-s|)$ for some function f.
   If it is not explicitly stated otherwise, stationarity means wide-sense stationarity.

Remark 1.7 1. It is important to note that neither does a strict-sense stationary process imply a wide-sense stationary process nor does a wide-sense stationary process implies a strict-sense stationary process.
2. Any stochastic process which is strict-sense stationary and has a finite variance is always wide-sense stationary.

Definition 1.21 **(Evolutionary Process)** An evolutionary stochastic process is a stochastic process that is not stationary (in both strict-sense and wide-sense).

Example 1.9 Consider uncorrelated RVs $\{X_n, n = 1, 2, \ldots\}$ having mean 0 and variance 1. Check whether this process is stationary or not.
Solution: For any n and m, consider,

$$C(n,m) = \text{cov}(X_n, X_m) = E[X_m X_n] = \begin{cases} 0, \; m \neq n; \\ 1, \; m = n. \end{cases}$$

Since covariance function depends only on the difference between n and m, $\{X_n, n = 1, 2, \ldots\}$ is a wide-sense stationary process. Furthermore, if X_n are identically distributed, then $\{X_n, n = 1, 2, \ldots\}$ is a strict-sense stationary process. In fact, if one chooses any finite positive value as variance of the RVs, then also, $\{X_n, n = 1, 2, \ldots\}$ is a strict-sense stationary process.

Example 1.10 Consider Example 1.7. Prove that, $\{X_i, i = 1, 2, \ldots\}$ is both wide-sense and strict-sense stationary processes.
Solution: The stochastic process $\{X_i, i = 1, 2, \ldots\}$ is such that $m(i) = E[X_i] = p$ and $E[X_i^2] = p$. Also,

$$\begin{aligned} C(i,j) &= E[X_i X_j] - E[X_i]E[X_j] \\ &= \begin{cases} 0, & i \neq j \\ p(1-p), & i = j. \end{cases} \end{aligned}$$

The above process is a wide-sense stationary process.

Now, for fixed positive integers n and h, consider the two n-dimensional random vectors $\{X_{i_1}, X_{i_2}, \cdots, X_{i_n}\}$ and $\{X_{i_1+h}, X_{i_2+h}, \cdots, X_{i_n+h}\}$. Since X_i are independent RVs, the joint distribution of the two random vectors $\{X_{i_1}, X_{i_2}, \cdots, X_{i_n}\}$ and $\{X_{i_1+h}, X_{i_2+h}, \cdots, X_{i_n+h}\}$ will be the product of the distributions of the RVs X_i's and X_{i+h}'s respectively. Since X_i are identically distributed RVs, the joint distribution for the two random vectors coincides. Hence, $\{X_i, i = 1, 2, \ldots\}$ is a strict-sense stationary process. In this case, $\{X_i, i = 1, 2, \ldots\}$ is a strict-sense as well as a wide-sense stationary process.

Example 1.11 Let $\{X_t, t \in T\}$ be a strict-sense stationary process with finite second-order moment and let, for any t, $Y_t = a + bt + X_t$ with $a \in R, b \neq 0$. Check whether $\{Y_t, t \in T\}$ is a wide-sense and strict-sense stationary process.
Solution: The function $m(t) = E[Y_t]$ is a function of t and hence $\{Y_t, t \in T\}$ is not a wide-sense stationary process. $\{Y_t, t \in T\}$ is not a strict-sense stationary process.

Example 1.12 (*White Noise Process*) Consider Example 1.9 with mean 0 and variance σ^2. Check whether this process is wide-sense stationary process or not.
Solution: All the three conditions of a wide-sense stationary stochastic process are satisfied and hence it's a wide-sense stationary process. This process is also called the white noise process.

Example 1.13 Let $\{X_t, t \geq 0\}$ be a stochastic process, where X_t is defined as

$$X_t = A\cos(\theta t) + B\sin(\theta t),$$

where, A and B are two uncorrelated RVs with mean 0 and variance σ^2, and θ is a constant. Check whether this process is a wide-sense stationary process or not.
Solution: For a fixed t, as mean of A and B is 0, $E[X_t] = 0$.

$$C(s, t) = E[X_t X_s] - E[X_t]E[X_s] = \sigma^2 \cos(\theta(t - s)),$$

which is a function of $(t - s)$, and $E[X_t^2] = \sigma^2$. Therefore, this process is a wide-sense stationary process.

1.4 Review of Standard Stochastic Processes

In this section, specific stochastic processes such as the Poisson process, Gaussian and Wiener process are presented. These stochastic processes are some examples of standard stochastic processes, each with its own properties and applications in various fields including finance, physics, biology, and engineering.

1.4.1 Poisson Process

Consider an event E that may happen at any time instant t and $N(t)$ is the number of occurrences of the event E in the time interval $(0, t]$. Then, $\{N(t), t \geq 0\}$ is a continuous parameter (or continuous-time) and non-negative valued stochastic process. Let $p_n(t) = P(N(t) = n)$.

Definition 1.22 **(Poisson Process)** Let $N(t)$ be the number of occurrences of the event E in the time interval $(0, t]$ of a random experiment. The stochastic process $\{N(t), t \geq 0\}$ is called the Poisson process, with intensity or rate $\lambda > 0$, if the following conditions are satisfied,

1. $N(0) = 0$ almost surely (a.s.).
2. $N(t)$ has stationary and independent increments.
3. For every $t > 0$, $N(t)$ follows a Poisson distribution with parameter λt.

The sample path of Poisson process is shown in Fig. 1.3. It begins at the origin, $N(0) = 0$ and is right continuous, non-decreasing and piece-wise constant with jumps of size 1 at the times of occurrence of the event. Poisson process is again discussed in detail in Chap. 5.

Remark 1.8 Poisson process is a continuous-time discrete state stochastic process with independent increment property. Hence, it is a Markov chain.

Example 1.14 Let $X(t)$ be the number of people immigrating into a territory on or before time t. Suppose that $\{X(t), t \geq 0\}$ follows the Poisson process with parameter $\lambda = 1$ per day. What is the probability that the elapsed time between the 100th and 101th immigration exceeds two days?
Solution: For fixed t, $X(t)$ follows Poisson distribution with parameter λt. Let T be the time of first immigration.

$$P(T > t) = P(X(t) = 0) = P(\text{no event takes place before } t) = e^{-\lambda t}.$$

The CDF of RV T is given by

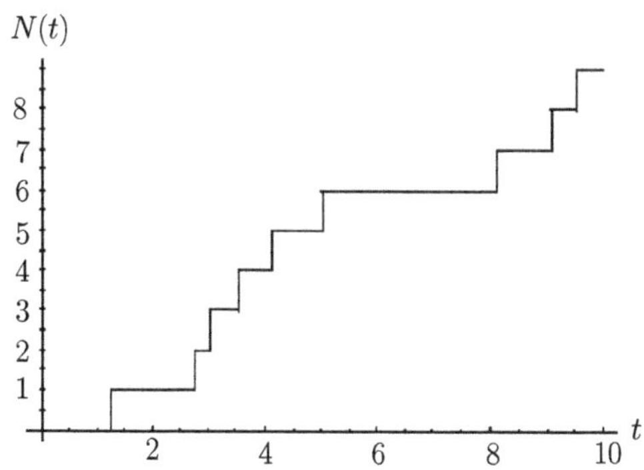

Fig. 1.3 Sample path of Poisson process with $\lambda = 0.5$

$$F_T(t) = P(T \le t) = 1 - P(T > t) = 1 - e^{-\lambda t}, \quad t \ge 0.$$

Thus, T follows an exponential distribution with parameter λ. Since, the Poisson process has independent increment property, time elapsed between 100th and 101th immigration is the same as the time of first immigration. Therefore, the required probability is given by

$$P(T > 2) = e^{-2}.$$

1.4.2 Gaussian Process

Definition 1.23 (**Gaussian Process**) A stochastic process $\{X_t, t \ge 0\}$ is called a Gaussian process if the distribution of each finite dimensional random vector is multivariate Gaussian (normal), i.e., if for all $t_1, t_2, \cdots, t_n$, and for all n, the distribution of $X = \left(X_{t_1}, X_{t_2}, \cdots, X_{t_n}\right)$ is multivariate normal.

The joint probability density function (PDF) of $(X_{t_1}, X_{t_2}, \ldots, X_{t_n})$ is given by

$$f_X(x) = \frac{1}{(2\pi)^{n/2}(det(\Sigma))^{1/2}} \exp\left[-\frac{1}{2}(X-\mu)\Sigma^{-1}(X-\mu)^{'}\right],$$

where

$$\begin{aligned} \mu &= (E[X_{t_1}], E[X_{t_2}], \ldots, E[X_{t_n}]) \\ \Sigma &= (\text{cov}(X_{t_i}, X_{t_j}); i, j = 1, 2, \ldots, n) \\ \text{cov}(X_{t_i}, X_{t_j}) &= E[(X_{t_i} - E[X_{t_i}])(X_{t_j} - E[X_{t_j}])]. \end{aligned}$$

Note that, two normally distributed RVs are independent if and only if these RVs are uncorrelated. Thus, n normally distributed RVs are independent if and only if its covariance matrix Σ is a diagonal matrix.

Example 1.15 Prove that if a Gaussian process $\{X_t, t \ge 0\}$ is wide-sense stationary, then it is strict-sense stationary.
Solution: To prove that $\{X_t, t \ge 0\}$ is a strict-sense stationary process, it should be shown that for all $t_1, t_2, \cdots, t_n \in \mathbb{R}^+$, for all n and for all $h > 0$, the joint distribution of $X = (X_{t_1}, X_{t_2}, \cdots, X_{t_n})$ is same as the joint distribution of $X^h = (X_{t_1+h}, X_{t_2+h}, \cdots, X_{t_n+h})$. As, $\{X_t, t \ge 0\}$ is a wide-sense stationary process,

$$m(t_i) = c \text{ (constant)}, \quad i = 1, 2, \ldots, n,$$

and

$$C(X_{t_i+h}, X_{t_j+h}) = C(X_{t_i}, X_{t_j}), \quad i, j = 1, 2, \ldots, n.$$

This implies that the mean vector and the covariance matrix of X is same as the mean vector and the covariance matrix of X^h, respectively. Hence, the joint distribution of X is same as the joint distribution of X^h, proving strict-sense stationarity.

1.4.3 Brownian Motion

Brownian motion (BM) is named after Robert Brown (21 December 1773 to 10 June 1858), a Scottish botanist who observed the random motion of particles suspended in a medium (a liquid or a gas) in the 1820s. Brownian motion was the subject of an important paper by German-born theoretical physicist, Albert Einstein (14 March 1879–18 April 1955) in 1905. The rigorous mathematical foundations for Brownian motion were first given by American mathematician Norbert Wiener (November 26, 1894–March 18, 1964) in 1923.

Definition 1.24 **(Brownian Motion)** A random process $\{W(t), t \geq 0\}$ with state space $\mathbb{R}$ is said to be a standard Brownian motion, also known as Wiener process, if it satisfies the following properties.

1. $W(0) = 0$ a.s..
2. The sample path of $\{W(t), t \geq 0\}$ is continuous on $[0, \infty)$.
3. It has stationary and independent increment.
4. For $0 \leq s < t$, $W(t) - W(s)$ is normally distributed with mean 0 and variance $t - s$.

A sample path of a Wiener process is shown in Fig. 1.4. Brownian motion is again discussed in detail in Chap. 11.

Remark 1.9 1. The covariance function $C(s, t)$ of a Brownian motion, for $s \leq t$,

$$\begin{aligned} C(s,t) &= E[(W(t) - E[W(t)])(W(s) - E[W(s)])] \\ &= E[W(t)W(s)] = E[(W(t) - W(s) + W(s))W(s)] = E\left[(W(s))^2\right] = s. \end{aligned}$$

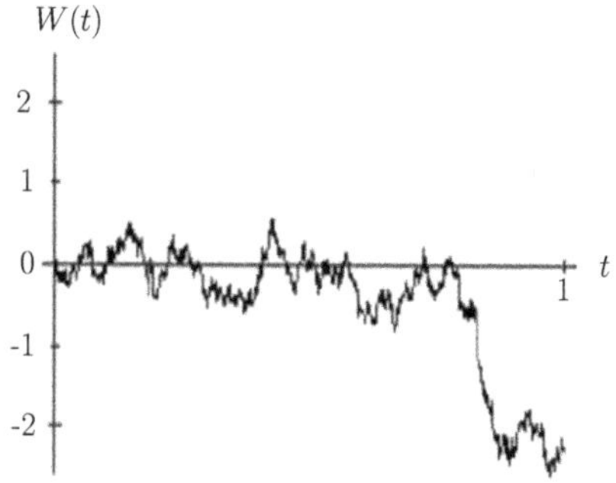

Fig. 1.4 Sample path of Wiener process

Hence, $C(s,t) = \min(s,t)$. Thus, the Wiener process is not a wide-sense stationary process.

2. The Wiener process is not a strict-sense stationary process as the variance of the Wiener process increases with time.
3. From the definition, $W(t+s) - W(s)$ is independent of the past, or alternatively, if $W(s) = x_0$, then no further knowledge of the values of $W(\tau)$ for $\tau < s$ has any effect on the knowledge of the probability distribution of $W(t+s) - W(s)$. Given $W(t)$, the future $W(t+h)$ for any $h > 0$ only depends on the future increment $W(t+h) - W(t)$ and this future is independent of the past. Hence, the Markov property is satisfied. Thus, $\{W(t), t \geq 0\}$ is a Markov process.

For more description, readers may refer to [1–11].

1.5 Exercise Problems

1. Trace the path of the following stochastic processes,
   (a) $\{X_k, k \in T\}$ where X_k is the waiting time of the kth student arriving at a bus stop and $T = \{1, 2, \ldots\}$.
   (b) $\{X_t, t \in T\}$ where X_t be the number of cars parked in a parking garage at a shopping mall at time t, and $T = [0, \infty)$.
   (c) $\{Y_t, t \in T\}$ where Y_t denotes the exchange rates of the Indian rupee (INR) against the US dollar (USD) at time t, and $T = [0, \infty)$.
   (d) $\{X_t, t \in T\}$ where X_t denotes the number of passengers in a bus at tth trip, and $T = \{1, 2, \ldots\}$.
2. Let $\{X_t, t \geq 0\}$ be a stochastic process with $X_t = A_0 + A_1 t + A_2 t^2$, where $A_i's$ are uncorrelated r.vs. with mean 0 and variance 1. Find the mean function and covariance function of X_t.
3. Let $\{N(t), t \geq 0\}$ be a Poisson process with rate λ. Find its covariance function $C(t_1, t_2)$.
4. Let Y_1, Y_2 be two IID RVs where each follows $Y_i \sim N(\mu, \sigma^2), i = 1, 2$. Define $X_t = Y_1 \cos(\lambda t) + Y_2 \sin(\lambda t), \quad -\infty < t < \infty$ where λ is a real constant. Show that $\{X_t, -\infty < t < \infty\}$ is a second-order process.
5. Suppose that $X_1, X_2, \ldots$ are IID RVs each having normal distribution with mean 0 and variance σ^2. Consider the stochastic process $\{S_n,\ n = 1, 2, \ldots\}$ where $S_n = \exp\left(\sum_{i=1}^{n} X_i - \frac{n}{2}\sigma^2\right)$. Find $E[S_n]$ for all n.
6. Let $\{X_t, 0 \leq t \leq T\}$ be a stochastic process such that $E[X_t] = 0$ and $E\left[X_t^2\right] = 1$ for all $t \in [0, T]$. Find the upper bound of $|E[X_t X_{t+h}]|$ for any $h > 0$ and $t \in [0, T-h]$.
7. Consider the process $\{X_t, t \geq 0\}$ with $X_t = A\cos(wt) + B\sin(wt)$ where w is a positive constant and A and B are uncorrelated RV with mean 0 and variance 1. Check whether $\{X_t, t \geq 0\}$ is wide-sense stationary or not.

8. In a communication system, the carrier signal at the receiver is modeled by the process $\{Y_t, t \geq 0\}$ which is defined as $Y_t = X_t \cos(2\pi w t + \Theta)$ where $\{X_t, t \geq 0\}$ is a zero-mean and wide-sense stationary process, and Θ is a uniform distributed RV in the interval $(-\pi, \pi)$ and w is a positive constant. Assume that, Θ is independent of the process $\{X_t, t \geq 0\}$. Is $\{Y_t, t \geq 0\}$ wide-sense stationary?
9. Consider Problem 4 with $P(Y_i = 1) = P(Y_i = -1) = \frac{1}{2}, i = 1, 2$ instead of normal distribution. Show that $\{X_t, -\infty < t < \infty\}$ is a second order stationary process which is not strict-sense stationary.
10. Prove that $\{X_t, -\infty < t < \infty\}$ is a second-order stationary process if for every number τ, the second-order process Y_t defined by $Y_t = X_{t+\tau}, -\infty < t < \infty$ has the same mean and covariance function as of X_t.
11. Consider the random telegraph signal, denoted by X_t, which jumps between two states, 0 and 1, according to the following rules. At time $t = 0$, the signal X_t start with equal probability for the two states, i.e., $P(X_0 = 0) = P(X_0 = 1) = 1/2$, and let the switching times be decided by a Poisson process $\{Y_t, t \geq 0\}$ with parameter λ independent of X_0. At time t, the signal

$$X_t = \frac{1}{2}\left(1 - (-1)^{X_0 + Y_t}\right), t > 0.$$

Is $\{X_t,\ t \geq 0\}$ wide-sense stationary.
12. Prove that every real-valued stochastic process $\{X_t, t \in T\}$ with independent increments is a Markov process.
13. Prove or disprove the following statement: If $\{X_t, t \in T\}$ is a Markov process, then $\{X_t, t \in T\}$ has independent increment property.
14. Let X and Y be two IID RVs each having uniform distribution on interval $[-\pi, \pi]$. Define, for $t \geq 0$, $Z_t = \sin(Xt + Y)$. Check whether $\{Z_t, t \geq 0\}$ is covariance stationary or not.
15. Let A be a positive RV that is independent of a strict-sense stationary random process $\{X_t, t \geq 0\}$. Show that $\{Y_t, t \geq 0\}$ where $Y_t = AX_t$ is also strict-sense stationary random process.
16. Let $\{X_t, t \geq 0\}$ be a stochastic process where the PMF of X_t is given by

$$P\{X_t = n\} = \begin{cases} \frac{at}{(1+at)} & n = 0 \\ \frac{(at)^{n-1}}{(1+at)^{n+1}} & n = 1, 2, \ldots. \end{cases}$$

Check whether $\{X_t, t \geq 0\}$ is covariance stationary or not.
17. Let $\{N(t), t \geq 0\}$ be a Poisson process with parameter λ. Show that

$$\lim_{t \to \infty} N(t) = \infty a.s..$$

Hint: Find the limit of $P\{N(k) \geq n\}$ as $k \to \infty$. Write $\{\lim_{t\to\infty} N(t) = \infty\}$ in terms of $\{N(k) \geq n\}$.

18. Let $\{N(t), t \geq 0\}$ be a Poisson process. Prove that the process $\{X(t), t \geq 0\}$, where

$$X(t) = N(t + L) - N(t), \ L > 0,$$

is covariance or wide-sense stationary.

19. Let $\{N(t), t \geq 0\}$ be a Poisson process with parameter λ. Show that

$$P\{N(t) \text{ is odd}\} = e^{(-\lambda t)} \sinh(\lambda t),$$
$$P\{N(t) \text{ is even}\} = e^{(-\lambda t)} \cosh(\lambda t).$$

Hint: Find the probability that $N(t) = 2n + 1$ and compare this to the n-th term of the Taylor expansion of $\sinh(\lambda t)$.

References

1. Bass, R.: Stochastic processes. Cambridge Series in Statistical and Probabilistic Mathematics (2011)
2. Brzezniak, Z., Tomasz, Z.: Basic stochastic processes: a course through exercises. Springer Science & Business Media (2000)
3. Castaneda, L.B., Arunachalam, V., Dharmaraja, S.: Introduction to Probability and Stochastic Processes with Applications. Wiley, New Jersey (2012)
4. Chung, K.L., Williams Ruth, J.: Introduction to Stochastic Integration, vol. 2. Birkhäuser, Boston (1990)
5. Chiang, C.L.: An Introduction to Stochastic Processes and their Applications. John Wiley, New York (1980)
6. Durrett, R. : Essentials of Stochastic Processes, 2nd edn. Springer, New York (2012)
7. Gusak, D.: Theory of Stochastic Processes. Springer
8. Mikosch T.: Elementary stochastic calculus with finance in view. World Scientific (1998)
9. Parzen, E.: Modern Probability Theory and its Applications, vol. 10. Wiley (1960)
10. Ross, S.M.: Introduction to Probability Models. Elsevier, 12/e, (2019)
11. Trivedi, K.S.: Probability and Statistics with Reliability, Queuing, and Computer Science Applications, 2nd edn. John Wiley & Sons (2001)

Chapter 2
Discrete-Time Markov Chains—Part I

A Markov process is a stochastic process used to model the events, where the system moves between states and the probability of transitioning from one state to the next state depends solely on the current state, with no influence from the previous visited states. Markov processes are categorized based on the nature of their state and parameter space. When the state space is discrete, the process is called a Markov chain (MC). If both the state and parameter space are discrete, it is specifically known as a discrete-time Markov chain (DTMC). In this chapter, the definition and basic properties of DTMC are presented. Also, limiting distribution, stationary distribution and steady-state analysis of the DTMCs are also considered. Moreover, this chapter and Chap. 3 both are devoted to an elaboration of the basic concepts and methods of DTMC.

2.1 Introduction to Markov Chains

Andrey Andreyevich Markov (14 June 1856 to 20 July 1922), a Russian mathematician, was the first to present the findings on these processes more than a century ago. Many researchers afterward worked on the expansions and implementations of his theory, including Kolmogorov and Erlang, among others. Roughly speaking, a Markov process is a random process with the Markov property. This is a basic description of relationships when it is thought of connections among RVs in stochastic processes. In short, the Markov property describes that the performance of a process in the future is solely determined by its present state, rather than its previous actions. In order to examine the process, the only key data from the past is the current condition of the process. Simply put, this means that the past has no influence on the future given the present. This trait is shared by a large number of systems that arise in reality, and hence, studying them is important. One can visualize the number of

D. Selvamuthu, *Introduction to Stochastic Processes*, University Texts in the Mathematical Sciences, https://doi.org/10.1007/978-981-97-6152-4_2

clients in the queue for service at a bank desk or the number of children a species has in a cycle, as an example.

Chains refer to a class of Markov processes in which the state space is discrete, and this nomenclature is used throughout in this book. In this chapter, DTMC are discussed, and after that continuous-time Markov chains (CTMCs) will be discussed.

The study of DTMCs are worthwhile from two important view points. Firstly, there are various concepts which can be presented at an elementary level. Secondly, there are large number of systems arising in practice that can be modeled by a DTMC. Thus, the subject has many useful applications.

The DTMC, also simply known as Markov chains (MCs), contain a state space of either $S = \{0, 1, \ldots, N\}$, which is termed as a finite Markov chain, or $S = \{0, 1, \ldots\}$, which is called a countable Markov chain. Depending on the context, the state 0 is included additionally in the state space.

2.2 Definition and Characterization

Within this section, are defined and characterized through state transitions.

Definition 2.1 (Discrete-time Markov chain) Let $\{X_n, n = 0, 1, \ldots\}$ be a sequence of RVs defined on a probability space $(\Omega, \mathfrak{F}, P)$. The stochastic process $\{X_n, n = 0, 1, \ldots\}$ with state space S is called a DTMC if, for all $i, j, i_0, i_1, \ldots, i_{n-1} \in S$, and for $n = 1, 2, \ldots$,

$$P\{X_{n+1} = j | X_n = i, X_{n-1} = i_{n-1}, \ldots, X_0 = i_0\} = P\{X_{n+1} = j | X_n = i\},$$

whenever the probability on the left-hand side is well-defined. The transition probability of X_{n+1} being in state j given that X_n is in state i is denoted by $p_{ij}(n, n+1)$ defined as

$$p_{ij}(n, n+1) = P\{X_{n+1} = j | X_n = i\}.$$

When $p_{ij}(n, n+1) = p_{ij}$ which is independent of n and p_{ij} is the probability that the state value undergoes a transition from state i to state j in one step, then DTMC is referred to as a time-homogeneous or simply homogeneous MC (or a Markov chain with stationary transition probabilities); otherwise, it is referred to as a non-homogeneous MC.

In other words, p_{ij} is the conditional probability that the DTMC is at time $n+1$ in state j, given that at time n it was in state i and is independent of time n.

The initial state distribution is the probability distribution of states at initial time 0, that is, the distribution of X_0, given below

$$P(X_0 = i) = \alpha_i, \quad \text{for all } \; i \in S,$$

where $\alpha_i \geq 0, i \in S$, and $\sum \alpha_i = 1$. Let the initial state distribution $\pi_0 = [\alpha_i]$ be the row vector whose elements are α_i.

In the study of Markov chains, the transition probabilities p_{ij} are fundamental entities. The situation of homogeneous Markov chains is a default one. Then, the stationary transition probabilities p_{ij} are known as one-step transition probabilities because they indicate the probability of transitioning from state i to state j in one step. In other words, if the MC is in state i at step $n-1$, then no matter how it got to state i it has probability p_{ij} of being in state j at the next step. The one-step transition probabilities p_{ij} satisfy

$$p_{ij} \geq 0, \forall\, i, j \in S \text{ and } \sum_{j \in S} p_{ij} = 1\, \forall\, i \in S.$$

These one-step transition probabilities may be written in the matrix form $\mathbf{P} = [p_{ij}]$ when $S = \{0, 1, \ldots\}$ as

$$\mathbf{P} = \begin{pmatrix} p_{00} & p_{01} & p_{02} & \cdots \\ p_{10} & p_{11} & p_{12} & \cdots \\ \vdots & \vdots & \vdots & \ddots \end{pmatrix}.$$

This matrix is called the transition probability matrix (TPM) of the Markov chain. Note that each row sum of $\mathbf{P}$ is one. A matrix with this property is called a stochastic matrix or Markov matrix. $\mathbf{P}$ is called a double stochastic matrix if both $\mathbf{P}$ and its transpose $\mathbf{P}^T$ are stochastic matrices.

Example 2.1 (Weather forecasting model) Suppose that the probability of rain tomorrow depends only on whether it is raining today and not on any prior weather conditions. This is known as the "memoryless property". Let α be the probability that it will rain tomorrow, given that it is raining today. Let β be the probability that it will not rain tomorrow, given that it is not raining today. Assume that, initially a day starts with no rain. Formulate this problem as a DTMC.

Solution: Let X_n be a random variable that indicates whether or not it will rain on the nth day. The parameter space $T = \{0, 1, \ldots\}$ is discrete and represents time. The state space is $S = \{0, 1\}$ where 0 denotes rain and 1 denotes no rain and is clearly discrete. The TPM of the MC $\{X_n, n = 0, 1, \ldots\}$ is given by

$$\mathbf{P} = \begin{pmatrix} \alpha & 1-\alpha \\ 1-\beta & \beta \end{pmatrix}.$$

The initial distribution is given by

$$P(X_0 = 1) = 1; \quad P(X_0 = 0) = 0.$$

Thus, $\{X_n, n = 0, 1, \ldots\}$ is a DTMC with the above given TPM and the initial distribution.

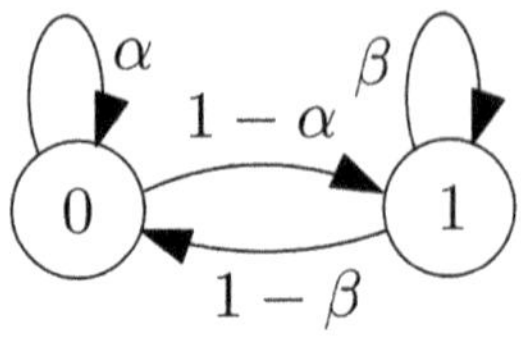

Fig. 2.1 State transition diagram for Example 2.2

2.2.1 State Transition Diagram

A *state transition diagram* is a graphical representation of the one-step transition probability matrix for the time-homogeneous DTMC. A state transition diagram is frequently used to visualize a Markov chain. The directed arcs between nodes represent one-step transitions between the states, while the nodes represent the states. Each transition is associated with a probability. Besides, they provide a good abstraction of a physical system under study. Such diagrams are simple to design. These diagrams can also be used for other purposes.

Example 2.2 Consider Example 2.1. Draw the state transition diagram for this DTMC.
Solution: The state transition diagram for this DTMC is shown in Fig. 2.1.

2.2.2 Sojourn Time Distribution

Definition 2.2 (Sojourn Time) Let $\{X_n, n = 0, 1, \ldots\}$ be a DTMC with state space $S = \{0, 1, \ldots\}$ and TPM $\mathbf{P} = [p_{ij}], i, j \in S$. For some $i \in S$, suppose $0 < p_{ii} < 1$. Let τ_i be the RV, known as sojourn time at state i, denoting the time taken for a change from state i into another state.

Theorem 2.1 *The PMF of the RV τ_i is given by*

$$P(\tau_i = n) = (1 - p_{ii})\, p_{ii}^{n-1}, n = 1, 2, \ldots$$

which is a geometric distribution with parameter $1 - p_{ii}$.

Proof For positive integers m, n

$$P\{\tau_i > m + n | X_0 = i\} = P\{\tau_i > m + n | X_m = i\} P\{\tau_i > m | X_0 = i\}.$$

That is,

$$F^c_{\tau_i}(m + n) = F^c_{\tau_i}(n) F^c_{\tau_i}(m), n = 1, 2, \ldots$$

where $F^c_{\tau_i}(n) = P\{\tau_i > n | X_0 = i\}$ if and only if $F_{\tau_i}(n) = (1 - p_{ii}) p_{ii}^{n-1}$, thus, the sojourn time τ_i has geometric distribution with parameter $1 - p_{ii}$.

The sojourn time is also known as the waiting time. The geometric distribution is the only discrete type random variable having memoryless property. It is a discrete analogue of the exponential distribution.

Example 2.3 (System with two components) A system has two components and one repairman. Assume that the probability of any one component breaking down on a given day is α. If the repair person is working on a component, the probability that he will complete the repair in a day is β. For simplicity, ignore the probability of a repair completion or a breakdown taking place except at the end of a day. Formulate this problem as a DTMC.
Solution: Let X_n be the number of components in operation at the end of the nth day. Assume that the behavior of $\{X_n, n = 0, 1, \ldots\}$ can be modeled as a Markov chain, and the state transition diagram for this Markov chain is shown in Fig. 2.2.

Example 2.4 (Barbershop problem) Consider the owner of a local one-chair barbershop. He is thinking of expanding the shop capacity because it seems there is too much demand and many people are turned away. From past information, during the time required to cut one person's hair, there may be 0, 1, or 2 arrivals with probabilities 0.3, 0.4, and 0.3, respectively. Assume that the shop has a fixed capacity of six people including the one whose hair is being cut. Further, any new arrival who finds six people in the barbershop is denied entry. Formulate this problem as a DTMC.
Solution: Let X_n be the number of people in the shop at the completion of the nth person's haircut. $\{X_n, n = 0, 1, \ldots\}$ is a Markov chain assuming IID arrivals. The state transition diagram is shown in Fig. 2.3.

Example 2.5 (Random Walk) A random walk is a mathematical process where an object takes a sequence of steps in random directions, often used to model unpredictable phenomena like stock prices or particle movements. For instance, consider

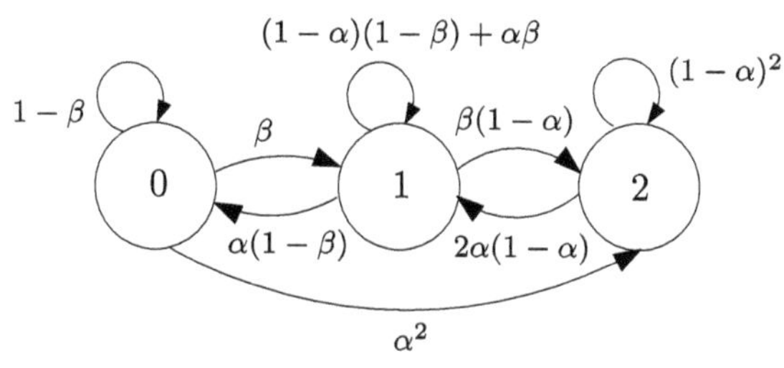

Fig. 2.2 State transition diagram for Example 2.3

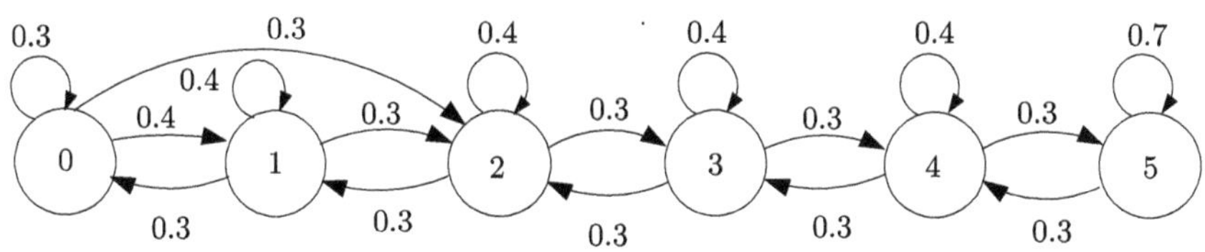

Fig. 2.3 State transition diagram for Example 2.4

the modification on Example 10.16. A particle performs a random walk on the state space $S = \{0, 1, \ldots, N\}$. Whenever it is at any position $k(k = 1, 2, \ldots, N-1)$, it either moves to $k-1$ with probability p or it moves to $(k-1)$ with probability q, or it remains at k with probability r $(p+q+r=1)$ in the next step. But as soon as it reaches 0, it remains there with probability α and it is reflected to state 1 with probability $1-\alpha$. Similarly, when the particle reaches state N, it remains there with probability β and it is reflected to state $N-1$ with probability $1-\beta$. Formulate this problem as a DTMC. Draw the state transition diagram for this DTMC.
Solution: Let X_n be the position of the particle after n steps. State transition diagram is shown in Fig. 2.4.

$\{X_n, n = 0, 1, \ldots\}$ is a DTMC whose TPM is given by

$$\mathbf{P} = \begin{pmatrix} \alpha & 1-\alpha & . & . & 0 \\ q & r & p & . & 0 \\ . & . & . & . & . \\ . & . & . & . & . \\ . & . & q & r & p \\ 0 & . & . & 1-\beta & \beta \end{pmatrix}.$$

If $\alpha = 0$, state 0 is a *reflecting barrier* and if $\alpha = 1$, state 0 is an *absorbing barrier*, if $0 < \alpha < 1$, state 0 is an elastic barrier. Similar is the case with state N.

Example 2.6 (Gambler's Ruin Problem) The Gambler's ruin problem is a problem in probability theory that models the situation of a gambler who repeatedly bets on a fair game (one with equal chances of winning or losing a fixed amount) until they either lose all the money or reach a certain wealth target. For instance, consider the modification in Example 2.5. When $\alpha = \beta = 1$, this problem can be considered as gambler's ruin problem when both the states 0 and N are absorbing states. That is, if the capital of the gambler increases to N rupees or ruin to 0, he/she quits playing. Let X_n be the amount the gambler has at the end of the nth play. Then, $\{X_n, n = 0, 1, \ldots\}$ is a stochastic process with state space $S = \{0, 1, \ldots, N\}$. The state transition diagram for this process is shown in Fig. 2.5.

For $i = 1, 2, \ldots N-1$,

$$p_{ij} = \begin{cases} p, & j = i+1 \\ q, & j = i-1 \\ 0, & j = i \\ 0, & \text{otherwise} \end{cases}$$

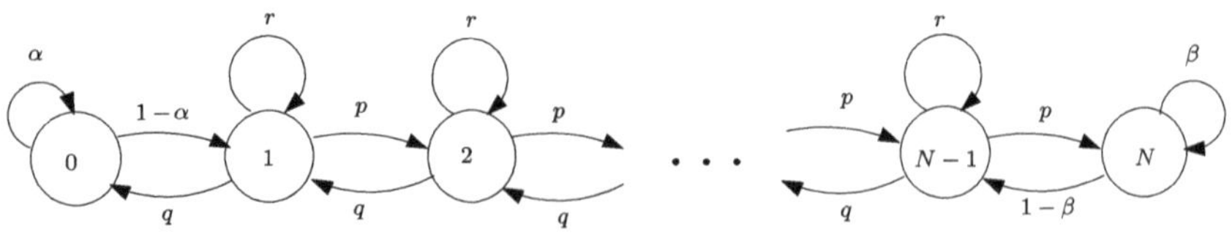

Fig. 2.4 State transition diagram for Example 2.5

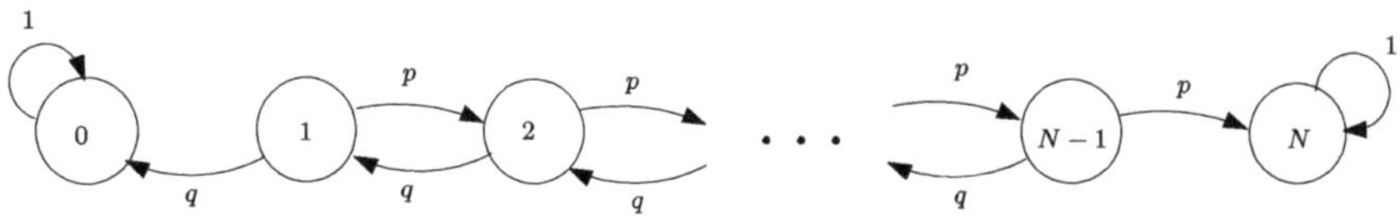

Fig. 2.5 State transition diagram for Example 2.6

and $p_{00} = p_{NN} = 1$ with $p + q = 1$.

Example 2.7 Consider a person playing a game in which a fair (unbiased) coin is tossed. The person earns a rupee every time the coin comes up heads. When the coin lands on tails, the player loses all of his/her previous gains and the game ends. Formulate this problem as a DTMC. Draw the state transition diagram for this DTMC.
Solution: Let X_n denote the capital of the person at the end of nth toss. However, $X_n = X_{n-1} + Y_n, n = 1, 2, \ldots$ where Y_n are IID RVs each with $P(Y_n = 0) = 1 - P(Y_n = 1) = 0.5, n = 1, 2, \ldots$ and independent on X_n,

$$X_{n+1} = \begin{cases} 0 & \text{with probability } \frac{1}{2} \\ X_n + 1 & \text{with probability } \frac{1}{2} \end{cases}$$

Thus, $\{X_n, n = 0, 1, \ldots\}$ forms a DTMC with state space {0,1, ...}. The state transition diagram for this DTMC is shown in Fig. 2.6.

Example 2.8 (Branching process problem) Consider a stochastic process in which each individual in generation n produces some random number of individuals in generation $n + 1$. This is according to a fixed probability distribution which does not vary from individual to individual. That is, the first generation of individuals is the collection of offsprings of a given individual. The next generation is formed by the offsprings of these individuals. Formulate this problem as a DTMC. Such a stochastic process is known as a *Branching process* which will be discussed later in Chap. 3.
Solution: Let X_n denote the number of individuals of the nth generation. Assume that the size of the zeroth generation is 1, hence, $X_0 = 1$. Let Y_i be the number of

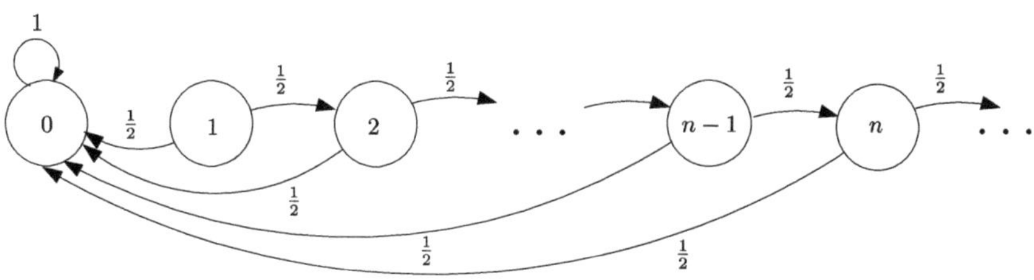

Fig. 2.6 State transition diagram for Example 2.7

offsprings of the ith individual of the nth generation. Suppose that $\{Y_i, i = 1, 2, \ldots\}$ are non-negative integer-valued IID RVs, each with probability mass function $p_j = P(Y_i = j)$, $j = 0, 1, \ldots$, and are independent of the size of the generation. Then

$$X_n = \sum_{i=1}^{X_{n-1}} Y_i, \quad n = 1, 2, \ldots$$

Thus, $\{X_n, n = 0, 1, \ldots\}$ is a DTMC with state space $S = \{0, 1, \ldots\}$.

2.2.3 *Chapman–Kolmogorov Equations*

The Chapman-Kolmogorov equation is an identity used to relate the joint probability distributions of various sets of coordinates on the stochastic process. The British mathematician Sydney Chapman (29 January 1888–16 June 1970) and the Russian mathematician Andrey Kolmogorov (25 April 1903–20 October 1987) independently derived an equation. It gives relationships between the n-step transition probability of the DTMC, which are defined by

$$p_{ij}^{(n)} = P\{X_{m+n} = j | X_m = i\},$$

for $i, j \in S$ and $m, n \geq 0$.

By analogy to the one-step case, the n-step transition probability matrix can be defined as $\mathbf{P}^{(n)} = [p_{ij}^{(n)}]$. Thus, $\mathbf{P}^{(1)} = \mathbf{P}$ and $\mathbf{P}^{(0)} = I$.

The higher-order transition probabilities $p_{ij}^{(n)}$ satisfy the following special case of Chapman–Kolmogorov equation which can be used to compute them.

Theorem 2.2 *Let $\{X_n, n = 0, 1, \ldots\}$ be a DTMC with state space S. For all $m, n \geq 0$ and all $i, j \in S$,*

$$p_{ij}^{(m+n)} = \sum_k p_{ik}^{(m)} p_{kj}^{(n)}. \tag{2.1}$$

Proof Noting that

$$\begin{aligned} p_{ij}^{(m+n)} &= \sum_{k=0}^{\infty} P\{X_{m+n} = j, X_m = k | X_0 = i\} \\ &= \sum_{k=0}^{\infty} P\{X_{m+n} = j | X_m = k, X_0 = i\} P\{X_m = k | X_0 = i\} \text{(by Markov Property)} \\ &= \sum_k p_{ik}^{(m)} p_{kj}^{(n)}. \end{aligned}$$

This result can be put in a matrix form as follows:

$$\mathbf{P}^{(m+n)} = \mathbf{P}^{(m)} \cdot \mathbf{P}^{(n)} = \mathbf{P}^{(n)} \cdot \mathbf{P}^{(m)}, \quad m, n \geq 0.$$

Equation (2.1) is known as the Chapman–Kolmogorov equation in DTMC. □

Remark 2.1 Note that beginning with

$$\mathbf{P}^{(2)} = \mathbf{P}^{(1+1)} = \mathbf{P} \cdot \mathbf{P} = \mathbf{P}^2$$

and continuing by induction, we can show that the n-step transition probability matrix $\mathbf{P}$ can be obtained by multiplying the matrix $\mathbf{P}$ by itself n times, i.e.,

$$\mathbf{P}^{(n)} = \mathbf{P}^{(n-1+1)} = \mathbf{P}^{(n-1)} \cdot \mathbf{P} = \mathbf{P}^{(n-2)} \cdot \mathbf{P} \cdot \mathbf{P} = \cdots = \mathbf{P}^n.$$

Hence,

$$\mathbf{P}^{(n)} = \mathbf{P}^n, \quad n = 2, 3, \ldots.$$

Example 2.9 Consider the simple weather forecasting model discussed in Example 2.1. Suppose that $\alpha = 0.7$ and $\beta = 0.6$, hence

$$\mathbf{P} = \begin{pmatrix} 0.7 & 0.3 \\ 0.4 & 0.6 \end{pmatrix}.$$

What is the probability that the weather will still be clear on 4th day from today, given that it is clear today?
Solution: It is required to find $P^{(4)}$,

$$\mathbf{P}^{(2)} = \mathbf{P} \cdot \mathbf{P} = \begin{pmatrix} 0.61 & 0.39 \\ 0.52 & 0.48 \end{pmatrix}$$

and

$$\mathbf{P}^{(4)} = \mathbf{P}^{(2)} \cdot \mathbf{P}^{(2)} = \begin{pmatrix} 0.5749 & 0.4251 \\ 0.5668 & 0.4332 \end{pmatrix}.$$

The required probability is $p_{11}^{(4)} = 0.4332$, and hence there is approximately a 57% chance that the weather will be clear on 4th day.

In order to compute unconditional probabilities, for instance in the last "What is the probability it will rain on Tuesday?". The initial state probability distribution is needed. Thus, once the transition probability matrix and the initial state distribution have been defined, the Markov chain is fully characterized.

2.2.3.1 Joint Probability Distribution

Now, the unconditional probabilities can be computed. The probability MC being in state j at a particular time n can be computed as

$$P(X_n = j) = \sum_{i \in S} P(X_0 = i) P\{X_n = j | X_0 = i\} = \sum_{i \in S} \alpha_i p_{ij}^{(n)}, \ j \in S$$

where $P(X_0 = i) = \alpha_i$ for all $i \in S$. Let $\pi_n = [\pi_n(j)]$ be a row vector whose elements are $\pi_n(j)$. Then $\pi_n = \pi_0 P^{(n)}, n = 1, 2, \ldots$. Similarly, the probability of a chain realization can be computed as follows:

$$\begin{aligned} P(X_0 = i_0, X_1 = i_1, \ldots, X_n = i_n) = & P(X_0 = i_0) \times P\{X_1 = i_1 | X_0 = i_0\} \\ & \times P\{X_2 = i_2 | X_0 = i_0, X_1 = i_1\} \\ & \cdots \times P\{X_n = i_n | X_0 = i_0, \ldots, X_{n-1} = i_{n-1}\}. \end{aligned}$$

The Markov property allows us to simplify the above equation as

$$\begin{aligned} P(X_0 = i_0, X_1 = i_1, \ldots, X_n = i_n) = & P(X_0 = i_0) \times P\{X_1 = i_1 | X_0 = i_0\} \\ & \times P\{X_2 = i_2 | X_1 = i_1\} \\ & \cdots \times P\{X_n = i_n | X_{n-1} = i_{n-1}\} \\ = & \alpha_{i_0} p_{i_0 i_1} p_{i_1 i_2} \cdots p_{i_{n-1} i_n}. \end{aligned}$$

Thus, the joint probability distribution of $X_r, X_{r+1}, \ldots, X_{r+n}$ can be computed in terms of the transition probabilities p_{ij} and the initial distribution π_0 of the Markov chain.

Example 2.10 In the simple weather forecasting model considered in Example 2.9, what is the probability that it will not rain on 4th day from today given that there was a 10% chance of rain today?
Solution: Required probability is calculated as

$$P(X_4 = 1) = \alpha_0 p_{01}^{(4)} + \alpha_1 p_{11}^{(4)} = 0.10 \times 0.4251 + 0.90 \times 0.4332 = 0.43239.$$

Example 2.11 Let $\{X_n, n \geq 0\}$ be a Markov chain with three states 0, 1, 2 and with TPM

$$\mathbf{P} = \begin{pmatrix} \frac{3}{4} & \frac{1}{4} & 0 \\ \frac{1}{4} & \frac{1}{2} & \frac{1}{4} \\ 0 & \frac{3}{4} & \frac{1}{4} \end{pmatrix}$$

and the initial distribution $P(X_0 = i) = \frac{1}{3}, \ i = 0, 1, 2$. Find

1. $P\{X_1 = 1 | X_0 = 2\}$
2. $P(X_2 = 2, X_1 = 1, X_0 = 2)$
3. $p_{01}^{(2)}$
4. $P(X_{102} = 1, X_{100} = 0)$.

Solution: The state transition diagram for this DTMC is shown in Fig. 2.7.
Required probabilities are given by

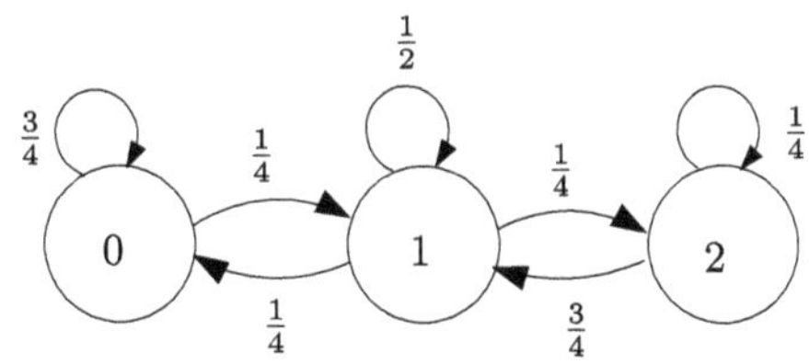

Fig. 2.7 State transition diagram for Example 2.11

1. $P\{X_1 = 1|X_0 = 2\} = \frac{3}{4}$.
2. $P(X_2 = 2, X_1 = 1, X_0 = 2) = \alpha_2 \times p_{21} \times p_{12} = \frac{1}{3} \times \frac{3}{4} \times \frac{1}{4} = \frac{1}{16}$.
3. $p_{01}^{(2)} = \frac{3}{4} \times \frac{1}{4} + \frac{1}{4} \times \frac{1}{2} = \frac{5}{16}$.
4. $P(X_{102} = 1, X_{100} = 0) = \alpha_0 \times p_{01}^{(2)} = \frac{5}{48}$.

2.3 Classification of States

Consider a Markov chain $\{X_n, n = 0, 1, \ldots\}$ over a state space $S = \{0, 1, \ldots\}$. The states are frequently categorized in distinct ways based on the system's fundamental features. It is possible to identify specific types of chains and examine their attributes in a unified manner using such classifications. The state transition diagrams for DTMC with finite state space and infinite state space are shown in Figs. 2.8 and 2.9, respectively, to understand the classification of states.

According to the definition, $p_{ii}^{(n)}$ is the probability of transitioning from state i to state i in n steps. Next, the so-called first return probabilities are defined by imposing the constraint that the process, having started at i, returns to i for the first time at the nth step.

2.3.1 *First Passage Probabilities*

In this subsection, we define first passage probabilities within the context of stochastic processes, examining the probability and timing of a process reaching a specific state for the first time.

Definition 2.3 (First Passage Probabilities) Let $\{X_n, n = 0, 1, \ldots\}$ be a DTMC with state space S and TPM $\mathbf{P}$. Let

$$f_{ii}^{(n)} = P\{X_n = i; X_m \neq i, m = 1, 2, \ldots, n-1|X_0 = i\},$$

be first return probability that the process, having started from state i, returns to state i for the first time at the nth step. In a similar way

$$f_{ij}^{(n)} = P\{X_n = j; X_m \neq j, m = 1, 2, \ldots, n-1|X_0 = i\},$$

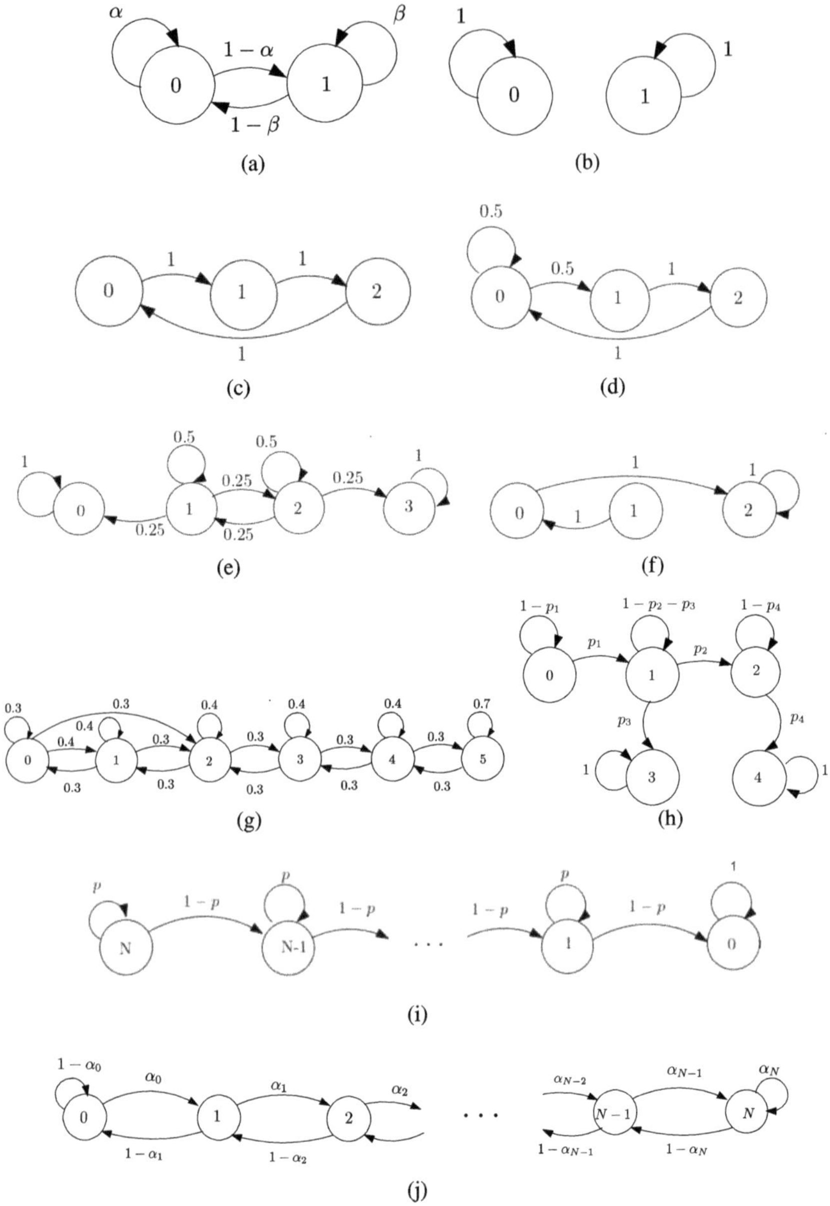

Fig. 2.8 State transition diagram of various finite state DTMCs

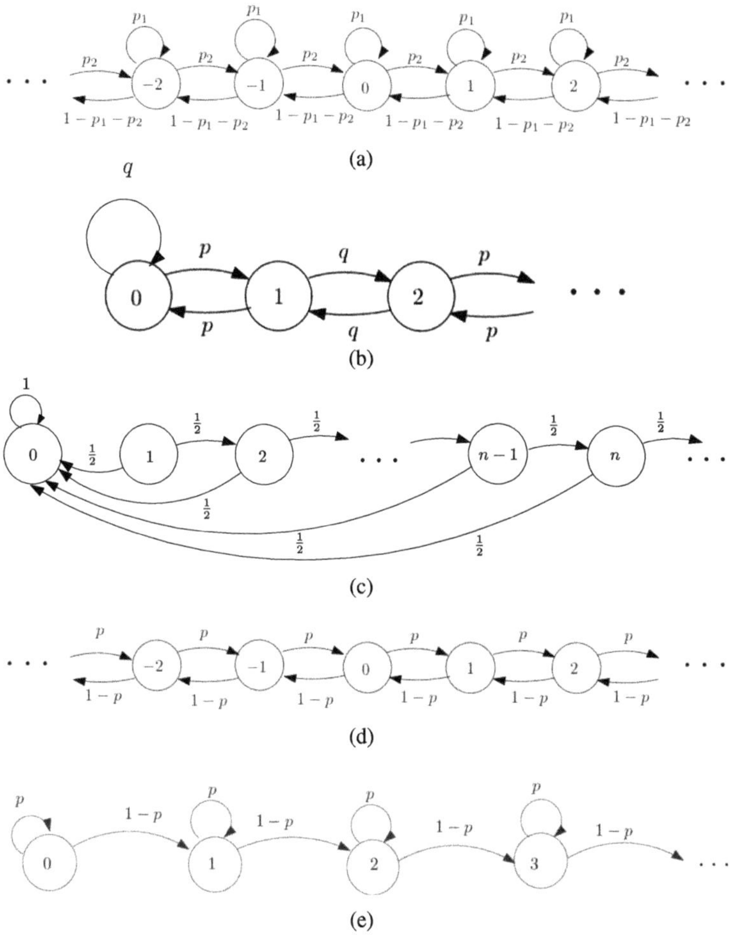

Fig. 2.9 State transition diagram of various infinite state DTMCs

gives the probability that the process started from state i reaches state j for the first time in exactly n steps. Clearly, $f_{ij}^{(0)} = 0$, for $i \neq j$ and $f_{ij}^{(1)} = p_{ij}$. These are known as the first passage probabilities.

Let F_{ij} denote the probability that starting from state i the system will ever reach state j regardless of number of steps. Clearly

$$F_{ij} = \sum_{n=1}^{\infty} f_{ij}^{(n)}.$$

For instance, from Fig. 2.8a, $F_{00} = \alpha + (1-\alpha)(1-\beta) + (1-\alpha)\beta(1-\beta) + \ldots = 1$ and $F_{01} = (1-\alpha) + \alpha(1-\alpha) + \alpha^2(1-\alpha) + \ldots = 1$.

Remark 2.2 Suppose $F_{ij} = 1$, then the system starting with state i is guaranteed to reach state j. In this case, $f_{ij}^{(n)}, n = 1, 2, \ldots$ is a probability distribution, and this yields the first passage time distribution for state j given that the system begins with state i.

Definition 2.4 (**Mean First Passage Time**) The mean first passage time from state i to state j is then calculated as follows:

$$\mu_{ij} = \sum_{n=1}^{\infty} n f_{ij}^{(n)}.$$

When $i = j$, the probability distribution of the *recurrence* times of state i is $f_{ii}^{(n)}, n = 1, 2, \ldots$ and $F_{ii} = 1$, implying that the return to state i is assured. The average mean recurrence time for state i is calculated as $\mu_{ii} = \sum_{n=1}^{\infty} n f_{ii}^{(n)}$ in this case.

For instance, from Fig. 2.8a,

$$\begin{aligned}
\mu_{01} &= 1(1-\alpha) + 2\alpha(1-\alpha) + 3\alpha^2(1-\alpha) + \cdots = \frac{1}{1-\alpha} \\
\mu_{00} &= \alpha + 2(1-\alpha)(1-\beta) + 3\beta(1-\alpha)(1-\beta) + \cdots \\
&= \alpha + (1-\alpha)(1-\beta)(2 + 3\beta + 4\beta^2 + \cdots) \\
&= \alpha + \frac{(1-\alpha)(2-\beta)}{1-\beta}.
\end{aligned}$$

2.3.2 *Recurrent and Transient States*

This subsection deals with the crucial concepts of recurrent and transient states. They are used to study the long-term behavior and stability of the Markov chain.

Definition 2.5 (**Recurrent State and Transient State**) If $F_{ii} = 1$, a state i is said to be recurrent (or persistent), and if $F_{ii} < 1$, it is said to be transient. If state i is recurrent, a MC starting from state i returns to state i with probability one. If state i is transient, a MC starting from state i has positive probability, i.e., $1 - F_{ii}$ of never returning to state i.

Definition 2.6 (**Positive and Null Recurrent States**) If $\mu_{ii} = \infty$, a recurrent state is said to be null recurrent, and if $\mu_{ii} < \infty$, then it is said to be positive (or non-null) recurrent.

For instance, from Fig. 2.8a, $F_{00} = 1$, implies state 0 is recurrent and

$$\mu_{00} = \alpha + \frac{(1-\alpha)(2-\beta)}{1-\beta}.$$

If $\beta = 1$ then $\mu_{00} < \infty$, state 0 is null recurrent, and if $\beta < 1$ then $\mu_{00} < \infty$, state 0 is positive (or non-null) recurrent.

Definition 2.7 **(Absorbing State)** A state i is said to be an absorbing state if $p_{ii} = 1$. An absorbing state is a special case of a positive recurrent state as $f_{ii}^{(1)} = p_{ii}^{(1)} = 1$, $f_{ii}^{(n)} = 0, n \geq 2$ gives $F_{ii} = 1$, $\mu_{ii} = 1$.

From Fig. 2.8b, $p_{00} = 1$, $p_{11} = 1$, and hence, both the states 0 and 1 are absorbing states. From Fig. 2.8e, states 0 and 3 are absorbing states. From Fig. 2.8h, states 3 and 4 are absorbing states. And, from Fig. 2.8i, state 0 is an absorbing state.

Remark 2.3 Note that null recurrent states can only occur in infinite state Markov chains. For example, the symmetric random walks in one (refer Example 2.30) and two dimensions are null recurrent.

Now, the simple results are stated below in the form of theorems.

Theorem 2.3 (First Entrance Theorem) *For all states $i, j \in S$,*

$$p_{ij}^{(n)} = \sum_{k=0}^{n} f_{ij}^{(k)} p_{jj}^{(n-k)}, \quad n \geq 1.$$

Proof

$$\begin{aligned} p_{ij}^{(n)} &= \sum_{k=0}^{n} P\{X_n = j, X_k = j, X_l \neq j \text{ for all } 0 \leq l \leq k-1 \mid X_0 = i\} \\ &= \sum_{k=0}^{n} P\{X_n = j \mid X_k = j\} P\{X_k = j, X_l \neq j \text{ for all } 0 \leq l \leq k-1 \mid X_0 = i\} \\ &= \sum_{k=0}^{n} p_{jj}^{(n-k)} f_{ij}^{(k)}, n \geq 1. \end{aligned}$$

Theorem 2.4 *Show that any state $i \in S$ is recurrent if and only if $\sum_{n=0}^{\infty} p_{ii}^{(n)} < \infty$. If state i is transient then*

$$\sum_{n=0}^{\infty} p_{ii}^{(n)} = \frac{1}{1 - F_{ii}}.$$

Proof Let

$$p_{ij}(s) = \sum_{n=0}^{\infty} p_{ij}^{(n)} s^n; \; F_{ij}(s) = \sum_{n=0}^{\infty} f_{ij}^{(n)} s^n.$$

Then,

$$
\begin{aligned}
p_{ij}(s) - \delta_{ij} &= \sum_{n=0}^{\infty} p_{ij}^{(n)} s^n - \delta_{ij} \\
&= \sum_{n=1}^{\infty} p_{ij}^{(n)} s^n \\
&= \sum_{n=1}^{\infty} \left[\sum_{k=0}^{n} p_{jj}^{(n-k)} f_{ij}^{(k)} \right] s^n \\
&= \sum_{k=0}^{\infty} f_{ij}^{(k)} \sum_{n=k}^{\infty} p_{jj}^{(n-k)} s^n \\
&= \left(\sum_{k=0}^{\infty} f_{ij}^{(k)} s^k \right) \left(\sum_{m=0}^{\infty} p_{jj}^{(m)} s^m \right), \quad m = n - k \\
&= F_{ij}(s) p_{jj}(s).
\end{aligned}
$$

Now,

$$
p_{ii}^{*} = \sum_{n=0}^{\infty} p_{ii}^{(n)} = p_{ii}(1) = \frac{1}{1 - F_{ii}(1)} = \frac{1}{1 - \sum_{n=0}^{\infty} f_{ii}^{(n)}} = \frac{1}{1 - F_{ii}}
$$

$$
= \begin{cases} \infty & \text{if } i \text{ is recurrent} \\ < \infty & \text{if } i \text{ is transient} \end{cases}.
$$

Therefore, state i is recurrent if and only if $\sum_{n=0}^{\infty} p_{ii}^{(n)} = \infty$, and when state is transient, then

$$
\sum_{n=0}^{\infty} p_{ii}^{(n)} = \frac{1}{1 - F_{ii}}.
$$

Some Important Results

Some results related to recurrent and transient states are as follows.

1. A recurring state is one of the states which is visited indefinitely. Assume that the recurrent state i is only visited $M < \infty$ times. There will be no more visits to state i after the Mth visit, because it is the last visit. This is incompatible with the fact that the chance of i being visited again after this time is $F_{ii} = 1$.
2. For transient states random number of visits is distributed as a geometric RV. Consider a MC which starts from transient state i. Then, with probability $1 - F_{ii} > 0$, the MC will never visit state i again. The probability that the chain visit state i, n more times is

$$
P\{n \text{ visits}\} = (F_{ii})^n (1 - F_{ii}), n = 0, 1, \ldots,
$$

which is the PMF of a modified geometric distribution with the expectation is $\frac{F_{ii}}{1-F_{ii}}$, which is a finite value.

3. The state i is said to be recurrent or transient depending on whether the series $\sum_{n=0}^{\infty} p_{ii}^{(n)}$ is divergent or convergent. For $n = 1, 2, \ldots,$ let

$$I_n = \begin{cases} 1, \text{ if } X_n = i; \\ 0, \text{ if } X_n \neq i, \end{cases}$$

be an indicator variable denoting whether the chain is in state i at the nth time point or not. Then, $\sum_{n=1}^{\infty} I_n$ is the total number of visits to state i, when the chain initiates in state i. Now, taking expectation,

$$E\Big[\sum_{n=1}^{\infty} I_n\Big] = \sum_{n=1}^{\infty} E[I_n | X_0 = i] = \sum_{n=1}^{\infty} P\{X_n = i | X_0 = i\} = \sum_{n=1}^{\infty} p_{ii}^{(n)} \text{ is } \infty \text{ or } < \infty$$

accordingly, as the state i is recurrent or transient.

4. If $p_{ij}^{(m)} > 0$ for some m and $p_{ji}^{(n)} > 0$ for some n. If state i is recurrent then state j is also recurrent.
Because, as i and j communicates, there exist positive integers m and n, such that $p_{ij}^{(m)} > 0$ and $p_{ji}^{(n)} > 0$. Hence, by Chapman–Kolmogorov equation,

$$p_{jj}^{(m+k+n)} \geq p_{ji}^{(n)} p_{ii}^{(k)} p_{ij}^{(m)}.$$

Summing over all possible values of k,

$$\sum_{k=1}^{\infty} p_{jj}^{(m+k+n)} \geq p_{ji}^{(n)} p_{ij}^{(m)} \sum_{k=0}^{\infty} p_{ii}^{(k)} = \infty.$$

as state i is recurrent. Thus, state j is also recurrent.

5. Let $G(i, j)$ denote the expected number of times the MC is in state j given that it starts in state i. Then,

$$G(i, j) = \frac{F_{ij}}{1 - F_{jj}}, \quad i, j \in S.$$

Further,

(i) If state j is a recurrent state, then $G(i, j) = \infty$.
(ii) If state j is a transient state, then $G(i, j)$ is finite for all $i \in S$.

These results describe the fundamental difference between a recurrent state and a transient state.

Now, the further classification of states such as aperiodic states, ergodic states, and closed class will be discussed.

2.3.3 Periodicity

Periodicity is one of the class properties and is defined as follows.

Definition 2.8 **(Periodicity)** The period d_i of a recurrent state i is defined as the greatest common divisor (gcd) of all m such that $p_{ii}^{(m)} > 0$. State i is said to be *aperiodic* if $d_i = 1$ and *periodic* if $d_i > 1$. If state i can never be revisited, then the period is defined as 0.

If every state has period 1, then the DTMC is said to be aperiodic. It can be proved that all the states in a communicating class have same period.

Example 2.12 Consider the Markov chain with TPM

$$\mathbf{P} = \begin{pmatrix} \frac{1}{2} & \frac{1}{2} \\ \frac{1}{3} & \frac{2}{3} \end{pmatrix},$$

and the corresponding state transition diagram is shown in Fig. 2.10. Find the periodicity of states 0 and 1.
Solution: The periodicity of state 0 is computed as

$$d_0 = \gcd\{n|p_{00}^{(n)} > 0\} = \gcd\{1, 2, \ldots\} = 1.$$

Similarly, the periodicity of state 1 is given by

$$d_1 = \gcd\{n|p_{11}^{(n)} > 0\} = \gcd\{1, 2, \ldots\} = 1.$$

Since the periodicity of the states 0 and 1 are 1, the Markov chain is aperiodic.

Example 2.13 Consider the Markov chain with TPM

$$\mathbf{P} = \begin{pmatrix} 0 & 0 & 1 \\ 1 & 0 & 0 \\ 0 & 0 & 1 \end{pmatrix},$$

and the corresponding state transition diagram is shown in Fig. 2.11. Find the period of the states 0,1, and 2.
Solution: Since states 0 and 1 are never revisited, the period of states 0 and 1 is 0. By the definition of periodicity , the periodicity of state 2 is computed as

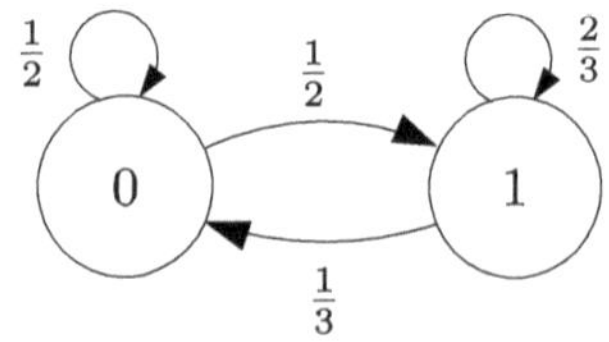

Fig. 2.10 State transition diagram for Example 2.12

Fig. 2.11 State transition diagram for Example 2.13

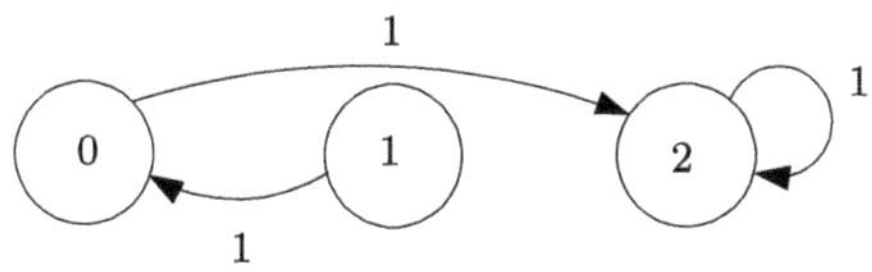

$$d_2 = \gcd\{n | p_{22}^{(n)} > 0\} = \gcd\{1, 2, \ldots\} = 1.$$

Further, the state 2 is an absorbing state. Hence, the periodicity of state 2 is 1.

2.3.4 *Irreducible Markov Chains*

In this subsection, the irreducible Markov chains in which each state can be reached from any other state are discussed in detail.

Definition 2.9 (Accessible, Communicating States, and Communicating Class) For some $n \geq 0$ if $p_{ij}^{(n)} > 0$, then state j *can be reached* or state j is *accessible* from state i; the relation is denoted by $i \rightarrow j$.

If two states are accessible from each other, then these two states are said to *communicate* with each other; it is denoted by $i \leftrightarrow j$. Communication between the states is an equivalence relation in the sense that it satisfies the three properties, i.e., reflexivity, symmetricity, and transitivity. Therefore, the communicating states in a Markov chain constitute equivalence classes called communicating classes.

Thus, the state space of a Markov chain may consist of a number of non-overlapping communicating classes. Note that, by identifying the communicating classes of the MC, each of which is relatively easy to understand, and which together give an understanding of the whole MC.

Definition 2.10 (Closed Class) A non-empty set $C \subseteq S$ is said to be closed if $\forall i \in C$ and $j \in S$ such that $i \leftrightarrow j$, then state $j \in C$. In simple words, exit from a closed class is not possible.

Equivalently, C is closed if and only if for all n,

$$p_{ij}^{(n)} = 0 \text{ for } i \in C, j \notin C.$$

If C is closed and suppose a MC starting in C, with probability one, then the MC will stay in C for ever. If state i is an absorbing state, then $\{i\}$ is closed.

If communicating class C is also closed, then every state in C is recurrent and C is called closed communicating class.

Definition 2.11 (Irreducible Markov chain) If the state space consists of only one closed communicating class, then the chain is said to be irreducible. That is, a Markov chain is *irreducible* if all the states are recurrent or all are transient; otherwise, the chain is called *reducible*.

Definition 2.12 (Regular Transition Probability Matrix) In any DTMC, a TPM $\mathbf{P}$ is regular if there exists n such that $\mathbf{P}^{(n)}$ has all strictly positive entries. A Markov chain with a regular transition probability matrix is irreducible.

Theorem 2.5 *In a finite irreducible Markov chain, all states are positive recurrent states.*

Proof A finite irreducible Markov chain must be positive or null recurrent using Corollary (2.1). Using Theorem (2.10) which will be proved later, it can be shown that positive or null recurrence is a class property. Thus, if one state is null recurrent, then all states are null recurrent. However, since $\sum_{j\in S} p_{ij}^{(n)} = 1$. As there are only finite number of states, it is impossible that $\lim_{n\to\infty} p_{ij}^{(n)} = 0$ for all $j \in S$. Thus, no state can be null recurrent. □

Thus, positive or null recurrence of an irreducible chain needs to be examined only when the Markov chain is not finite.

Remark 2.4
1. In a chain, recurrence (either positive or null) and transient are class properties. That is, all states in a communicating class are either recurrent (positive or null) states or all are transient states. The result that the recurrence and transience are class properties is an obvious consequence of the last result.
2. In a finite state Markov chain, all states can not be transient. In a finite Markov chain, some states must be recurrent states, in fact in an irreducible finite Markov chain all states are recurrent since there is only one equivalence class and recurrence is a class property. It can also be shown that all states in a finite Markov chain cannot be null recurrent.
3. An irreducible DTMC is a MC whose state space is irreducible, i.e., a MC in which every state leads back to itself and also to every other state. In such a MC, it is necessarily either all states are transient or recurrent.

From Fig. 2.8a, c, f, and j, the chain is irreducible, and hence, all states are recurrent. And, from Fig. 2.8d, the chain is reducible, and hence, states 0 and 1 are transient and 2 is recurrent. For instance, from Fig. 2.8g, the chain is reducible, but it has two closed communicating classes named as {0,1,2} and {3,4}. All the states are recurrent.

Example 2.14 Consider a Markov chain with $S = \{0, 1, 2, 3\}$ and with TPM

$$\mathbf{P} = \begin{pmatrix} 0 & 0 & \frac{1}{2} & \frac{1}{2} \\ 1 & 0 & 0 & 0 \\ 0 & 1 & 0 & 0 \\ 0 & 1 & 0 & 0 \end{pmatrix}.$$

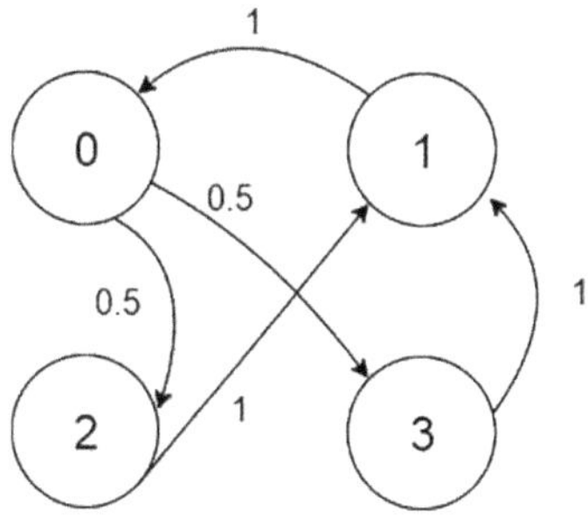

Fig. 2.12 State transition diagram for Example 2.14

The corresponding state transition diagram is shown in Fig. 2.12. Classify the states for this DTMC.

Solution: In this problem,

$$F_{00} = f_{00}^{(3)} = (p_{02}p_{21} + p_{03}p_{31})p_{10} = (0.5 \times 1 + 0.5 \times 1)1 = 1.$$

Also,

$$\mu_{00} = 3 \times f_{00}^{(3)} + 0 = 3.$$

Hence, state 0 is a positive recurrent. All states communicate with each other. Hence, the chain is irreducible and, therefore, all states must be positive recurrent.

Example 2.15 Consider a finite Markov chain with $S = \{0, 1, 2, 3, 4\}$ and TPM

$$\mathbf{P} = \begin{pmatrix} 1/2 & 1/2 & 0 & 0 & 0 \\ 1/2 & 1/2 & 0 & 0 & 0 \\ 0 & 0 & 1/2 & 1/2 & 0 \\ 0 & 0 & 1/2 & 1/2 & 0 \\ 1/4 & 1/4 & 0 & 0 & 1/2 \end{pmatrix}.$$

The corresponding state transition diagram is shown in Fig. 2.13. Classify the states of this DTMC.

Solution: This Markov chain consists of three communicating classes $\{0, 1\}$, $\{2, 3\}$, and $\{4\}$. While the first two classes $\{0, 1\}$ and $\{2, 3\}$ are recurrent, the third one $\{4\}$ is transient. Also note that classes $\{0, 1\}$ and $\{2, 3\}$ are closed communicating classes.

Example 2.16 Consider a Markov chain with $S = \{0, 1, 2, 3, 4\}$ and TPM

$$\mathbf{P} = \begin{pmatrix} 1 & 0 & 0 & 0 & 0 \\ 0 & 1 & 0 & 0 & 0 \\ 0 & 0 & 1/2 & 1/2 & 0 \\ 0 & 0 & 1/2 & 1/2 & 0 \\ 1/4 & 1/4 & 0 & 0 & 1/2 \end{pmatrix}.$$

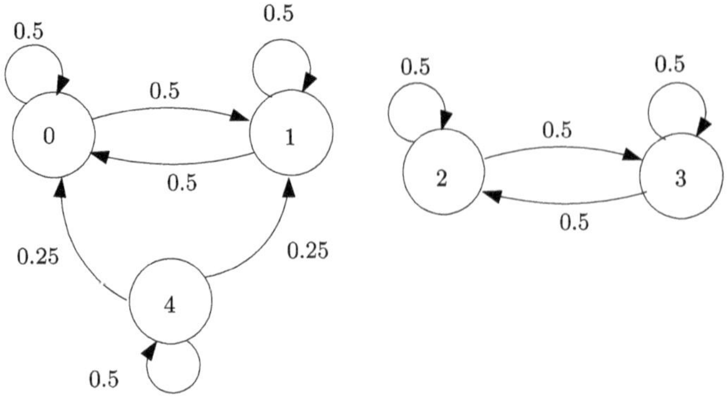

Fig. 2.13 State transition diagram for Example 2.15

Fig. 2.14 State transition diagram for Example 2.16

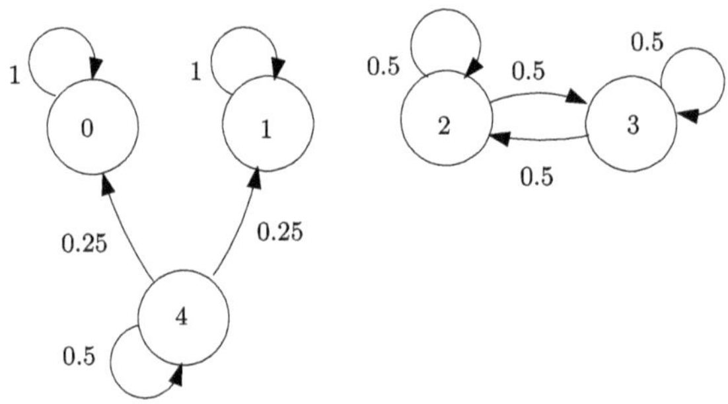

The corresponding state transition diagram is shown in Fig. 2.14. Classify the states of this DTMC.

Solution: This Markov chain consists of four communicating classes {0}, {1}, {2, 3}, and {4}. The first two classes are recurrent each having one absorbing state, the third one is recurrent, and the fourth one is transient.

Example 2.17 The Markov chain with TPM

$$\mathbf{P} = \begin{pmatrix} \frac{3}{4} & \frac{1}{4} & 0 \\ \frac{1}{4} & \frac{1}{2} & \frac{1}{4} \\ 0 & \frac{3}{4} & \frac{1}{4} \end{pmatrix}$$

is irreducible (Fig. 2.15).

Fig. 2.15 State transition diagram for Example 2.17

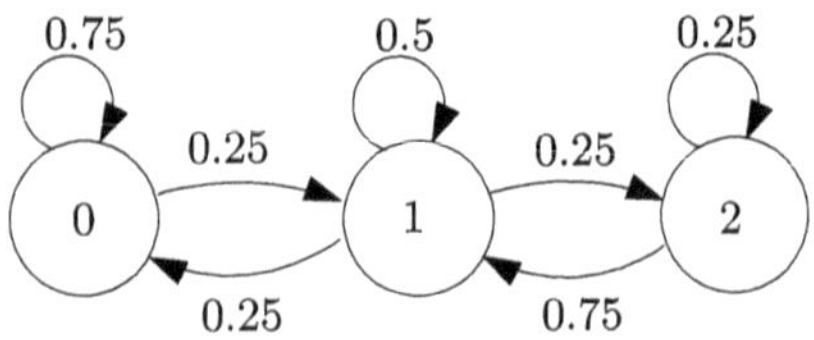

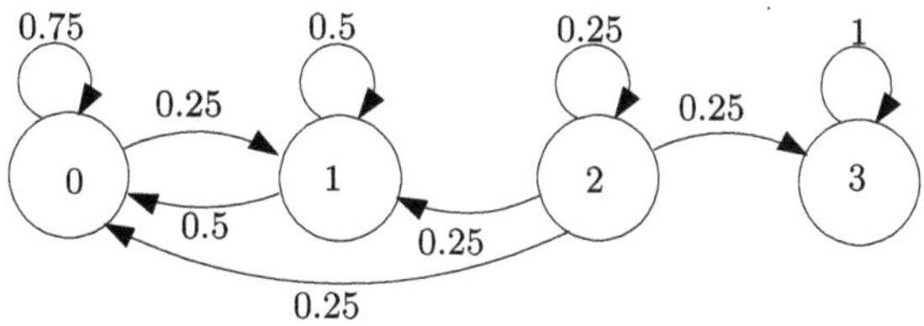

Fig. 2.16 State transition diagram for Example 2.18

Example 2.18 The Markov chain with $S = \{0, 1, 2, 3\}$ and with TPM

$$\mathbf{P} = \begin{pmatrix} \frac{3}{4} & \frac{1}{4} & 0 & 0 \\ \frac{1}{2} & \frac{1}{2} & 0 & 0 \\ \frac{1}{4} & \frac{1}{4} & \frac{1}{4} & \frac{1}{4} \\ 0 & 0 & 0 & 1 \end{pmatrix}$$

is reducible. It has three communicating classes $\{0, 1\}$, $\{2\}$, and $\{3\}$ (Fig. 2.16).

2.3.5 *Ergodicity*

Ergodicity is the time average of a function over a realization or sample, and it is related to the space average. That is, whenever a stochastic process is ergodic, the temporal average is the same for every starting points.

Definition 2.13 (**Ergodic Property**) An aperiodic and positive recurrent state of an irreducible Markov chain is called an *ergodic state* . An irreducible Markov chain is ergodic if all its states are irreducible, aperiodic and positive recurrent. Ergodicity is also a class property.

From Fig. 2.8a, f, and j, the chains are irreducible, positive recurrent, and aperiodic, and hence the chains are ergodic. From Fig. 2.8b, d, e, g–i the chains are reducible, whereas from Fig. 2.8c, the chain with all the states having period 3, and hence these chains are non-ergodic. Similarly, from Fig. 2.9a, b, and d, the chains are ergodic based on the proper choice of the transition probabilities, whereas from Fig. 2.9c, the chain is reducible and hence non-ergodic and whereas from Fig. 2.9e, the chain with all the states are transient states, and hence the chain is non-ergodic.

Example 2.19 Consider the three-state Markov chain having TPM $\mathbf{P} = \begin{pmatrix} 0 & 2/3 & 1/3 \\ 1/2 & 0 & 1/2 \\ 1/2 & 1/2 & 0 \end{pmatrix}$ with $S = \{1, 2, 3\}$. Check whether this chain is ergodic.

Solution: The state transition diagram is shown in Fig. 2.17. The chain is irreducible. Further, the states are aperiodic and positive recurrent, and hence the chain is ergodic.

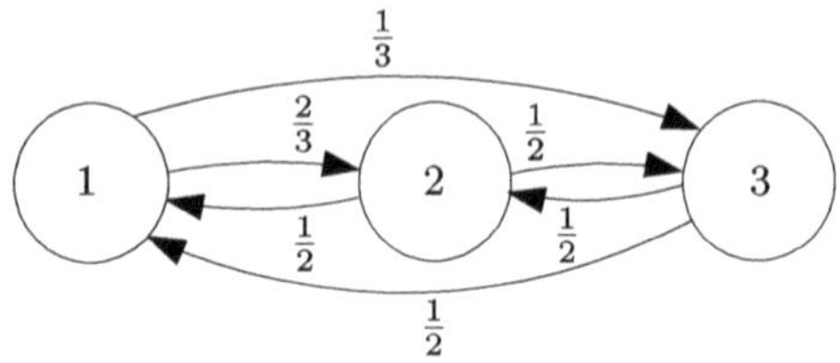

Fig. 2.17 State transition diagram for Example 2.34

2.4 Limiting Distribution

Now we discuss the limiting and stationary distribution of Markov chains. A sufficient condition for the existence of stationary distribution will also be stated.

Consider the TPM

$$\mathbf{P} = \begin{pmatrix} 0.7 & 0.3 \\ 0.4 & 0.6 \end{pmatrix}$$

and using the Chapman—Kolmogorov equation examine examine the powers of this TPM.

$$\mathbf{P}^{(2)} = \begin{pmatrix} 0.61 & 0.39 \\ 0.52 & 0.48 \end{pmatrix}, \quad \mathbf{P}^{(4)} = \begin{pmatrix} 0.5749 & 0.4251 \\ 0.5668 & 0.4332 \end{pmatrix}, \quad \mathbf{P}^{(8)} \approx \begin{pmatrix} 0.572 & 0.428 \\ 0.570 & 0.430 \end{pmatrix}.$$

It can be observed that rows in the above matrix become approximately same. For example, both $p_{00}^{(8)}$ and $p_{10}^{(8)}$ are approximately same. The probability of going into state 0 becomes nearly independent of the initial state as time passes (here, after 8 time steps). Indeed, the n-step transition probabilities $p_{ij}^{(n)}$ approach a value, that is independent of the initial state i under specific conditions as $n \to \infty$. Another way to put it is that when n is large enough, the chain's probabilistic behavior becomes independent of the initial state, i.e., $P\{X_n = j | X_0 = i\} = P(X_n = j)$.

It is now investigated whether being independent of the initial state i, a Markov chain may reach a steady or stable state after a large number of transitions. In other words, under what conditions, if any, does $p_{ij}^{(n)}$ as n approaches ∞ tend to a limit which is independent of the starting state i? The property of the limiting distribution of $p_{ij}^{(n)}$ being independent of the starting state i is known as ergodicity, which is discussed later in this chapter, and the Markov chain becomes ergodic as a result. The transition probabilities settle down and become stable when such constraints exist. After that, the MC exhibits some long-run regularity features.

Definition 2.14 **(Limiting Distribution)** Let $\{X_n, n = 0, 1, \ldots\}$ be a DTMC with state space S. The probability distribution $\pi = [\pi_i]_{i \in S}$ is called the limiting distribution of the Markov chain if

$$\pi_j = \lim_{n\to\infty} P\{X_n = j | X_0 = i\} \text{ for all } i, j \in S,$$

and

$$\sum_{j \in S} \pi_j = 1.$$

By the above definition, when a limiting distribution exists, it does not depend on the initial state ($X_0 = i$), thus

$$\pi_j = \lim_{n \to \infty} P(X_n = j), \text{ for all } j \in S.$$

Example 2.20 Let $\{X_n, n = 0, 1, \ldots\}$ be a DTMC with state space $S = \{0, 1\}$. For the DTMC with TPM

$$\mathbf{P} = \begin{pmatrix} 0 & 1 \\ 1 & 0 \end{pmatrix},$$

find the limiting distribution of the DTMC, if it exists.
Solution: For any n

$$\mathbf{P}^{(2n)} = \begin{pmatrix} 1 & 0 \\ 0 & 1 \end{pmatrix}, \quad \mathbf{P}^{(2n+1)} = \begin{pmatrix} 0 & 1 \\ 1 & 0 \end{pmatrix}.$$

Note that the values of each $p_{ij}^{(n)}$ oscillate between 0 and 1 and, therefore, it doesn't converge to single limit as $n \to \infty$. Hence, the limiting distribution does not exist.

Example 2.21 Let $\{X_n, n = 0, 1, \ldots\}$ be a DTMC with state space $S = \{0, 1\}$. For the DTMC with TPM

$$\mathbf{P} = \begin{pmatrix} 0.7 & 0.3 \\ 0.4 & 0.6 \end{pmatrix},$$

find the limiting distribution, if it exists.
Solution: The limit of $\mathbf{P}^{(n)}$ (as given above by continuing the limiting procedure) is given by

$$\lim_{n \to \infty} \mathbf{P}^{(n)} = \begin{pmatrix} 0.5714 & 0.4286 \\ 0.5714 & 0.4286 \end{pmatrix}.$$

Thus, (0.5714, 0.4286) is the limiting distribution of the DTMC.

Example 2.22 Suppose $\{X_n, n = 0, 1, \ldots\}$ is a DTMC with state space $S = \{1, 2, 3\}$ and the following one-step TPM:

$$\mathbf{P} = \begin{pmatrix} 0.20 & 0.30 & 0.50 \\ 0.10 & 0.00 & 0.90 \\ 0.55 & 0.00 & 0.45 \end{pmatrix}.$$

Find the limiting distribution of the DTMC, if it exists.

Solution: Different values of n the n-step transition probability matrix is given by

$$n=2:\mathbf{P}^{(2)}=\begin{pmatrix}0.3450 & 0.0600 & 0.5950\\ 0.5150 & 0.0300 & 0.4550\\ 0.3575 & 0.1650 & 0.4775\end{pmatrix},$$

$$n=4:\mathbf{P}^{(4)}=\begin{pmatrix}0.3626 & 0.1207 & 0.5167\\ 0.3558 & 0.1069 & 0.5373\\ 0.3790 & 0.1052 & 0.5158\end{pmatrix},$$

$$n=10:\mathbf{P}^{(10)}=\begin{pmatrix}0.3704 & 0.1111 & 0.5185\\ 0.3703 & 0.1111 & 0.5186\\ 0.3704 & 0.1111 & 0.5185\end{pmatrix},$$

$$n\geq 11:\mathbf{P}^{(n)}=\begin{pmatrix}0.3704 & 0.1111 & 0.5185\\ 0.3704 & 0.1111 & 0.5185\\ 0.3704 & 0.1111 & 0.5185\end{pmatrix}.$$

It can be observed that the probability mass function of X_n approaches

$$\pi=[0.3704, 0.1111, 0.5185].$$

Theorem 2.6 (Limiting Distribution) *Let $\{X_n, n=0,1,\ldots\}$ be a DTMC with state space $S=\{1,2,\ldots,N\}$. If a limiting distribution π exists, it satisfies*

$$\pi_j=\sum_{i=1}^{N}\pi_i p_{ij},\quad j\in S, \tag{2.2}$$

and

$$\sum_{j=1}^{N}\pi_j=1. \tag{2.3}$$

Proof Conditioning on X_n and using the law of total probability,

$$\mathrm{P}\left(X_{n+1}=j\right)=\sum_{i=1}^{N}\mathrm{P}\left(X_n=i\right)p_{ij},\quad j\in S. \tag{2.4}$$

Now, let n tend to ∞ on both the right- and left-hand sides. Then, assuming that the limiting distribution exists,

$$\lim_{n\to\infty}\mathrm{P}\left(X_n=j\right)=\lim_{n\to\infty}\mathrm{P}\left(X_{n+1}=j\right)=\pi_j,\quad j\in S.$$

Substituting the above result in (2.4), (2.2) is obtained. Equation (2.3) follows since π is a probability mass function. Further, (2.2) can be written in matrix form as

$$\pi = \pi \mathbf{P},$$

and is called the balance equations or the steady-state equations and (2.4) is called the normalizing equation.

The above theorem is illustrated with the following example.

Example 2.23 Consider Example 2.22. Find the limiting distribution of this DTMC using the above theorem.
Solution: Let $\pi = [\pi_1, \ \pi_2, \ \pi_3]$, then by the above theorem,

$$\pi \mathbf{P} = \pi \text{ and } \sum_{i=1}^{3} \pi_i = 1$$

$$\begin{aligned}
0.2\pi_1 + 0.1\pi_2 + 0.55\pi_3 &= \pi_1 \\
0.3\pi_1 + 0.0\pi_2 + 0.00\pi_3 &= \pi_2 \\
0.5\pi_1 + 0.9\pi_2 + 0.45\pi_3 &= \pi_3.
\end{aligned}$$

Solving the above system of linear equations,

$$\pi_2 = 0.30\pi_1, \quad \pi_3 = \frac{0.77}{0.55}\pi_1.$$

Using $\sum_{i=1}^{3} \pi_i = 1$,

$$\pi_1 = 0.3704, \quad \pi_2 \quad = 0.1111, \quad \pi_3 = 0.5185.$$

Example 2.24 Consider a DTMC $\{X_n, n \geq 0\}$ with state space $\{1, 2, 3\}$ and transition matrix

$$\mathbf{P} = \begin{pmatrix} 0.20 & 0.80 & 0 \\ 0.10 & 0.90 & 0 \\ 0 & 0 & 1 \end{pmatrix}.$$

Find the limiting distribution, if it exists.
Solution: Computing the matrix powers $\mathbf{P}^{(n)}$ for increasing values of n,

$$\lim_{n \to \infty} \mathbf{P}^{(n)} = \begin{pmatrix} 0.1111 & 0.8889 & 0 \\ 0.1111 & 0.8889 & 0 \\ 0 & 0 & 1.0000 \end{pmatrix}.$$

It can be observed that every row of limiting value of $\mathbf{P}^{(n)}$ is not the same. This implies that there doesn't exist unique limiting distribution which is independent of the initial distribution.

2.5 Stationary Distribution

In this section, the concept of a stationary distribution for a Markov chain is discussed.

Definition 2.15 (Stationary Distribution) Consider a DTMC $\{X_n, n = 0, 1, \ldots\}$ with state space S and TPM $\mathbf{P} = [p_{ij}]$. A probability distribution $[\pi_i]$ is called *stationary* (or *invariant*) for the given Markov chain, if it satisfies

$$\pi_j = \sum_i \pi_i p_{ij}, \quad j \in S$$

such that $\pi_i \geq 0, \sum_i \pi_i = 1$. In vector notations, this is written as $\pi \mathbf{P} = \pi, \pi \mathbf{e} = 1$, with π as a row vector and $\mathbf{e}$ as a column vector of ones. Stationary distribution is also called as steady-state distribution.

Once a Markov chain has reached a distribution π such that $\pi P = \pi$, it will stay there. If $\pi P = \pi$, the distribution π become an equilibrium distribution. Equilibrium does not imply that the values of X_n and X_{n+1} are identical. It implies that the distribution of X_{n+1} and X_n are identical. An equilibrium distribution goes by a number of other names. Equilibrium distributions are also referred to as invariant or stationary if π is one of them.

If the initial state is chosen according to the stationary distribution, i.e., if

$$P(X_0 = j) = \pi_j, \quad j \in S,$$

then, using the total probability theorem,

$$P(X_1 = j) = \sum_{k \in S} P(X_0 = k) p_{kj} = \sum_{k \in S} \pi_k p_{kj} = \pi_j.$$

Similarly, for all n

$$P(X_n = j) = \pi_j, \quad j \in S.$$

Consequently, all subsequent states will have the same distribution.

While having a unique stationary distribution is desired in many applications, not all Markov chains have one. Following is an illustration of a Markov chain without a stationary distribution. It will be shown one in which stationary distribution does exists but the probability distribution over the states does not eventually converge to it in steady state (or to equilibrium). To guarantee the existence of a stationary distribution to which the Markov chain converges in steady-state, some conditions will be applied based on these examples.

Example 2.25 Consider a simple two-state Markov chain with states 0 and 1, where the chain is invariant for all time periods $n \geq 0$. As a result, $\mathbf{P} = \begin{pmatrix} 1 & 0 \\ 0 & 1 \end{pmatrix}$ gives the transition probability matrix for this Markov chain. The stationary distribution is

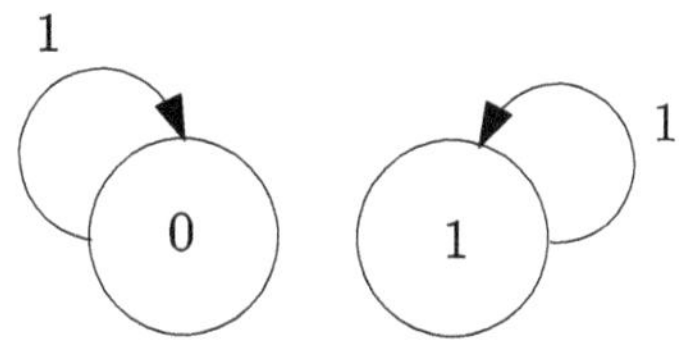

Fig. 2.18 State transition diagram for Example 2.25

not unique because $\pi \mathbf{P} = \pi$ holds true for any distribution π. In this illustration, the Markov chain is constructed such that if it began in one state, it would end up staying there forever. A Markov chain without a unique stationary distribution is one in which one state cannot be reached from another. Figure 2.18 contains the state transition diagram.

Note that the Markov chain given in the example above is reducible. Let us consider another example.

Example 2.26 Consider Example 2.20, a simple two-state Markov chain with states 0 and 1 that alternates between the two states for time intervals $n \geq 0$. As a result, $\mathbf{P} = \begin{pmatrix} 0 & 1 \\ 1 & 0 \end{pmatrix}$. The state transition diagram is given in Fig. 2.19.

Discuss the stationary distribution for this Markov chain.

Solution: The stationary distribution is obtained by solving $\pi \mathbf{P} = \pi, \pi \mathbf{e} = 1$ which gives $\pi = (1/2, 1/2)$. However, as demonstrated in the aforementioned Example 2.25, the system does not converge to this stationary distribution starting from any initial condition. The state of this Markov chain flips between 0 and 1 on a regular basis, which prevents it from converging to the stationary distribution.

It is easy to see that

$$\mathbf{P}^{(2n-1)} = \begin{pmatrix} 0 & 1 \\ 1 & 0 \end{pmatrix} \text{ and } \mathbf{P}^{(2n)} = \begin{pmatrix} 1 & 0 \\ 0 & 1 \end{pmatrix} \text{ for } n \in \mathbb{N}.$$

This means that the values of each $p_{ij}^{(n)}$ oscillate between 0 and 1 and, therefore, cannot converge to anything when $n \to \infty$. In other words, there is no limiting distribution. The matrix $\mathbf{P}^{(n)}$ never becomes a matrix with both rows identical as n gets large. That is, limiting distribution does not exist, even though stationary distribution exists. A more subtle reason for the non-existence of limiting distributions can be dependence on the initial conditions.

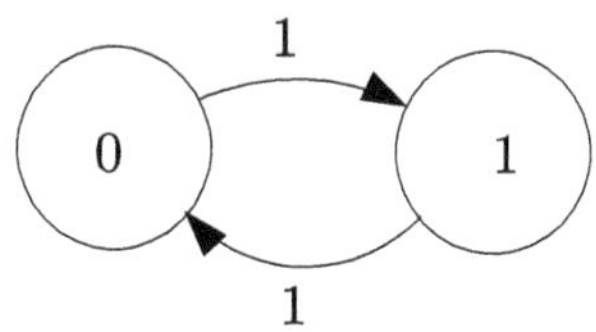

Fig. 2.19 State transition diagram for Example 2.26

Now, a theorem is presented as follows which is related to limiting distribution of Markov chain.

Theorem 2.7 (Basic limit theorem of renewal theory) *Let $\{f_n\}$ be a sequence such that $f_n \geq 0, \sum f_n = d$ and d be the greatest common divisor of those n for which $f_n > 0$. Let $\{u_n\}$ be another sequence such that $u_0 = 1$ and $u_n = \sum_{k=1}^{n} f_k u_{n-k}$ $(n \geq 1)$. Then*

$$\lim_{n\to\infty} u_n = \frac{d}{\mu},$$

where $\mu = \sum_{n=1}^{\infty} n f_n$, the limit being zero when μ is ∞, i.e., $\lim_{n\to\infty} u_n = 0$.

Remark 2.5 The proof of Theorem 2.7 is presented for $d = 1$ and the proof for $d > 1$ is left as an exercise.

Proof It can be proved that $u_n < 1$. Let $\limsup\limits_{n\to\infty} u_n = a$. Therefore, a subsequence u_{n_k} can be chosen such that $u_{n_k} \to a$ and without loss of generality $u_{n_k+j} \to v_j$ as $k \to \infty$ for all $j \in \mathbb{R}$. Now pass the limit in $u_n = \sum\limits_{k=1}^{n} f_k u_{n-k}$ and by using the dominated convergence theorem (see Appendix A.2)

$$v_n = \sum_{j=1}^{\infty} f_j v_{n-j} \quad \text{for } n \in (-\infty, \infty). \tag{2.5}$$

In particular,

$$v_0 = \sum_{j=1}^{\infty} f_j v_{-j}.$$

Since, $\limsup\limits_{n\to\infty} u_n = a$ which implies $v_j \leq a$ for all j. Let $S = \{n : f_n > 0\}$ then $v_k = a$ for $k \in S$, and using (2.5) it can be proved that $v_k = a$ for $k = k_1 + k_2 + \cdots + k_l$. Since the greatest common divisor of S is 1, there exists k_0 such that for $k \geq k_0$, $v_k = a$, and again by using (2.5) it can be concluded that $v_j = a$ for all $j \in \mathbb{R}$. For $n = 1, \ldots, N$

$$\begin{aligned} u_1 + u_2 + \cdots + u_N = (f_1 + f_2 + \cdots + f_N) + (f_1 + f_2 + \cdots + f_{N-1})u_1 \\ + \cdots + (f_1 + f_2 + \cdots + f_{N-k})u_k + \cdots + f_1 p_{N-1}. \end{aligned}$$

Let, $T_k = \sum_{i=k}^{\infty} f_k$, $T_1 = 1$ and $\sum_{k=0}^{\infty} T_k = \mu$. Therefore,

$$\sum_{k=1}^{N} T_k u_{N-k+1} = \sum_{j=1}^{N} f_j.$$

However, along the subsequence $N = n_k$, $u_{N-j} \to a$ for every j, if $\sum_{k=0}^{\infty} T_k = \mu < \infty$, then by dominated convergence theorem (see Appendix A.2), $\mu a = 1$ and,

$\limsup_{n\to\infty} u_n = a = \frac{1}{\mu}$. Further, by using the same argument, it can be also concluded that $\liminf_{n\to\infty} u_n = a = \frac{1}{\mu}$ which implies

$$\lim_{n\to\infty} u_n = \frac{1}{\mu}.$$

If $\sum_{k=0}^{\infty} T_k = \mu = \infty$ then by Fatou's Lemma (see Appendix A.2) $\lim_{N\to\infty} u_N = a = 0$.

The following theorem will be used to obtain some important results as explained below.

Theorem 2.8 *Let **P** be the TPM of a given MC.*

1. *Suppose state i is positive recurrent, then as $n \to \infty$*

   (i) state i is periodic with period d, then

$$p_{ii}^{(n)} \to \frac{d}{\mu_{ii}},$$

   (ii) state i is aperiodic, then

$$p_{ii}^{(n)} \to \frac{1}{\mu_{ii}}.$$

2. *Suppose state i is null recurrent, then as $n \to \infty$,*

$$p_{ii}^{(n)} \to 0.$$

Proof Let state i be recurrent. Then $\mu_{ii} = \sum n f_{ii}^{(n)}$ is defined. From the first entrance theorem (Theorem 2.3),

$$p_{ii}^{(n)} = \sum_{k=0}^{n} f_{ii}^{(k)} p_{ii}^{(n-k)}.$$

Identifying the quantities above with those of the basic limit theorem of renewal theory, the required result is obtained. □

Theorem 2.9 *If state j is positive recurrent and aperiodic, then for every i*

$$\lim_{n\to\infty} p_{ij}^{(n)} = \frac{F_{ij}}{\mu_{jj}},$$

and if state j is null recurrent, then for every i

$$\lim_{n\to\infty} p_{ij}^{(n)} = 0.$$

Proof Let state j be recurrent. Then $\mu_{ij} = \sum nf_{ij}^{(n)}$ is defined. From the first entrance theorem (Theorem 2.3),

$$p_{ij}^{(n)} = \sum_{k=0}^{(n)} f_{ij}^{(k)} p_{jj}^{(n-k)}.$$

Identifying the quantities above with those of the basic limit theorem of renewal theory, the required result is obtained. □

Corollary 2.1 *In a finite Markov chain, not all states can be null recurrent.*

Proof Let all the states be null recurrent. Then for all $j \in S$, $\lim_{n\to\infty} p_{ij}^{(n)} = 0 \,\forall\, i \in S$. Also, $\sum_{j\in S} p_{ij}^{(n)} = 1$, which is a contradiction.

Theorem 2.10 *All states in an irreducible chain belong to the same type, namely transient, null recurrent, or positive recurrent. All states are either aperiodic or periodic, with the latter having all of them the same period.*

Proof Since the chain is irreducible, every state is accessible from every other state. If $i, j \in S$, then $p_{ij}^{(N)} = a > 0$ for some $N \geq 1$ and $p_{ji}^{(M)} = b > 0$ for some $M \geq 1$. This implies that

$$p_{jk}^{(n+m)} = p_{jk}^{(m+n)} = \sum_r p_{jr}^{(m)} p_{rk}^{(n)} \geq p_{jr}^{(m)} p_{rk}^{(n)}, \text{ for each } r.$$

Here

$$p_{ii}^{(n+N+M)} \geq p_{ij}^{(N)} p_{jj}^{(n)} p_{ji}^{(M)} = abp_{jj}^{(n)}$$

and

$$p_{jj}^{(n+N+M)} \geq p_{ji}^{(M)} p_{ii}^{(n)} p_{ij}^{(N)} = abp_{ii}^{(n)}.$$

From the above, it is clear that the two series $\sum p_{ii}^{(n)}$ and $\sum p_{jj}^{(n)}$ converge or diverge together. Thus, the two states i and j are either both transient or both recurrent.

Suppose that state i is a null recurrent, then $p_{ii}^{(n)} \to 0$ as $n \to \infty$, then from the above, $p_{jj}^{(n)} \to 0$ as $n \to \infty$, thus state j is also null recurrent. Suppose that state i is a positive recurrent and has period t, then $p_{ii}^{(n)} > 0$ whenever n is a multiple of t. Now

$$p_{ii}^{(N+M)} \geq p_{ij}^{(N)} p_{ji}^{(M)} = ab > 0,$$

so that $(N + M)$ is a multiple of t. Again, from the above for n, a multiple of t,

$$p_{jj}^{(n+N+M)} \geq abp_{ii}^{(n)} > 0.$$

Thus, $(n + N + M)$ is a multiple of t, and therefore, t is the period of state j also.

Example 2.27 Consider a two-state Markov chain with state space $S = \{0, 1\}$. Let the matrix $\mathbf{P}$ be given by

$$\mathbf{P} = \begin{pmatrix} p_1 & 1 - p_1 \\ 1 - p_2 & p_2 \end{pmatrix}.$$

Classify the states of this Markov chain. Also, discuss whether the limiting distribution and stationary distribution exist or not.
Solution: Consider the four cases:

(a) When $0 < p_1 < 1$ and $0 < p_2 < 1$.
Here, both the states 0 and 1 are aperiodic and positive recurrent. There is one closed communicating class given by $\{0,1\}$, i.e., equal to S, and hence, the chain is irreducible. These properties imply that the chain is ergodic.
(b) When $0 < p_1 < 1$ and $p_2 = 1$.
Here, the state 0 is a transient state and state 1 is positive recurrent which is also absorbing. There is one closed communicating class given by $\{1\}$, i.e., not same as S, and hence, the chain is reducible.
(c) When $p_1 = 1$ and $0 < p_2 < 1$.
Similar discussion as in (b) gives the state 0 is positive recurrent which is also absorbing and state 1 is transient. Hence, the chain is reducible.
(d) When $p_1 = 1$ and $p_2 = 1$. Here, both the states 0 and 1 are aperiodic and positive recurrent. There are two closed communicating classes given as $\{0\}$,$\{1\}$, and hence, the chain is reducible.

Example 2.28 Consider a three-state Markov chain with state space $S = \{0, 1, 2\}$. Let the matrix $\mathbf{P}$ be given by

$$\mathbf{P} = \begin{pmatrix} 0 & 1 & 0 \\ 0 & 0 & 1 \\ 1 & 0 & 0 \end{pmatrix}.$$

Classify the states of this Markov chain. Also, discuss whether limiting distribution and stationary distribution exist or not.
Solution: The state transition diagram is shown in Fig. 2.20.

Here, $p_{ii}^{(3n)} > 0$ for all $i \in S$ and $n \in \mathbb{N}$, and hence the period is 3 for all the states. Hence, the states are not aperiodic. All the states are positive recurrent, and there is one closed communicating class same as S, and hence, chain is irreducible.

As shown in Example 2.26, the stationary distribution is obtained by solving $\pi\mathbf{P} = \pi, \pi\mathbf{e} = 1$ which gives $\pi = (\frac{1}{3}, \frac{1}{3}, \frac{1}{3})$, but the system does not converge to it. Hence, the limiting distribution doesn't exist.

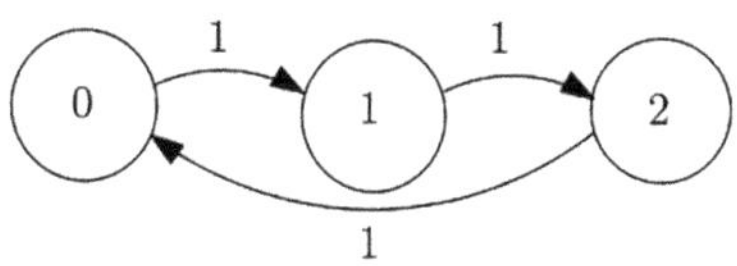

Fig. 2.20 State transition diagram for Example 2.28

Example 2.29 Consider the previous example with the matrix $\mathbf{P}$ be given by

$$\mathbf{P} = \begin{pmatrix} 0 & 1 & 0 \\ 0 & p & 1-p \\ 1 & 0 & 0 \end{pmatrix}$$

where $0 \le p \le 1$. Classify the states of this Markov chain. Also, discuss whether limiting distribution and stationary distribution exist or not.
Solution: Based on the values of p, there will be three cases.

1. When $p = 0$, this Markov chain is the same as in the previous example.
2. When $p = 1$, states 0 and 2 are transient states and state 1 is absorbing state. Hence, the limiting distribution for this Markov chain is $v = (0, 1, 0)$. The stationary distribution is obtained by solving $\pi\mathbf{P} = \pi$ with $\sum_i \pi_i = 1$ and is given by $\pi = (0, 1, 0)$.
3. When $0 < p < 1$, the state transition diagram is shown in Fig. 2.21. This chain is aperiodic, positive recurrent, and irreducible. Hence, the limiting distribution is same as the stationary distribution. By solving $\pi\mathbf{P} = \pi$ with $\sum_i \pi_i = 1$, $\pi = \left(\frac{1-p}{3-2p}, \frac{1}{3-2p}, \frac{1-p}{3-2p}\right)$.

Example 2.30 Consider an infinite state Markov chain $\{X_n, n = 0, 1, \ldots\}$ with state space $S = \{\ldots, -2, -1, 0, 1, 2, \ldots\}$. Here $p_{ii+1} = p$ and $p_{ii-1} = 1 - p$ for some $0 < p < 1$ and all $i \in S$. Classify the states of this Markov chain.
Solution: The state transition diagram is shown in Fig. 2.22. Take $q = 1 - p$. $\{X_n, n = 0, 1, \ldots\}$ is referred as a random walk.

Suppose $X_0 = i$. Then the transition probability can be computed.

$$P\{X_n = j | X_0 = i\} = \binom{n}{\frac{n+j-i}{2}} p^{\frac{n+j-i}{2}} q^{\frac{n-j+i}{2}}$$

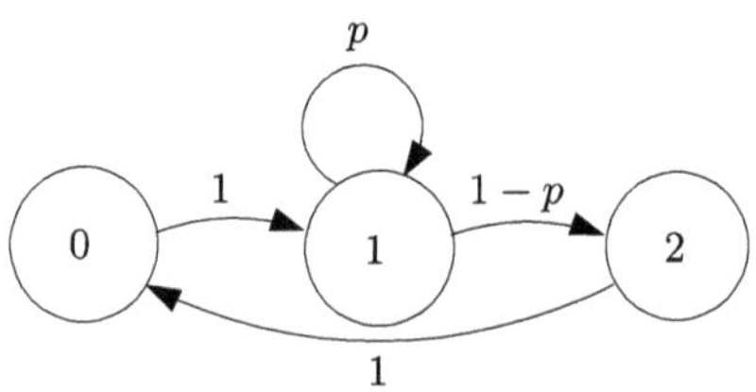

Fig. 2.21 State transition diagram for Example 2.29

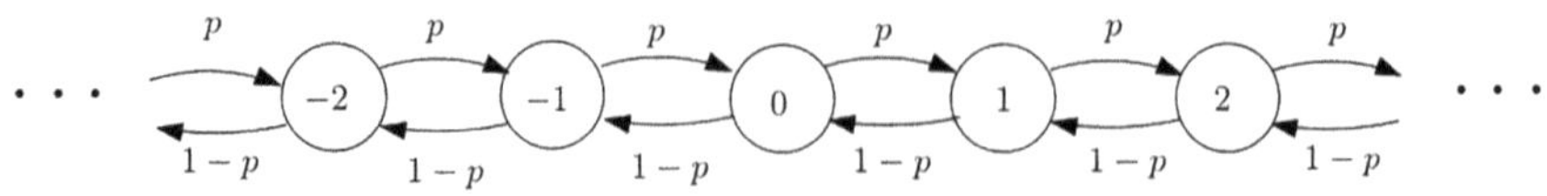

Fig. 2.22 State transition diagram for Example 2.30

for $n + j - i$ is an even non-negative integer, otherwise $P\{X_n = j | X_0 = i\} = 0$. For instance $i = j = 0$, we have

$$p_{00}^{(2n)} = \binom{2n}{n} p^n q^n.$$

In the symmetric random walk, $p = q = \frac{1}{2}$. In this case, all the states are recurrent states and the closed communicating class is same as the state space, and hence, the chain is irreducible. Here, all the states are periodic with period 2. On the other hand, when $p \neq q$, all the states are transient.

Example 2.31 Consider an infinite state Markov chain with state space $S = \{\ldots, -2, -1, 0, 1, 2, \ldots\}$. Here, $p_{ii} = p_1$, $p_{ii+1} = p_2$ and $p_{ii-1} = 1 - p_1 - p_2$ for some $0 < p_1, p_2 < 1$ and all $i \in S$. Classify the states of this Markov chain.
Solution: The state transition diagram is shown in Fig. 2.23.

All the states are recurrent states and aperiodic and the closed communicating class is same as state space. Hence, the chain is irreducible.

Example 2.32 Consider a finite state Markov chain with state space $\{1, 2, 3\}$ and TPM

$$\mathbf{P} = \begin{pmatrix} 0.2 & 0.4 & 0.4 \\ 0.4 & 0.2 & 0.4 \\ 0.4 & 0.4 & 0.2 \end{pmatrix}.$$

Classify the states of this Markov chain. Also, discuss whether limiting distribution and stationary distribution exist or not.
Solution: The state transition diagram is shown in Fig. 2.24.

Clearly the DTMC is irreducible and aperiodic.

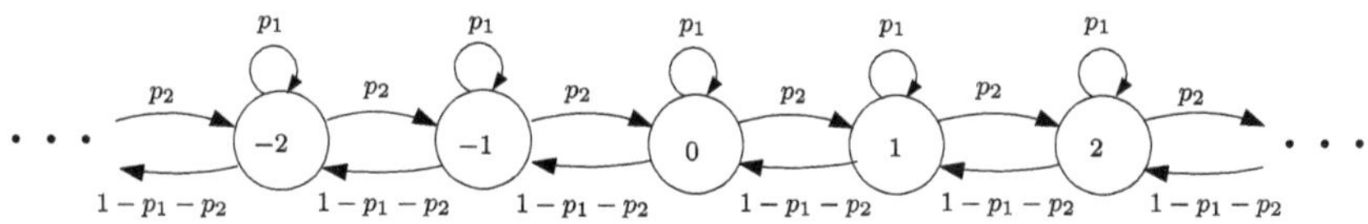

Fig. 2.23 State transition diagram for Example 2.31

Fig. 2.24 State transition diagram for Example 2.32

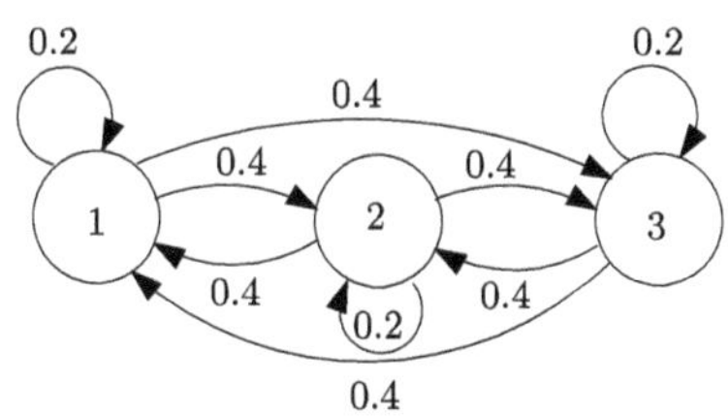

$$\mathbf{P}^{(20)} = \begin{pmatrix} 0.333 & 0.333 & 0.333 \\ 0.333 & 0.333 & 0.333 \\ 0.333 & 0.333 & 0.333 \end{pmatrix}.$$

From here, it can be observed that chain is positive recurrent. In this example, limiting distribution is the same as unique steady-state or stationary distribution.

In the situation of finite irreducible chains, the following theorem provides a sufficient condition for the existence of unique stationary, equivalently steady-state distribution π.

Theorem 2.11 (Ergodic Theorem) *Consider, a finite irreducible, aperiodic Markov chain with TPM $\boldsymbol{P} = [p_{ij}]$, where the limits*

$$\pi_j = \lim_{n\to\infty} p_{ij}^{(n)}$$

exist and are independent of the initial state i, i.e., regardless of the initial distribution of the chain, the distribution of X_n approaches $\pi = [\pi_j]$ as $n \to \infty$. The limits π_j are such that $\pi_j \geq 0, \sum_j \pi_j = 1$. Furthermore, the limiting probability distribution $[\pi_j]$ is identical with the stationary distribution for the given chain. That is, it satisfies

$$\pi_j = \sum_i \pi_i p_{ij}, \quad \sum_j \pi_j = 1.$$

Proof Since the states are aperiodic and positive recurrent, for each pair j, k, $\lim\limits_{n\to\infty} p_{jk}^{(n)}$ exists and equal to F_{jk}/μ_k using Theorem 2.9.

If state j is recurrent, then for every k that can be reached from state j, $F_{kj} = 1$. By the above result, since k is recurrent, $F_{jk} = 1$, such that

$$\pi_k = \lim_{n\to\infty} p_{jk}^{(n)} = \frac{1}{\mu_k} > 0$$

and is independent of j.

$$p_{jk}^{(n+m)} = \sum_i p_{ji}^{(n)} p_{ik}^{(m)}, \forall m,$$

$$\pi_k = \lim_{n\to\infty} p_{jk}^{(m+n)} = \lim_{n\to\infty} \sum_i p_{ji}^{(n)} p_{ik}^{(m)} \geq \sum_i \lim_{n\to\infty} p_{ji}^{(n)} p_{ik}^{(m)} = \sum_i \pi_i p_{ik}^{(m)}.$$

Suppose, if possible, that $\pi_k > \sum\limits_i \pi_i p_{ik}^{(m)}$. Summing over k,

$$\sum_k \pi_k > \sum_k \sum_i \pi_i p_{ik}^{(m)} = \sum_i \pi_i$$

which is impossible. Hence, the equality holds in the above and $\pi_k = \sum_i \pi_i p_{ik}^{(m)}$ for all $m \geq 1$. In particular, for $m = 1, \pi_k = \sum_i \pi_i p_{ik}$, and for large $m, \pi_k = \sum_i \pi_i \pi_k = \pi_k \sum_i \pi_i$. Hence $\sum_i \pi_i = 1$. This shows that $\{\pi_i\}$ is a probability distribution. This distribution is unique.

The distribution given in the ergodic theorem above is known as stationary or equilibrium distribution, and the probabilities are known as stationary or equilibrium probabilities. The row vector π is also sometimes called as invariant probability measure of the TPM **P**. In view of this result, the stationary distribution of a finite irreducible, aperiodic chain is obtained by the unique solution of the system of equations in the theorem.

Example 2.33 Consider Example 2.24. Discuss the limiting distribution for the Markov chain. Further, find the stationary distribution for this Markov chain, if it exists.

Solution: Since the states are aperiodic and positive recurrent and the finite Markov chain is reducible, the above ergodic theorem is not satisfied. Hence, a unique limiting distribution does not exist which is independent of initial distribution.

For the stationary distribution of this Markov chain, we solve $\pi \mathbf{P} = \pi$ such that $\pi_i \geq 0$ and $\sum_{i=1}^{3} \pi_i = 1$, i.e.,

$$\begin{pmatrix} \pi_1 & \pi_2 & \pi_3 \end{pmatrix} \begin{pmatrix} 0.2 & 0.8 & 0 \\ 0.1 & 0.9 & 0 \\ 0 & 0 & 1 \end{pmatrix} = \begin{pmatrix} \pi_1 & \pi_2 & \pi_3 \end{pmatrix} \text{ and } \pi_1 + \pi_2 + \pi_3 = 1.$$

After simplification, we get

$$\begin{aligned} \pi_1 &= 0.2\pi_1 + 0.1\pi_2 \\ \pi_2 &= 0.8\pi_1 + 0.9\pi_2 \\ \pi_1 + \pi_2 + \pi_3 &= 1. \end{aligned}$$

Solving these equations, we get $\pi_2 = 8\pi_1$ and $\pi_3 = 1 - 9\pi_1$ which implies the stationary distributions are $\begin{pmatrix} \pi_1 & 8\pi_1 & 1 - 9\pi_1 \end{pmatrix}$ for $0 \leq \pi_1 \leq \frac{1}{9}$.

The distribution of chain at each step is same, i.e., $P(X_n = i) = \pi_i$ which implies π_i the stationary distribution for the chain $\{X_n, n = 0, 1, \ldots\}$. Now another concept which is closely related to the stationary distribution is the invariant measure $\mu = (\mu_i)$ for the chain if $0 < \mu_i < 1$ and $\mu = \mu \mathbf{P}$.

The converse of the above theorem also holds (without proof):

Theorem 2.12 *If a chain is irreducible and aperiodic and if there exists a unique stationary distribution* $\pi = [\pi_i]$ *for the chain, then the chain is ergodic and* $\pi_i = 1/\mu_{ii}$.

Hence, ergodicity is a necessary and sufficient condition for the existence of π in such chains.

Example 2.34 Consider Example 2.19. Obtain the stationary distribution if it exists.
Solution: The chain is irreducible, and the states are aperiodic positive recurrent, and hence chain is ergodic. Solving the system of equations, $\pi_1 = 1/3, \pi_2 = 10/27, \pi_3 = 8/27$ are obtained as the stationary distribution of the chain.

The previous results for a finite MC can be applied to chains with an countably infinite state space. It is worth noting that the states of such a chain might not even be a single closed set. The states do not belong to any closed set, including S, when $p_{ij} = 1$ for $j = i + 1, i = 0, 1, \ldots$; and zero otherwise. A more sensitive classification of states when working with Markov chains with countable state space—transient, null, and positive recurrent. Positive recurrence is an extra need in the case of a countable MC, whereas irreducibility (or some form of reducibility) and aperiodicity are sufficient for ergodicity of a finite MC. The following theorem, which is stated below, is crucial in the theory of countable Markov chains.

Theorem 2.13 (General Ergodic Theorem) *Consider, an irreducible, aperiodic Markov chain $\{X_n, n = 0, 1, \ldots\}$ with state space $S = \{0, 1, \ldots\}$ and having TPM $\boldsymbol{P} = [p_{ij}]$.*
If the chain is null recurrent or transient then $\lim_{n\to\infty} p_{ij}^{(n)} = 0$.
If the chain is positive recurrent then the limits $\lim_{n\to\infty} p_{ij}^{(n)} = \pi_j$ exist and are independent of the initial state i. The limits π_j are such that $\pi_i \geq 0, \sum_i \pi_i = 1$. The probabilities π_i's are given by the solution of system of equations

$$\pi_j = \sum_i \pi_i p_{ij}, \sum_i \pi_i = 1.$$

Proof For every n and M, $1 = \sum_{i=0}^{\infty} p_{ji}^{(n)} \geq \sum_{i=0}^{M} p_{ji}^{(n)}$. Letting $n \to \infty$, it is obtained $1 \geq \sum_{i=0}^{M} \pi_i$, for every M. Thus, $\sum_{j=0}^{\infty} \pi_j \leq 1$. Now $p_{1j}^{(n+1)} \geq \sum_{k=0}^{M} p_{1k}^{(n)} p_{kj}$, if we let $n \to \infty$, it is obtained $\pi_j \geq \sum_{k=0}^{M} \pi_k p_{kj}$. Next, since the left-hand side is independent of M. However, $M \to \infty$ gives

$$\pi_j \geq \sum_{k=0}^{\infty} \pi_k p_{kj} \,. \tag{2.6}$$

Multiplying by p_{ji}, then summing on j and using (2.6), yields $\pi_j \geq \sum_{k=0}^{\infty} \pi_k p_{kj}^{(2)}$ thus in general, $\pi_j \geq \sum_{k=0}^{\infty} \pi_k p_{kj}^{(n)}$ for any n. Suppose strict inequality holds for some j. Adding these inequalities with respect to j, we get

$$\sum_{j=0}^{\infty} \pi_j > \sum_{j=0}^{\infty} \sum_{k=0}^{\infty} \pi_k p_{kj}^{(n)} = \sum_{k=0}^{\infty} \pi_k \sum_{j=0}^{\infty} p_{kj}^{(n)} = \sum_{k=0}^{\infty} \pi_k,$$

which is a contradiction. Thus, $\pi_j = \sum_{k=0}^{\infty} \pi_k p_{kj}^{(n)}$ for all n. Since $\sum \pi_k$ converges and $p_{kj}^{(n)}$ is uniformly bounded, letting $n \to \infty$, it can be concluded that

$$\pi_j = \sum_{k=0}^{\infty} \pi_k \lim_{n\to\infty} p_{kj}^{(n)} = \pi_j \sum_{k=0}^{\infty} \pi_k, \quad \text{for every } j.$$

Thus, $\sum_{j=0}^{\infty} \pi_j = 1$ since $\pi_j > 0$ by positive recurrence. Suppose the sequence $x = \{x_n\}$ satisfies the same relations as π. Then

$$x_k = \sum_{j=0}^{\infty} x_j p_{jk} = \sum_{j=0}^{\infty} x_j p_{jk}^{(n)},$$

and if let $n \to \infty$ as before,

$$x_k = \sum_{j=0}^{\infty} x_j \lim_{n\to\infty} p_{jk}^{(n)} = \pi_k \sum_{j=0}^{\infty} x_j = \pi_k.$$

Example 2.35 (Birth-Death chains) Consider a DTMC $\{X_n; n = 1, 2, \ldots\}$ with state space $S = \{0, 1, \ldots\}$ with TPM $\mathbf{P} = [p_{ij}]_{i,j\in S}$ and as long as $|i - j| > 1$, the transition probabilities fulfill $p_{ii+1} = p_i > 0, i \geq 0$, $p_{ii-1} = q_i > 0, i \geq 1$, and $p_{ii} = 0$ elsewhere. As a result, $p_{00} = 1 - p_0$ and $p_{ii} = 1 - p_i - q_i$, where $i \geq 1$. When a system changes state from i to $i + 1$, it is considered as a birth whereas when the system changes state from $i + 1$ to i it is considered as a death. Note that one birth or death can only take place in one unit of time due to the restriction on the transition probabilities.

$$\mathbf{P} = \begin{pmatrix} 1-p_0 & p_0 & 0 & 0 & \ldots \\ q_1 & 1-p_1-q_1 & p_1 & 0 & \ldots \\ 0 & q_2 & 1-p_2-q_2 & p_2 & \ldots \\ \vdots & \vdots & \vdots & \vdots & \ddots \end{pmatrix}.$$

Classify the states of this Markov chain. Also, discuss whether limiting distribution and stationary distribution exist or not.

Solution: The state transition diagram is shown in Fig. 2.25. Based on the values of p_i and q_j, the following three cases arise.

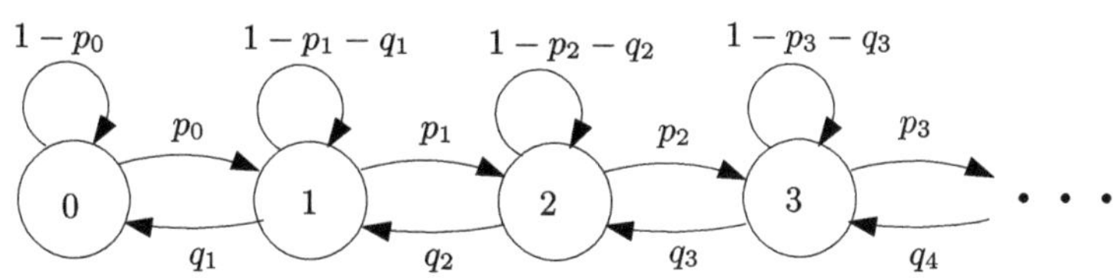

Fig. 2.25 State transition diagram for Example 2.35

1. Case 1. When $p_i = 0, i = 0, 1, \ldots$ and $q_j > 0, j = 1, 2, \ldots$. State 0 is absorbing state and states $1, 2, \ldots$ are transient states. Then $\lim_{n\to\infty} p_{ij}^{(n)} = \begin{cases} 1, & j = 0 \\ 0, & j \neq 0 \end{cases}$.
2. Case 2. When $p_i > 0, i = 0, 1, \ldots$ and $q_j = 0, j = 1, 2, \ldots$. All states are transient states. $\lim_{n\to\infty} p_{ij}^{(n)} = 0$, for all $j \in S$.
3. Case 3. When $p_i > 0, i = 0, 1, \ldots$ and $q_j > 0, j = 1, 2, \ldots$. All states are recurrent and aperiodic. The Markov chain is aperiodic and irreducible.

Then, by applying Theorem 2.13, the limiting distribution exists and is independent of initial distribution. The equations for the stationary probabilities take the form

$$\pi_0 = (1 - p_0)\pi_0 + q_1\pi_1$$
$$\pi_i = p_{i-1}\pi_{i-1} + (1 - p_i - q_i)\pi_i + q_{i+1}\pi_{i+1}, \qquad i = 1, 2, \ldots.$$

From the first equation, $p_0\pi_0 = q_1\pi_1$. By induction, it can be verified that $p_i\pi_i = q_{i+1}\pi_{i+1}$ for $i = 0, 1, \ldots$. For convenience, define $\rho_i = p_i/q_{i+1}$. Then, $\pi_{i+1} = \rho_i\pi_i$ and iterating this,

$$\pi_i = \pi_0 \prod_{j=0}^{i-1} \rho_j, \text{ and therefore, } \pi_0 = \frac{1}{1 + \sum_{i=1}^{\infty} \prod_{j=0}^{i-1} \rho_j}.$$

If $\sum_{i=1}^{\infty} \prod_{j=0}^{i-1} \rho_j < \infty$, then π_0 is positive and all the states are positive recurrent. If this sum of products is infinite, then no state is positive recurrent. If ρ_j is bounded above 1, say $\rho_j \leq 1 - \epsilon$ for some fixed $\epsilon > 0$ and all sufficiently large j, then this sum of products will converge and the states will be positive recurrent.

For the simple birth-death chain where $p_i = p, q_i = q, \rho_j = \rho = p/q$ for all j. When $\rho < 1$, the above calculations yield $\pi_i = (1 - \rho)\rho^i$ for $i = 0, 1, \ldots$. The chain is positive recurrent. The chain is null recurrent when $\rho = 1$ and transient when $\rho > 1$. Suppose the state space is finite, and the chain is positive recurrent for all values of ρ.

Example 2.36 Consider $\{X_n, n = 1, 2, \ldots\}$ a birth-death chain with state space $S = \{0, 1, \ldots\}$ such that

$$p_i = \begin{cases} p & i = 0, 2, \ldots \\ q & i = 1, 3, \ldots \end{cases}$$

and

$$q_i = \begin{cases} q & i = 0, 2, \ldots \\ p & i = 1, 3, \ldots \end{cases}$$

where $0 < p < 1$ and $q = 1 - p$. Determine whether chain is transient or recurrent. Classify the states of this Markov chain. Also, discuss whether limiting distribution and stationary distribution exist or not.

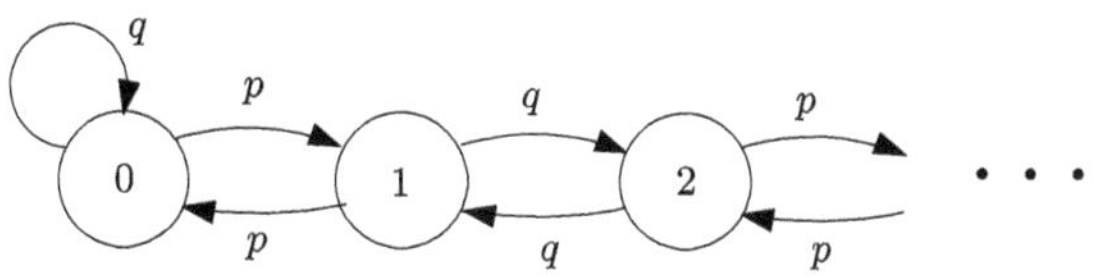

Fig. 2.26 State transition diagram for Example 2.36

Solution: The state transition diagram is shown in Fig. 2.26.

Clearly, the Markov chain is irreducible and aperiodic. Define

$$f_{ii}^{(n)} = P\{X_n = i; X_m \neq i, m = 1, 2, \ldots, n-1 | X_0 = i\},$$

then,

$$\begin{aligned} f_{00}^1 &= q; \; f_{00}^2 = p^2; \; f_{00}^3 = 0; \; f_{00}^4 = p^2q^2; \\ f_{00}^5 &= 0; \; f_{00}^6 = p^4q^2; \; f_{00}^7 = 0; \; f_{00}^8 = p^4q^4; \\ f_{00}^9 &= 0; \; f_{00}^{10} = p^6q^4; \; f_{00}^{11} = 0; \; f_{00}^{12} = p^6q^6 \end{aligned}$$

$$F_{00} = q + p^2 + p^2q^2(1 + p^2q^2 + \ldots) + p^4q^2(1 + p^2q^2 + \cdots) = q + p^2 + \frac{p^2q^2(1+p^2)}{1-p^2q^2}.$$

Since, $F_{00} < 1$ which implies state 0 is transient and since for an irreducible Markov chain transience is class property, each state is transient and hence, this chain is transient. By Theorem 2.13, $\lim_{n\to\infty} p_{ij}^{(n)} = 0$.

The equations for the stationary probabilities take the form:

$$\begin{aligned} \pi_0 &= q\pi_0 + p\pi_1 \\ \pi_1 &= p\pi_0 + q\pi_2. \end{aligned}$$

From the first equation, $\pi_0 = \pi_1$ and using this in second equation, $\pi_1 = \pi_2$. Continuing in similar way $\pi_0 = \pi_1 = \pi_2 = \ldots$. Also, $\sum_{i=0}^{\infty} \pi_i = 1$which is not possible in this case for any value of π_i. Hence stationary distribution does not exist.

For more discussions, readers may refer to [1–5]

2.6 Exercise Problems

1. The one-step transition probability matrix of DTMC $\{X_n, n = 0, 1, \ldots\}$ with state space $\{1, 2, 3\}$ is

$$\mathbf{P} = \begin{pmatrix} 0.1 & 0.5 & 0.4 \\ 0.6 & 0.2 & 0.2 \\ 0.3 & 0.4 & 0.3 \end{pmatrix}$$

and the initial distribution is $\pi_0 = (0.7, 0.2, 0.1)$. Find

(a) $P(X_2 = 3)$
(b) $P(X_3 = 2, X_2 = 3, X_1 = 3, X_0 = 2)$
(c) $P\{X_{30} = 3 | X_{27} = 1\}$.

2. For a Markov chain $\{X_n, n = 0, 1, \ldots\}$ with state space $E = \{0, 1, 2, 3, 4\}$ and transition probability matrix $\mathbf{P}$ given below, classify the states of the chain. Also determine the closed communicating classes.

$$\text{(a)}\quad \mathbf{P} = \begin{pmatrix} \frac{1}{2} & 0 & 0 & \frac{1}{2} & 0 \\ \frac{1}{2} & \frac{1}{2} & 0 & 0 & 0 \\ \frac{1}{4} & \frac{1}{2} & 0 & 0 & \frac{1}{4} \\ 0 & 0 & 0 & 1 & 0 \\ 0 & 0 & \frac{1}{2} & \frac{1}{4} & \frac{1}{4} \end{pmatrix}. \qquad \text{(b)}\quad \mathbf{P} = \begin{pmatrix} 0 & \frac{1}{3} & 0 & \frac{1}{3} & \frac{1}{3} \\ \frac{1}{3} & \frac{1}{3} & 0 & \frac{1}{3} & 0 \\ 0 & 0 & \frac{2}{3} & 0 & \frac{1}{3} \\ \frac{1}{4} & \frac{1}{4} & 0 & \frac{1}{4} & \frac{1}{4} \\ 0 & 0 & \frac{1}{3} & 0 & \frac{2}{3} \end{pmatrix}.$$

3. Assume $\{Y_n, n = 0, 1, \ldots\}$ is a sequence of IID random variables taking values on $\mathbf{N} \cup \{0\}$, with probability distribution $P(Y_n = i) = a_i$ $(i = 0, 1, \ldots)$. Let $X_{n+1} = (X_n - 1)^+ + Y_n$. Prove that $\{X_n, n = 0, 1, \ldots\}$ is a Markov chain and write down its transition probability matrix.
4. Consider an inventory model. Let Y_n be a quantity for the daily need, X_n be inventory quantity at nth day and (s, S) is inventory storage. That means, when the inventory X_n is not more than s, it should be stocked to the inventory S; when the inventory X_n is more than s, it should be stocked. Assume $\{Y_n, n = 0, 1, \ldots\}$ is a sequence of IID RVs with probability distribution $P(Y_n = i) = a_i$ $(i = 0, 1, \ldots)$. Fix $0 < s < S < \infty$. Let

$$X_{n+1} = \begin{cases} X_n - Y_{n+1}, \text{ if } s < X_n \le S; \\ S - Y_{n+1}, \text{ if } X_n \le s. \end{cases}$$

Prove that $\{X_n, n = 0, 1, \ldots\}$ is a Markov chain and write down its transition probability matrix.
5. Let X_0 be an integer-valued random variable, $P(X_0 = 0) = 1$, that is independent of the IID sequence $Z_1, Z_2, \ldots$, where $P(Z_n = 1) = p$, $P(Z_n = -1) = q$ and $P(Z_n = 0) = 1 - (p + q)$. Let $X_n = \max\{0, X_{n-1} + Z_n\}$, $n = 1, 2, \ldots$. Prove that $\{X_n, n = 0, 1, \ldots\}$ is a DTMC. Write the one-step TPM or draw the state transition diagram for this Markov chain.
6. A particle starts at the origin and moves to and fro on a straight line. At any move it jumps either 1 unit to the right or 1 unit to the left each with probability $1/3$ or $1/3$ respectively. With probability $1/3$, the particle may stay at the same position in any move. Model this process as a stochastic process. Write the stochastic process with state space and parameter space. Verify whether this process satisfies Markov property.
7. Suppose that a machine can be in two states $\{0, 1\}$ where 0 implies working and 1 implies out of order on a day. The probability that a machine is working on a particular day depends on the state of the machine during two previous

days. Specifically assume that $P\{X_{n+1} = 0 | X_{n-1} = j, X_n = k\} = q_{jk} \quad j, k = 0$, where X_n denotes the state of the machine on nth day.

(a) Show that $\{X_n, n = 1, 2, \ldots\}$ is not a DTMC.
(b) Define a new state space for the problem by taking the pairs (j, k) where j and k are 0 or 1. It is said that machine is in state (j, k) on day n if the machine is in state j on day $(n - 1)$ and in state k on nth day. Show that in this case the state space of the system is a DTMC.
(c) It is given that the machine was working on Monday and Tuesday. Find the probability that it will be working on Thursday?

8. Consider the chain with transition matrix

$$\mathbf{P} = \begin{pmatrix} 1/6 & 1/3 & 1/2 & 0 \\ 1/2 & 1/2 & 0 & 0 \\ 1/6 & 1/3 & 1/2 & 0 \\ 0 & 1/6 & 1/3 & 1/2 \end{pmatrix}.$$

Identify the closed class.

9. Consider a Markov Chain with state space $S = \{0, 1, 2, 3, 4\}$ and transition probability matrix

$$\mathbf{P} = \begin{pmatrix} 1 & 0 & 0 & 0 & 0 \\ 0 & 0.25 & 0.75 & 0 & 0 \\ 0 & 0.5 & 0.5 & 0 & 0 \\ 0.25 & 0.25 & 0 & 0.25 & 0.25 \\ 0 & 0 & 0 & 0.5 & 0.5 \end{pmatrix}.$$

(a) Classify the states of the chain.
(b) Determine the stationary distribution for states 1 and 2.
(c) For the transient states, calculate μ_{ij}, the expected number of visits to transient state j, given that the process started in a transient state i.
(d) Find $P\{X_5 = 2 \mid X_3 = 1\}$.

10. Show that if a Markov Chain is irreducible and $P_{ii} > 0$ for some state i, then the chain is aperiodic.
11. Consider the Markov chain $\{X_n, n = 0, 1, 2\}$ with $S = \{0, 1, 2\}$ and transition probability matrix given as:

$$\mathbf{P} = \begin{pmatrix} \frac{1}{4} & \frac{1}{2} & \frac{1}{4} \\ \frac{1}{3} & 0 & \frac{2}{3} \\ \frac{1}{2} & 0 & \frac{1}{2} \end{pmatrix}.$$

(a) Classify the chain as irreducible, aperiodic, and find the stationary distribution.
(b) Consider a DTMC with transition probability matrix $\begin{pmatrix} 0.4 & 0.6 & 0 \\ 0.4 & 0 & 0.6 \\ 0 & 0.4 & 0.6 \end{pmatrix}$. Find the stationary distribution for this Markov chain.

(c) Assume $X_0 = 1$ and let R be the first time that the chain returns to state 1, i.e., $R = min\{m \geq 1 : X_n = 1\}$. Find $E[R|X_0 = 1]$.

12. A mathematics professor has 2 umbrellas. He keeps one of them at home and the other in the office. Every morning, when he leaves home, he checks the weather and takes an umbrella with him if it rains. In case both the umbrellas are in the office, he gets wet. The same procedure is repeated in the afternoon when he leaves the office to go home. The professor lives in a tropical region, and therefore, the chance of rain in the afternoon is higher than in the morning; it is $1/5$ in the afternoon and $1/20$ in the morning. Whether it rains or not is independent of whether it rained the last time he checked. On day 0, there is an umbrella at home, and 1 in the office. Note that, there are two trips each day. What is the expected number of days that will pass before the professor gets wet? What is the probability that the first time he gets wet it is on his way home from the office?
13. Consider a DTMC on non-negative integers where the chain, moves from i, to $i + 1$ with probability p, to state 0 with probability $1 - p$, and then return to i with probability $0 < p < 1$. Show that this DTMC is irreducible and recurring and that it has a unique stationary distribution, also, find the stationary distribution.
14. A random walker walks on the integers. At each step, he moves one-step right with probability $\frac{1}{3}$, two steps left with probability $\frac{1}{3}$, and stays on his place with probability $\frac{1}{3}$. What is the probability that the walker will return to his original location after three steps?
15. Show that the birth and death chain is recurrent if and only if

$$\sum_{k=0}^{\infty} \gamma_k = \infty$$

where, $\gamma_0 = 1$ and

$$\gamma_k = \frac{q_1 \cdots q_k}{p_1 \cdots p_k}, \qquad k = 1, 2, \ldots$$

$$q_i = p_{ii-1}, \qquad i = 1, 2, \ldots$$

$$p_i = p_{ii+1}, \qquad i = 1, 2, \ldots.$$

16. Show that an irreducible birth and death chain on $S = \{0, 1, \ldots\}$ is recurrent if and only if

$$\sum_{k=0}^{\infty} \frac{q_1 \cdots q_k}{p_1 \cdots p_k} = \infty$$

where

$$q_i = p_{i\,i-1} \text{ and } p_i = p_{i\,i+1}, \quad i = 1, 2, \ldots$$

17. Consider the birth and death chain on {0, 1, ...} with

$$p_k = \frac{k+2}{2(k+1)} \text{ and } q_k = \frac{k}{2(k+1)}, \; k \geq 0$$

determine whether the chain is recurrent or transient.
18. Prove that $\mu_{ij} = \frac{F_{ij}}{1-F_{ij}}$, $i, j \in S$.
19. Prove that if i is a recurrent state and suppose $p_{ij}^{(n)} > 0$ for some n, then state j is recurrent and $F_{ij} = F_{ji} = 1$.
20. Define

$$e_{jk}^{(n)} = P\{X_n = k, X_m \neq j, 0 < m < n | X_0 = j\}$$

denotes the probability that the chain starting from state j, the chain leads to state k at the nth step without hitting j in the middle. Prove that, if $p_{jk}^{(n)} > 0$, then there exists $m \leq n$ such that $e_{jk}^{(n)} > 0$.
21. Let $\{X_n, n = 0, 1, \ldots\}$ be a DTMC. Prove that the distribution of X_n is independent of n if and only if the initial distribution π is a stationary distribution.
22. Consider a DTMC model arising in an insurance problem. To compute insurance or pension premiums for professional diseases such as silicosis, it is needed to compute the average degree of disability at pre-assigned time periods. Suppose that, m degrees of disability $S_1, S_2, \ldots, S_m$ are retained. Assume that an insurance policyholder can go from degree S_i to degree S_j with a probability p_{ij}. This strong assumption leads to the construction of the DTMC model in which $\mathbf{P} = [p_{ij}]$ is the one-step transition probability matrix related to the degree of disability. Using real observations recorded in India, the following transition matrix $\mathbf{P}$ is considered :

$$\mathbf{P} = \begin{pmatrix} 0.90 & 0.10 & 0 & 0 & 0 \\ 0 & 0.95 & 0.05 & 0 & 0 \\ 0 & 0 & 0.90 & 0.05 & 0.05 \\ 0 & 0 & 0 & 0.90 & 0.10 \\ 0 & 0 & 0.05 & 0.05 & 0.90 \end{pmatrix}$$

(a) Classify the states of the chain as transient, positive recurrent, or null recurrent along with period.
(b) Find the limiting distribution for the degree of disability.

23. Consider a DTMC with state space $S = \{0, 1, \ldots\}$ and define transition probabilities as $p_{jj+2} = v_j$ and $p_{j0} = 1 - v_j$, for $j \in S$. Find the transition probability matrix of Markov chain and classify the states of this chain.
24. Consider an irreducible DTMC with state space $S = \{0, 1, \ldots, m\}$. Assume that, $\mathbf{P} = [p_{ij}]$ is a double stochastic matrix, i.e., $\sum_{k=1}^{m} p_{ki} = \sum_{k=1}^{m} p_{ik} = 1$ for each

$i = 1, 2, \ldots, m$. Prove that the stationary distribution of this Markov chain is the uniform distribution on $\{1, 2, \ldots, m\}$.

25. Consider the simple random walk on a circle. Assume that K odd number of points labeled $0, 1, \ldots, K - 1$ are arranged on a circle clockwise. From i, the walker moves to $i + 1$ (with K identified with 0) with probability p $(0 < p < 1)$ and to $i - 1$ (with -1 identified with $K - 1$) with probability $1 - p$. Find the steady-state distribution for this random walk, if it exists.

References

1. Castaneda, L.B., Arunachalam, V., Dharmaraja, S.: Introduction to Probability and Stochastic Processes with Applications. Wiley, New Jersey (2012)
2. Conolly, B.W., Parthasarathy, P.R., Dharmaraja, S.: A chemical queue. Math. Sci. **21**, 83–91 (1997)
3. Ross, S.M.: Introduction to Probability Models. Elsevier, 12/e, (2019)
4. Stockmayer, W.H., Gobush, W., Norvich, R.: Local-jump models for chain dynamics. Pure Appl. Chem. **26**(3–4), 537–544 (1971)
5. Trivedi, K.S.: Probability and Statistics with Reliability, Queuing, and Computer Science Applications, 2nd edn. John Wiley & Sons (2001)

Chapter 3
Discrete-Time Markov Chains—Part II

In this chapter, the reducible Markov chains and time-reversed Markov chains will be discussed. Finally, discrete-time branching processes are presented.

3.1 Reducible Markov Chains

In the previous chapter, the limiting properties of irreducible Markov chains are discussed. Similar behavior for certain types of reducible chains is examined in this chapter.

In general, the different types of reducible Markov chains are classified as follows:

1. Finite chains with single closed class.
2. Absorbing Markov chains.
3. Chain with more than one closed class.

The state transition diagram of various types of reducible Markov chains is shown in Fig. 3.1.

3.1.1 *Finite Chains with Single Closed Class*

Consider finite Markov chains with a closed class C of interconnected states and a state space S. Assume that C is also aperiodic. In this case, the ergodic theorem is stated below.

Theorem 3.1 (Ergodic Theorem) *Consider a finite Markov chain* $\{X_n, n = 0, 1, \ldots\}$ *with aperiodic states. Let* **P** *be the TPM of the given* m*-state chain with state space* S*, and* **N** *be the transition (submatrix) of transitions among the* $k(\leq m)$ *members of the closed class* C*. Let* π_1 *be the stationary distribution corresponding to the*

D. Selvamuthu, *Introduction to Stochastic Processes*, University Texts in the Mathematical Sciences, https://doi.org/10.1007/978-981-97-6152-4_3

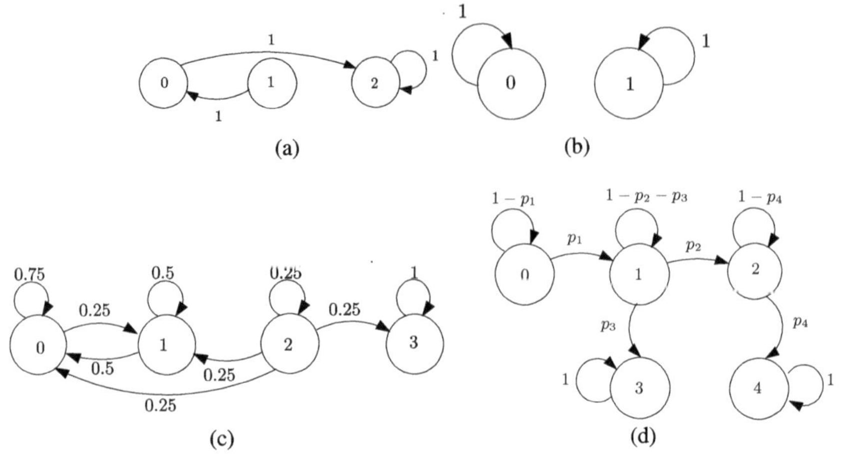

Fig. 3.1 State transition diagram of various reducible DTMCs

stochastic submatrix **N**. *If* $\pi = (\pi_1, \mathbf{0})$, *then* $\mathbf{P}^{(\mathbf{n})} \to \mathbf{e}\pi$ *as* $n \to \infty$. *In other words, element-wise* π *is the stationary distribution corresponding to the matrix* **P** *which is also limiting distribution.*

Proof Let $\mathbf{P} = \begin{pmatrix} \mathbf{N} & \mathbf{0} \\ \mathbf{R} & \mathbf{M} \end{pmatrix}$ where **N** is the $k \times k$ sub-matrix of transitions among k states of the closed class C, **R** is $k \times m - k$ sub-matrix of transitions of $m - k$ transient states to k states of closed class C and **M** is the $m - k \times m - k$ sub-matrix of transitions among $m - k$ transient states. If the MC is considered only upto k states, then its TPM is given by **N**. By general ergodic theorem, the stationary distribution exists, then $\mathbf{N}^{(\mathbf{n})} \to \pi_1$. When j is transient, $\lim_{n\to\infty} p_{ij}^{(n)} = 0$. Hence, $\mathbf{M}^{(\mathbf{n})}$ and $\mathbf{R}^{(\mathbf{n})} \to 0$. Thus, $\mathbf{P}^{(\mathbf{n})} \to \mathbf{e}\pi$ as $n \to \infty$.

Example 3.1 Consider a reducible chain with $S = \{0, 1, 2, 3\}$ and TPM $\mathbf{P} = \begin{pmatrix} \mathbf{N} & \mathbf{0} \\ \mathbf{R} & \mathbf{M} \end{pmatrix}$, where states 2 and 3 form a single closed class C and states 0 and 1 are transient forming set T. Also it is given that

$$\mathbf{N} = \begin{pmatrix} \frac{1}{2} & \frac{1}{2} \\ 1 & 0 \end{pmatrix}, \quad \mathbf{R} = \begin{pmatrix} \frac{1}{2} & \frac{1}{2} \\ \frac{7}{10} & 0 \end{pmatrix}, \quad \mathbf{M} = \begin{pmatrix} 0 & 0 \\ \frac{3}{10} & 0 \end{pmatrix}.$$

Draw the state transition diagram for this MC and find the stationary distribution for this Markov chain.

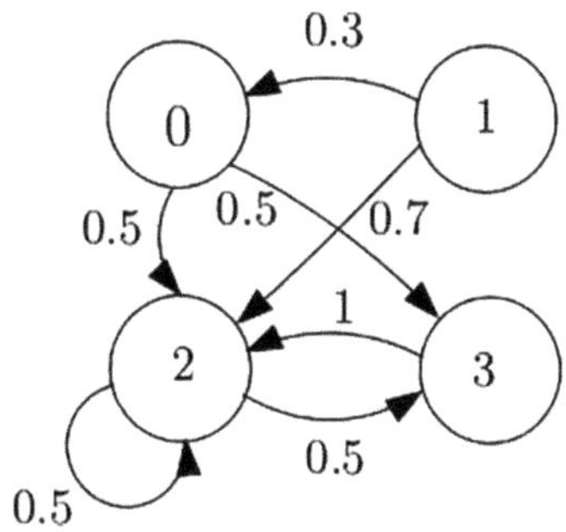

Fig. 3.2 State transition diagram for Example 3.1

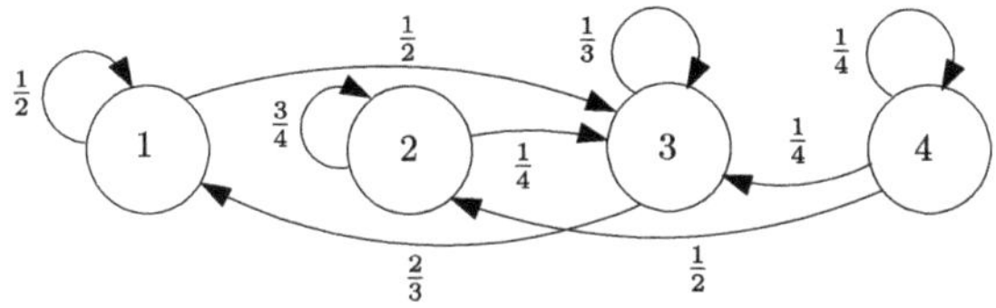

Fig. 3.3 State transition diagram for Example 3.2

Solution: The state transition diagram is shown in Fig. 3.2. Solving $\pi_1 \mathbf{N} = \boldsymbol{\pi}_1$, and $\pi_1 \mathbf{e} = 1$ for π_1, we get $\pi_1 = \left(\frac{2}{3}, \frac{1}{3}\right)$. Then, $\mathbf{N}^{(n)} \to \mathbf{e}\pi_1$. Thus,

$$\mathbf{P}^{(n)} \to \mathbf{e}\pi, \text{ where } \pi = \left(0, 0, \frac{2}{3}, \frac{1}{3}\right).$$

3.1.2 Absorbing Markov Chains

Assume that there are $m - k$ transient states T in a given MC with finite state space having m states, and that all the k recurring states in the Markov chain is absorbing. Taking the absorbing states first and then the transitory states, the states can then be reorganised. The transient matrix can be represented in the canonical form $\mathbf{P} = \begin{pmatrix} \mathbf{I} & 0 \\ \mathbf{R} & \mathbf{M} \end{pmatrix}$, where $\mathbf{M}$ denotes the $m - k \times m - k$ sub-matrix of transitions between the members of the set of transient states T, and $\mathbf{I}$ is the $k \times k$ unit matrix, denotes the transitions between the members of the set of absorbing states.

$$\mathbf{P}^{(n)} = \begin{pmatrix} \mathbf{I} & \mathbf{0} \\ \mathbf{R}^{(n)} & \mathbf{M}^{(n)} \end{pmatrix}.$$

As $n \to \infty$, $\mathbf{M}^{(n)} \to 0$ as $\lim_{n \to \infty} p_{ij}^{(n)} = 0, i, j \in T$. If the chain begins with $i \in T$, it will remain in T for a finite number of steps before entering the set of absorbing states, where it will remain indefinitely. One of the intriguing quantities to investigate in these scenarios is the probability of absorption.

Theorem 3.2 *Let a_{ik} denote the probability that the chain starting with a transient state i eventually gets absorbed in an absorbing state k. If the absorption probability matrix is denoted by $\mathbf{A} = [a_{ik}], i \in T, k \in S \setminus T$, then*

$$\boldsymbol{G} = (\boldsymbol{I} - \boldsymbol{M})^{-1}, \quad \boldsymbol{A} = (\boldsymbol{I} - \boldsymbol{M})^{-1}\boldsymbol{R} = \boldsymbol{GR}.$$

Proof Firstly, let us prove that $(\mathbf{I} - \mathbf{M})$ has an inverse by proving that the homogeneous system

$$(\mathbf{I} - \mathbf{M})x = 0,$$

has a unique solution $x = 0$. Indeed, $(\mathbf{I} - \mathbf{M})x = 0$ is equivalent to $x = \mathbf{M}x$ and

$$x = \mathbf{M}x = \mathbf{M}^{(2)}x = \cdots = \mathbf{M}^{(n)}x = \cdots$$

But $\mathbf{M}^{(n)}x \to 0$. Hence, $x = 0$. Now consider

$$\mathbf{I} - \mathbf{M}^{(n+1)} = (\mathbf{I} - \mathbf{M})\left(\mathbf{I} + \mathbf{M} + \mathbf{M}^{(2)} + \cdots + \mathbf{M}^{(n)}\right).$$

Multiplying (to the left) both sides by $\mathbf{G} = (\mathbf{I} - \mathbf{M})^{-1}$:

$$\mathbf{G} - \mathbf{G}\mathbf{M}^{(n+1)} = \mathbf{I} + \mathbf{M} + \mathbf{M}^{(2)} + \cdots + \mathbf{M}^{(n)}.$$

Therefore, letting $n \to \infty$,

$$\mathbf{G} = \mathbf{I} + \mathbf{M} + \mathbf{M}^{(2)} + \cdots = \sum_{n=0}^{\infty} \mathbf{M}^{(n)}.$$

Now, conditioning on the first step, for any $i \in T$ and any $j \in S \setminus T$,

$$a_{ij} = p_{ij} + \sum_{k \in T} p_{ik} a_{kj}$$

Thus,

$$\mathbf{A} = \mathbf{R} + \mathbf{M}\,\mathbf{A}.$$

That is,

$$(\mathbf{I}\text{-}\mathbf{M})\mathbf{A} = \mathbf{R}.$$

Therefore

$$\mathbf{A} = (\mathbf{I}\text{-}\mathbf{M})^{-1}\mathbf{R} = \mathbf{G}\,\mathbf{R}.$$

The matrix $\mathbf{G} = (\mathbf{I}\text{-}\mathbf{M})^{-1}$ is known as the fundamental matrix. It plays a crucial role, say, in finding out the mean time for absorption in some states. The cases where there is more than one closed class of states in a reducible Markov chain can be handled in a similar way.

Example 3.2 Let us consider a Markov chain with states space $S = \{0, 1, 2, 3\}$ and probability transition matrix

$$\mathbf{P} = \begin{pmatrix} 0 & \frac{1}{2} & 0 & \frac{1}{2} \\ \frac{1}{2} & 0 & \frac{1}{2} & 0 \\ 0 & 0 & 1 & 0 \\ 0 & 0 & 0 & 1 \end{pmatrix}.$$

If the chain, starts from state 0, determine the expected number of visits that the chain makes to state 0 before getting absorbed.

Solution: The state transition diagram is shown in Fig. 3.4.

The closed classes are given by $C_1 = \{2\}$ and $C_2 = \{3\}$. The set of transient states $T = \{0, 1\}$. That is, the transient states are 0 and 1 and the absorbing states are 2 and 3. Recognizing matrix **P** (with the order of states 2-3-0-1):

$$\mathbf{P} = \begin{pmatrix} 1 & 0 & 0 & 0 \\ 0 & 1 & 0 & 0 \\ 0 & \frac{1}{2} & 0 & \frac{1}{2} \\ \frac{1}{2} & 0 & \frac{1}{2} & 0 \end{pmatrix}.$$

That is,

$$\mathbf{R} = \begin{pmatrix} 0 & \frac{1}{2} \\ \frac{1}{2} & 0 \end{pmatrix}, \quad \mathbf{M} = \begin{pmatrix} 0 & \frac{1}{2} \\ \frac{1}{2} & 0 \end{pmatrix}$$

$$\begin{aligned} \mathbf{G} = (\mathbf{I}\text{-}\mathbf{M})^{-1} &= \left(\begin{pmatrix} 1 & 0 \\ 0 & 1 \end{pmatrix} - \begin{pmatrix} 0 & \frac{1}{2} \\ \frac{1}{2} & 0 \end{pmatrix} \right)^{-1} \\ &= \begin{pmatrix} \frac{4}{3} & \frac{2}{3} \\ \frac{2}{3} & \frac{4}{3} \end{pmatrix}. \end{aligned}$$

The expected number of visits that the chain makes to state 0 before getting absorbed is given by $\mu_{00} = \frac{4}{3}$.

Example 3.3 Consider a five-state Markov chain with state space $S = \{0, 1, 2, 3, 4\}$. Let the matrix **P** be given by

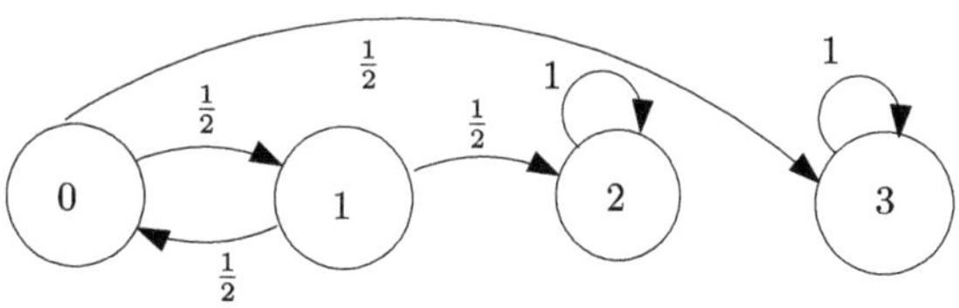

Fig. 3.4 State transition diagram for Example 3.2

$$\mathbf{P} = \begin{pmatrix} 1-p_1 & p_1 & 0 & 0 & 0 \\ 0 & 1-p_2-p_3 & p_2 & p_3 & 0 \\ 0 & 0 & 1-p_4 & 0 & p_4 \\ 0 & 0 & 0 & 1 & 0 \\ 0 & 0 & 0 & 0 & 1 \end{pmatrix}, \quad p_i > 0,\ i = 1, 2, 3, 4.$$

Classify the states of this Markov chain. Also, find the probability of absorption in states 3 and 4.

Solution: The state transition diagram is shown in Fig. 3.5.

Here, p_1, p_2, p_3, p_4 are all greater than 0. The states 0, 1, and 2 are transient states, and 3 and 4 are recurrent aperiodic states. There are two closed communicating classes given by $\{3\}$ and $\{4\}$. Hence, the chain is reducible.

Here,

$$\mathbf{M} = \begin{pmatrix} 1-p_1 & p_1 & 0 \\ 0 & 1-p_2-p_3 & p_2 \\ 0 & 0 & 1-p_4 \end{pmatrix},$$

$$\mathbf{R} = \begin{pmatrix} 0 & 0 \\ p_3 & 0 \\ 0 & p_4 \end{pmatrix}.$$

The fundamental matrix is given by:

$$\mathbf{G} = (\mathbf{I}\text{-}\mathbf{M})^{-1} = \begin{pmatrix} \frac{1}{p_1} & \frac{1}{p_2+p_3} & \frac{p_2}{(p_2+p_3)p_4} \\ 0 & \frac{1}{p_2+p_3} & \frac{p_2}{(p_2+p_3)p_4} \\ 0 & 0 & \frac{1}{p_4} \end{pmatrix}.$$

Hence, the probability of absorption in states 3 and 4 is given in matrix form as:

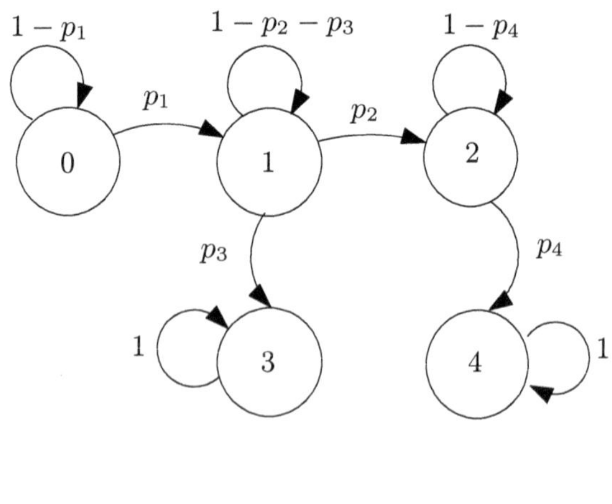

Fig. 3.5 State transition diagram for Example 3.3

$$\mathbf{A} = \mathbf{G}\,\mathbf{R} = \begin{pmatrix} \frac{p_3}{p_2+p_3} & \frac{p_2}{p_2+p_3} \\ \frac{p_3}{p_2+p_3} & \frac{p_2}{p_2+p_3} \\ 0 & 1 \end{pmatrix}.$$

Example 3.4 (Gambler's ruin problem) Two gamblers A and B, wager on successive independent tosses of an unbiased coin that will result in a heads-up outcome with probability p. If the coin lands on heads, gambler A wins a rupee from gambler B, and if it lands on tails, gambler B wins a rupee from gambler A. The total amount of rupees wagered by the two players, let's say N, remains constant as a result. The game is over as soon as either player declares bankruptcy, that is, runs out of money. Let's say that player A starts out with a fortune of i, $0 \leq i \leq N$. The amount of money that gambler A has after the nth toss is X_n. If $X_n = 0$, gambler A loses everything, and the game is over. If $X_n = N$, gambler B loses everything he had, and the game is over. Otherwise, the game continues. For this Markov chain, write the one-step transition probability matrix and draw the state transition diagram. Find the probability that gambler A goes with amount N given that he started with i amount.
Solution: Here,$\{X_n, n = 0, 1, \ldots\}$ forms a reducible Markov chain with state 0 and state N as absorbing states (Fig. 3.6).

Define

$$g(i) = P\{A \text{ goes with amount } N | X_0 = i\}, \quad 0 \leq i \leq N.$$

Clearly, $g(0) = 0$ and $g(N) = 1$. The function $g(i)$ satisfies

$$g(i) = pg(i+1) + (1-p)g(i-1), \quad 1 \leq i \leq N-1.$$

The above equation can be rewritten as

$$g(i+1) - g(i) = \frac{1-p}{p}(g(i) - g(i-1)), \quad 1 \leq i \leq N-1.$$

For $i = 1$,

$$g(2) - g(1) = \frac{1-p}{p}(g(1) - g(0)) = \frac{1-p}{p}g(1).$$

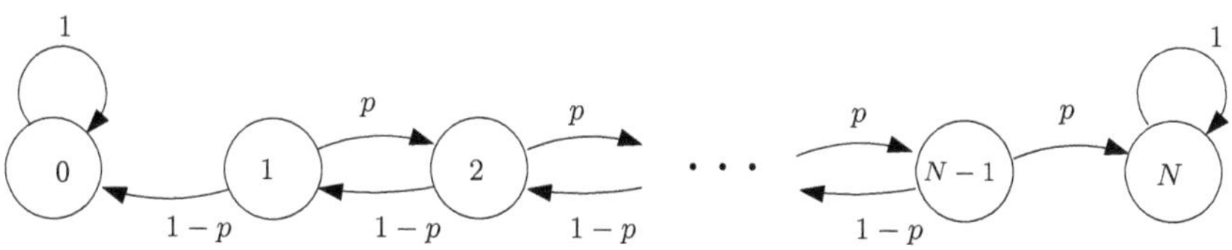

Fig. 3.6 State transition diagram for Example 3.4

For $i = 2$,

$$g(3) - g(2) = \frac{1-p}{p}(g(2) - g(1)) = \left(\frac{1-p}{p}\right)^2 g(1).$$

Hence,

$$g(i) - g(i-1) = \left(\frac{1-p}{p}\right)^{i-1} g(1), \quad 1 \le i \le N.$$

Adding all,

$$g(i) - g(1) = \sum_{j=1}^{i-1} \left(\frac{1-p}{p}\right)^j g(1).$$

$$g(i) = \sum_{j=0}^{i-1} \left(\frac{1-p}{p}\right)^j g(1), \quad 1 \le i \le N.$$

Using $g(N) = 1$,

$$1 = g(N) = \sum_{j=0}^{N-1} \left(\frac{1-p}{p}\right)^j g(1).$$

Hence,

$$g(1) = \frac{1}{\sum_{j=0}^{N-1} \left(\frac{1-p}{p}\right)^j}.$$

This implies,

$$g(i) = \frac{\sum_{j=0}^{i-1} \left(\frac{1-p}{p}\right)^j}{\sum_{j=0}^{N-1} \left(\frac{1-p}{p}\right)^j}, \quad 1 \le i \le N.$$

Therefore,

$$g(i) = P\{A \text{ goes with amount } N | X_0 = i\} = \begin{cases} \frac{1-\left(\frac{1-p}{p}\right)^i}{1-\left(\frac{1-p}{p}\right)^N}, & p \neq \frac{1}{2} \\ \frac{i}{N}, & p = \frac{1}{2} \end{cases}$$

3.1.3 Chain with More Than One Closed Class

Consider a finite reducible chain with more than one irreducible closed class of persistent states and some transient states and transition probability matrix **P**. To fix our ideas suppose that there are two irreducible closed classes C_1 and C_2 of positive recurrent states with transition sub matrices $\mathbf{N_1}$ and $\mathbf{N_2}$, respectively. Then,

$$\mathbf{P} = \begin{pmatrix} \mathbf{N_1} & \mathbf{0} & \mathbf{0} \\ \mathbf{0} & \mathbf{N_2} & \mathbf{0} \\ \mathbf{R_1} & \mathbf{R_2} & \mathbf{M} \end{pmatrix}$$

where $\mathbf{M}$ is the (sub) matrix of transitions among the m_1 transient states (which form a set T). Let $C = C_1 \cup C_2$, such that $S = C \cup T$. To find the probability of absorption of a transient state. Consider a new Markov chain with transition probabilities

$$\bar{p}_{ij} = \begin{cases} p_{ij}, i \in T, j \in S \\ \delta_{ij}, i \in C, j \in S \end{cases}.$$

The new chain then becomes an absorbing chain with the same transient states. The transition matrix becomes

$$\tilde{\mathbf{P}} = \begin{pmatrix} \mathbf{I} & \mathbf{0} & \mathbf{0} \\ \mathbf{0} & \mathbf{I} & \mathbf{0} \\ \mathbf{R_1} & \mathbf{R_2} & \mathbf{M} \end{pmatrix} = \begin{pmatrix} \mathbf{I} & \mathbf{0} \\ \mathbf{R} & \mathbf{M} \end{pmatrix}$$

where $\mathbf{R}$ is the augmented matrix having the same number of rows as $\mathbf{M}$ and with columns of $\mathbf{R_1}$ augmented by those of $\mathbf{R_2}$. The absorption probabilities can be computed in the same manner from

$$\mathbf{A} = [a_{ik}] = \mathbf{G}\,\mathbf{R}$$

where $\mathbf{G} = (\mathbf{I}\text{-}\mathbf{M})^{-1}$ is the fundamental matrix.

Example 3.5 Consider the Markov chain having a transition probability matrix.

$$\mathbf{P} = \begin{pmatrix} 1/3 & 0 & 2/3 & 0 & 0 & 0 \\ 0 & 1/2 & 1/4 & 0 & 1/4 & 0 \\ 2/5 & 0 & 3/5 & 0 & 0 & 0 \\ 0 & 1/4 & 1/4 & 1/4 & 0 & 1/4 \\ 0 & 0 & 0 & 0 & 1/2 & 1/2 \\ 0 & 0 & 0 & 0 & 3/4 & 1/4 \end{pmatrix}.$$

The chain consists of two classes of positive recurrent aperiodic states $C_1 = \{1, 3\}$, $C_2 = \{5, 6\}$ and transient states 2 and 4. Find the stationary distribution of this Markov Chain if the chain starts from (a) state 3 and (b) state 5.

Also, find the probability of absorption of state 2 and state 4.

Solution:

$$\mathbf{N_1} = \begin{pmatrix} \frac{1}{3} & \frac{2}{3} \\ \frac{2}{5} & \frac{3}{5} \end{pmatrix}, \mathbf{N_2} = \begin{pmatrix} \frac{1}{2} & \frac{1}{2} \\ \frac{1}{2} & \frac{1}{2} \end{pmatrix}.$$

Solving $\pi_1 \mathbf{N_1} = \pi_1$, we get $\pi_1 = \left(\frac{3}{8}, \frac{1}{8}\right)$. Similarly solving $\pi_2 \mathbf{N_2} = \pi_2$, we get $\pi_2 = \left(\frac{1}{2}, \frac{1}{2}\right)$.

(a) If chain starts from state 3 then, $\pi = \left(\frac{3}{8}, 0, \frac{1}{8}, 0, 0, 0\right)$.

(b) If chain starts from state 5 then, $\pi = \left(0, 0, 0, 0, \frac{1}{2}, \frac{1}{2}\right)$.

Now,

$$\mathbf{R_1} = \begin{pmatrix} 0 & \frac{1}{4} \\ 0 & \frac{1}{4} \end{pmatrix}, \mathbf{R_2} = \begin{pmatrix} \frac{1}{4} & 0 \\ 0 & \frac{1}{4} \end{pmatrix}, \mathbf{M} = \begin{pmatrix} \frac{1}{2} & 0 \\ \frac{1}{4} & \frac{1}{4} \end{pmatrix}.$$

$$(\mathbf{I} - \mathbf{M})^{-1} = \begin{pmatrix} \frac{1}{2} & 0 \\ \frac{1}{6} & \frac{1}{3} \end{pmatrix}, \mathbf{R} = \begin{pmatrix} 0 & \frac{1}{4} & \frac{1}{4} & 0 \\ 0 & \frac{1}{4} & 0 & \frac{1}{4} \end{pmatrix}.$$

The probability of absorption

$$\mathbf{A} = \mathbf{GR} = \begin{pmatrix} 0 & \frac{1}{2} & \frac{1}{2} & 0 \\ 0 & \frac{1}{2} & \frac{1}{6} & \frac{1}{3} \end{pmatrix}.$$

Hence both State 2 and State 4 have equal probability of absorption in class C_1 and C_2.

3.2 Reversed and Time-Reversible Markov Chains

A "stationary Markov chain" is the one whose initial state distribution is the same as its stationary distribution or one that has run indefinitely. It supposedly has attained "stationarity". In their general form, time non-homogeneous Markov chain is not stationary. Note that a time non-homogeneous Markov chain cannot be stationary. Only stationary chains can run in homogeneous systems over time.

Definition 3.1 **(The reversed Markov chain)** Consider a stationary, ergodic Markov chain $\{X_n, \ n = 0, 1, \ldots\}$ with transition probability matrix $\mathbf{P}$ and stationary distribution $\{\pi_i, i \in S\}$. Now consider the chain in reverse order, i.e., $X_{m+1}, X_m, X_{m-1}, X_{m-2}, \ldots$. This is called the reversed chain.

Theorem 3.3 *The reversed chain is also a Markov chain.*

Proof Consider a portion of the forward Markov chain

$$\ldots, X_{m-2}, X_{m-1}, X_m, X_{m+1}, X_{m+2}, X_{m+3}, \ldots$$

and suppose that X_{m+1} is the present state. Then, by using the Markov property for the forward chain, the future $X_{m+2}, X_{m+3}, \ldots$ is independent of the past $\ldots, X_{m-1}, X_m$. Since independence is the symmetric property, i.e., if X is independent of Y , then Y is independent of X, therefore the past $\ldots, X_{m-1}, X_m$ is independent of the future $X_{m+2}, X_{m+3}, \ldots$. Hence, in terms of the reversed chain, the past which is independent of the future:

$$P\{X_m = j|X_{m+1} = i, X_{m+2}, \ldots\} = P\{X_m = j|X_{m+1} = i\} = q_{ij} \text{ (say)}.$$

The transition probabilities of the reversed Markov chain can be obtained as follows.

$$q_{ij} = P\{X_m = j|X_{m+1} = i\} = \frac{P(X_m = j)P\{X_{m+1} = i|X_m = j\}}{P(X_{m+1} = i)} = \frac{\pi_j p_{ji}}{\pi_i},$$

where the fact that the forward chain is running in stationarity is used.

Definition 3.2 (**Time-reversible Markov chain**) An ergodic Markov chain is time-reversible if $q_{ij} = p_{ij}$ for all states i and j.

Some results with trivial proofs are stated below:
Results:

1. A Markov chain is said to be time-reversible if $\pi_i p_{ij} = \pi_j p_{ji}$ for all states i and j.
2. For a time-reversible Markov chain, the proportion of transitions $i \to j$ is equal to the proportion of transitions $j \to i$.
   Since the time reversibility condition given in the above result is

$$P\{X_n = i|X_{n+1} = j\} = P\{X_n = j|X_{n+1} = i\}$$

   for any n sufficiently large that stationarity applies, but $P\{X_n = i|X_{n+1} = j\}$ is the proportion of transitions that move $i \to j$ and $P\{X_n = j|X_{n+1} = i\}$ is for transitions $j \to i$.
3. If a π can be found such that $\pi_i p_{ij} = \pi_j p_{ji}$ with $\sum_i \pi_i = 1$ for all states i, j, then the process is said to be reversible and π is the stationary distribution of the chain.
   Proof: Suppose x_i such that $\sum_i x_i = 1$. Then,

$$\sum_i x_i p_{ij} = \sum_i x_j p_{ji} = x_j \sum_i p_{ji} = x_j.$$

   Thus, it is shown that the x_j satisfies the equations defining a stationary distribution. Hence proved.

Example 3.6 A random walk on the integers (or a subset of integers, as in this case) with finite state space $S = \{0, 1, 2, \ldots, N\}$ moves either one step left or one-step

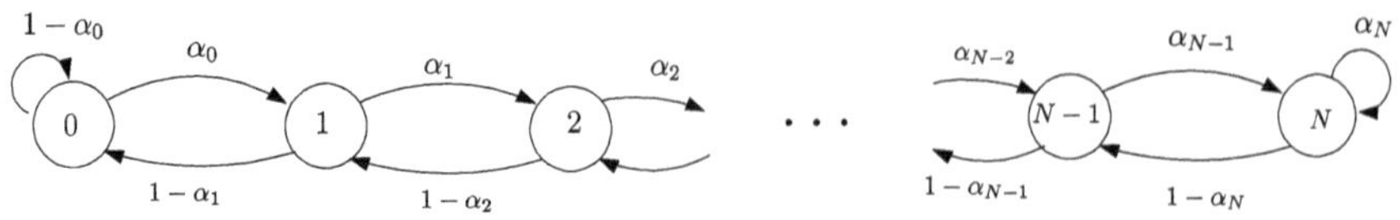

Fig. 3.7 State transition diagram for Example 3.6

right during each time step. The transition probabilities given as are $p_{ii+1} = \alpha_i = 1 - p_{ii-1}, i = 1, 2, \ldots, N-1$, $p_{01} = \alpha_0 = 1 - p_{00}$, $p_{NN} = \alpha_N = 1 - p_{NN-1}$. Find the stationary distribution of this Markov chain.

Solution: The state transition diagram is shown in Fig. 3.7. A reversible Markov chain, is a random walk. If a process jumps right from position i, $0 < i < N$, there is only one way to return to state i, which is via $i+1$, hence it must have previously jumped left from $i+1$. As the chain is positive recurrent, for every jump to the right at i, there must be a jump left from $i+1$. This implies that the forward flux (rate) from state i to $i+1$ is equal to the backward flux from state $i+1$ to i. This balanced flux is a key indicator of a time-reversible process.

Since the process is time-reversible, the stationary distribution can be obtained from the reversibility conditions

$$
\begin{aligned}
\pi_{\mathbf{0}}\alpha_0 &= \pi_{\mathbf{1}}(1-\alpha_1)\\
\pi_{\mathbf{1}}\alpha_1 &= \pi_{\mathbf{2}}(1-\alpha_2)\\
&\vdots\\
\pi_{\mathbf{i}}\alpha_i &= \pi_{\mathbf{i+1}}(1-\alpha_{i+1})\\
&\vdots\\
\pi_{\mathbf{N-1}}\alpha_{N-1} &= \pi_{\mathbf{N}}(1-\alpha_N)
\end{aligned}
$$

with solution

$$
\begin{aligned}
\pi_{\mathbf{1}} &= \frac{\pi_{\mathbf{0}}\alpha_0}{1-\alpha_1}\\
\pi_{\mathbf{2}} &= \frac{\pi_{\mathbf{0}}\alpha_0\alpha_1}{(1-\alpha_1)(1-\alpha_2)}\\
&\vdots\\
\pi_{\mathbf{j}} &= \frac{\pi_{\mathbf{0}}\alpha_0\alpha_1\cdots\alpha_{j-1}}{(1-\alpha_1)(1-\alpha_2)\cdots(1-\alpha_j)}.
\end{aligned}
$$

Then, use the condition $\sum_{i}^{N} \pi_{\mathbf{i}} = 1$ to obtain

$$\pi_0 = \left[1 + \sum_{j=1}^{N} \frac{\alpha_0 \alpha_1 \cdots \alpha_{j-1}}{(1-\alpha_1)(1-\alpha_2)\cdots(1-\alpha_j)}\right]^{-1}.$$

Next, a theorem giving a necessary and sufficient condition for an ergodic MC to be time-reversible is stated and the proof is left as an exercise.

Theorem 3.4 *An ergodic MC with $p_{ij} = 0$, whenever $p_{ji} = 0$ is time-reversible if and only if any path from state i to state i has the same probability as the reverse path. In other words,*

$$p_{ii_1} p_{i_1 i_2} \cdots p_{i_k i} = p_{ii_k} p_{i_k i_{k-1}} \cdots p_{i_1 i},$$

for all states $i, i_1, \ldots, i_k$, $k = 1, 2, \ldots$.

Queueing theory research has long focused on using this set of concepts to analyze queueing systems and queueing networks. It generates a large number of simple results for systems that at first glance seem to be extremely complex. A rapid conclusion in this direction is that any birth-death chain with a steady-state (equilibrium) probability distribution is reversible.

Example 3.7 Consider M balls, distributed in two urns. At each time step, you randomly select one ball from the total of M balls and move it from the urn where it currently resides to the other urn. Let X_n be the number of balls in urn 1. Prove that this chain has a reversible distribution

Solution: The transition probabilities are:

$$\begin{aligned}
p_{01} &= 1, \\
p_{MM-1} &= 1, \\
p_{ii-1} &= \frac{i}{M}, \ 1 \le i \le M, \\
p_{ii+1} &= \frac{M-i}{M}, \ 1 \le i \le M-1.
\end{aligned}$$

The invariant measure put each ball at random into one of the two urns, as switching any ball between the two urns does not alter this assignment. Thus, π_i is binomial distributed with parameters M and 0.5,

$$\pi_i = \binom{M}{i} \frac{1}{2^M}, \quad 0 \le i \le M.$$

The following equations need to be verified in order to check the time reversibility

$$\pi_0 p_{01} = \pi_1 p_{10},$$
$$\pi_i p_{ii+1} = \pi_{i+1} p_{i+1i},\ 1 \le i \le M-1,$$
$$\pi_i p_{ii-1} = \pi_{i-1} p_{i-1i},\ 1 \le i \le M-1,$$
$$\pi_M p_{MM-1} = \pi_{M-1} p_{M-1M}.$$

It can be easily observed by putting the values of p_{ij} that this Markov chain is time-reversible.

3.3 Discrete-Time Branching Processes

Several families in Victorian England felt in the mid-nineteenth century that their family names might go extinct. Was it merely irrational fear that drove them to this conclusion, or was it something more serious? Sir Francis Galton submitted the following query when they decided to ask around.
"How many male children (on average) must each generation of a family have in order for the family name to continue in perpetuity?"

London-based mathematician H. W. Watson (25 February 1827–11 January 1903) in 1874 presented a solution to the "problem of the extinction of families" which is as follows: Let $p_0, p_1, \ldots$ be the odds that the number of sons of a man has resp. $0, 1, \ldots$, and each son have the same probability for his own sons, and so on. What are the chances that the male line will extinct after r generations, and what are the chances that any given number of descendants in the male line will die in any given generation?

Watson gave the solution "every family will die out, even when the population size, on average, increases from one generation to the next". Francis Galton prefaced to the solution by Watson and Galton recognized that a first step in studying the hypothesis would be to determine the probability that an ordinary family will disappear using fertility data for the whole population. The mathematical model of Galton and Watson is known as a Galton–Watson process. In 1938, A. Kolmogorov also treated the above problem and determined the asymptotic form of the probability that the family is still in existence after a large finite number of generations.

In this section, the results on extinction and survival probabilities expressed in terms of the mean and the generating function of a random variable whose distribution models the branching process are presented.

Definition 3.3 (Branching Process) The branching process is a system of particles (individuals, cells, molecules, etc.) that produce a random number of offsprings at random times during their lives or immediately after they die.

Definition 3.4 (General Branching Processes) The general branching processes are those in which individuals can reproduce offspring during their lifetime.

Let Z_n denotes the number of individuals in the population in the nth generation. Assume that the population starts from one individual, i.e., at $n = 0$, $Z_0 = 1$. At $n = 1$, that one individual produces some random number of offspring Z_1 and dies. Similarly, at $n = 2$, each Z_1 individual will produce some random number of offspring and die. The total number of offspring produced at $n = 2$ is represented by Z_2. The distribution of the number of offspring produced is assumed to be the same for each individual in every generation and independent of the number of individuals in the generation and the number of offspring others have. If at any $n = k$, $Z_k = 0$, then $Z_m = 0\ \forall m \geq k$, i.e., the population is extinct.

Definition 3.5 (**Classical Branching Process**) A stochastic process $\{Z_n, n = 0, 1, \ldots\}$ is called a classical branching process if it assumes the production of offsprings in the above manner. Let Y_i (or $Y_{i,n}$) be the number of offspring of the ith individual of the nth generation. Suppose that, $\{Y_i, i = 1, 2, \ldots\}$ are non-negative integer valued IID RVs with probability mass function $p_j = P(Y_i = j), j = 0, 1, \ldots$ and independent of the size of the generation. Then

$$Z_n = \sum_{i=1}^{Z_{n-1}} Y_i, \quad n = 1, 2, \ldots$$

and $\{Z_n, n = 0, 1, \ldots\}$ is a DTMC.

Example 3.8 Consider that a cell either produces a single copy with a probability of 0.99 or does not reproduce at all with a probability of 0.01. While the living population might survive for a while to the next period, what is the probability of surviving till the nth generation? What is the probability of extinction?
Solution: The probability of surviving till nth generation is $(\frac{99}{100})^n$, since the population would need to reproduce successfully, with probability 0.99 for n generations. This goes to 0 as n becomes very large, therefore the probability of extinction for this population is 1.

The concept of type space is one of the most important concepts in branching process theory. The type space is the set of all conceivable types of particles involved in the process, which can be limited, countably infinite, or in a continuum. A given type of particle can create particles of a different type. Type transitions, as well as the type space, are restricted, resulting in distinct features of the resulting processes.

3.3.1 Galton–Watson Process

In this section, Galton–Watson Process is presented, a foundational stochastic model widely used in understanding population dynamics and branching phenomena.

Definition 3.6 (**Galton–Watson Process**) Let $\{Z_n, n = 0, 1, \ldots\}$ be a classical branching process. The distribution of particle lifetime τ has a significant impact on the behaviour and analysis of the process. When $\tau = 1$, the method is known as the Galton-Watson process.

If τ is an exponential distributed random variable, then the resulting process is called a Markov branching process. If τ is an arbitrary non-negative random variable, the resulting process is called an "age-dependent" or Bellman–Harris process.

Consider particles that can produce more of the same type of particles. The first generation is made up of particles from the 0-th generation, the second generation is made up of particles from the first generation, and so on. The number of particles in the 0th, 1st, ... generations is denoted by $Z_0, Z_1, \ldots$. Furthermore, the following assumptions are made:

1. If the size of the nth generation is known, the probability distribution governing the subsequent generations is independent of the sizes of the earlier generations (i.e., generations preceding the nth). Therefore it becomes a discrete-time Markov chain $\{Z_n, n = 0, 1, \ldots\}$. It is assumed that the transition probabilities of the Markov chain remain constant over time. That is $\{Z_n, n = 0, 1, \ldots\}$ is a time-homogeneous DTMC. Note that the number of offsprings produced by a particle is independent of how many other particles are present. Also, it is assumed that $Z_0 = 1$.
2. If $Z_n = 0$ then $P(Z_{n+1} = 0) = 1$. The probability distribution of Z_1 is given by

$$P(Z_1 = k) = p_k, \quad k = 0, 1, \ldots, \quad \sum_{k=0}^{\infty} p_k = 1$$

where p_k is the probability that a particle existing in the nth generation has k offsprings in the $(n+1)$-th generation.

3.3.1.1 One-Step Transition Probabilities

Assume that p_k is independent of the generation number n. The conditional distribution of Z_{n+1} given $Z_n = k$ is appropriate to the assumption that different particles reproduce independently; that is, Z_{n+1} follows the distribution of the sum of k independent RVs, each distributed like Z_1. Since $\{Z_n, n = 0, 1, \ldots\}$ is a time-homogeneous DTMC, the one-step transition probabilities are given by

$$p_{ij} = P\{Z_{n+1} = j | Z_n = i\} = P(Y_1 + Y_2 + \ldots + Y_i = j), \quad i, j = 1, 2, \ldots \quad (3.1)$$

where $Y_1, Y_2, \ldots, Y_i$ are independent RVs, each distributed like Z_1.

Example 3.9 Consider a Galton–Watson process $\{Z_n, n = 0, 1, \ldots\}$ with offspring distribution $\{p_k, k = 0, 1, 2\}$. Assume that $p_0 = 1/5$, $p_1 = 3/5$ and $p_2 = 1/5$. Calculate $P\{Z_2 = 3 | Z_1 = 2\}$, $P\{Z_3 = 2 | Z_2 = 3\}$ and $P(Z_2 = 2)$.

Solution:

$$\begin{aligned} P(Z_2=3|Z_1=2) &= P(Y_1+Y_2=3) \\ &= P(Y_1=0, Y_2=3)+P(Y_1=1, Y_2=2) \\ &\quad +P(Y_1=2, Y_2=1) \\ &\quad +P(Y_1=3, Y_2=0) \\ &= P(Y_1=0)P(Y_2=3)+P(Y_1=1)P(Y_2=2) \\ &\quad +P(Y_1=2)P(Y_2=1)+P(Y_1=3)P(Y_2=0) \\ &= p_0p_3+p_1p_2+p_2p_1+p_3p_0 \\ &= 0+\frac{3}{25}+\frac{3}{25}+0=\frac{6}{25}. \end{aligned}$$

Similarly,

$$\begin{aligned} P(Z_3=2|Z_2=3) &= P(Y_1+Y_2+Y_3=2) \\ &= p_0p_0p_2+p_0p_2p_0+p_2p_0p_0+p_1p_1p_0+p_1p_0p_1+p_0p_1p_1 \\ &= 3p_0(p_1^2+p_0p_2). \end{aligned}$$

Now,

$$\begin{aligned} P(Z_2=2) &= P\{Z_2=2|Z_1=0\}P(Z_1=0)+P\{Z_2=2|Z_1=1\}P(Z_1=1) \\ &\quad +P\{Z_2=2|Z_1=2\}P(Z_1=2) \\ &= 0+P(Y_1=2)P(Z_1=1)+P(Y_1+Y_2=2)P(Z_1=2) \\ &= p_2p_1+(p_0p_2+p_1p_1+p_2p_0)p_2 \\ &= p_2(p_1+p_0p_2+p_1p_1+p_2p_0). \end{aligned}$$

Example 3.10 The first generation of particles is the collection of offsprings of a given particle. The next generation is formed by the offsprings of these members. Assume that the probability that a particle has k offsprings (splits into k parts) is p_k, where $p_0=0.4$, $p_1=0.3$, $p_2=0.3$. Further, assume that the particles act independently and identically irrespective of the generation. Find the probability that there is no particle in the second generation.
Solution:

$$\begin{aligned} P(Z_2=0) &= P\{Z_2=0|Z_1=0\}P(Z_1=0)+P\{Z_2=0|Z_1=1\}P(Z_1=1) \\ &\quad +P\{Z_2=0|Z_1=2\}P(Z_1=2) \\ &= P(Z_1=0)+P(Y_1=0)P(Z_1=1)+P(Y_1+Y_2=0)P(Z_1=2) \\ &= p_0+p_0p_1+p_0^2p_2 \\ &= 0.568. \end{aligned}$$

3.3.1.2 Probability Generating Function

Let $\{Y_i, i = 1, 2, \ldots\}$ be IID random variables each having the same distribution as Z_1. Let $H(z)$ be the probability generating function of Y_i.

$$H(z) = \sum_k P(Y_i = k)z^k = \sum_k p_k z^k, \quad | z | \le 1$$

where z is a complex variable. Let $H_n(z)$ be the probability generating function of Z_n.

$$H_n(z) = \sum_k P(Z_n = k)z^k, \quad | z | \le 1$$

where z is a complex variable.

Since $Z_0 = 1$, $H_0(z) = z$, and

$$\begin{aligned} H_1(z) &= \sum_k P(Z_1 = k)z^k \\ &= \sum_k P(Z_1 = k|Z_0 = 1)z^k \\ &= \sum_k p_k z^k \\ &= H(z). \end{aligned}$$

Theorem 3.5 *$H_n(z) = H_{n-1}(H(z))$ and $H_n(z) = H(H_{n-1}(z))$.*

Proof For $n = 1, 2, \ldots$,

$$\begin{aligned} P(Z_n = k) &= \sum_i P(Z_n = k \mid Z_{n-1} = i)P(Z_{n-1} = i) \\ &= \sum_i P(Y_1 + Y_2 + \cdots + Y_i = k)P(Z_{n-1} = i). \end{aligned}$$

Now,

$$\begin{aligned} H_n(z) &= \sum_k P(Z_n = k)z^k \\ &= \sum_k \left[\sum_i P(Z_n = k \mid Z_{n-1} = i)P(Z_{n-1} = i) \right] z^k \\ H_n(z) &= \sum_i P(Z_{n-1} = i) \left[\sum_k P(Y_1 + Y_2 + \cdots + Y_i = k)z^k \right]. \end{aligned}$$

Since $\{Y_i, i = 1, 2, \ldots\}$ are IID random variables and PGF of Y_i is $H(z)$, then the PGF of $Y_1 + Y_2 + \cdots + Y_i$ is $[H(z)]^i$. Hence,

$$H_n(z) = \sum_i P(Z_{n-1} = i)[H(z)]^i = H_{n-1}(H(z)).$$

$H_1(z) = H(z) = H(H_0(z))$ since $H_0(z) = z$. Suppose for $n = j$, $H_j(z) = H(H_{j-1}(z))$, then

$$\begin{aligned} H_{j+1}(z) &= H_j(H(z)) \\ &= H(H_{j-1}(H(z))) \\ &= H(H_j(z)). \end{aligned}$$

By induction, it is proved.

Theorem 3.6 *Let $\mu = E[Z_1]$ and $\sigma^2 = Var(Z_1)$. Then*

$$E[Z_n] = \mu^n,$$

and

$$Var(Z_n) = \begin{cases} \frac{\mu^{n-1}(\mu^n-1)\sigma^2}{\mu-1}, & \mu \neq 1 \\ n\sigma^2 & \mu = 1 \end{cases}.$$

Proof By Theorem 3.5, $H_n(z) = H_{n-1}(H(z))$. Then differentiating and putting $z = 1$, $H_n^{'}(1) = H_{n-1}^{'}(H(1))H^{'}(1)$, $H(1) = 1$, $H^{'}(1) = \mu$ and for $n = 1, 2, \ldots$

$$H_n^{'}(1) = H_{n-1}^{'}(1)\mu = \cdots = \mu^n.$$

Hence $E[Z_n] = \mu^n$.

Remark 3.1 1. If $\mu = 1$, then $E[Z_n] \to 1$ and $\text{Var}[Z_n] \to \infty$ as $n \to \infty$.
2. If $\mu < 1$, then $E[Z_n] \to 0$ and $\text{Var}[Z_n] \to 0$ as $n \to \infty$.
3. If $\mu > 1$, then $E[Z_n] \to \infty$ and $\text{Var}[Z_n] \to \infty$ as $n \to \infty$.

Criticality
A very important classification is based on the mean progeny count $\mu = E[Z_1]$.

$$E[Z_n] = \mu^n.$$

Therefore, in the expected value sense, the process grows geometrically if $\mu > 1$, stays constant if $\mu = 1$, and decays geometrically if $\mu < 1$. These three cases are called supercritical, critical, and subcritical, respectively:

$$\begin{aligned} &\mu > 1, \ \text{supercritical} \ \ E[Z_n] \uparrow \infty \\ &\mu = 1, \ \text{critical} \ \ E[Z_n] = 1 \\ &\mu < 1, \ \text{subcritical} \ \ E[Z_n] \downarrow 0. \end{aligned}$$

Example 3.11 A population begins with a single individual. In each generation, each individual in the population dies with probability $\frac{1}{2}$ or doubles with probability $\frac{1}{2}$. Let Z_n denote the number of individuals in the population in the nth generation. Find the mean and variance of Z_n.
Solution: Let Y be a RV with $P(Y = 0) = \frac{1}{2}$, $P(Y = 2) = \frac{1}{2}$. Let $\mu = E[Y] = 1$ and $\sigma^2 = \text{Var}[Y] = 1$. Then $E[Z_n] = \mu^n = 1$, and $\text{Var}[Z_n] = n\sigma^2 = n$.

Example 3.12 Consider the Galton–Watson process $\{Z_n, n = 0, 1, \ldots\}$ with offspring distribution $\{p_k\}$. Assume that $p_0 = 1/5$, $p_1 = 3/5$ and $p_2 = 1/5$. Find $E[Z_n]$ and $\text{Var}[Z_n]$.
Solution: Since $\mu = E[Z_1] = 1$. Hence, this process is a critical Galton–Watson process. $E[Z_1^2] = \frac{7}{5}$, then $\sigma^2 = \frac{2}{5}$. Hence, for $n = 1, 2, \ldots$

$$E[Z_n] = 1 \text{ and } \text{Var}(Z_n) = \frac{2n}{5}.$$

3.3.1.3 Probability of Extinction

If $Z_n = 0$ for some value $n \geq 1$, then $Z_m = 0$ for all $m \geq n$ and also

$$P\{Z_m = 0 | Z_n = 0\} = 1, \quad \text{for all } m \geq n.$$

A branching process ends when $P(Z_n = 0) = 1$, for some value of $n = 1, 2, \ldots$.

Theorem 3.7 *If $\mu \leq 1$, then the probability of extinction is one. If $\mu > 1$, then the probability of extinction is the positive root less than unity of the equation $H(z) = z$.*

Proof Let d_n be the probability that extinction occurs on or before the n-th generation, i.e., $d_n = P(Z_n = 0)$. Since, $d_0 = 0$.

$$\begin{aligned} d_{n+1} &= P(Z_{n+1} = 0) \\ &= P\{Z_{n+1} = 0 \mid Z_n = 0\}P(Z_n = 0) + P\{Z_{n+1} = 0 \mid Z_n \neq 0\}P(Z_n \neq 0) \\ &= 1 \times P(Z_n = 0) + P\{Z_{n+1} = 0 \mid Z_n \neq 0\}P(Z_n \neq 0) \\ &\geq P(Z_n \neq 0) = d_n. \end{aligned}$$

If

$$0 = d_0 \leq d_1 \leq \cdots \leq 1,$$

i.e., $\{d_n, n = 0, 1, \ldots\}$ is an increasing and bounded above sequence. There exists $d = \lim_{n\to\infty} d_n$, $d \in [0, 1]$ where d is the probability of extinction of the process.

$$
\begin{aligned}
d_m &= P(Z_m = 0) \\
&= \sum_j P\{Z_m = 0 \mid Z_1 = j\} P(Z_1 = j) \\
&= \sum_j P\{Z_m = 0 \mid Z_1 = j\} p_j.
\end{aligned}
$$

Let Y_i be the number of offspring (in m^{th} generation) of the ith individual in first generation.

$$
\begin{aligned}
P(Z_m = 0 \mid Z_1 = j) &= P(Y_1 + Y_2 + \cdots + Y_j = 0) \\
&= P(Y_1 = 0) P(Y_2 = 0) \cdots P(Y_j = 0) \\
&= [P(Z_{m-1} = 0)]^j \\
&= d_{m-1}^j.
\end{aligned}
$$

Hence,

$$
d_m = \sum_j d_{m-1}^j p_j = H(d_{m-1}).
$$

Since $d_m \to d$ when $m \to \infty$, the value d satisfies the equation $z = H(z)$. Note that the solution to equation $z = H(z)$ represents intersection of the graphs of $y = z$ and $y = H(z)$.

Remark 3.2 1. In the case $\mu > 1$, both extinction and survival are possible with nonzero probabilities. Then,

$$
P(Z_1 = i | \text{extinct}) = \frac{P(Z_1 = i \text{ and extinct})}{P(\text{extinct})} = \frac{P(Y = i) d^i}{H(d)}.
$$

Denote the last expression by $P(\widehat{Y} = i)$. Observe that $E[\widehat{Y}] < 1$. Hence, the distribution of Galton–Watson process with Y, conditioned on the extinction is the same as the distribution of the Galton–Watson process with $\widehat{Y}$.

2. In the case $1 < \mu < \infty$,

$$
\frac{Z_n}{\mu^n} \to X \ a.s. \, n \to \infty,
$$

for some positive random variable X.

Example 3.13 Consider the Galton–Watson process $\{Z_n, n = 0, 1, \ldots\}$ with offspring distribution $\{p_k\}$. Assume that $p_0 = 1/5$, $p_1 = 3/5$ and $p_2 = 1/5$. Find the probability of extinction.

Solution: Since $\mu = E[Z_1] = 1$.

$$
H(z) = \sum_k p_k z^k = \frac{1}{5} + \frac{3}{5} z + \frac{1}{5} z^2.
$$

For instance,

$$P(Z_2 = 2) = \frac{H_2''(0)}{2!},$$

using

$$H_2(z) = H(H(z)) = \frac{1}{5} + \frac{3}{5}H(z) + \frac{1}{5}[H(z)]^2.$$

Since $\mu = 1$, by Theorem 3.7, the probability of extinction is one.

Example 3.14 Suppose Z_i's are binomial random variables with parameters 3 and p. Find the probability of extinction.

Solution: Here, $E[Z] = 3p$. When $3p > 1$, by Theorem 3.7, the probability of extinction is the smallest solution of $z = H(z)$ which is smaller than 1. Now,

$$z = H(z) \implies z = (q + pz)^3.$$

That is,

$$p^3z^3 + 3p^2qz^2 + (3pq^2 - 1)z + q^3 = 0.$$

It is known that $z = 1$ is a solution, it is rewritten as

$$(z - 1)(p^3z^2 + (3p^2q + p^3)z - q^3) = 0.$$

Suppose $p = 0.5$, $E[Z] = \frac{3}{2} > 1$. Then, $z^2 + 4z - 1 = 0$ gives $d = 0.236$.

For more discussions, readers may refer to [1–5]

3.4 Exercise Problems

1. Let, $\{X_n, n = 0, 1, \ldots\}$ be a time-homogeneous DTMC with state space $S = \{0, 1, 2, 3, 4\}$ and one-step transition probability matrix

$$\mathbf{P} = \begin{pmatrix} 1 & 0 & 0 & 0 & 0 \\ 0.5 & 0 & 0.5 & 0 & 0 \\ 0 & 0.5 & 0 & 0.5 & 0 \\ 0 & 0 & 0.5 & 0 & 0.5 \\ 0 & 0 & 0 & 0 & 1 \end{pmatrix}.$$

   (a) Classify the states of the chain as transient, positive recurrent, or null recurrent.
   (b) When $P(X_0 = 2) = 1$, find the expected number of times the Markov chain visits state 1 before being absorbed.
   (c) When $P(X_0 = 1) = 1$, find the probability that the Markov chain gets absorbed in state 0 .

2. A rumor-spreading paradigm is one method of information dissemination over a network. Assume that there are 5 hosts connected to the network. One host starts off by sending a message. Every round, the message is sent by one host to another host who is selected independently and uniformly at random from the other 4 hosts. When all hosts have received the message, the process ends. Create a discrete-time model of this process with Markov chains

   (a) X_n be state of host ($i = 1, 2, \ldots, 5$) who received the message at the end of nth round.
   (b) Y_n be number of hosts having the message at the end of nth round.

   Find one-step transition probability matrix for the above DTMC. Classify the states of the chains as transient, positive recurrent, or null recurrent.

3. Consider a Markov Chain with state space $S = \{0, 1, 2, 3, 4\}$ and transition probability matrix

$$\begin{pmatrix} 1 & 0 & 0 & 0 & 0 \\ 0 & \frac{3}{4} & \frac{1}{4} & 0 & 0 \\ 0 & \frac{3}{4} & \frac{1}{4} & 0 & 0 \\ \frac{1}{4} & \frac{1}{4} & 0 & \frac{1}{4} & \frac{1}{4} \\ 0 & 0 & 0 & \frac{1}{2} & \frac{1}{2} \end{pmatrix}.$$

   (a) Classify the states of the chain.
   (b) Determine the stationary distribution for states 1 and 2.
   (c) For the transient states, calculate μ_{ij}, the expected number of visits to transient state j, given that the process started in a transient state i.
   (d) Find $P(X_5 = 2 \mid X_3 = 1)$.

4. Let us consider a Markov chain with states space $S = \{0, 1, 2, 3, 4\}$ and probability transition matrix

$$\mathbf{P} = \begin{pmatrix} 0 & \frac{1}{4} & 0 & \frac{3}{4} & 0 \\ \frac{1}{2} & 0 & \frac{1}{2} & 0 & 0 \\ 0 & 0 & 1 & 0 & 0 \\ 0 & 0 & \frac{1}{3} & \frac{2}{3} & 0 \\ 0 & 0 & 0 & 0 & 1 \end{pmatrix}.$$

   (a) If the chain, starts from state 0, find the probability of absorption in state 3 and state 5.
   (b) If the chain, starts from state 3, determine the expected number of visits that the chain makes to state 3 before getting absorbed.

5. Let $p_{00} = 1$ and, for $j > 0$, $p_{jj} = p$, $p_{jj-1} = q$ where $p + q = 1$, define the transition probability matrix of DTMC.

   (a) Find $f_{j0}^{(n)}$, the probability that absorption takes place exactly at the nth step given that the initial state is j.
   (b) Find the expectation of this distribution.

6. Prove that

$$Var(Z_n) = \begin{cases} \frac{\mu^{n-1}(\mu^n-1)\sigma^2}{\mu-1}, & \mu \neq 1 \\ n\sigma^2 & \mu = 1 \end{cases}.$$

7. Let $\{Z_n, n = 0, 1, \ldots\}$ be a classical branching process such that the distribution of $\{Y_i\}$ follows the geometric distribution with $p = 0.5$. Find the probability of extinction.
8. Think about doing a rabbit mating experiment. The evolution of a specific gene that is present in both types G and g is compared. A rabbit has two genes, either GG (dominant), Gg (hybrid, where the order is immaterial and gG is same to Gg) or gg (recessive). When two rabbits are mated, the progeny has an equal chance of receiving a gene from each parent. The progeny will either be dominant with probability $\frac{1}{2}$ or hybrid with probability $\frac{1}{2}$ if a dominant (GG) with a hybrid (Gg) are mate. Begin by mating a hybrid with a rabbit of the specified character (GG,Gg, or gG). Through several generations, the procedure is repeated, always mating with a hybrid, and the progeny is mated with another hybrid. Start with a hybrid rabbit. Let μ_n be the probability distribution of the character of the rabbit of the n-th generation. Then, compute $\mu_3(gg)$.
9. Assume that a virus can exist in three different strains and that, in each generation, it either remains the same or mutates to a different strain with probability p that is selected at random. What is the likelihood that the strain in generation n will be the same as generation 0?
10. Assume that a bacteria reproduce by cell division, and after one unit of time, the bacterium dies with probability $\frac{1}{6}$, remains the same with probability $\frac{1}{3}$ or splits into two with probability $\frac{1}{2}$. Assume initially at $n = 0$ there are 1000 bacteria. Then

    (a) Derive the generating function for the distribution of the size of the population at time $n \in \mathbb{N}_0$.
    (b) Compute the probability of extinction for the population.
    (c) Let m_n be the maximum possible number of bacteria at time n. Find m_n and compute the probability that there are exactly m_n bacteria in the population at time n, where $n \in \mathbb{N}_0$.
    (d) Given that there are 1000 bacteria in the population at time $n = 50$, compute the expected number of bacteria at time $n = 51$.

References

1. Athreya K.B., Ney, P.E.: Branching Processes. Springer-Verlay, New Jersey (1972)
2. Castaneda, L. B., Arunachalam, V., Dharmaraja, S.: Introduction to Probability and Stochastic Processes with Applications. Wiley, New Jersey (2012)

3. Harris, T.E.: The Theory of Branching Processes. Springer-Verlag, Berlin, and Prentice Hall, Englewood Cliffs, New Jersey (1963)
4. Ross, S.M.: Introduction to Probability Models. Elsevier, 12/e (2019)
5. Trivedi, K.S.: Probability and Statistics with Reliability, Queuing, and Computer Science Applications, 2nd edn. John Wiley & Sons (2001)

Chapter 4
Continuous-Time Markov Chains

Depending on the state space and parameter space, Markov processes can be categorised into several groups. A Markov chain is a Markov process with discrete state space. A Markov chain with discrete parameter space is known as a DTMC, which was explored in the previous two chapters. After learning about the DTMC, one would be interested in studying a continuous-time Markov chain in which transitions can occur at any time. It is a very useful class of stochastic processes and is abbreviated as CTMC. For example, the number of customers in a coffee shop and the number of planes at airport gates are both instances of continuous-time Markov chains (CTMC). The limiting and stationary distributions of the CTMCs are discussed in this chapter. Poisson process is an example of a discrete state continuous-time Markov process and will be studied in detail in Chap. 5.

4.1 Definition and Characterization

This section discusses the continuous-time Markov chains, which are dynamic systems evolving continuously in time. Here, we define and explore these Markov chains, examining their continuous-state transitions and fundamental properties.

Definition 4.1 A discrete-state continuous parameter stochastic process $\{X(t), t \geq 0\}$ with state space S is called a CTMC if for all $0 \leq t_0 < t_1 < \ldots < t_n < t$ and for all states $x_0, x_1, \ldots, x_n \in S$ and $x \in \mathbb{R}$, the conditional distribution satisfies, for $n = 1, 2, \ldots$

$$P\{X(t) \leq x | X(t_0) = x_0, X(t_1) = x_1, \ldots, X(t_n) = x_n\} = P\{X(t) \leq x | X(t_n) = x_n\}.$$

The Markov chain is called time-homogeneous if $P\{X(t+s) = j \mid X(s) = i\}$ for any $s \geq 0$, does not depend on s, and this probability denoted by $P_{ij}(t)$ is known

D. Selvamuthu, *Introduction to Stochastic Processes*, University Texts in the Mathematical Sciences, https://doi.org/10.1007/978-981-97-6152-4_4

as the stationary transition probability of transition from state i to state j in a time interval of length t. $\mathbf{P}(t) = [P_{ij}(t)]$ is called transition probability matrix for CTMC.

For a time-homogeneous CTMC, the following notations are given for further discussions. Suppose the state space $S = \{0, 1, \ldots\}$, and the random variable $X(t)$ has the probability mass function

$$\pi_j(t) = P(X(t) = j), \quad j \in S.$$

Since $\pi_j(t) \geq 0$, for $j \in S$ and $\sum_{j \in S} \pi_j(t) = 1$. Let

$$\pi(t) = [\pi_0(t), \pi_1(t), \ldots].$$

Initial state probability vector is given as follows:

$$\pi(0) = [\pi_0(0), \pi_1(0), \ldots].$$

Given $\pi_i(0)$ and P_{ij}(t), for $j \in S$,

$$\begin{aligned}\pi_j(t) &= P(X(t) = j) \\ &= \sum_{i \in S} P\{X(t) = j | X(0) = i\} P(X(0) = i) \\ &= \sum_{i \in S} \pi_i(0) P_{ij}(t).\end{aligned}$$

4.1.1 Sojourn Time Distribution

Let $\{X(t), t \geq 0\}$ be a time-homogeneous CTMC with state space S. Assume that anytime the Markov chain enters state i in S, the length of the time spent in state i is a random variable, denoted by H_i and is referred to as the holding time in state i. Whenever the holding time expires, the process transitions into state j with transition probability at time t is $P_{ij}(t)$, which is independent of the past, and so on.

Now, the holding time or sojourn time distribution for the time-homogeneous CTMCs is provided as follows:

Theorem 4.1 *Consider a time-homogeneous CTMC* $\{X(t), t \geq 0\}$. *At time* $t = 0$, *given the initial state probability vector* $\pi(0)$, *the sojourn time in the state i before moving into another state has exponential distribution with parameter* λ_i.

Proof Let H_i be a RV, referred to as the sojourn time, representing the time taken to transition from state i to a different state. Define

$$F_{H_i}(t) = P\{H_i \leq t | X(0) = i\}, t \geq 0.$$

Then, for $s, t \geq 0$

$$P\{H_i > s + t | X(0) = i\} = P\{H_i > s + t | X(s) = i\} P\{H_i > s | X(0) = i\}$$
$$F^c_{H_i}(s + t) = F^c_{H_i}(t) F^c_{H_i}(s)$$

where

$$F^c_{H_i}(t) = P\{H_i > t | X(0) = i\}.$$

Hence for t taking values in $[0, \infty)$, $0 < F^c_{H_i}(t) < 1$, and it satisfies the Cauchy functional equation, namely

$$F^c_{H_i}(t) = \exp(-\lambda_i t), \quad t > 0 \text{ for some } \lambda_i > 0.$$

The above result is valid for some $\lambda_i > 0$, depending on i. Each time state i is visited, the chain spends, on average, $E[H_i] = \frac{1}{\lambda_i}$ units of time before moving on. Here, λ_i is called the transition intensity from state i or transition rate at state i. □

Since the sojourn time H_i has an exponential distribution with parameter λ_i for any state i, the sojourn time distribution is exponential for all state i except the absorbing state.

Example 4.1 Consider a two state time-homogeneous Markov chain $\{X(t), t \geq 0\}$. The state space $S = \{0, 1\}$, where 0 is the down or non-operational state and 1 is the up or operational state. Time spent in state 1 or the time spent by the system in moving from state 1 to state 0 is exponential distributed with parameter λ and time spent in state 0 or time spent in moving from state 0 to state 1 is exponential distributed with parameter μ. The state transition diagram is shown in Fig. 4.1.

Example 4.2 Consider a three state time-homogeneous CTMC $\{X(t), t \geq 0\}$ with state space $S = \{1, 2, 3\}$. The state transition diagram is shown in Fig. 4.2.

For $i \neq j$, the rate λ_{ij} denotes that the system spends exponential distributed time with parameter λ_{ij} in state i before transition into state j.

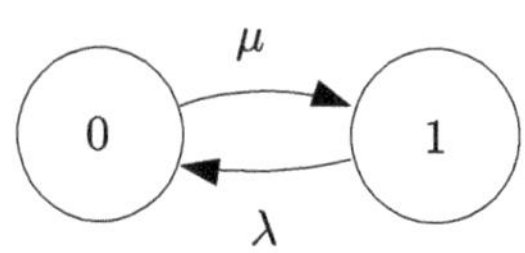

Fig. 4.1 State transition diagram for Example 4.1

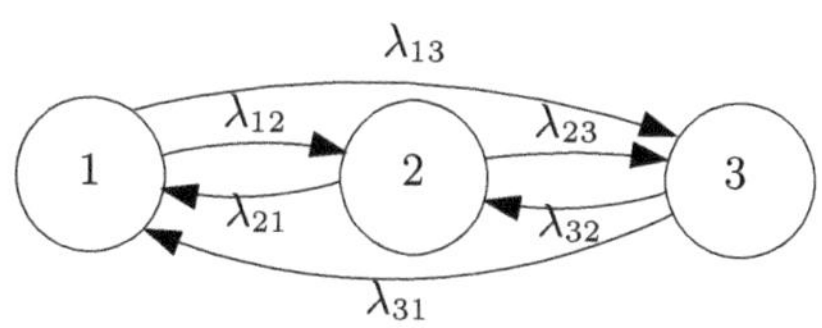

Fig. 4.2 State transition diagram for Example 4.2

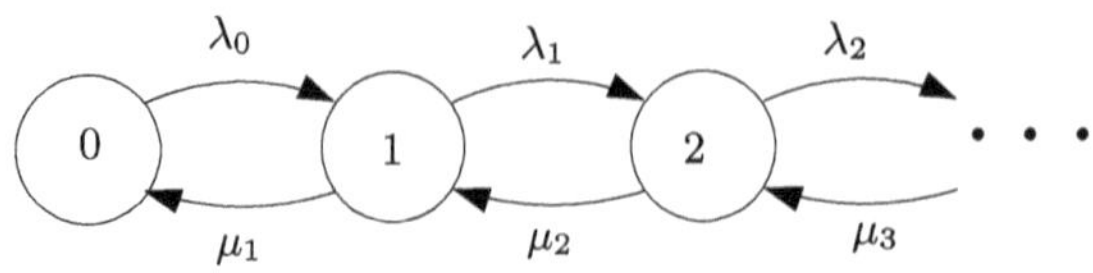

Fig. 4.3 State transition diagram for Example 4.3

Example 4.3 Consider a CTMC $\{X(t), t \geq 0\}$ with the state space $S = \{0, 1, \ldots\}$ and the state transition diagram is shown in Fig. 4.3. The time spent in state 0 is exponential distributed with the parameter λ_0. In state 1, the system can spend exponential distributed time before traveling to state 2 with parameter λ_1, and can spend exponential distributed time before moving to state 0 with parameter μ_1, both of which are independent of one another. As a result, the time spent in state 1 is the minimum of the two independent exponential distributed times and hence, the time spent in state 1 become exponential distribution with the parameter $\lambda_1 + \mu_1$.

4.1.2 Chapman–Kolmogorov Equations

The Chapman–Kolmogorov equations are one of the most important equations and the starting point for derivations of many results for the time-homogeneous CTMC. These equations define the relationships between the transition probabilities of higher order in terms of lower ones. Note that,

$$\begin{aligned} P_{ij}(t+s) &= P\{X(t+s) = j | X(0) = i\} \\ &= \sum_{k \in S} P\{X(t+s) = j, X(t) = k | X(0) = i\} \\ &= \sum_{k \in S} P\{X(t+s) = j | X(t) = k, X(0) = i\} P\{X(t) = k | X(0) = i\}, \end{aligned}$$

which by using the Markov property, becomes,

$$P_{ij}(t+s) = \sum_{k \in S} P_{ik}(t) P_{kj}(s), \qquad \forall\, i, j \in S \;\; t \geq 0, s \geq 0.$$

These above equations are referred as Chapman–Kolmogorov equations (C-K equations) for a time-homogeneous CTMC.

4.1.3 Infinitesimal Generator Matrix

Define for $i, j \in S$

$$q_{ij} = \frac{d}{dt} P_{ij}(t)\Big|_{t=0}, \quad \text{for } i \neq j$$
$$q_{ii} = \frac{d}{dt} P_{ii}(t)\Big|_{t=0}.$$

That is,

$$q_{ij} = \lim_{\Delta t \to 0} \frac{P_{ij}(\Delta t) - P_{ij}(0)}{\Delta t}, \quad i \neq j$$
$$q_{ii} = \lim_{\Delta t \to 0} \frac{P_{ii}(\Delta t) - P_{ii}(0)}{\Delta t}.$$

Since $P_{ij}(0) = 0$, for $i \neq j$ and $P_{ii}(0) = 1$, clearly $q_{ij} \geq 0$, for $i \neq j$ and $q_{ii} \leq 0$ for any i. Also

$$P_{ij}(\Delta t) = q_{ij}\Delta t + o(\Delta t), \quad i \neq j$$
$$P_{ii}(\Delta t) = 1 + q_{ii}\Delta t + o(\Delta t),$$

where $\frac{o(\Delta t)}{\Delta t} \to 0$ as $\Delta t \to 0$. From the definition, q_{ij} can be interpreted as the instantaneous rate of change of state, from state i to state j and represents the parameters of the CTMC called as transition intensities. Since $\sum_j P_{ij}(\Delta t) = 1$,

$$\sum_j q_{ij} = 0.$$

Hence, $q_{ii} = -\sum_{j \neq i} q_{ij}$. Thus, $-q_{ii}$ can be interpreted as the instantaneous rate of leaving state i. This tells us that the sojourn time in state i, or the amount of time the system waits before transitioning to a new state from state i, is exponential distributed with rate $-q_{ii}$.

Thus, using these transition intensities a matrix $\mathbf{Q} = [q_{ij}]$ can be constructed such that,

1. $q_{ij} \geq 0, \quad i \neq j$
2. $q_{ii} = -\sum_{j \neq i} q_{ij}$.

Hence, the row sum of each row of $\mathbf{Q}$ is zero. This matrix $\mathbf{Q}$ is known as the infinitesimal generator matrix (or the transition rate matrix).

Definition 4.2 (Embedded Markov Chain) Let $\{X(t), t \geq 0\}$ be a continuous-time discrete-state stochastic process. Let τ_n denote the time epoch at which the nth transition occurs, it is then observed that $X_n = X(\tau_n^+)$ is the state right after the nth transition. The sequence of RVs $X_0, X_1, \ldots$ consecutively, represents the state visited at nth transition, and the process moves from one state to another state according to the one-step transition probabilities, $p_{ij} = P\{X_{n+1} = j | X_n = i\}$. Therefore, $\{X_n, n = 0, 1, \ldots\}$ defines the underlying DTMC, called the embedded Markov chain (EMC) of $\{X(t), t \geq 0\}$. The transition matrix $[p_{ij}]$, and the holding time rates or sojourn time rates $-q_{ii}$, completely determines the CTMC.

Example 4.4 *(Embedded DTMC or Poisson Process)* Consider the Poisson process $\{N(t), t \geq 0\}$, where $N(t)$ denotes the state of the system at time t. Let $X_n = N(t_n^+)$, $n = 1, 2, \ldots$ where t_n^+ is the time after the nth transition in the system. The embedded Markov chain $\{X_n\}$ satisfies $X_n = n, n = 0, 1, \ldots$. The transition from state n to $n+1$ occurs with probability 1. The transition probability matrix of $\{X_n, n = 0, 1, \ldots\}$ is

$$\mathbf{P} = \begin{pmatrix} 0 & 1 & 0 & 0 & 0 & \cdots \\ 0 & 0 & 1 & 0 & 0 & \cdots \\ 0 & 0 & 0 & 1 & 0 & \cdots \\ 0 & 0 & 0 & 0 & 1 & \cdots \\ \vdots & \vdots & \vdots & \vdots & \vdots & \vdots \end{pmatrix}.$$

4.1.4 Classification of States

The states of a CTMC can be described in a similar way as in a DTMC. Given that $X(0) = i$, let T_{ii} be the amount of (continuous) time until the chain re-visits state i. The initial transition in the chain will occur at time H_i (sojourn time in state i), therefore $T_{ii} \geq H_i$.

1. If $P\{X(t) = j|X(0) = i\} = P_{ij}(t) > 0$ for some $t \geq 0$, state j is said to be reachable from state i in a CTMC. State i and state j are said to communicate, if state j is reachable from state i and state i is reachable from state j.
2. A state i is referred to as recurrent if the ultimate return to that state is the certain event, or if $P(T_{ii} < \infty) = 1$. Positive recurrence is the property of a state i where the mean time of return is finite, i.e., $E[T_{ii}] < \infty$. In addition to being recurrent, a state i is said to be null recurrent if $E[T_{ii}] = \infty$, meaning that it will never return.
3. Given a state i, if the ultimate return has probability less than 1, the state is called transient.
4. A state $i \in S$ is called an absorbing state if $q_{ij} = 0$ for all $j \neq i$, i.e., $q_{ii} = 1$.
5. If $\sum_{j \neq i} q_{ij} < \infty$, then the state i is called stable or regular. CTMC is regular means that, the number of transitions from each state in finite time is finite with probability 1, i.e., a regular CTMC has no explosions.
6. If $\displaystyle\sum_{j \neq i} q_{ij} = \infty$, then the state i is called an instantaneous state. The sojourn time in an instantaneous state is zero.
7. The state space of a CTMC can be partitioned into a set of transient states and closed sets of recurrent states.
8. A CTMC is irreducible if each state is accessible from every other state. Observe that states i and j communicate in a CTMC if and only if they do so for the embedded DTMC $\{X_n, n = 0, 1, \ldots\}$. They communicate in continuous-time if

and only if they do this at transition epochs. A CTMC is irreducible if and only if its embedded DTMC is irreducible.

9. A CTMC is ergodic if it is irreducible and positive recurrent.

Example 4.5 Consider Fig. 4.4a. Each of the two states, 0 and 1, is accessible from the other, and hence they are communicating states. This Markov chain is an irreducible CTMC since it only contains two communicating states.

Example 4.6 Consider the Fig. 4.4b. In this, state 0 doesn't communicate with any other state. Hence, this Markov chain is reducible. There are two communicating classes in this Markov Chain, $\{\{0\}, \{1, 2\}\}$. State 0 is an absorbing state and states 1 and 2 are transient states.

Example 4.7 Consider Fig. 4.4c, in this if $\lambda_{ij} > 0\, \forall i,\, j$, then all states are accessible from all other states. Therefore, this Markov Chain is irreducible. This is a finite CTMC, hence not all states can be transient. Thus, all states in this chain are recurrent.

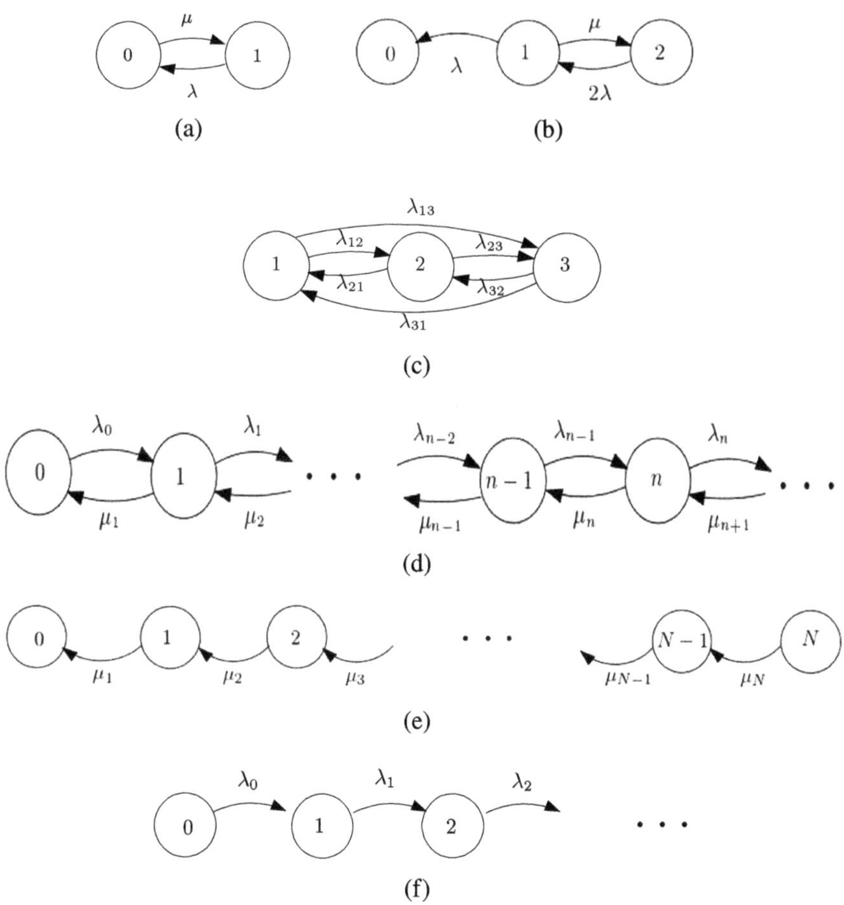

Fig. 4.4 State transition of CTMC models

Example 4.8 Consider Fig. 4.4d. In this, all states communicate with each other. Hence, the Markov chain is irreducible. Further, all states are recurrent.

Example 4.9 Consider the Fig. 4.4e. In this chain, state 0 is an absorbing state and the remaining states are transient.

Example 4.10 Consider Fig. 4.4f. In this chain, if the system starts from any state, it is not possible to return to the same state. Hence, this Markov chain is reducible. All the states are transient states.

Example 4.11 Consider a system with two machines. Machine 1 has a failure rate λ_1 and machine 2 has failure rate λ_2. There is only one service man and the repair rate of each of the failed machine is μ. The system shuts off when both machines fail. That is, there is no more repair when the system shutdowns. In this instance, $\{X_t, t \geq 0\}$ is a CTMC with a state space $S = \{0, I_1, I_2, 2\}$ where 0 indicates that both units have failed, I_1 indicates that machine 1 is working while machine 2 has failed, I_2 indicates that machine 1 has failed and machine 2 is working, and 2 indicates that both units are working. The state transition diagram is shown in Fig. 4.5.

The corresponding infinitesimal generator matrix is given by:

$$\mathbf{Q} = \begin{pmatrix} 0 & 0 & 0 & 0 \\ \lambda_1 & -(\lambda_1 + \mu) & 0 & \mu \\ \lambda_2 & 0 & -(\lambda_2 + \mu) & \mu \\ 0 & \lambda_2 & \lambda_1 & -(\lambda_1 + \lambda_2) \end{pmatrix}.$$

Here, state 0 is the absorbing state as $q_{00} = 0$. States I_1, I_2 and 2 are transient.

Example 4.12 An aircraft has two engines, one engine in each wing. The aircraft can land using at least one engine. Assume that the failure time of each engine follows exponential distribution with parameter λ_1 and λ_2, respectively, to complete a mission, and that engine failures are independent. Write the infinitesimal generator matrix for this Markov chain.

Solution: Let $X(t)$ be the state of the working condition of the two engines at time t. The state space is $S = \{(1, 1), (1, 0), (0, 1), (0, 0)\}$. The state transition diagram for this Markov chain is shown in Fig. 4.6 .

The infinitesimal generator matrix for this Markov chain is given by

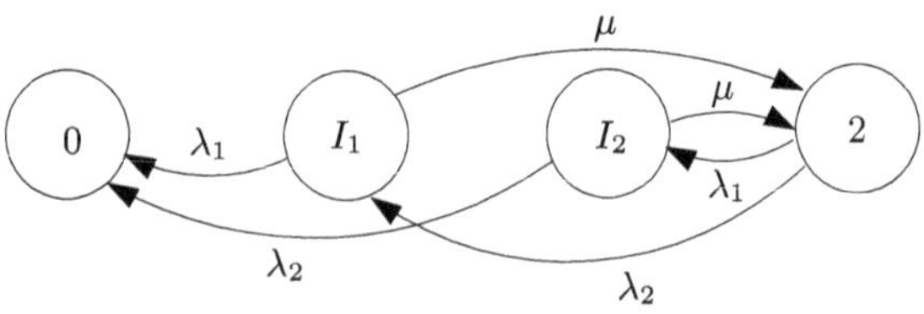

Fig. 4.5 State transition diagram for Example 4.10

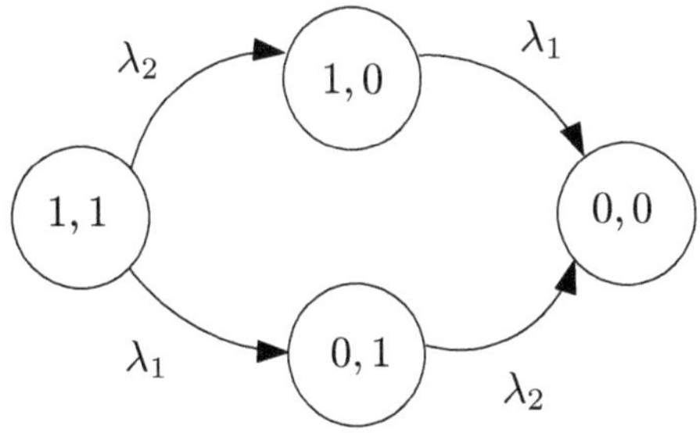

Fig. 4.6 State transition diagram for Example 4.12

$$\mathbf{Q} = \begin{pmatrix} -(\lambda_1+\lambda_2) & \lambda_2 & \lambda_1 & 0 \\ 0 & -\lambda_1 & 0 & \lambda_1 \\ 0 & 0 & -\lambda_2 & \lambda_2 \\ 0 & 0 & 0 & 0 \end{pmatrix}.$$

In this Markov chain, the states (1, 1), (1, 0) and (0, 1) are transient states. The state (0, 0) is an absorbing state.

4.1.5 Kolmogorov Equations

There are two sets of differential equations, namely Kolmogorov forward and Kolmogorov backward equations. The main difference between these two sets is the way these are evolved in time. The forward equations are constructed by moving forward in time, whereas the backward equations are constructed by moving backward in time. Consider the Chapman–Kolmogorov equations

$$P_{ij}(t+s) = \sum_{k \in S} P_{ik}(t) P_{kj}(s).$$

Differentiate with respect to s,

$$\frac{d}{ds} P_{ij}(t+s) = \sum_{k \in S} P_{ik}(t) \frac{d}{ds} P_{kj}(s).$$

Letting $s \to 0$,

$$P'_{ij}(t) = \sum_{k \in S} P_{ik}(t) q_{kj}.$$

The differential equation is given as

$$\mathbf{P}'(t) = \mathbf{P}(t)\mathbf{Q},$$

with initial condition $\mathbf{P}(0) = \mathbf{I}$. The above equation is known as the forward Kolmogorov equation.

Similarly, by differentiating Chapman–Kolmogorov equations with respective to t at $t = 0$,

$$P'_{ij}(t) = \sum_{k \in S} q_{ik} P_{kj}(t).$$

In matrix form,

$$\mathbf{P}'(t) = \mathbf{QP}(t), \mathbf{P}(0) = \mathbf{I},$$

where the matrix $\mathbf{P(t)} = [P_{ij}(t)]$. The above equation is known as the backward Kolmogorov equation. Thus,

$$\mathbf{P}'(t) = \mathbf{P}(t)\mathbf{Q}, \mathbf{P}(0) = \mathbf{I}$$
$$\text{and } \mathbf{P}'(t) = \mathbf{QP}(t), \mathbf{P}(0) = \mathbf{I}$$

are the forward and backward Kolmogorov equations, respectively, which can be solved to get state transition probabilities $\mathbf{P}(t)$, given the infinitesimal generator $\mathbf{Q}$ of a CTMC.

To simply remember which Kolmogorov equation is forward and which is backward, keep in mind that $\mathbf{P}(t)\mathbf{Q}$ on the right hand side of the forward Kolmogorov equation is in alphabetical order, whilst $\mathbf{QP}(t)$ on the right hand side of the backward Kolmogorov equation is in reverse (backward) alphabetical order. In the presence of a finite state space, both forward and backward Kolmogorov equations are always fully specified. There may be technical issues with an infinite state space since there may be infinitely many transitions in finite time.

But, both equations have a matrix-exponential solution. It has a formal solution of the form

$$\mathbf{P}(t) = \mathbf{P}(0)e^{\mathbf{Q}t}, \qquad \mathbf{P}(0) = \mathbf{I},$$

where

$$e^{\mathbf{Q}t} = \mathbf{I} + \mathbf{Q}t + \frac{1}{2}(\mathbf{Q}t)^2 + \cdots = \sum_{i=0}^{\infty} \frac{1}{i!}(\mathbf{Q}t)^i.$$

The transition probabilities can be expressed as a matrix-exponential function of the rate matrix $\mathbf{Q}$, i.e.,

$$\mathbf{P}(t) = e^{\mathbf{Q}t} = \sum_{i=0}^{\infty} \frac{1}{i!}(\mathbf{Q}t)^i.$$

This matrix exponential is a unique solution to both the forward and backward Kolmogorov equations with initial condition $\mathbf{P}(0) = \mathbf{I}$. There are various numerical recipes for estimating $e^{\mathbf{Q}t}$ to any desired level of accuracy. For any square matrix $\mathbf{A}$, it is known that $e^{\mathbf{A}} = \lim_{n \to \infty} \left(\mathbf{I} + \frac{\mathbf{A}}{n}\right)^n$ thus, one can choose large n such that $e^{\mathbf{Q}t} \approx \left(\mathbf{I} + \frac{\mathbf{Q}t}{n}\right)^n$.

Also, consider

$$\pi_i(t+s) = P(X(t+s)=i) = \sum_{k\in S} \pi_k(t)P_{ki}(s).$$

Differentiate with respect to s,

$$\frac{d}{ds}\pi_i(t+s) = \sum_{k\in S} \pi_k(t)\frac{d}{ds}P_{ki}(s).$$

Letting $s \to 0$,

$$\pi_i'(t) = \sum_{k\in S} \pi_k(t)q_{ki},$$
$$\pi'(t) = \pi(t)\mathbf{Q}.$$

The above equation is also known as the forward Kolmogorov equation. This equation is also known as master equation which describes the temporal behavior of conditional probabilities for Markov processes.

Similarly, the backward Kolmogorov equation can be obtained as

$$\pi_i'(t) = \sum_{k\in S} q_{ik}\pi_k(t),$$
$$\pi'(t) = \mathbf{Q}\pi(t).$$

Thus,

$$\pi'(t) = \pi(t)\mathbf{Q},$$
$$\text{and } \pi'(t) = \mathbf{Q}\pi(t),$$

are the form of forward and backward Kolmogorov equations, respectively. These can be solved to get state probability distribution $\pi(t)$ at time t given initial state probability distribution $\pi(0)$ and infinitesimal generator $\mathbf{Q}$ for a CTMC.

The conditional probabilities, $P_{ij}(t) = P\{X(s+t) = j|X(s) = i\}$ are called the transition probabilities of the CTMC for all states i and j and for all times $s > 0$ and $t > 0$. The homogeneous transition probabilities are given as:

$$P\{X(s+t) = j|X(s) = i\} = P\{X(t) = j|X(0) = i\} = P_{ij}(t),$$

for all states i and j and for all times $s > 0$ and $t > 0$, where the independence of future states from s characterizes the homogeneous property.

Example 4.13 Consider Example 4.1. Find the time-dependent solution of this stochastic process.

Solution: Based on the given assumption,

$$P_{01}(\Delta t) = \mu \Delta t + o(\Delta t), \quad P_{00}(\Delta t) = 1 - \mu \Delta t + o(\Delta t),$$

$$P_{10}(\Delta t) = \lambda \Delta t + o(\Delta t), \quad P_{11}(\Delta t) = 1 - \lambda \Delta t + o(\Delta t).$$

Hence, the **Q** matrix is given by,

$$\mathbf{Q} = \begin{pmatrix} q_{00} & q_{01} \\ q_{10} & q_{11} \end{pmatrix} = \begin{pmatrix} -\mu & \mu \\ \lambda & -\lambda \end{pmatrix}.$$

Observe that the eigenvalues of **Q** are 0 and $-(\lambda + \mu)$. The forward Kolmogorov equations are given by,

$$\begin{pmatrix} P'_{00}(t) & P'_{01}(t) \\ P'_{10}(t) & P'_{11}(t) \end{pmatrix} = \begin{pmatrix} P_{00}(t) & P_{01}(t) \\ P_{10}(t) & P_{11}(t) \end{pmatrix} \times \begin{pmatrix} -\mu & \mu \\ \lambda & -\lambda \end{pmatrix}$$

or

$$\begin{aligned} P'_{i0}(t) &= -\mu P_{i0}(t) + \lambda P_{i1}(t), \\ P'_{i1}(t) &= \mu P_{i0}(t) - \lambda P_{i1}(t). \end{aligned}$$

For $i = 1$,

$$P'_{11}(t) = \mu P_{10}(t) - \lambda P_{11}(t).$$

Suppose that the initial conditions are $P_{10}(0) = 0$, $P_{11}(0) = 1$. Since $P_{10}(t) + P_{11}(t) = 1$, the differential equation for $P_{11}(t)$ becomes

$$P'_{11}(t) = -(\lambda + \mu) P_{11}(t) + \mu.$$

Solving the above differential equation with the initial condition $P_{11}(0) = 1$,

$$P_{11}(t) = \frac{\mu}{\lambda + \mu} + \frac{\lambda}{\lambda + \mu} e^{-(\lambda+\mu)t}.$$

Using $P_{10}(t) + P_{11}(t) = 1$,

$$P_{10}(t) = \frac{\lambda}{\lambda + \mu} - \frac{\lambda}{\lambda + \mu} e^{-(\lambda+\mu)t}.$$

Similarly, solving differential equation corresponding to $i = 0$, $P_{00}(t)$ and $P_{01}(t)$. Hence, transition probability matrix is obtained as:

$$\mathbf{P}(t) = \begin{pmatrix} P_{00}(t) & P_{01}(t) \\ P_{10}(t) & P_{11}(t) \end{pmatrix} = \begin{pmatrix} \frac{\lambda}{\lambda+\mu} + \frac{\mu}{\lambda+\mu} e^{-(\lambda+\mu)t} & \frac{\mu}{\lambda+\mu} - \frac{\mu}{\lambda+\mu} e^{-(\lambda+\mu)t} \\ \frac{\lambda}{\lambda+\mu} - \frac{\lambda}{\lambda+\mu} e^{-(\lambda+\mu)t} & \frac{\mu}{\lambda+\mu} + \frac{\lambda}{\lambda+\mu} e^{-(\lambda+\mu)t} \end{pmatrix}. \tag{4.1}$$

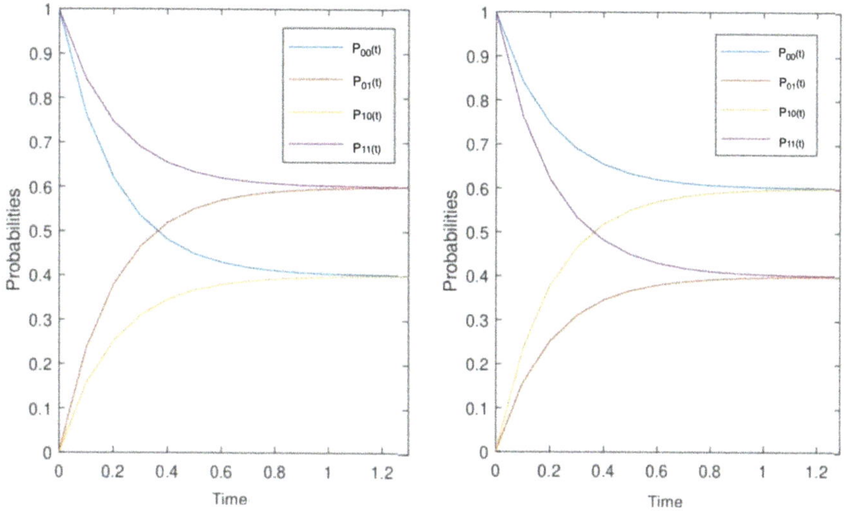

(a) $\lambda = 2, \mu = 3$ (b) $\lambda = 3, \mu = 2$

Fig. 4.7 Transition probabilities for different λ and μ for Example 4.13

Since the eigenvalues of matrix $\mathbf{Q}$ are 0 and $-(\lambda + \mu)$, the first term of transition probabilities corresponds to the eigenvalue 0. The time-dependent probabilities are shown in Fig. 4.7 for two sets of values of λ and μ.

Example 4.14 Consider a three state time-homogeneous Markov chain $\{X(t), t \geq 0\}$ with the state space $S = \{0, 1, 2\}$. The state transition diagram for this process is shown in Fig. 4.8.

Assume that, $P(X(0) = 0) = 1$. Obtain the time-dependent probabilities for this stochastic process.

Solution: The $\mathbf{Q}$ matrix is given by,

$$\mathbf{Q} = \begin{pmatrix} q_{00} & q_{01} & q_{02} \\ q_{10} & q_{11} & q_{12} \\ q_{20} & q_{21} & q_{22} \end{pmatrix} = \begin{pmatrix} -\lambda & \lambda & 0 \\ \mu & -(\mu + \lambda) & \lambda \\ 0 & \mu & -\mu \end{pmatrix}.$$

The forward Kolmogorov equations are given by,

Fig. 4.8 State transition diagram for Example 4.14

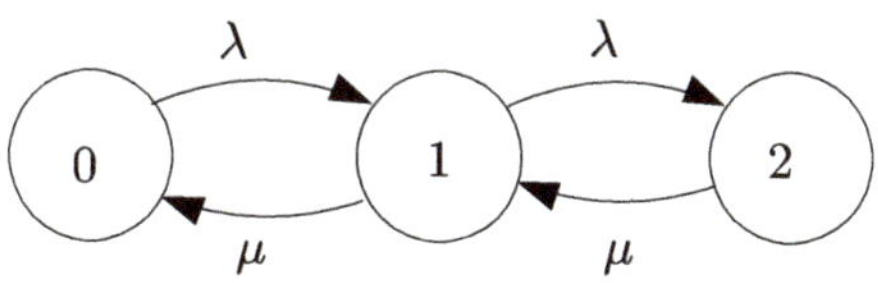

$$\pi_0'(t) = -\lambda\pi_0(t) + \mu\pi_1(t), \tag{4.2}$$
$$\pi_1'(t) = \lambda\pi_0(t) - (\mu+\lambda)\pi_1(t) + \mu\pi_2(t), \tag{4.3}$$
$$\pi_2'(t) = \lambda\pi_1(t) - \mu\pi_2(t). \tag{4.4}$$

The initial conditions are given by $\pi_0(0) = 1, \pi_1(0) = \pi_2(0) = 0$. Let $\widehat{\pi_j}(s)$ be the Laplace transform (see Appendix A.6) of $\pi_j(t),\ j = 0, 1, 2$ where $\widehat{\pi_j}(s) = \int_0^\infty e^{-st}\pi_j(t)dt,\ j = 0, 1, 2$. Taking Laplace transform of Equations (4.2–4.4)

$$s\widehat{\pi_0}(s) - 1 = -\lambda\widehat{\pi_0}(s) + \mu\widehat{\pi_0}(s), \tag{4.5}$$
$$s\widehat{\pi_1}(s) = \lambda\widehat{\pi_0}(s) + \mu\widehat{\pi_2}(s) - (\lambda+\mu)\widehat{\pi_1}(s), \tag{4.6}$$
$$s\widehat{\pi_2}(s) = \lambda\widehat{\pi_1}(s) - \mu\widehat{\pi_2}(s). \tag{4.7}$$

Solving (4.5–4.7), taking inverse Laplace transform and finally after simplification, the time-dependent probabilities are given by

$$\pi_0(t) = \frac{1}{1+\rho+\rho^2} - \frac{\rho\sqrt{\rho}e^{-(\lambda+\mu-\sqrt{\lambda\mu})t}}{2\rho-2(1+\rho)\sqrt{\rho}} + \frac{\rho\sqrt{\rho}e^{-(\lambda+\mu+\sqrt{\lambda\mu})t}}{2\rho+2(1+\rho)\sqrt{\rho}},$$
$$\pi_1(t) = \frac{\rho}{1+\rho+\rho^2} + \frac{\rho(-\rho+\sqrt{\rho})e^{-(\lambda+\mu-\sqrt{\lambda\mu})t}}{2\rho-2(1+\rho)\sqrt{\rho}} - \frac{\rho(\rho+\sqrt{\rho})e^{-(\lambda+\mu+\sqrt{\lambda\mu})t}}{2\rho+2(1+\rho)\sqrt{\rho}},$$
$$\pi_2(t) = \frac{\rho^2}{1+\rho+\rho^2} + \frac{\rho^2 e^{-(\lambda+\mu-\sqrt{\lambda\mu})t}}{2\rho-2(1+\rho)\sqrt{\rho}} + \frac{\rho^2 e^{-(\lambda+\mu+\sqrt{\lambda\mu})t}}{2\rho+2(1+\rho)\sqrt{\rho}}.$$

Note that, the above results are obtained by a different method in Sect. 6.2.2.3. The time-dependent probabilities $P_{ij}(t), i, j = 0, 1, 2$, are shown in Fig. 4.9 for two sets of values of λ and μ.

Example 4.15 Consider the CTMC system with state space given S = {2, 1, 0} and the state transition diagram shown in Fig. 4.10. The state 2 means both the units are working. The states 1 and 0 means one unit is working and no unit is working respectively. Assume that, initially, the system is in state 2. Find the reliability of the system at any time t.
Solution: The infinitesimal generator matrix for the given CTMC is

$$\mathbf{Q} = \begin{pmatrix} -2\lambda & 2\lambda & 0 \\ \mu & -(\mu+\lambda) & \lambda \\ 0 & 0 & 0 \end{pmatrix}.$$

Using $\mathbf{P}'(t) = \mathbf{P}(t)\mathbf{Q}$,

$$\begin{pmatrix} P_{22}'(t) & P_{21}'(t) & P_{20}'(t) \\ P_{12}'(t) & P_{11}'(t) & P_{10}'(t) \\ P_{02}'(t) & P_{01}'(t) & P_{00}'(t) \end{pmatrix} = \begin{pmatrix} P_{22}(t) & P_{21}(t) & P_{20}(t) \\ P_{12}(t) & P_{11}(t) & P_{10}(t) \\ P_{02}(t) & P_{01}(t) & P_{00}(t) \end{pmatrix} \begin{pmatrix} -2\lambda & 2\lambda & 0 \\ \mu & -(\mu+\lambda) & \lambda \\ 0 & 0 & 0 \end{pmatrix}.$$

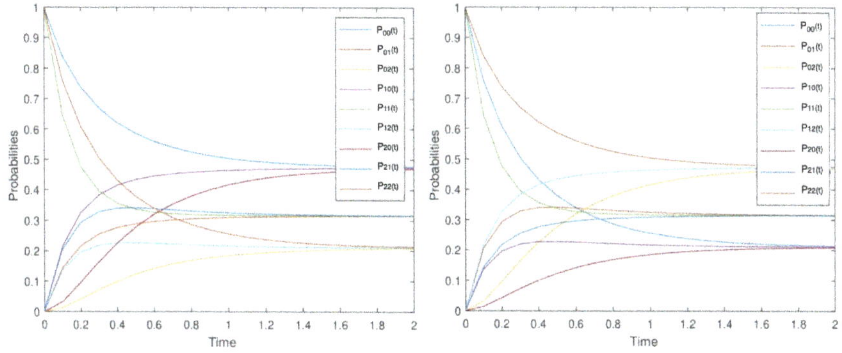

(a) $\lambda = 2, \mu = 3$ (b) $\lambda = 3, \mu = 2$

Fig. 4.9 Transient probabilities for different λ and μ for Example 4.14

Fig. 4.10 State transition diagram for Example 4.15

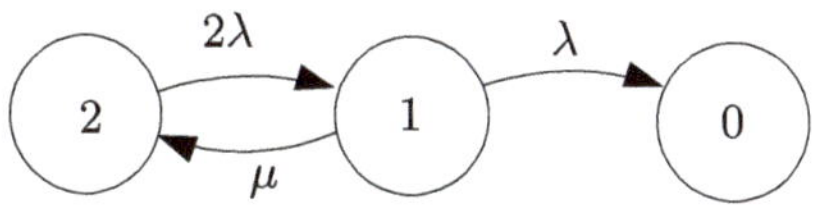

Using, $\pi'(t) = \pi(t)\mathbf{Q}$ where the vector $\pi(t) = (\pi_2(t), \pi_1(t), \pi_0(t))$. The system of linear equations is given by

$$\begin{aligned}
\pi_2'(t) &= -2\lambda\pi_2(t) + \mu\pi_1(t), \\
\pi_1'(t) &= 2\lambda\pi_2(t) - (\lambda + \mu)\pi_1(t), \\
\pi_0'(t) &= \lambda\pi_1(t).
\end{aligned}$$

with the initial conditions $\pi_2(0) = 1, \pi_1(0) = 0, \pi_0(0) = 0$. Using Laplace transforms the above system of equations reduces to

$$\begin{aligned}
s\widehat{\pi_2}(s) &= 1 - 2\lambda\widehat{\pi_2}(s) + \mu\widehat{\pi_1}(s), \\
s\widehat{\pi_1}(s) &= 2\lambda\widehat{\pi_2}(s) - (\lambda + \mu)\widehat{\pi_1}(s), \\
s\widehat{\pi_0}(s) &= \lambda\widehat{\pi_1}(s),
\end{aligned}$$

where $\widehat{\pi_j}(s) = \int_0^\infty e^{-st}\pi_j(t)dt,\ j = 0, 1, 2$. Solving the above system of algebraic equations,

$$\widehat{\pi_0}(s) = \frac{2\lambda^2}{s(s^2 + (3\lambda + \mu)s + 2\lambda^2)}.$$

Let $s^2 + (3\lambda + \mu)s + 2\lambda^2 = (s + s_1)(s + s_2)$. Then the preceding expression can be rearranged such that

$$\begin{aligned}\widehat{\pi_0}(s) &= \frac{2\lambda^2}{s} \cdot \frac{1}{s_1 - s_2}\left(\frac{1}{s+s_2} - \frac{1}{s+s_1}\right) \\ &= \frac{2\lambda^2}{s_1 - s_2}\left(\frac{1}{s_2}\left(\frac{1}{s} - \frac{1}{s+s_2}\right) - \frac{1}{s_1}\left(\frac{1}{s} - \frac{1}{s+s_1}\right)\right),\end{aligned}$$

where

$$s_1, s_2 = \frac{(3\lambda + \mu) \pm \sqrt{\lambda^2 + 6\lambda\mu + \mu^2}}{2}.$$

Inverting the Laplace transform $\widehat{\pi_0}(s)$,

$$\pi_0(t) = \frac{2\lambda^2}{s_1 - s_2}\left(\frac{1}{s_2}\left(1 - e^{-s_2 t}\right) - \frac{1}{s_1}\left(1 - e^{-s_1 t}\right)\right).$$

However in a CTMC, reliability is a metric that can be used to analyze the probability of a system or component performing its intended function without failure over a specified period of time. In this problem, the reliability of the system is given by

$$R(t) = 1 - \pi_0(t).$$

The time dependent probabilities $\pi_j(t)$, $j = 0, 1$ and 2 are shown in the Fig. 4.11 for varying values of λ and μ.

Using $s_1 s_2 = 2\lambda^2$ and simplifying the expression, we get the reliability as

$$R(t) = \frac{2\lambda^2}{s_1 - s_2}\left(\frac{e^{-s_2 t}}{s_2} - \frac{e^{-s_1 t}}{s_1}\right).$$

For the fixed values of λ and μ, the reliability $R(t)$ is shown in Fig. 4.12.

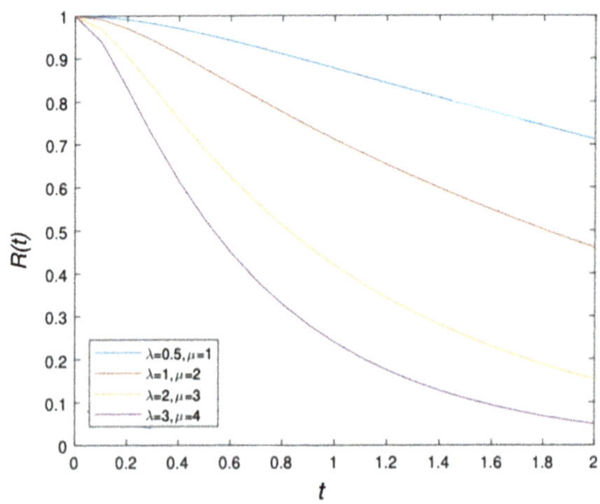

Fig. 4.11 Reliability function

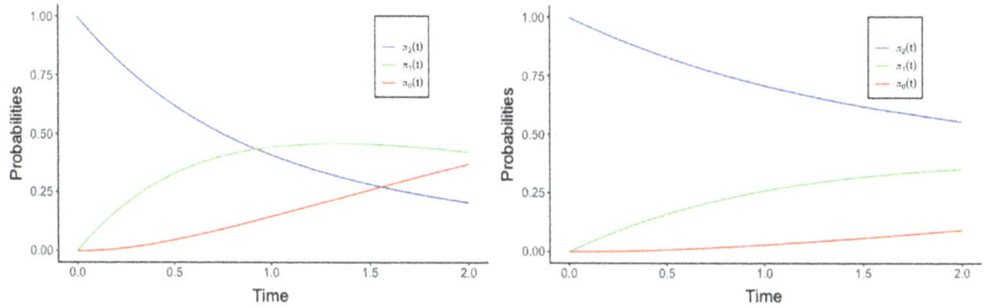

(a) $\lambda = 0.5, \mu = 0.2$ (b) $\lambda = 0.2, \mu = 0.3$

Fig. 4.12 Transient probabilities for different λ and μ

Mean time to failure (MTTF) is one of the key performance indicators that measures the mean time the system functions before its first non-repairable failure. The MTTF of the system is given by

$$\begin{aligned}\text{MTTF} = \int_0^{\infty} R(t)dt &= \frac{2\lambda^2}{s_1 - s_2}\left[\frac{1}{s_2^2} - \frac{1}{s_1^2}\right] = \frac{2\lambda^2\,(s_1 + s_2)}{s_1^2 s_2^2} \\ &= \frac{2\lambda^2(3\lambda + \mu)}{\left(2\lambda^2\right)^2} = \frac{3}{2\lambda} + \frac{\mu}{2\lambda^2}.\end{aligned}$$

Note that the MTTF of the two-component parallel redundant system, in the absence of a repair facility (i.e., $\mu = 0$), is equal to the first term, $3/(2\lambda)$, in the above expression. Therefore, the effect of a repair facility is to increase the mean life by $\mu/\left(2\lambda^2\right)$, or by a factor

$$\frac{\mu/2\lambda^2}{3/2\lambda} = \frac{\mu}{3\lambda}.$$

Example 4.16 Let $\{X(t), t \geq 0\}$ be a CTMC with state space $S = \{1, 2, 3\}$ and

$$\mathbf{Q} = \begin{pmatrix} -2 & 1 & 1 \\ 1 & -1 & 0 \\ 2 & 1 & -3 \end{pmatrix}.$$

Find the time-dependent probabilities for this system when the system is in state 1 initially.

Solution: Eigenvalues of $\mathbf{Q}$ are 0, -2 and -4 and the eigenvectors corresponding to the eigenvalues are given by $\begin{bmatrix} 1 \\ 1 \\ 1 \end{bmatrix}$, $\begin{bmatrix} 1 \\ -1 \\ 1 \end{bmatrix}$ and $\begin{bmatrix} -3/5 \\ 1/5 \\ 1 \end{bmatrix}$ respectively. Since, $\mathbf{P}(t) = e^{\mathbf{Q}t}$ which implies $\mathbf{Q}$ can be written as $\mathbf{Q} = SDS^{-1}$ where D is the diagonal matrix with diagonal elements as eigenvalues of $\mathbf{Q}$ and S is the matrix with column vector as eigenvector corresponding to each eigenvalues. Thus,

$$
\begin{aligned}
\mathbf{P}(t) = e^{\mathbf{Q}t} &= Se^{Dt}S^{-1} \\
&= \begin{bmatrix} 1 & -3/5 & 1 \\ -1 & 1/5 & 1 \\ 1 & 1 & 1 \end{bmatrix} \begin{bmatrix} e^{-2t} & 0 & 0 \\ 0 & e^{-4t} & 0 \\ 0 & 0 & 1 \end{bmatrix} \begin{bmatrix} 1/4 & -1/2 & 1/4 \\ -5/8 & 0 & 5/8 \\ 3/8 & 1/2 & 1/8 \end{bmatrix} \\
&= \begin{bmatrix} \frac{3}{8} + \frac{1}{4}e^{-2t} + \frac{3}{8}e^{-4t} & \frac{1}{2} - \frac{1}{2}e^{-2t} & \frac{1}{8} + \frac{1}{4}e^{-2t} - \frac{3}{8}e^{-4t} \\ \frac{3}{8} - \frac{1}{4}e^{-2t} - \frac{1}{8}e^{-4t} & \frac{1}{2} + \frac{1}{2}e^{-2t} & \frac{1}{8} - \frac{1}{4}e^{-2t} + \frac{1}{8}e^{-4t} \\ \frac{3}{8} + \frac{1}{4}e^{-2t} - \frac{5}{8}e^{-4t} & \frac{1}{2} - \frac{1}{2}e^{-2t} & \frac{1}{8} + \frac{1}{4}e^{-2t} + \frac{5}{8}e^{-4t} \end{bmatrix}.
\end{aligned}
$$

Hence, time-dependent probabilities for this system when the system is in state 1 initially is given by

$$P_{11}(t) = \frac{3}{8} + \frac{1}{4}e^{-2t} + \frac{3}{8}e^{-4t}, \tag{4.8}$$

$$P_{12}(t) = \frac{1}{2} - \frac{1}{2}e^{-2t}, \tag{4.9}$$

$$P_{13}(t) = \frac{1}{8} + \frac{1}{4}e^{-2t} - \frac{3}{8}e^{-4t}. \tag{4.10}$$

Note that, this problem is revisited in Example 4.22.

Example 4.17 Consider Example 4.12. Assume that both the engines are working at time 0. The mission reliability of the aircraft is obtained as follows.

The forward Kolmogorov equations are given by

$$\pi'(t) = \pi(t)\mathbf{Q},$$

where $\pi(t) = [\pi_{11}(t), \pi_{10}(t), \pi_{01}(t), \pi_{00}(t)]$ and

$$\mathbf{Q} = \begin{pmatrix} -(\lambda_1 + \lambda_2) & \lambda_1 & \lambda_2 & 0 \\ 0 & -\lambda_2 & 0 & \lambda_2 \\ 0 & 0 & -\lambda_1 & \lambda_1 \\ 0 & 0 & 0 & 0 \end{pmatrix}.$$

Given that, both the engines are in working condition at time 0. Hence, the initial condition is given by $\pi_{11}(0) = 1$. The solution of the above system of linear differential equations can be obtained by the Laplace transform method. Let $\widehat{\pi}_j(s)$ denote the Laplace transform of $\pi_j(t)$. Taking Laplace transform in the above equations,

$$
\begin{aligned}
s\widehat{\pi}_{11}(s) - 1 &= -(\lambda_1 + \lambda_2)\widehat{\pi}_{11}(s), \\
s\widehat{\pi}_{10}(s) &= \lambda_1\widehat{\pi}_{11}(s) - \lambda_2\widehat{\pi}_{10}(s), \\
s\widehat{\pi}_{01}(s) &= \lambda_2\widehat{\pi}_{11}(s) - \lambda_1\widehat{\pi}_{01}(s), \\
s\widehat{\pi}_{00}(s) &= \lambda_2\widehat{\pi}_{10}(s) + \lambda_1\widehat{\pi}_{01}(s).
\end{aligned}
$$

Taking inverse Laplace transform, the time-dependent probabilities are obtained which are given by

$$\pi_{11}(t) = e^{-(\lambda_1+\lambda_2)t},$$
$$\pi_{10}(t) = e^{-\lambda_2 t} - e^{-(\lambda_1+\lambda_2)t},$$
$$\pi_{01}(t) = e^{-\lambda_1 t} - e^{-(\lambda_1+\lambda_2)t},$$
$$\pi_{00}(t) = 1 - e^{-\lambda_1 t} - e^{-\lambda_2 t} + e^{-(\lambda_1+\lambda_2)t}.$$

Hence, the required mission reliability of the aircraft, denoted by $R(t)$ is given by

$$R(t) = 1 - \pi_{00}(t) = e^{-\lambda_1 t} + e^{-\lambda_2 t} - e^{-(\lambda_1+\lambda_2)t}.$$

Example 4.18 **Reliability of Voice over Internet Protocol (VoIP) system**
Consider a VoIP system given in [7]. This system is considered initially in state U. State U is a system state in which the entire system is operating as intended. If the primary server fails in this state and the system's control automatically changes to the standby server, the system switches from state U to state U_S, where the standby server is in control. The damaged server is then repaired.

The system switches from state U to state F_{S_1}, where this change is done manually if the primary server fails but the system control does not automatically switch to the backup server. The system switches from state F_{S_1} to state U_S following the manual transition to the standby server. Assume that the automated switching system malfunctions with probability p, and that control is automatically shifted from the failing server to the backup server with probability $1 - p$. The system shifts from state U to state U_R, where the call routing is quickly shifted to the hot standby router if one of the routers breaks while the system is in state U. The broken router is then repaired.

The system switches from state U to state U_P, when it utilizes power from the UPS if the electric power source is interrupted while it is in state U. The damaged electrical supply is then repaired. System can undergo rejuvenation if it is in condition U. The control of the system is automatically transferred to the standby server whenever the system has to be rejuvenated.

The system then switches from state U to state R_{jS}, where the backup server is in charge. However, it is possible that the automatic switching mechanism will also malfunction while the system undergoes rejuvenation. The system then transitions from state U to R_{jF}, when the entire process must be completed manually. The system switches from state R_{jF} to state R_{jS} following the manual switching. Similarly, the transition from one state to another state takes place.

For simplicity, assume that the states are numbered as $U \equiv 0$, $U_S \equiv 1$, $U_R \equiv 2$, $U_P \equiv 3$, $R_{jS} \equiv 4$, $F_{S_1} \equiv 5$, $F_{S_2} \equiv 6$, $F_{S_3} \equiv 7$, $F_{S_4} \equiv 8$, $U_{SR} \equiv 9$, $U_{SP} \equiv 10$, $U_{RP} \equiv 11$, $R_{jF} \equiv 12$, $U_{SRP} \equiv 13$ and $D \equiv 14$.

Next, the reliability and performance metrics are presented. Let $\pi_i(t)$ be the time-dependent probability of each state $i, i \in S$ of the CTMC shown in Fig. 4.13. The time-dependent probabilities are obtained by solving the following system of difference-differential equations for the CTMC:

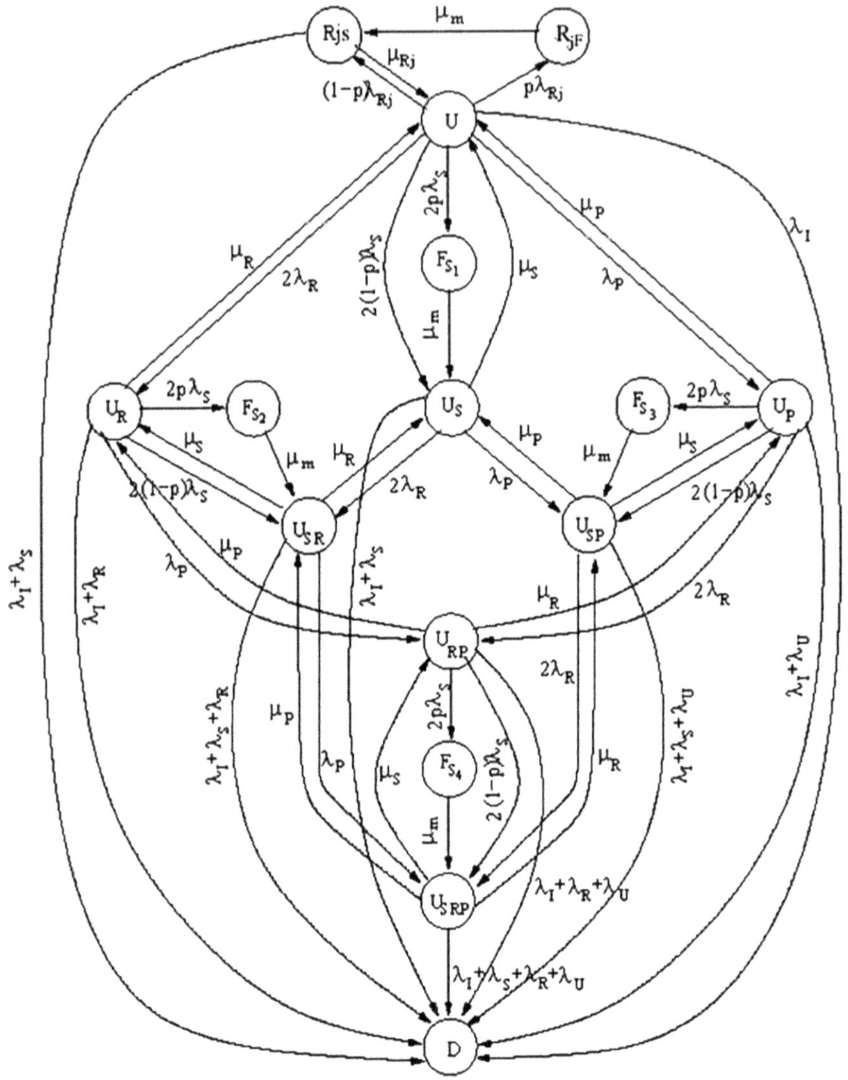

Fig. 4.13 State transition diagram for VoIP reliability model

$$
\begin{aligned}
\pi_0'(t) &= -\left[2(1-p)\lambda_s + 2p\lambda_s + 2\lambda_R + \lambda_P + (1-p)\lambda_{R_j} + p\lambda_{R_j} + \lambda_I\right]\pi_0(t) \\
&\quad +\mu_S\pi_1(t) + \mu_R\pi_2(t) + \mu_P\pi_3(t) + \mu_{R_j}\pi_4(t), \\
\pi_1'(t) &= 2(1-p)\lambda_s\pi_0(t) - \left(\mu_S + 2\lambda_R + \lambda_p + \lambda_I + \lambda_s\right)\pi_1(t) \\
&\quad +\mu_m\pi_s(t) + \mu_R\pi_9(t) + \mu_p\pi_{10}(t), \\
\pi_2'(t) &= 2\lambda_R\pi_0(t) - \left(\mu_R + 2p\lambda_S + 2(1-p)\lambda_s + \lambda_p + \lambda_I + \lambda_R\right) \\
&\quad +\mu_S\pi_9(t) + \mu_P\pi_{11}(t), \\
\pi_3'(t) &= \lambda_p\pi_0(t) - (\mu_P + 2p\lambda_s + 2(1-p)\lambda_s + 2\lambda_R + \lambda_I + \lambda_U) \\
&\quad +\mu_S\pi_{10}(t) + \mu_R\pi_{11}(t), \\
\pi_4'(t) &= (1-p)\lambda_{R_j}\pi_0(t) - \left(\mu_{R_j} + \lambda_I + \lambda_{\mathrm{s}}\right)\pi_4(t) + \mu_m\pi_{12}(t), \\
\pi_5'(t) &= 2p\lambda_{\mathrm{s}}\pi_0(t) - \mu_m\pi_5(t), \\
\pi_6'(t) &= 2p\lambda_s\pi_2(t) - \mu_m\pi_6(t), \\
\pi_7'(t) &= 2p\lambda_s\pi_3(t) - \mu_m\pi_7(t), \\
\pi_8'(t) &= 2p\lambda_s\pi_0(t) - \mu_{\mathrm{m}}\pi_8(t), \\
\pi_9'(t) &= 2\lambda_R\pi_1(t) + 2(1-p)\lambda_S\pi_2(t) + \mu_{\mathrm{m}}\pi_6(t) \\
&\quad -\left(\mu_R + \mu_s + \lambda_p + \lambda_s + \lambda_R + \lambda_I\right)\pi_9(t) + \mu_p\pi_{13}(t), \\
\pi_{10}'(t) &= \lambda_p\pi_1(t) + 2(1-p)\lambda_s\pi_3(t) + \mu_{\mathrm{m}}\pi_7(t) \\
&\quad -\left(\mu_p + \mu_S + \lambda_I + \lambda_U + 2\lambda_R + \lambda_s\right)\pi_{10}(t) + \mu_R\pi_{13}(t), \\
\pi_{11}'(t) &= \lambda_p\pi_2(t) + 2\lambda_R\pi_3(t) - (2p\lambda_{\mathrm{S}} + 2(1-p)\lambda_{\mathrm{S}} \\
&\quad +\mu_p + \mu_R + \lambda_U + \lambda_R + \lambda_I)\pi_{11}(t) + \mu_S\pi_{13}(t), \\
\pi_{12}'(t) &= 2p\lambda_S\pi_{11}(t) - \mu_{\mathrm{m}}\pi_{12}(t), \\
\pi_{13}'(t) &= \lambda_P\pi_9(t) + 2\lambda_R\pi_{10}(t) + 2(1-p)\lambda_S\pi_{11}(t) + \mu_m\pi_{12}(t) \\
&\quad -\left(\mu_p + \mu_R + \mu_S + \lambda_S + \lambda_R + \lambda_U + \lambda_I\right)\pi_{13}(t), \\
\pi_{14}'(t) &= \lambda_I\pi_0(t) + (\lambda_I + \lambda_s)\pi_1(t) + (\lambda_I + \lambda_U)\pi_2(t) + (\lambda_1 + \lambda_U)\pi_3(t) \\
&\quad +(\lambda_I + \lambda_S)\pi_4(t) + (\lambda_I + \lambda_s + \lambda_R)\pi_9(t) + (\lambda_I + \lambda_s + \lambda_U)\pi_{10}(t) \\
&\quad +(\lambda_I + \lambda_U + \lambda_R)\pi_{11}(t) + (\lambda_I + \lambda_s + \lambda_U + \lambda_R)\pi_{13}(t).
\end{aligned}
$$

Assuming that the initial state of the CTMC is state 0, hence, $\pi_0(0) = 1$ and $\pi_i(0) = 0$ for $i = 1, 2, \ldots, 14$, the technique of Laplace transform can be applied to obtain $\pi_i(t)\ \forall\ \ i = 0, 1, \ldots, 14$. Since state 14 is the complete down state, $\pi_{14}(t)$ gives the probability that the system has failed at or before time t. Then, the reliability of the system is given by

$$R(t) = 1 - \pi_{14}(t).$$

4.2 Limiting Distribution

In this section, the limiting distribution for the time-homogeneous CTMC is defined, followed by some examples. The long-term behavior of the CTMCs and the proportion of times that the Markov chain spends in each state as t becomes large are discussed.

Definition 4.3 Consider the time-homogeneous CTMC $\{X(t), t \geq 0\}$ with state space S. The limiting probabilities of the CTMC, denoted by υ_j, $j \in S$, are given by

$$\upsilon_j = \lim_{t\to\infty} P_{ij}(t) = \lim_{t\to\infty} P\{X(t) = j|X(0) = i\} \text{ for } i \in S, j \in S,$$

provided the limit exist. Here, $\lim_{t\to\infty} P_{ij}(t)$ expressed by υ_j, is independent of i, because it is not reliant on the initial state i, satisfies

$$\upsilon_j \geq 0 \text{ and } \sum_j \upsilon_j = 1.$$

Example 4.19 Consider Example 4.13. Find the limiting distribution, if it exists.
Solution: Taking limit $t \to \infty$ in Equation (4.1), the limiting distribution can be obtained. That is,

$$\lim_{t\to\infty} \mathbf{P(t)} = \begin{pmatrix} \frac{\lambda}{\lambda+\mu} & \frac{\mu}{\lambda+\mu} \\ \frac{\lambda}{\lambda+\mu} & \frac{\mu}{\lambda+\mu} \end{pmatrix}. \tag{4.11}$$

Thus, the limiting distribution vector $\mathbf{v} = (\upsilon_0, \upsilon_1)$ is given by

$$\mathbf{v} = \left(\frac{\lambda}{\lambda+\mu}, \frac{\mu}{\lambda+\mu}\right).$$

4.3 Stationary Distribution

Suppose $\{X(t), t \geq 0\}$ is a time-homogeneous CTMC with state space $S = \{0, 1, \ldots\}$, with probability distribution $\pi_j(t) = P(X(t) = j)$ for $j \in S$. Here, $\pi_j(t)$ is time-dependent for the state $j \in S$. These time-dependent or transient solutions are required for a thorough explanation of the CTMC. However, obtaining such answers is fraught with difficulties. Furthermore, in many practical scenarios, it is necessary to understand the behaviour in the steady state, that is, when the system attain a stationary distribution. This section deals with the concept of stationary distribution, throwing light on their fundamental role in modeling long-lasting stochastic processes.
It is easier and convenient to determine

$$\pi_j = \lim_{t\to\infty} P(X(t) = j), \ j \in S$$

provided the limit exists. When this limit exists, it is said that the system has reached steady state; and the corresponding probability distribution vector $\pi = [\pi_n]$, is called stationary distribution.

Definition 4.4 (Stationary Distribution) Consider time-homogeneous CTMC $\{X(t), t \geq 0\}$ with state space S and the infinitesimal generator matrix $\mathbf{Q}$. The vector $\pi = [\pi_0, \pi_1, \ldots]$ is called a stationary distribution of a CTMC if it satisfies $\pi\mathbf{Q} = \mathbf{0}$ with

1. $\pi_j \geq 0 \; \forall \; j \in S$.
2. $\sum_{j \in S} \pi_j = 1$.

The stationary distribution is also known as the steady-state distribution and equilibrium distribution.

Remark 4.1 1. For an irreducible, positive recurrent CTMC, the stationary distribution exists and it is unique. The steady-state vector $\pi = [\pi_0, \pi_1, \ldots]$ is uniquely determined by,

$$\pi\mathbf{Q} = \mathbf{0}, \quad \sum_{j \in S} \pi_j = 1.$$

Thus, for an irreducible CTMC with all positive recurrent states, limiting distribution and stationary distribution coincide.

2. Recall that periodic irreducible finite state DTMC has a single stationary distribution that is not a limiting distribution. If $d > 0$ is the period of DTMC, the probabilities of transitions as d grows.
Note that a CTMC cannot be periodic and that it admits a stationary distribution. Moreover, there are limiting distributions that are not stationary, and there are stationary distributions that are not limiting distributions.

Example 4.20 Consider an embedded Markov chain $\{X_n, n = 0, 1, \ldots\}$ of a CTMC $\{X(t), t \geq 0\}$ where $X_n = X(\tau_n^+)$ and τ_n denotes the time epoch at which the nth transition occurs. Verify that any stationary distribution for the EMC is also a stationary distribution, and vice versa.
Solution: $\mathbf{P}$ is the one-step transition probability matrix and $\mathbf{Q}$ is the infinitesimal generator matrix. Take $q \geq \max_i |q_{ii}|, i \in S$. Hence, $\mathbf{P} = \mathbf{I} + \frac{1}{q}\mathbf{Q}$. Now,

$$\begin{aligned}
\pi\mathbf{P} = \pi &\Leftrightarrow \pi\left(\mathbf{I} + \frac{1}{q}\mathbf{Q}\right) = \pi \\
&\Leftrightarrow \pi + \frac{\pi}{q}\mathbf{Q} = \pi \\
&\Leftrightarrow \pi\mathbf{Q} = \mathbf{0}.
\end{aligned}$$

Example 4.21 Consider Example 4.19. Find the stationary distribution of this CTMC.
Solution: In this CTMC, the matrix $\mathbf{Q}$ is given by

$$\mathbf{Q} = \begin{pmatrix} -\mu & \mu \\ \lambda & -\lambda \end{pmatrix}.$$

The above CTMC is irreducible having all the recurrent states, and hence, by the previous remark the stationary distribution exists and is unique. Stationary distribution is given by

$$\pi \mathbf{Q} = \mathbf{0}, \quad \sum_{j \in S} \pi_j = 1.$$

That is,

$$(\pi_0, \pi_1) \begin{pmatrix} -\mu & \mu \\ \lambda & -\lambda \end{pmatrix} = (0, 0).$$

Thus,

$$-\mu \pi_0 + \lambda \pi_1 = 0 \Rightarrow \pi_1 = \frac{\mu}{\lambda} \pi_0.$$

Since, $\pi_0 + \pi_1 = 1$,

$$\Rightarrow \left(1 + \frac{\mu}{\lambda}\right) \pi_0 = 1$$

$$\Rightarrow \pi_0 = \frac{\lambda}{\lambda + \mu}, \quad \pi_1 = \frac{\mu}{\lambda + \mu}.$$

The limiting distribution of the CTMC obtained in Example 4.11 coincides with the stationary distribution obtained here.

Example 4.22 Consider the CTMC in Example 4.16 with the $\mathbf{Q}$ matrix as

$$\mathbf{Q} = \begin{pmatrix} -2 & 1 & 1 \\ 1 & -1 & 0 \\ 2 & 1 & -3 \end{pmatrix}.$$

Find the steady-state distribution for this CTMC, if it exists.

Solution: Every state can communicate with every other state and hence the given CTMC is irreducible and finite and hence, all states are positive recurrent.

To find the steady-state solution, solve $\pi \mathbf{Q} = 0$ with $\sum_{j=1}^{3} \pi_j = 1$. That is,

$$\begin{aligned} -2\pi_1 + \pi_2 + 2\pi_3 &= 0, \\ \pi_1 - \pi_2 + \pi_3 &= 0, \\ \pi_1 - 3\pi_3 &= 0, \\ \pi_1 + \pi_2 + \pi_3 &= 1. \end{aligned}$$

Solving the above equations, $\pi_1 = \frac{3}{8}$, $\pi_2 = \frac{4}{8}$, $\pi_3 = \frac{1}{8}$.
Further, as $t \to \infty$ in (4.8–4.10),

$$P_{11}(t) \to \frac{3}{8}, \quad P_{12}(t) \to \frac{4}{8}, \quad P_{13}(t) \to \frac{1}{8}.$$

Note that, these limiting probabilities will be the same for all i. Thus, the limiting distribution vector $\mathbf{v} = [v_1, v_2, v_3]$ is given by

$$\mathbf{v} = \left(\frac{3}{8}, \frac{4}{8}, \frac{1}{8}\right).$$

Example 4.23 Consider Example 4.15. Find the limiting distribution and stationary distribution, if they exist.
Solution: Since state 0 is the absorbing state, hence as $t \to \infty$,

$$\mathbf{v} = [v_0, v_1, v_2] = [0, 0, 1].$$

Similarly, the stationary distribution π is

$$\pi_0 = 1, \quad \pi_1 = 0, \quad \pi_2 = 1.$$

Example 4.24 DRX mechanism is an energy-saving mechanism for the user equipment (UE). A Markov modeling is presented for the performance analysis of power saving in the DRX mechanism. For the model in detail, reader can refer to [4]. The states are given as follows:

- Idle period: It is the state in which UE is active waiting for packets to provide service.
- Working period: It is the state in which the UE is active, i.e., providing service to the packets.
- Short cycle: It is a state in which the UE turns off most of its components. This short cycle is repeated a predetermined number of times (say N) if no data packet arrives and ends otherwise.
- Long sleep: It is a state in which UE enters on the expiration of the short cycle if no data packet arrives.

Let $\{(N(t), S(t)), t \geq 0\}$ represent the two dimensional CTMC where $\{N(t), t \geq 0\}$ denotes the number of packets in the system at time t and $\{S(t), t \geq 0\}$ represents the state of the system at time t with state space

$$\Omega = \{(i, j); i = 0, 1, \ldots; j = 0, 1, \ldots, N, N + 1\}.$$

The state $(0, 0)$ represents that the system is in an idle period. The state $(i, 0), i = 1, 2, \ldots$ represents that the system is in working period providing service to ith packet and the state $(0, j), j = 1, 2, \ldots, N$ represents the system is in jth short cycle. The state $(i, N + 1)$ represents the system is in long sleep with i number of packets. In both idle and working periods, the UE is active to provide service to the packets.
Let $P_{ij}(t) = P\{N(t) = i, S(t) = j\}, i = 0, 1, \ldots$ and $j = 0, 1, \ldots, N + 1$ be the probability that the server is in state j with i number of packets in the system at any time t. Then, $P_{ij}(t)$ satisfies the following Kolmogorov forward equations:

$$P'_{00}(t) = -(\beta+\lambda)P_{00}(t) + \mu P_{10}(t) + (\nu-\lambda)P_{0N+1}(t), \tag{4.12}$$

$$P'_{10}(t) = -(\mu+\lambda)P_{10}(t) + \lambda P_{00}(t) + \mu P_{20}(t) + \lambda(P_{01}(t) + P_{02}(t) + \ldots + P_{0N}(t)) + (\nu-\lambda)P_{1(N+1)}(t), \tag{4.13}$$

$$P'_{i0}(t) = -(\mu+\lambda)P_{i0}(t) + \lambda P_{(i-1)0}(t) + \mu P_{(i+1)0}(t) + (\nu-\lambda)P_{i(N+1)}(t), \quad i = 2, 3, \ldots, \tag{4.14}$$

$$P'_{01}(t) = -\alpha P_{01}(t) + \beta P_{00}(t), \tag{4.15}$$

$$P'_{0j}(t) = -\alpha P_{0j}(t) + (\alpha-\lambda)P_{0(j-1)}(t), \; j = 2, 3, \ldots, N, \tag{4.16}$$

$$P'_{0(N+1)}(t) = -\nu P_{0(N+1)}(t) + (\alpha-\lambda)P_{0N}(t), \tag{4.17}$$

$$P'_{i(N+1)}(t) = -\nu P_{i(N+1)}(t) + \lambda P_{(i-1)(N+1)}(t), i = 1, 2, \ldots. \tag{4.18}$$

Assume that the system is initially empty, i.e., $P_{00} = 1$. The conditions for the existence of the system are $\lambda \leq \alpha$ and $\lambda < \nu$. The steady-state probabilities of the system are obtained from the time-dependent system size probabilities. It is known that, $\lim_{t\to\infty} P_{ij}(t) = \pi_{ij}$ and $\lim_{t\to\infty} P'_{ij}(t) = 0$. . Hence, using these the steady-state probabilities of UE being in different states, i.e., $(0, 0)$, $(0, j)$, $(i, N+1)$ and $(n, 0)$ for $n = 1, 2, \ldots; 1 \leq j \leq N; i \geq 0$ are given by

$$\pi_{0j} = \frac{\beta(\alpha-\lambda)^{j-1}}{\alpha^j}\pi_{00}, \quad j = 1, 2, \ldots, N,$$

$$\pi_{i(N+1)} = \frac{\beta\lambda^i(\alpha-\lambda)^N}{\nu^{i+1}\alpha^N}\pi_{00}, \quad i = 0, 1, \ldots$$

$$\pi_{n0} = A_n\pi_{00}, \quad n = 1, 2, \ldots,$$

where

$$A_n = \left[\frac{\mu b^{n+1}}{(\mu-\lambda)} + \frac{\beta b^{n-1}((\alpha-\lambda)^N - \alpha^N)}{\alpha^N(\mu-\lambda)}\right]\left[b^{n+1} - b^{n-1}\right] + \frac{\beta b^n(\alpha-\lambda)^N}{(\mu-\lambda)\alpha^N}$$
$$\times\left[\frac{(\nu-\lambda)b^{n+2}\left(1-\left(\frac{\lambda}{\nu b^2}\right)^{n+1}\right)}{(\nu b^2-\lambda)} - b^n\left(1-\left(\frac{\lambda}{\nu}\right)^{n+1}\right) + \left(\frac{\lambda}{\nu}\right)^{n+1}(b^{-n} - b^n)\right],$$

where

$$\pi_{00} = \frac{1}{1 + \sum_{i=1}^{\infty} A_i + \sum_{j=1}^{N} \frac{\beta(\alpha-\lambda)^{j-1}}{\alpha^j} + \sum_{i=0}^{\infty} \frac{\beta\lambda^i(\alpha-\lambda)^N}{\nu^{i+1}\alpha^N}}.$$

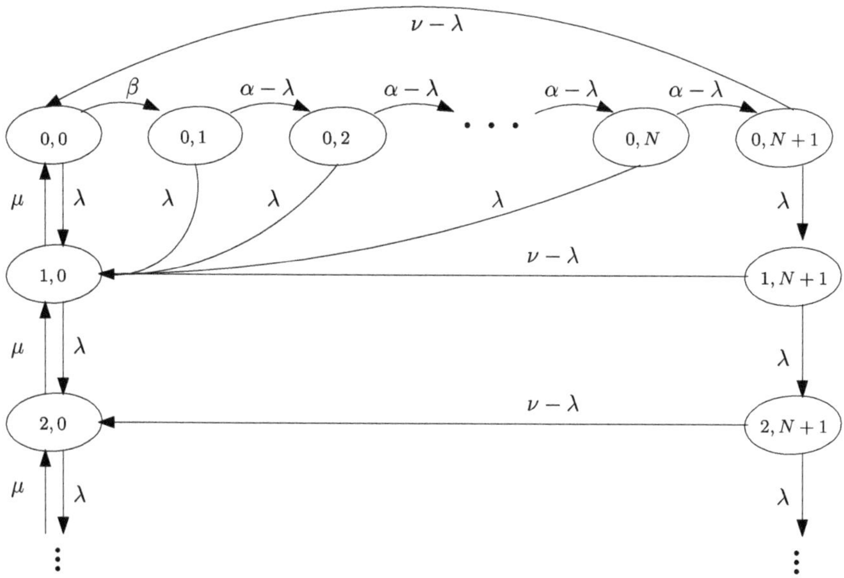

Fig. 4.14 State transition diagram for the proposed Markov model

The condition for the existence of these steady-state probabilities is $\lambda < \mu$. The proof is left as an exercise problem for the reader. The time-dependent probabilities are presented in [4].

Later, this scenario will be discussed in Chap. 8.

4.4 Continuous-Time Branching Processes

In Chap. 3, Sect. 3.3.1 considered the Galton-Watson process, where $\{Z_n, n = 0, 1, \ldots\}$ denote the number of individuals in the population in the nth generation in which the lifetime of each particle is one unit of time. Hence, these lifetimes are assumed to follow the DTMC. A natural generalization is to allow these lifetimes to be continuous type random variables. Instead of the DTMC $\{Z_n, n = 0, 1, \ldots\}$, it will be considered a continuous-time stochastic process $\{Z(t), t \geq 0\}$ where $Z(t)$ is the number of particles at time t. This stochastic process in general will not be Markov process, unless the lifetimes are assumed to be independent exponentially distributed random variables. In this section, the Markov branching process in continuous-time will be studied in detail.

4.4.1 Bellman–Harris Branching Processes

The Bellman–Harris branching process is more general than the branching processes considered in the earlier chapter. Now, it is assumed that, lifetimes of particles are non-negative random variables with arbitrary distributions. Suppose $Z(t)$ denote the number of particles at time t. This stochastic process $\{Z(t), t \geq 0\}$ is so-called age-dependent process which is generally non-Markov, but two of its special cases arc Markov: the Galton–Watson process and the age-dependent branching process with exponential lifetimes. Note that the Bellman-Harris process is more difficult to analyze, but it has many properties similar to these two processes (Fig. 4.15).

4.4.1.1 Markov Branching Process

Let $Z(t)$ be number of particles at time t. Let

$$\delta_{1k} + a_k h + o(h), \quad k = 0, 1, \ldots,$$

represent the probability that a single particle will split producing k particles during a small time interval $(t, t + h)$ of length h. δ_{1k} denotes the Kronecker delta function. Assume that $a_1 \leq 0, a_k \geq 0$ for $k = 0, 2, 3, \ldots$ and $\sum_{k=0}^{\infty} a_k = 0$. Further postulate that individual particles act independent of each other, always governed by the infinitesimal probabilities. Note that it is also assuming time-homogeneity

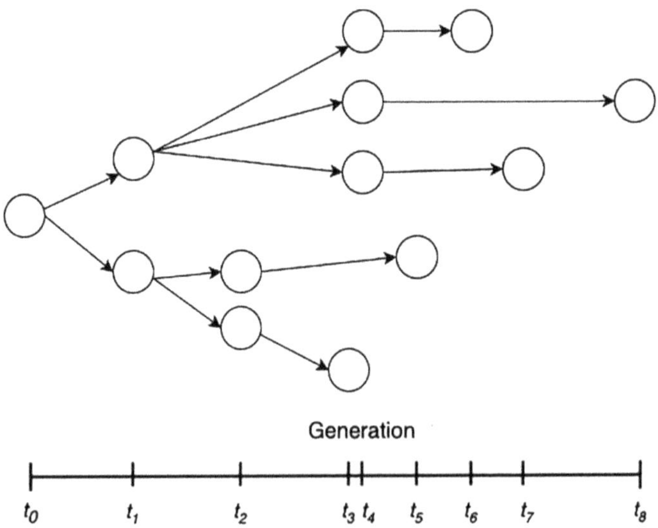

Fig. 4.15 Sample path of a branching process

for the transition probabilities since a_k is not a function of the time at which the conversion or splitting occurs. Each particle lives a random length of time following an exponential distribution with mean $1/\lambda = a_0 + a_2 + a_3 + \cdots$. On completion of i-th lifetime, it produces a random number D of descendants of like particles. The probability mass function of D is

$$P(D = k) = \frac{a_k}{a_0 + a_2 + a_3 + \ldots}, \quad k = 0, 2, \ldots .$$

The lifetime and progeny distribution of separate individuals are independent and identically distributed. Using the independence assumptions,

$$P\{Z(t+h) = n + k - 1 | Z(t) = n\} = na_k h + o(h), \quad k = 0, 2, \ldots$$

and

$$P\{Z(t+h) = n | Z(t) = n\} = 1 - na_1 h + o(h).$$

Definition 4.5 A continuous-time Markov process $\{Z(t), t \geq 0\}$ with the state space $S = \{0, 1, \ldots\}$ is said to be a continuous-time Markov branching process if the transition probabilities $P_{ij}(t)$ satisfy

$$\sum_{j=0}^{\infty} P_{ij}(t) z^j = \left[\sum_{j=0}^{\infty} P_{1j}(t) z^j \right]^i \quad \text{for all } i \geq 0 \text{ and } \mid z \mid \leq 1.$$

4.4.1.2 Probability Generating Function

Let

$$P_{ij}(t) = P\{Z(t+s) = j | Z(s) = i\}, \quad i, j = 0, 1, \ldots, t, s \geq 0$$

and let

$$\phi(t; z) = \sum_{j=0}^{\infty} P_{1j}(t) z^j,$$

the PGF for $P_{1j}(t)$.

Theorem 4.2 *The PGF for $P_{1j}(t)$, $\phi(t; z)$, satisfies:*

$$\phi(t + v; z) = \phi(t; \phi(v; z)).$$

This is the continuous-time analog of Theorem 3.5, in the case of discrete-time branching processes.

Proof Since individual particles act independently, the fundamental relation

$$\sum_j P_{ij}(t)z^i = \left[\sum_j P_{1j}(t)z^j\right]^i = [\phi(t; z)]^i .$$

The formula means that the population $Z(t; i)$ involving in time t from i initial parents is same, in the sense of probability, as the combined sum of i population each having one initial parents. Also, this formula characterizes and distinguishes branching processes from other continuous-time Markov chains. By the time-homogeneity, the Chapman–Kolmogorov equations take the form

$$P_{ij}(t+v) = \sum_k P_{ik}(t)P_{kj}(v).$$

Now,

$$\begin{aligned}[\phi(t+v; z)]^i &= \sum_j P_{ij}(t+v)z^j \\ &= \sum_j \sum_k P_{ik}(t)P_{kj}(v)z^j \\ &= \sum_k P_{ik}(t) \sum_j P_{kj}(v)z^j \\ &= \sum_k P_{ik}(t)\,[\phi(v; z)]^k \\ &= [\phi(t; \phi(v; z))]^i .\end{aligned}$$

When $i = 1$, the results are obtained. □

Theorem 4.3 *Let* $u(z) = \sum_k a_k z^k$. *Then* $\phi(t; z)$ *satisfies:*

$$\frac{\partial \phi(t; z)}{\partial t} = \frac{\partial \phi(t; z)}{\partial z} u(z),$$

$$\frac{\partial \phi(t; z)}{\partial t} = u(\phi(t; z))$$

with the initial condition

$$\phi(0; z) = \sum_j P_{1j}(0)z^j = z.$$

Proof Part 1: Since

$$\phi(h; z) = \sum_j P_{1j}(h)z^j$$

$$
\begin{aligned}
&= \sum_j \left(\delta_{1j} + a_j h + o(h)\right) z^j \\
&= z + h \sum_j a_j z^j + o(h) \\
&= z + hu(z) + o(h).
\end{aligned}
$$

Now,

$$\phi(t+h; z) = \phi(t; \phi(h; z)) = \phi(t; z + hu(z) + o(h)).$$

By Taylor's theorem, expand the right-hand side with respect to the second variable

$$\phi(t+h; z) = \phi(t; z) + \frac{\partial \phi(t; z)}{\partial z} hu(z) + o(h).$$

Hence

$$\frac{\phi(t+h; z) - \phi(t; z)}{h} = \frac{\partial \phi(t; z)}{\partial z} u(z) + \frac{o(h)}{h}.$$

Taking $h \to 0^+$,

$$\frac{\partial \phi(t; z)}{\partial t} = \frac{\partial \phi(t; z)}{\partial z} u(z).$$

This is a partial differential equation for the function of two variables $\phi(t; z)$ with the initial condition

$$\phi(0; z) = \sum_j P_{1j}(0) z^j = z.$$

Part 2: Now consider,

$$\phi(v+h; z) = \phi(v; \phi(h; z)).$$

By Taylor's theorem,

$$\phi(v+h; z) = \phi(v; z) + hu(\phi(v; z)) + o(h).$$

Hence

$$\frac{\phi(v+h; z) - \phi(v; z)}{h} = u(\phi(v; z)) + \frac{o(h)}{h}.$$

Taking $h \to 0^+$, then substituting $v = t$,

$$\frac{\partial \phi(t; z)}{\partial t} = u(\phi(t; z)).$$

This is an ordinary differential equation with initial condition

$$\phi(0; z) = \sum_j P_{1j}(0) z^j = z.$$

4.4.1.3 Mean of $Z(t)$

Consider

$$\frac{\partial \phi(t;z)}{\partial t} = \frac{\partial \phi(t;z)}{\partial z} u(z).$$

By differentiating with respect to z and interchanging the order of differentiation on the left side,

$$\frac{\partial^2 \phi(t;z)}{\partial t \partial z} = u(z)\frac{\partial^2 \phi(t;z)}{\partial z^2} + u'(z)\frac{\partial \phi(t;z)}{\partial z}.$$

If $z = 1$, $u(1) = \sum_k a_k = 0$, and then,

$$\frac{\partial m(t)}{\partial t} = u'(1)m(t),$$

where

$$m(t) = E[Z(t)] = \left.\frac{\partial \phi(t;z)}{\partial z}\right|_{z=1}.$$

But, since $Z(0) = 1$, then $m(0) = 1$ and the solution is

$$m(t) = e^{u'(1)t}.$$

4.4.1.4 Probability of Extinction

Definition 4.6 The extinction probability, q, is defined by

$$q = \lim_{t\to\infty} P_{10}(t).$$

Now, it is assumed that $a_0 > 0$, otherwise extinction is impossible. It is enough to consider the case where the process starts with a single individual at time zero. In fact, with $z = 0$

$$P_{i0}(t) = [P_{10}(t)]^i.$$

Now, $P_{i0}(t)$ is non-decreasing in t,

$$\begin{aligned} P_{i0}(t+v) &= [\phi(t+v;0)]^i \\ &= [\phi(t;\phi(v;0))]^i \\ &\geq [\phi(t;0)]^i \\ &= P_{i0}(t). \end{aligned}$$

4.4.1.5 Derivation of Probability of Extinction

Let t_0 be a fixed positive number. Consider the discrete-time process

$$Z(0), Z(t_0), Z(2t_0), \ldots, Z(nt_0), \ldots$$

where $Z(t)$ is the population size at time t.
Assume that, the population size at $t = 0$ is one. Since $Z(t)$ is assumed to be a Markov process, the discrete process $Y_n = Z(nt_0)$ will be a Markov chain which is also a discrete-time branching process. By the hypothesis of homogeneity of the probability function of $Z(t)$ and

$$\sum_j P_{ij}(t)s^i = \left[\sum_j P_{1j}(t)z^j\right]^i = [\phi(t; z)]^i .$$

$$\begin{aligned}\sum_k P\{Y_{n+1} = k|Y_n = i\}z^k &= E\left[z^{Y_{n+1}}|Y_n = i\right]\\ &= E\left[z^{Z((n+1)t_0)}|Z(nt_0) = i\right]\\ &= E\left[z^{Z(t_0)}|Z(0) = i\right]\\ &= [\phi(t_0; z)]^i\\ &= \left[E\left[s^{Z(t_0)}|Z(0) = 1\right]\right]^i\\ &= \left[E\left[s^{Y_1}|Y_0 = 1\right]\right]^i .\end{aligned}$$

This shows that $\{Y_n, n = 0, 1, \ldots\}$ is a branching process.

The probability generating function of the number of offspring of a single individual in this process is $\phi(t_0; z)$. By Theorem 3.7, the probability of extinction for the Y_n process is the smallest non-negative root of the equation

$$\phi(t_0; z) = z.$$

But

$$\begin{aligned}P(Y_n = 0 \text{ for some } n) &= \lim_{n\to\infty} P(Y_n = 0)\\ &= \lim_{n\to\infty} P(Z(nt_0) = 0)\\ &= \lim_{t\to\infty} P(Z(t) = 0)\\ &= q.\end{aligned}$$

The extinction probability q of the continuous-time branching process $Z(t)$ is the smallest non-negative root of the equation

$$\phi(t_0; z) = z,$$

where t_0 is any positive number.

Hence, expect that it should also be able to calculate q from an equation that does not depend on time.

Theorem 4.4 *The probability of extinction q is the smallest non-negative root of the equation*

$$u(z) = 0.$$

Hence, $q = 1$ if and only if $u^{'}(1) \leq 0$.

Proof Since q satisfies

$$\phi(t_0; z) = z,$$

for any $t_0 > 0$,

$$\frac{\phi(v + h; z) - \phi(v; z)}{h} = u(\phi(v; z)) + \frac{o(h)}{h}.$$

If $z = q$, then

$$\frac{q - q}{h} = u(q) + \frac{o(h)}{h}.$$

Hence, for any $h > 0$,

$$0 = u(q) + \frac{o(h)}{h}$$

Taking $h \to 0^+$, obtain

$$u(q) = 0.$$

Since

$$u^{''}(z) = \sum_{k=1}^{\infty} a_k k(k-1) z^{k-2} \geq 0,$$

$u(s)$ is convex in the interval $[0, 1]$. As $u(1) = 0$ and $u(0) = a_0 > 0$, $u(z)$ may have at most one zero in the interval $(0, 1)$. According to whether $u^{'}(1) \leq 0$ or $u^{'}(1) > 0$ holds, the cases $q = 1$ or $q < 1$ respectively. Note that $E[Z(t_0)] = E[Y]$ if and only if $u^{'}(1) > 0$. This means that for the discrete-time branching process $Z(nt_0), n = 0, 1, \ldots$ ($t_0 > 0$ fixed), extinction occurs with probability less than one, and therefore the same is true for the process $Z(t)$. Here, the probability of extinction q is the smallest zero of $u(s)$ in $[0, 1]$. In a similar manner it is concluded that, if $u^{'}(1) \leq 0$, q must be equal to one. In either case q is the smallest non-negative root of $u(z) = 0$.

4.4.1.6 Limit Theorem

If $u'(1) = 0$ and $u''(1) < \infty$, then

$$P\{Z(t) > 0|Z(0) = 1\} \approx \frac{2}{tu''(t)}, \quad t \to \infty$$

and

$$\lim_{t\to\infty} P\left(\frac{2Z(t)}{tu''(1)} > \lambda|Z(t) \neq 0\right) = e^{-\lambda}, \quad \lambda > 0.$$

If $u'(1) > 0$ and $u''(1) < \infty$, then

$$\frac{Z(t)}{e^{tu'(1)}},$$

has a limit distribution as $t \to \infty$.

Remark 4.2 The Kolmogorov equations are usually difficult to solve, and hence the distribution of $Z(t)$ can rarely be obtained explicitly. However, there are two important stochastic processes where they can be solved, namely the Yule process and the birth-death process.

Example 4.25 (Yule process) Consider the Yule process $\{Z(t), t \geq 0\}$ with PGF $\phi(z) = z^2$. That is, each particle lives an exponential distributed lifetime and then splits into two particles. Find $\phi(t; z)$.
Solution: The backward Kolmogorov equations is given by

$$\frac{\partial}{\partial t}\phi(t; z) = a[\phi^2(t; z) - \phi(t; z)] \text{ with } \phi(0; z) = z.$$

The solution is

$$\phi(t; z) = ze^{-at}[1 - (1 - e^{-at})z]^{-1}.$$

Now, some other important branching processes are briefed as follows.

1. **Bellman–Harris Processes**
Consider a classical branching process in which offsprings are born at the time of the parents' death. Consider the number of particles living at time t to be $Z(t)$. The process is known as a "age-dependent" or Bellman–Harris process when the distribution of particle lifespan τ is an arbitrary non-negative random variable. Assume that every particle reproduces and dies independently. This model generalizes the birth-death process in two ways: first, individual particle lifespans do not have to have an exponential distribution, and second, several particles can be born. Because the process $Z(t)$ is not a Markov process, it is normally analyzed using renewal theory.
2. **Bellman–Harris Processes with Disasters**
Consider the population model which follows a Bellman–Harris process. At random times, disasters trouble the population and each particle alive at the time of

the disaster survives with probability p. The survival of any particle is assume to be independent of the survival of any other particle. Measure of interest is limiting behavior when extinction does not occur.

3. **Bellman–Harris Processes with Immigration**
Consider a Bellman–Harris process $\{Z(t), t \geq 0\}$ where $Z(t)$ denotes the population size at time t. In addition, it is allowed the sudden appearance into the system of newly born particles called immigrants. Immigrants are assumed to arrive in groups of various sizes with the probability of n immigrants in a group immigrating at time t given by $p_n(t), n = 1, 2, \ldots$. Once these particles arrive, they reproduce and die according to the Bellman-Harris process. Measures of interest are the mean, the limiting distribution and the asymptotic behavior. This branching process is widely used to describe growth and decay of biological populations.

For more discussions, readers may refer to [1–14].

4.5 Exercise Problems

1. Consider a time-homogeneous CTMC $\{X(t), t \geq 0\}$ with the infinitesimal generator matrix $\mathbf{Q} = \begin{pmatrix} -3 & 2 & 1 \\ 1 & -2 & 1 \\ 1 & 4 & -5 \end{pmatrix}$ and initial distribution $(0, 0, 1)$. Find $P(\tau > t)$ where τ denotes the first transition time of the Markov chain.
2. Consider an individual running a tiny auto body shop. A Poisson process with rate λ describes the arrival process of vehicles in need of repair. The man begins by bumping one car, painting it, and then moving on to the next. The length of time to bump a car is exponential with mean $\frac{1}{\mu_1}$ and length of time to paint a car is exponential with mean $\frac{1}{\mu_2}$. Model the shop as a Markov process $\{X(t), t \geq 0\}$, where $X(t) = (i, j)$ if there are $i > 0$ cars in the shop and the car being repaired is in stage j. The stage $j = 0$ means the bumping stage and $j = 1$ means the printing stage. If the shop is empty, let $X(t) = (0, 0)$. Determine the infinitesimal generator matrix $\mathbf{Q}$.
3. Let's say a business employs four operators to handle a single phone number. A busy signal will be heard if someone calls while all four operators are occupied. Let $X(t)$ be the number of active operators at time t. Assume that each individual service process and the arrival processes are Poisson processes with rates λ and μ, respectively. Determine the forward Kolmogorov equations for the Markov Process and the infinitesimal generator matrix $\mathbf{Q}$.
4. In orbit, there are two communication satellites. Each satellite has an exponential lifetime with a mean of $\frac{1}{\mu}$. A replacement is sent up if one fails. The preparation and sending up of a replacement takes an exponential amount of time, with mean $\frac{1}{\lambda}$. Let X_t represents the number of satellites in orbit at the given moment. Consider $\{X(t), t \geq 0\}$ as a CTMC with the state space $\{0, 1, 2\}$. Show that the

infinitesimal generator matrix is given by

$$\mathbf{Q} = \begin{pmatrix} -\lambda & \lambda & 0 \\ \mu & -(\lambda+\mu) & \lambda \\ 0 & 2\mu & -2\mu \end{pmatrix}.$$

Write down the Kolmogorov forward and backward equations for the above process.

5. Consider a stochastic process $\{X(t), t \geq 0\}$ which take the value 0 and 1 with probabilities $p_0(t)$ and $p_1(t)$, respectively. Also

$$\mathrm{P}\{X(t+\Delta t) = 1 | X(t) = 0\} = \alpha \Delta t + o(\Delta t)$$

and

$$\mathrm{P}\{X(t+\Delta t) = 0 | X(t) = 1\} = \beta \Delta t + o(\Delta t)$$

As $\Delta t \to 0$, $o(\Delta t) \to 0$ and α and β are positive constants. Assume $p_0(0) = 1$.

(a) Write the forward Kolmogorov equations for the Markov process $\{X(t), t \geq 0\}$.
(b) Derive the transient or time-dependent probability distribution of the process.

6. Suppose that the probability of an accident happening in time t to $t + \Delta t$ is $(\nu + j\lambda)\Delta t + o(\Delta t)$ given that $j \geq 0$ accidents have occurred previous to time t, for $\nu, \lambda > 0$. Let $X(t)$ denote number of accidents happen up to and till time t.

(a) Write the forward Kolmogorov equations for the Markov process $\{X(t), t \geq 0\}$.
(b) Prove that

$$P_{0n+1}(t) = \frac{\nu(\nu+\lambda)\dots(\nu+n\lambda)}{(n+1)!\lambda^{n+1}} e^{-\nu t} \left(1 - e^{-\lambda t}\right)^{n+1}, \quad n = 0, 1, \dots.$$

7. Two categories of jobs, designated as 1 and 2, arrive at a processor according to independent Poisson processes with rates λ_1 and λ_2 respectively. Let $X(t)$ represent the kind of the latest job arrival prior to time t. Show that that $\{X(t), t \geq 0\}$ constitutes a CTMC and describe its transition rates. Show that $P_{ij}(t) = a + be^{-(\lambda_1+\lambda_2)t}$, and identify the coefficients a and b. Prove that $\{X(t), t \geq 0\}$ is ergodic and determine its stationary distribution.
8. Consider a CTMC $\{X(t), t \geq 0\}$ with state space $\{0, 1, 2\}$ and

$$\mathbf{Q} = \begin{pmatrix} -1 & 1 & 0 \\ 0 & -1 & 1 \\ 1 & 0 & -1 \end{pmatrix}.$$

(a) Find its transition probability matrix $\mathbf{P}(t)$.
(b) Show that the limiting distribution of $\mathbf{P}(t)$ as $t \to \infty$ is $\pi = (1/3, 1/3, 1/3)$ and

$$\frac{2}{3}e^{-3t/2} \le \sup_i \sum_j \mid P_{ij}(t) - \pi_j \mid \le \frac{4}{3}e^{-3t/2}.$$

9. Consider a CTMC $\{X(t), t \ge 0\}$ with

$$\mathbf{Q} = \begin{pmatrix} -3 & 1 & 2 \\ 2 & -4 & 2 \\ 2 & 1 & -3 \end{pmatrix}.$$

Prove that its transition probability matrix is

$$\mathbf{P}(t) = \frac{1}{5}\begin{pmatrix} 2+3e^{-5t} & 1-e^{-5t} & 2-2e^{-5t} \\ 2-2e^{-5t} & 1+4e^{-5t} & 2-2e^{-5t} \\ 2-2e^{-5t} & 1-e^{-5t} & 2+3e^{-5t} \end{pmatrix},$$

hence, the limiting distribution of $\mathbf{P}(t)$ is $\pi = (2/5, 1/5, 2/5)$ and

$$\sup_i \sum_j \mid P_{ij}(t) - \pi_j \mid = \frac{8}{5}e^{-5t}.$$

10. Consider a reversible CTMC $\{X(t), t \ge 0\}$ with state space S. Suppose $\mathbf{P}(t) = [P_{ij}(t)]$ has stationary distribution $\boldsymbol{\pi}$. Then, prove that $\lim_{t\to\infty} P_{ij}(t) = \pi_j$.
11. Let $\{X(t), t \ge 0\}$ be a CTMC. Fix $s > 0$, define $Y(t) = X(s-t)$. Then, prove that,

(a) $\{Y(t), 0 \le t \le s\}$ is a CTMC.
(b) $\{Y(t), 0 \le t \le s\}$ has homogeneous transition probability matrix if and only if $\{X(t), t \ge 0\}$ is further stationary, that is the distribution of $X(t)$, $m_i = P(X(t) = i)$ is independent of t. Moreover, show that the transition probability matrix of $\{Y(t), 0 \le t \le s\}$ is given by $P^*_{ij}(t) = m_j P_{ij}(t)/m_i$ where $[P_{ij}(t)]$ is the transition probability matrix of $\{X(t), t \ge 0\}$

References

1. Athreya, K.B., Ney, P.E.: Branching Processes. Springer-Verlay, New Jersey (1972)
2. Castaneda, L.B., Arunachalam, V., Dharmaraja, S.: Introduction to Probability and Stochastic Processes with Applications. Wiley, New Jersey (2012)
3. Dharmaraja, S.: Transient solution of a two-processor system. Math. Comput. Model. **32**, 1117–1123 (2000)
4. Dharmaraja, S., Aggarwal, A., Sudhesh, R.: Analysis of energy saving in user equipment in LTE-A using stochastic modelling. Telecommun. Syst. **80**, 123–140 (2022)
5. Dharmaraja, S., Di Crescenzo, A., Giorno, V., Nobile, A.G.: A continuous-time Ehrenfest model with catastrophes and its jump-diffusion approximation. J. Stat. Phys. **161**, 326–345 (2015)
6. Dharmaraja, S., Kumar, R.: Transient solution of a Markovian queuing model with heterogeneous servers and catastrophes. Opsearch **52**, 810–826 (2015)
7. Gupta, V., Dharmaraja, S.: Reliability and performance modeling of VoIP system with multiple component failure. Int. J. Reliab. Safety **7**, 58–74 (2013)
8. Haring, G., Marie, R., Puigjaner, R., Trivedi, K.S.: Loss formulas and their applications to optimization for cellular network. IEEE Trans. Veh. **50**(3), 664–673 (2001)
9. Harris, T.E.: The Theory of Branching Processes. Springer-Verlag, Berlin, and Prentice Hall, Englewood Cliffs, New Jersey (1963)
10. Jindal, V., Dharmaraja, S., Trivedi, K.S.: Markov modeling approach for survivability analysis of cellular networks. Int. J. Perform. Eng. **7**(5), 429 (2011)
11. Kijima, M. : (1997). Markov processes for stochastic modeling, vol. 6. CRC Press
12. Peng, N.F.: Spectral representations of the transition probability matrices for continuous time finite Markov chains. J. Appl. Prob. **33**, 28–33 (1996)
13. Ross, S.M.: Introduction to Probability Models. Elsevier, 12/e (2019)
14. Trivedi, K.S.: Probability and Statistics with Reliability, Queuing, and Computer Science Applications, 2nd edn. John Wiley & Sons (2001)

Chapter 5
Birth-Death Processes

In Chap. 4, the characterization of CTMCs and their limiting and stationary distributions were discussed in detail. Birth-Death Processes (BDPs), whose state space is the state of non-negative integers, are a subclass of Markov chains with continuous-time parameters. These processes are distinguished by the fact that if a transition happens, it leads to a neighbouring state. When a stochastic process consists of the absorption and emission of photons or particles, the excitation and de-excitation of atoms or nuclei, or of electrons in semiconductors, the birth and death of individuals, the arrival and departure of customers, it is characterized by birth and death (one-step or generation-recombination) processes.

In this chapter, a detailed study on BDPs is presented. Further, some special BDPs such as pure birth and pure death processes, are also discussed. Also another important stochastic process, namely Poisson process, the basic concepts and properties of the Poisson process both time-homogeneous and in-homogeneous have been provided.

Finally, a section on continuous-time branching processes, in particular, the Bellman-Harris branching process is presented.

5.1 Birth-Death Processes

In this section, we explore birth-death processes, fundamental models capturing state transitions over time. Their versatile applications across diverse fields are presented.

D. Selvamuthu, *Introduction to Stochastic Processes*, University Texts in the Mathematical Sciences, https://doi.org/10.1007/978-981-97-6152-4_5

5.1.1 Characterization

A birth-death process, $\{X(t), t \geq 0\}$, is a CTMC with the state space $S = \{0, 1, \ldots\}$ and with stationary transition probabilities, i.e.,

$$P_{ij}(t) = \mathrm{P}\{X(t+s) = j | X(s) = i\}, \quad i, j \in S$$

which do not depend on s, as given below. For $i, j \in S$ and as $h \downarrow 0$, $P_{ij}(t)$ is such that

1. $P_{ii+1}(h) = \lambda_i h + o(h), \; i = 0, 1, \ldots.$
2. $P_{ii-1}(h) = \mu_i h + o(h), \; i = 1, 2, \ldots.$
3. $P_{ii}(h) = 1 - (\lambda_i + \mu_i)h + o(h), \; i = 0, 1, \ldots.$
4. $P_{ij}(h) = o(h), \; |i - j| > 1.$
5. $P_{ij}(0) = \delta_{ij}$ (Kronecker delta).

The parameters λ_i and μ_i are called the *birth and death rates*, respectively, which are given by

$$\lambda_i \geq 0, \; i \in S, \; \mu_0 = 0 \text{ and } \mu_i \geq 0, \; i \in S \backslash \{0\}.$$

In Point 1 and 2, we assume that if the process starts at state i, then in a small interval of time the probabilities of going one-step up or down are essentially proportional to the length of the interval. In Point 4, we assume that transitions of more than one step are not allowed. When the birth or death rates depends on state i, then the process $\{X(t), t \geq 0\}$ is called a state-dependent BDP.

Definition 5.1 (*Birth-Death Processes*) A CTMC $\{X(t), t \geq 0\}$ with the state space $\{0, 1, \ldots\}$ is called a birth-death process (BDP) if there exist rates $\lambda_i \; (> 0), \; i = 0, 1, \ldots$ and $\mu_i \; (> 0), \; i = 1, 2, \ldots$ such that its possible transition rates are given by

$$\begin{aligned} q_{ii+1} &= \lambda_i, \quad i = 0, 1, \ldots \\ q_{ii-1} &= \mu_i, \quad i = 1, 2, \ldots \\ q_{ii} &= -(\lambda_i + \mu_i), \quad i = 1, 2, \ldots \\ q_{00} &= -\lambda_0 \\ q_{ij} &= 0 \text{ for } |i - j| > 1. \end{aligned}$$

Thus, if the process is in state $i \neq 0$ at any time instant then the only transitions possible from state i is either to state $i + 1$ (called a birth) or to state $i - 1$ (called a death), with respective instantaneous transition rates λ_i and μ_i, respectively. Moreover, no death is possible when the process is in state 0.

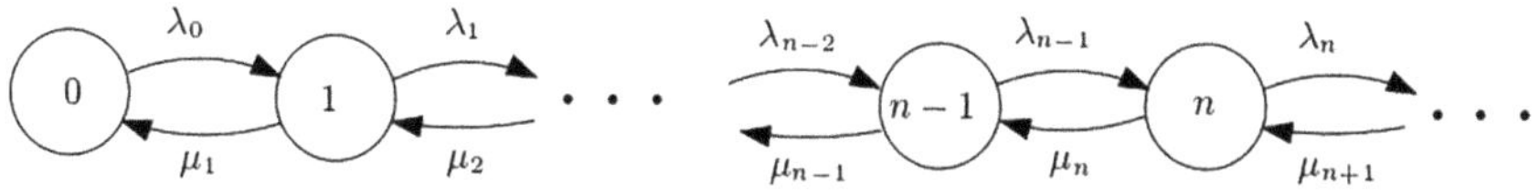

Fig. 5.1 State transition diagram for the BDP

For $t \geq 0$, define the transition probability matrix $\mathbf{P}(t)$ such that

$$\mathbf{P}(t) = \{P_{ij}(t),\ \ i,\ j \in S\}$$

satisfies the system of forward difference differential equations, called the Kolmogorov forward equations, which are given as follows

$$\begin{aligned} P'_{i0}(t) &= -\lambda_0 P_{i0}(t) + \mu_1 P_{i1}(t), i = 0, 1, \ldots \\ P'_{ij}(t) &= \lambda_{j-1} P_{ij-1}(t) - (\lambda_j + \mu_j) P_{ij}(t) + \mu_{j+1} P_{ij+1}(t), \quad (5.1) \\ i, j &= 1, 2, \ldots. \end{aligned}$$

The state transition diagram of a birth-death process is shown in Fig. 5.1.

The corresponding infinitesimal generator matrix is

$$\mathbf{Q} = [q_{ij}] = \begin{pmatrix} -\lambda_0 & \lambda_0 & 0 & 0 \ldots \\ \mu_1 & -(\lambda_1 + \mu_1) & \lambda_1 & 0 \ldots \\ 0 & \mu_2 & -(\lambda_2 + \mu_2) & \lambda_2 \ldots \\ \vdots & \vdots & \vdots & \vdots \ \vdots \end{pmatrix}.$$

Example 5.1 Consider a birth-death process $\{Z(t), t \geq 0\}$ with birth rates $\lambda_i = i\lambda, i = 0, 1, \ldots$ and death rates $\mu_n = i\mu, i = 1, 2, \ldots$. Find $\phi(t; z) = \sum_{j=0}^{\infty} P_{1j}(t) z^j$, the PGF for $P_{1j}(t)$. For details, refer Section 4.4.1.2.

Solution: Here, $a = \lambda + \mu, \phi(z) = (\mu + \lambda z^2) a^{-1}$. The Kolmogorov backward equation becomes

$$\frac{\partial}{\partial t}\phi(t; z) = [\lambda\phi^2(t; z) - (\lambda + \mu)\phi(t; z) + \mu] \ \text{ with } \phi(0; z) = z.$$

The solution is

$$\phi(t; z) = \frac{\mu(z - 1) - e^{(\mu - \lambda)t}(\lambda z - \mu)}{\lambda(z - 1) - e^{(\mu - \lambda)t}(\lambda z - \mu)}.$$

5.1.2 Spectral Representation

There is a close connection between BDPs and Orthogonal Polynomials (OPs).The $P_{mn}(t)$'s admit the integral representation

$$P_{mn}(t) = \frac{\lambda_0\lambda_1\ldots\lambda_{n-1}}{\mu_1\mu_2\ldots\mu_n}\int_0^\infty e^{-xt}Q_m(x)Q_n(x)\mathrm{d}\alpha(x), \tag{5.2}$$

where $\{Q_n(x)\}$ are orthogonal polynomials over the interval $[0, \infty)$ with respect to the spectral measure $d\alpha(x)$. Here, $\mathrm{d}\alpha(x)$ is a positive probability measure and $Q_n(x)$'s are generated by

$$Q_0(x) = 1, \quad Q_1(x) = (\lambda_0 + \mu_0 - x)/\lambda_0,$$
$$-xQ_n(x) = \lambda_n Q_{n+1}(x) - (\lambda_n + \mu_n)Q_n(x) + \mu_n Q_{n-1}(x), \quad n = 1, 2, \ldots$$

and satisfy the orthogonality relation

$$\int_0^\infty Q_n(x)Q_m(x)d\alpha(x) = \frac{\mu_1\mu_2\ldots\mu_n}{\lambda_0\lambda_1\ldots\lambda_{n-1}}\delta_{mn}.$$

When $\mu_0 = 0$, it can be shown by induction that $Q_n(0) = 1$ for all $n \geq 0$.

Equation (5.2) is remarkable in the sense that the dependence on t in the right-hand side is restricted entirely to the monotone decreasing exponential term e^{-xt} and that the dependence on m and n is factored as the product $Q_m(x)Q_n(x)$. In view of formula (5.2), the problem of finding the transition probability functions of a BDP, from the birth and death rates, amounts to finding the polynomials $Q_n(x)$ and the measure with respect to which they are orthogonal from the parameter in the recurrence relation. However, it will often be difficult to find the transition probability functions of a BDP explicitly.

5.1.3 Time-Dependent Analysis

Through recurrence relations, birth and death models are frequently used to calculate the equilibrium distributions of reversible reactions. Because the equations involved are significantly simpler, very basic approaches may be applied, and steady state analysis has been widely used. However, when the time horizon of processes is finite, such steady state evaluations are incorrect in many applicable scenarios. Steady state metrics of system performance simply do not make sense in many prospective applications since the system may never reach equilibrium.

There are numerous cases where steady-state results do not match actual events. The time-dependent analysis help us understand how a system behaves when its parameters are changed. Furthermore, transient analysis is useful in obtaining optimal solutions that lead to system control. Furthermore, transient solutions exist for a broader class of problems than steady state solutions, which exist solely under convergence conditions. As a result, transitory analysis is critical if relevant solutions in real-world situations are to be obtained.

There are several approaches to study the transient behavior of BDPs analytically. The most common approach employs generating functions and/or Laplace transformations. The transient system size probabilities are difficult to obtain analytically and often impossible. Closed form solutions are in general preferred since they provide deep insight into the transient behavior of these processes. In practical problems the underlying birth and death rates are state-dependent. The difficulty is compounded in the transient analysis of such models. Hence it is pertinent to develop techniques to solve the resulting birth and death equations numerically therefore as to gain an insight into the behavior of various system characteristics.

Because of their interesting convergence properties and connections with many branches of mathematics, such as number theory, special functions, differential equations, moment problems, and orthogonal polynomials, approximations using Continued Fractions (CFs) have a prominent place in mathematical literature. This CF technique is utilized to discover approximate values for time-dependent probabilities, mean and variance of system size for BDPs that do not permit closed form analytical solutions.

Let

$$\pi_n(t) = P(X(t) = n), n = 0, 1, \ldots$$

and

$$\pi(t) = [\pi_0(t), \pi_1(t), \ldots].$$

Since

$$\pi_n(t) = \sum_{i=0}^{\infty} \pi_i(0) P_{in}(t), n = 0, 1, \ldots$$

which in matrix form is

$$\pi(t) = \pi(0)\mathbf{P}(t).$$

Now, multiplying Kolmogorov forward equation by $\pi(0)$ gives

$$\pi(0)\mathbf{P}'(t) = \pi(0)\mathbf{P}(t)\mathbf{Q}$$

or

$$\pi'(t) = \pi(t)\mathbf{Q}.$$

This in expanded form is given by

$$\begin{aligned} \pi_0'(t) &= -\lambda_0 \pi_0(t) + \mu_1 \pi_1(t), \\ \pi_n'(t) &= \lambda_{n-1}\pi_{n-1}(t) - (\lambda_n + \mu_n)\pi_n(t) + \mu_{n+1}\pi_{n+1}(t), \quad n = 1, 2, \ldots. \end{aligned} \tag{5.3}$$

The above system of difference-differential equations can be solved with the initial condition to obtain the time-dependent probabilities.

Example 5.2 *(Unrestricted random walk (URW))* Consider the most basic linear, unrestricted random walk that can be constructed, where the time intervals between steps have a negative exponential distribution. Regardless of how many steps have occurred, it is of interest in things like the particle's location $N(t)$ at time t (rather than its position N_n at the nth step), the first passage returns to zero, the sojourn times, and other results. Such a walk is generated by a simple queueing model which will be discussed in Chap. 6. The queueing context physically restricts $N(t)$ to non-negative values, but if this immense barrier is lifted (allowing $N(t)$ to take on negative values as well), for a "randomized random walk" in which $N(t)$ denotes the position at time t of a particle that is moving on the x-axis in accordance with the walk's rules. The walk is supposed to be performed by the traditional particle, which may take only unit steps to the right and left on the unrestricted x-axis. Its position $N(t)$ at time t is governed by two independent Poisson processes with parameters λt and μt for motion to the right and left respectively. Assume $P(N(0) = 0) = 1$. Let $\pi_n(t) = P(N(t) = n)$ for the positive and negative integers $-\infty < n < \infty$. For this BDP, one obtains

$$\frac{d\pi_n(t)}{dt} = \lambda\pi_{n-1}(t) - (\lambda + \mu)\pi_n(t) + \mu\pi_{n+1}(t), n = 0, \pm 1, \ldots \tag{5.4}$$

with $\pi_0(0) = 1$. The state transition diagram for this randomized random walk is shown in Fig. 5.2.

The solution for the above system of equations (5.4) is given by

$$\pi_n(t) = e^{-(\lambda+\mu)t} \rho^{n/2} I_n \left(2t\sqrt{\lambda\mu}\right), \quad n = 0, \pm 1, \pm 2, \ldots, \tag{5.5}$$

where $\rho = \frac{\lambda}{\mu}$ and $I_n(.)$ is the modified Bessel function of the first kind. Also, show that

$$\pi_n(t) = e^{-(\lambda+\mu)t} \sum_{r \geq 0} \frac{(\lambda t)^{n+r} (\mu t)^r}{(n+r)!r!}, \tag{5.6}$$

$$\pi_{-n}(t) = e^{-(\lambda+\mu)t} \sum_{r \geq 0} \frac{(\mu t)^{n+r} (\lambda t)^r}{(n+r)!r!}, \quad n = 0, 1, \ldots. \tag{5.7}$$

Solution: Define the PGF for $\pi_n(t)$

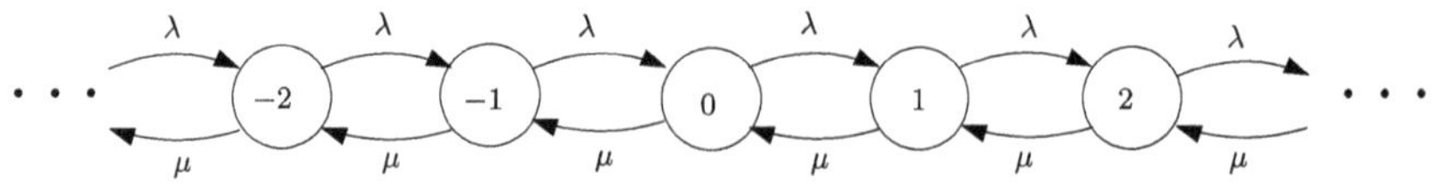

Fig. 5.2 State transition diagram for the randomized random walk

$$\phi(t;z) = \sum_{n=-\infty}^{\infty} \pi_n(t) z^n. \tag{5.8}$$

Multiplying (5.4) by z^n and summing over $n(-\infty < n < \infty)$,

$$\frac{\partial \phi}{\partial t} = \lambda z \phi - (\lambda + \mu)\phi + \frac{\mu \phi}{z}. \tag{5.9}$$

Let the Laplace transform of $\phi(t;z)$ be $\widehat{\phi}(s;z)$. Taking the Laplace transform on (5.9),

$$\begin{aligned} z(s\widehat{\phi}(s;z) - 1) &= (\lambda z^2 - (\lambda+\mu)z + \mu)\widehat{\phi}(s;z) \\ \Longrightarrow \quad \widehat{\phi}(s;z) &= \frac{z}{(sz + (\lambda+\mu)z - \mu - \lambda z^2)}. \end{aligned}$$

The inverse Laplace transform of the above equation is

$$\begin{aligned} \phi(t;z) &= e^{-at} e^{-(\lambda z + \frac{\mu}{z})t} \\ &= e^{-at} e^{-\frac{2\sqrt{\lambda\mu}t}{2}\left(\sqrt{\rho}z + \frac{1}{\sqrt{\rho}z}\right)}, \end{aligned} \tag{5.10}$$

where $a = \lambda + \mu$.

Referring to the generating function of $I_n(.)$, (5.10) yields

$$\phi(t;z) = e^{-at} \sum_{n=-\infty}^{\infty} I_n(2\sqrt{\lambda\mu}t)(\sqrt{\rho}z)^n.$$

Extraction of the coefficients of z^n in $\phi(t;z)$ gives (5.5) by using the property $I_n(.) = I_{-n}(.)$. Writing $I_n(y)$ in series form as,

$$I_n(y) = \sum_{m=0}^{\infty} \frac{(y/2)^{n+2m}}{m!(n+m)!}.$$

Equation (5.5) can be written as for $n = 0, 1, \ldots,$

$$\pi_n(t) = e^{-(\lambda+\mu)t} \sum_{r=0}^{\infty} \frac{(\lambda t)^{n+r}(\mu t)^r}{(n+r)!r!},$$

$$\pi_{-n}(t) = e^{-(\lambda+\mu)t} \sum_{r=0}^{\infty} \frac{(\mu t)^{n+r}(\lambda t)^r}{(n+r)!r!}.$$

Here the interpretation of $\pi_n(t)$ is the probability that the position of particle is n at time t. But in our model, $\pi_n(t)$ is interpreted as the average length of the nth bond at time t. In this model, which may be thought of as a BDP on an infinite state space, it is investigated the transient behavior of diatomic systems with infinite beads. It is useful to distinguish between the two scenarios-unrestricted random walk, or RRW, and restricted random walk, or queueing system-based on whether the state space is a collection of integers or a set of non-negative integers. In the section that follows, first the transient behavior of the unrestricted random walk model is examined.

Example 5.3 *(Randomized random walk)* Consider a freely-joint chain of regularly alternating kinds of beads with the structure ... ABABAB ... joined by bonds of the same length. Label a particular bead zero, and designate those to one side of it with positive integers and those to the other side with negative ones. These two kinds of beads have alternating jump rates λ and μ respectively, i.e., the forward and backward jump rates for even labeled beads are λ and μ, respectively, and these rates are reversed for odd labeled beads. The state-transition diagram for this model is shown in Fig. 5.3.

Here, the local-jump process can be described as a rotation of bead i about an axis passing through beads $i-1$ and $i+1$ as defined in the previous chapter. Without loss of generality, assume that, the average length of nth bond at any time t lies between 0 and 1. Let $\pi_n(t)$ be the average length of the nth bond at time t. $\pi_0(0) = 1$. The forward Kolmogorov equations for this model become

$$\begin{aligned}\frac{d\pi_{2n}(t)}{dt} &= \mu\pi_{2n-1}(t) - (\lambda+\mu)\pi_{2n}(t) + \lambda\pi_{2n+1}(t),\\ \frac{d\pi_{2n+1}(t)}{dt} &= \lambda\pi_{2n}(t) - (\lambda+\mu)\pi_{2n+1}(t) + \mu\pi_{2n+2}(t),\\ n &= 0, \pm 1, \pm 2, \ldots.\end{aligned} \tag{5.11}$$

The time-dependent average length of bond vectors in the RRW are given by

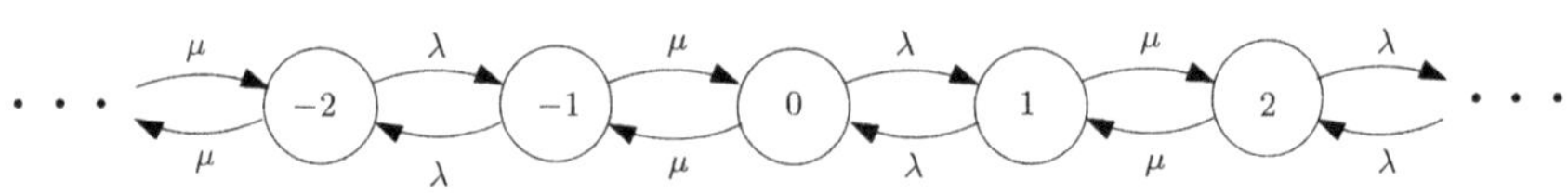

Fig. 5.3 State-transition diagram for unrestricted random walk model

$$\pi_{2n+1}(t) = e^{-at} \sum_{\nu=n}^{\infty} \frac{(\mu t)^{2\nu+1}}{(2\nu+1)!} \sum_{r=n}^{\nu} \binom{\nu}{r-n} \binom{\nu+1}{r+1} \rho^{2r+1-n}, \tag{5.12}$$

$n = 0, 1, \ldots$

$$\pi_{2n+1}(t) = e^{-at} \sum_{\nu=n}^{\infty} \frac{(\lambda t)^{2\nu+1}}{(2\nu+1)!} \sum_{r=n}^{\nu} \binom{\nu}{r-n} \binom{\nu+1}{r+1} \rho^{n-2r-1}, \tag{5.13}$$

$n = -1, -2, \ldots$

$$\pi_{2n}(t) = e^{-at} \sum_{\nu=n}^{\infty} \frac{(\mu t)^{2\nu}}{(2\nu)!} \sum_{r=n}^{\nu} \binom{\nu}{r} \binom{\nu}{r-n} \rho^{2\nu-2r+n}, \tag{5.14}$$

$n = 0, \pm 1, \pm 2, \ldots.$

When $\lambda = \mu$ in (5.13), the formula reduces to modified Bessel functions of the first kind (see Eqs. (5.6) and (5.7)) and hence the right-hand sides can be described as *generalized modified Bessel functions of the first kind*. Specifically, $\pi_0(t)$ can be expressed as

$$\pi_0(t) = e^{-at} \frac{d}{dt} \int_0^t I_0(as) I_0(b(t-s))ds,$$

or

$$\pi_0(t) = e^{-at} \left\{ b \int_0^t I_0(as) I_1(b(t-s))ds + I_0(at) \right\}, \tag{5.15}$$

where $a = \lambda + \mu$ and $b = \lambda - \mu$ (refer to Example 5.2 for further analysis).

Example 5.4 *(Restricted random walk (RRW))* It is interesting to investigate the behavior of the local-jump process in the infinite chain whose leftmost bead is labeled as zero while those to its right are labeled with positive integers. This model turns-out to be the restricted random walk or queueing system. The system state is increased by one upon an arrival to the queue (equivalent to the particle taking a step in the positive direction) and decreased by one when service to a customer is completed and the customer leaves the queue.

Let $N(t)$ denote the state of the local-jump process at time t. Here, the contrast with Example 5.2 is that the rules governing the jump process depend on whether $N(t)$ is even or odd, as defined earlier. It is supposed, unless stated otherwise, that $N(0) = 0$. Let

$$\pi_n(t) = P(N(t) = n), \quad n = 0, 1, \ldots. \tag{5.16}$$

The differential-difference equations corresponding to (5.11) are

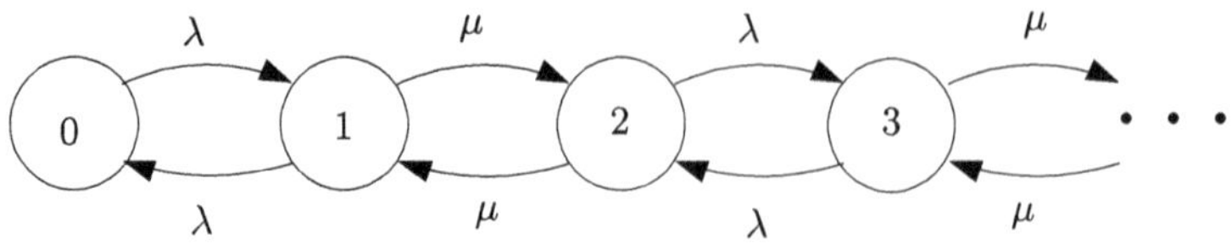

Fig. 5.4 State-transition diagram for restricted random walk model

$$\frac{\mathrm{d}\pi_0(t)}{\mathrm{d}t} + \lambda\pi_0(t) = \lambda\pi_1(t),$$
$$\frac{\mathrm{d}\pi_{2n}(t)}{\mathrm{d}t} + (\lambda+\mu)\pi_{2n}(t) = \mu\pi_{2n-1}(t) + \lambda\pi_{2n+1}(t),$$
$$n = 1, 2, \ldots$$
$$\frac{\mathrm{d}\pi_{2n+1}(t)}{\mathrm{d}t} + (\lambda+\mu)\pi_{2n+1}(t) = \lambda\pi_{2n}(t) + \mu\pi_{2n+2}(t),$$
$$n = 0, 1, \ldots. \tag{5.17}$$

The state-transition diagram is shown in Fig. 5.4.

The transient probabilities of this system are given by

$$\pi_0(t) = \frac{e^{-at}}{a-b} aA(at) - bA(-bt) - ab\int_0^t A(as)A(-b(t-s))ds, \tag{5.18}$$

where $A(t) = I_0(t) + I_1(t)$, $a = \lambda + \mu$ and $b = \lambda - \mu$. Similarly, the other system state probabilities $\pi_n(t)$ can be obtained in terms of the modified Bessel functions of the first kind.

5.1.3.1 Continued Fraction Methodology

In this section, the approximate transient system size probability values for the BDPs are obtained using the CF approach. The importance of CF techniques is discussed in Appendix A.8.

Define

$$L_{r-1} = \prod_{k=0}^{r-1} \lambda_k,\ r = 1, 2, \ldots; \quad M_r = \prod_{k=1}^{r} \mu_k,\ r = 1, 2, \ldots$$

with $L_{-1} = M_0 = 1$. For notational convenience, use $P_n(t)$ for $P_{mn}(t)$. Let

$$f_r(s) = (-1)^r M_r \int_0^\infty e^{-st} P_r(t)\mathrm{d}t, \quad r = 1, 2, \ldots$$

and

$$f_0(s) = \widehat{P}_0(s).$$

Taking Laplace transforms of the system of equations and assuming that there are zero units initially in the system, $f_0(s)$ simplifies to the following expression,

$$f_0(s) = \frac{1}{s + \lambda_0 + \frac{f_1(s)}{f_0(s)}}.$$

Similarly,

$$\frac{f_r(s)}{f_{r-1}(s)} = \frac{-\lambda_{r-1}\mu_r}{s + \lambda_r + \mu_r + \frac{f_{r+1}(s)}{f_r(s)}}, \quad r = 1, 2, \ldots .$$

This results in the CF expression

$$f_0(s) = \frac{1}{s + \lambda_0} - \frac{\lambda_0\mu_1}{s + \lambda_1 + \mu_1} - \frac{\lambda_1\mu_2}{s + \lambda_2 + \mu_2} - \cdots . \tag{5.19}$$

Let the nth approximate of (5.19) be given by $\dfrac{A_n(s)}{B_n(s)}$ where

$$\begin{aligned}
A_1(s) &= 1,\\
A_2(s) &= s + \lambda_1 + \mu_1,\\
&\vdots\\
A_n(s) &= (s + \lambda_{n-1} + \mu_{n-1})A_{n-1}(s) - \lambda_{n-2}\mu_{n-1}A_{n-2}(s),
\end{aligned} \tag{5.20}$$
$n = 3, 4, \ldots,$

and

$$\begin{aligned}
B_1(s) &= s + \lambda_0,\\
B_2(s) &= (s + \lambda_1 + \mu_1)B_1(s) - \lambda_0\mu_1,\\
&\vdots\\
B_n(s) &= (s + \lambda_{n-1} + \mu_{n-1})B_{n-1}(s) - \lambda_{n-2}\mu_{n-1}B_{n-2}(s),
\end{aligned} \tag{5.21}$$
$n = 3, 4, \ldots .$

Then

$$f_0(s) \simeq \frac{A_n(s)}{B_n(s)}, \tag{5.22}$$

where $A_n(s)$ and $B_n(s)$ are given by (5.20) and (5.21), respectively, and can be written in tridiagonal determinant form as follows

$$A_n(s) = \begin{vmatrix} s+\lambda_1+\mu_1 & 1 & & \\ \lambda_1\mu_2 & s+\lambda_2+\mu_2 & 1 & \\ & & \ddots & \\ & & \lambda_{n-2}\mu_{n-1} & s+\lambda_{n-1}+\mu_{n-1} \end{vmatrix}_{(n-1)\times(n-1)},$$

$$B_n(s) = \begin{vmatrix} s+\lambda_0 & 1 & & \\ \lambda_0\mu_1 & s+\lambda_1+\mu_1 & 1 & \\ & & \ddots & \\ & & \lambda_{n-2}\mu_{n-1} & s+\lambda_{n-1}+\mu_{n-1} \end{vmatrix}_{n\times n}. \quad (5.23)$$

Note that $B_n(s)$ given by (5.23) is clearly zero when $-s$ is an eigenvalue of the matrix

$$\mathbf{F_n} = \begin{pmatrix} \lambda_0 & \mu_1 & & & \\ \lambda_0 & \lambda_1+\mu_1 & \mu_2 & & \\ & \lambda_1 & \lambda_2+\mu_2 & \mu_3 & \\ & & & \ddots & \\ & & & \lambda_{n-2} & \lambda_{n-1}+\mu_{n-1} \end{pmatrix}_{n\times n}.$$

This matrix is quasi-symmetric and can be transformed into a real symmetric matrix $\mathbf{E_n}$ by a similarity transformation $\mathbf{E_n} = \mathbf{D_n^{-1}F_nD_n}$ where

$$\mathbf{D_n} = \text{Diag}\{1, \sqrt{L_0M_1}, \sqrt{L_1M_2}, \ldots, \sqrt{L_{n-2}M_{n-1}}\}.$$

The matrix thus obtained is given by

$$\mathbf{E_n} = \begin{bmatrix} \lambda_0 & \sqrt{\lambda_0\mu_1} & & & \\ \sqrt{\lambda_0\mu_1} & \lambda_1+\mu_1 & \sqrt{\lambda_1\mu_2} & & \\ & \sqrt{\lambda_1\mu_2} & \lambda_2+\mu_2 & \sqrt{\lambda_2\mu_3} & \\ & & & \ddots & \\ & & & \sqrt{\lambda_{n-2}\mu_{n-1}} & \lambda_{n-1}+\mu_{n-1} \end{bmatrix}_{n\times n}.$$

This is a real symmetric positive definite tri-diagonal matrix with non-zero sub-diagonal elements, and therefore the eigenvalues are real and distinct (Wilkinson (1965)). These eigenvalues are denoted by $s_1^{(n)}, s_2^{(n)}, \ldots, s_n^{(n)}$, that is, $-s_1^{(n)}, -s_2^{(n)}, \ldots, -s_n^{(n)}$ are the roots of $B_n(s)$. Similarly the roots of $A_n(s)$ are negative, real and distinct and are denoted by $-z_1^{(n)}, -z_2^{(n)}, \ldots, -z_{n-1}^{(n)}$.

Using partial fractions, (5.22) can be expressed as

$$f_0(s) \simeq \sum_{j=1}^{n} \frac{\prod_{r=1}^{n-1}(z_r^{(n)} - s_j^{(n)})}{(s + s_j^{(n)})\prod_{r=1,r\neq j}^{n}(s_r^{(n)} - s_j^{(n)})}.$$

Upon inversion,

$$P_0(t) \simeq \sum_{j=1}^{n} \frac{\prod_{r=1}^{n-1}(z_r^{(n)} - s_j^{(n)})}{\prod_{r=1,r\neq j}^{n}(s_r^{(n)} - s_j^{(n)})} e^{-s_j^{(n)}t}.$$

Murphy and O'Donohoe [12] found the probabilities $P_r(t)$ for the system, starting initially with $m(> 0)$ units, by using the properties of continued fractions. These probabilities are given by

$$P_r(t) \simeq \sum_{j=1}^{k} H_j^{(r)} e^{-s_j^{(k)}t}, \quad r = 0, 1, \ldots, \tag{5.24}$$

where

$$H_j^{(r)} = \frac{B_r(-s_j^{(k)})B_m(-s_j^{(k)})A_k(-s_j^{(k)})}{\left(\prod_{i=0}^{m-1}\lambda_i\right)\left(\prod_{i=1}^{r}\mu_i\right)\left(\prod_{i=1,i\neq j}^{k}(s_i^{(n)} - s_j^{(n)})\right)},$$

and

$$k = \begin{cases} m + n \text{ for } r \leq m \\ r + n \text{ for } r \geq m \end{cases}.$$

5.1.4 Steady-State Analysis

For steady-state probabilities denoted by π_n, allowing $t \to \infty$ in Eq. (5.3)

$$\pi_n = \lim_{t\to\infty} P(X(t) = n), \quad n = 0, 1, \ldots$$

With the choice of λ_n, μ_n such that BDP is irreducible and positive recurrent,

$$\begin{aligned} 0 &= -\lambda_0\pi_0 + \mu_1\pi_1 \\ 0 &= \lambda_{n-1}\pi_{n-1} - (\lambda_n + \mu_n)\pi_n + \mu_{n+1}\pi_{n+1}, \quad n = 1, 2, \ldots \end{aligned}$$

Solving the above system of equations,

$$\pi_n = \frac{\lambda_{n-1}\lambda_{n-2}\ldots\lambda_0}{\mu_n\mu_{n-1}\ldots\mu_1}\pi_0, \quad n = 1, 2, \ldots$$

and using $\sum_{n=0}^{\infty}\pi_n = 1$, π_0 as follows

$$\pi_0 = \frac{1}{1 + \frac{\lambda_0}{\mu_1} + \frac{\lambda_0 \lambda_1}{\mu_1 \mu_2} + \cdots},$$

provided the denominator in the above expression is finite. If the denominator is infinite, then $\pi_n = 0$ for all n. The steady-state or stationary distribution does not exist.

5.2 Finite Birth-Death Process

Finite BDP is a BDP with finite state space $\{0, 1, \ldots, N\}$. No birth is possible if system is in state N at any time instant. Thus, $\lambda_N = 0$ and state transition diagram is shown in Fig. 5.5.

Example 5.5 The incoming traffic follows the Poisson process with a rate of λ in a parking lot with N spaces, but only while there are open spots. The occupancy times have an exponential distribution with mean $1/\mu$. Let $\{X(t), t \geq 0\}$ be the number of occupied parking spaces at time t.

(a) Determine infinitesimal generator matrix Q and the forward Kolmogorov equations for the BDP $\{X(t), t \geq 0\}$.
(b) Determine the equilibrium probability distribution of the process.

Solution:

(a) The infinitesimal generator matrix **Q** is given by

$$\mathbf{Q} = \begin{pmatrix} -\lambda & \lambda & \cdot & \cdot\ \cdot & \cdot & \cdot \\ \mu & -(\lambda+\mu) & \lambda & \cdot\ \cdot & \cdot & \cdot \\ \cdot & \mu & -(\lambda+\mu) & \lambda\ \cdot & \cdot & \cdot \\ \cdot & \cdot & \cdot & \cdot\ \cdot & \cdot & \cdot \\ \cdot & \cdot & \cdot & \cdot\ \mu & -(\lambda+\mu) & \lambda \\ \cdot & \cdot & \cdot & \cdot\ \cdot & \mu & -\mu \end{pmatrix}_{(N+1)\times(N+1)}.$$

Forward Kolmogorov equation for the BDP $\{X(t), t \geq 0\}$ is given by

$$\begin{aligned} P'_{m0}(t) &= -\lambda P_{m0}(t) + \mu P_{m1}(t), \\ P'_{mn}(t) &= \lambda P_{mn-1}(t) - (\lambda+\mu) P_{mn}(t) + \mu P_{mn+1}(t), \ 1 \leq n \leq N-1, \\ P'_{mN}(t) &= \lambda P_{mN-1}(t) - \mu P_{mN}(t). \end{aligned}$$

(b) Let equilibrium probability distribution of the process be $\pi = [\pi_0, \pi_1, \ldots, \pi_N]$ where π is the solution of $\pi \mathbf{Q} = 0$. The equilibrium probabilities are given as

$$\pi_n = \frac{\lambda^n}{\mu^n} \pi_0, \quad n = 1, 2, \ldots, N$$

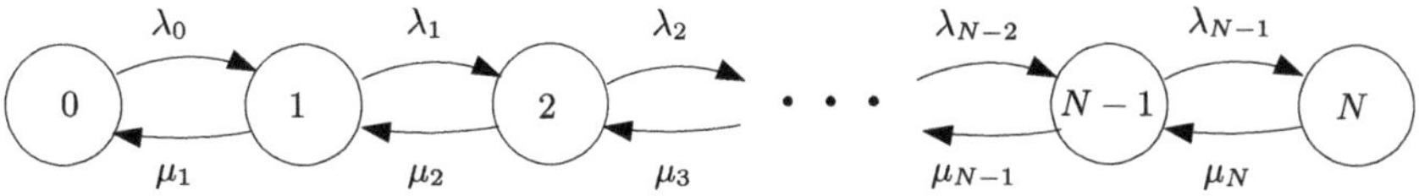

Fig. 5.5 State transition diagram for the finite BDP

where

$$\pi_0 = \frac{1}{1 + \sum_{i=1}^{N} \left(\frac{\lambda}{\mu}\right)^i} = \frac{(\lambda + \mu)\,\mu^N}{\lambda^{N+1} - \mu^{N+1}}.$$

Hence, the equilibrium probabilities are given by $\pi_n = \dfrac{\lambda^n}{\mu^n}\left(\dfrac{(\lambda + \mu)\,\mu^N}{\lambda^{N+1} - \mu^{N+1}}\right)$ for $n = 1, 2, \ldots, N$.

5.2.1 *Transient Solution*

A finite or truncated BDP on $\{0, 1, \ldots, N\}$ satisfies a finite system of equations:

$$\begin{aligned}
P'_{m0}(t) &= -\lambda_0 P_{m0}(t) + \mu_1 P_{m1}(t),\\
P'_{mn}(t) &= \lambda_{n-1} P_{mn-1}(t) - (\lambda_n + \mu_n) P_{mn}(t) + \mu_{n+1} P_{mn+1}(t),\ 1 \le n \le N-1,\\
P'_{mN}(t) &= \lambda_{N-1} P_{mN-1}(t) - \mu_N P_{mN}(t).
\end{aligned} \tag{5.25}$$

The infinitesimal generator $\mathbf{Q}$ of this finite BDP is given by

$$\mathbf{Q} = \begin{pmatrix}
-\lambda_0 & \lambda_0 & \cdot & \cdot & \cdot & \cdot & \cdot \\
\mu_1 & -(\lambda_1 + \mu_1) & \lambda_1 & \cdot & \cdot & \cdot & \cdot \\
\cdot & \mu_2 & -(\lambda_2 + \mu_2) & \lambda_2 & \cdot & \cdot & \cdot \\
\cdot & \cdot & \cdot & \cdot & \cdot & \cdot & \cdot \\
\cdot & \cdot & \cdot & \cdot & \mu_{N-1} & -(\lambda_{N-1} + \mu_{N-1}) & \lambda_{N-1} \\
\cdot & \cdot & \cdot & \cdot & \cdot & \mu_N & -\mu_N
\end{pmatrix}. \tag{5.26}$$

Note that the above matrix has real and distinct eigenvalues (Brown [1], Peng [14]).

There exists a number of methods to find exact transient solution of finite BDPs. Some of them are listed below.

1. Murphy and O'Donohoe [12] have discussed the method of finding $P_{mn}(t)$ with the initial number m in the system.

   Define two sets of polynomials recursively as follows:

$$\begin{aligned} A_1(s) &= 1 \\ A_2(s) &= s + \lambda_1 + \mu_1 \\ A_n(s) &= (s + \lambda_{n-1} + \mu_{n-1})A_{n-1}(s) - \lambda_{n-2}\mu_{n-1}A_{n-2}(s), \quad n = 3, 4, \ldots, N \\ A_{N+1}(s) &= (s + \mu_N)A_N(s) - \lambda_{N-1}\mu_N A_{N-1}(s), \end{aligned} \tag{5.27}$$

and

$$\begin{aligned} B_1(s) &= s + \lambda_0 \\ B_2(s) &= (s + \lambda_1 + \mu_1)B_1(s) - \lambda_0\mu_1 \\ B_n(s) &= (s + \lambda_{n-1} + \mu_{n-1})B_{n-1}(s) - \lambda_{n-2}\mu_{n-1}B_{n-2}(s), \quad n = 3, 4, \ldots, N \\ B_{N+1}(s) &= (s + \mu_N)B_N(s) - \lambda_{N-1}\mu_N B_{N-1}(s). \end{aligned} \tag{5.28}$$

Then, the solution of (5.25) is given, for $0 \le m,\ n \le N$, by

$$P_{mn}(t) = \sum_{j=0}^{N} H_j^{(n)} e^{-s_j t}, \tag{5.29}$$

where

$$H_j^{(n)} = \frac{B_n(-s_j)B_m(-s_j)A_{N+1}(-s_j)}{\left(\prod_{k=0}^{m-1}\lambda_k\right)\left(\prod_{k=1}^{n}\mu_k\right)\left(\prod_{i=0,i\neq j}^{N}(s_i - s_j)\right)},$$

and the s_j's are the roots of $B_{N+1}(s)$.

2. Rosenlund [16] has expressed the transient solution of (5.25) in terms of a sequence of polynomials $C_n(s)$, $(n = 0, 1, \ldots, N)$. Define these polynomials recursively as

$$\begin{aligned} &C_0(s) = 1, \quad C_1(s) = s + \lambda_0, \\ &C_r(s) = (s + \lambda_{r-1} + \mu_{r-1})C_{r-1}(s) - \lambda_{r-2}\mu_{r-1}C_{r-2}(s), 2 \le r \le N + 1. \end{aligned}$$

Further, set $D_{mm} = 1$ and

$$D_{mn} = \begin{cases} \mu_{n+1}\cdots\mu_m, & n < m \\ \lambda_m\cdots\lambda_{n-1}, & n > m. \end{cases} \tag{5.30}$$

Then, the solution of (5.25) is given, for $0 \le m,\ n \le N$, by

$$P_{mn}(t) = \pi_n + \sum_{r=1}^{N} \frac{D_{mN}D_{Nn}C_n(-\beta_r)C_m(-\beta_r)\exp(-\beta_r t)}{C_N(-\beta_r)C'_{N+1}(-\beta_r)} \tag{5.31}$$

where

$$\pi_n = \frac{D_{mN} D_{Nn} C_n(0) C_m(0)}{C_N(0) \left(\prod_{r=1}^{N} \beta_r\right)}, \tag{5.32}$$

$$C'_{N+1}(-\beta_r) = -\beta_r \prod_{k=1, k \neq r}^{N} (\beta_k - \beta_r), \quad r = 1, 2, \ldots, N \tag{5.33}$$

and the β_r's are the roots of $C_{N+1}(s)$.

3. Chiang [3] has obtained the solution of the system of Eq. (5.25) in simple form by using matrix methods. He derived

$$P_{mn}(t) = \sum_{l=0}^{N} \frac{A'_{mn}(\rho_l)}{\prod_{i=0, i \neq l}^{N} (\rho_l - \rho_i)} e^{\rho_l t}, \quad m, n = 0, 1, \ldots, N \tag{5.34}$$

where $A'_{mn}(\rho_l)$ is the (n, m)th cofactor of the characteristic matrix $\mathbf{Q}$ which is defined in (5.26) and the ρ_l's are eigenvalues of $\mathbf{Q}$.

4. Van Doorn [21] has given the spectral representation of the transition probability function $P_{mn}(t)$. This spectral representation is an explicit representation as a mixture of exponentials. In this connection, define the polynomials $Q_n(x), n \in S$ associated with the birth-death parameters λ_n and μ_n, by the recurrence relations

$$\begin{aligned} Q_0(x) &= 1 \\ -x Q_0(x) &= -\lambda_0 Q_0(x) + \lambda_0 Q_1(x) \\ -x Q_n(x) &= \mu_n Q_{n-1}(x) - (\lambda_n + \mu_n) Q_n(x) + \lambda_n Q_{n+1}(x), \quad 0 < n < N \end{aligned} \tag{5.35}$$

and the polynomial $Q_{N+1}(x)$ of degree $N + 1$, defined by

$$-x Q_N(x) = \mu_N Q_{N-1}(x) - \mu_N Q_N(x) + Q_{N+1}(x). \tag{5.36}$$

These polynomials are known to be orthogonal. It is well known that $Q_{N+1}(x)$ has $N + 1$ distinct real zeros $x_0 < x_1 < \ldots < x_N$.
The spectral representation of $P_{mn}(t), \quad m, n \in S$ is given by

$$P_{mn}(t) = \pi_n \sum_{k=0}^{N} \exp(-x_k t) Q_m(x_k) Q_n(x_k) \rho_k \tag{5.37}$$

where

$$\rho_k = \frac{1}{\sum_{i=0}^{N} Q_i^2(x_k) \pi_i}. \tag{5.38}$$

5. Nikiforov et al. [13] have derived the transient solution of (5.25) in terms of orthogonal polynomials of a discrete variable. Define the polynomials recursively as follows.

$$xy_n(x) = \lambda_{n-1}y_{n-1}(x) - (\lambda_n + \mu_n)y_n(x) + \mu_n y_n(x) \tag{5.39}$$

with

$$y_{-1}(x) = y_{N+1}(x) = 0, \;\; y_0(x) = 1.$$

These polynomials $y_n(x)$ satisfy orthogonality relations of the form

$$\sum_{i=0}^{N} y_n(x_i)y_m(x_i)\rho_i = \delta_{m,n}d_n^2 \tag{5.40}$$

where ρ_i and d_n^2 are positive constants. Then, the solution of (5.25) can be written as follows, for $0 \le m, \; n \le N$,

$$P_{mn}(t) = \sum_{i=0}^{N} P_{mn}(0)\frac{y_n(x_i)y_m(x_i)}{d_m^2}e^{x_i t}\rho_i \tag{5.41}$$

where x_i are the roots of the equation $y_{N+1}(x) = 0$.

6. Kijima [9] has presented the transient solution of a finite BDP in terms of eigenvalues of the corresponding infinitesimal generator $\mathbf{Q}$ defined in (5.26).
   Let $\theta_j, \;\; j = 0, 1, \ldots, N$, be the eigenvalues of $\mathbf{Q}$ and suppose that $\mathbf{Q}$ can be written as

$$\mathbf{Q} = \mathbf{M}\theta_{\mathbf{D}}\mathbf{M}^{-1}$$

   where $\mathbf{M}$ is a non-singular matrix and $\theta_{\mathbf{D}}$ is the diagonal matrix with the θ_j's as diagonal elements. Then the solution can be written as follows:

$$\begin{aligned}\mathbf{P}(t) = \exp\{\mathbf{Q}t\} &= \sum_{n=0}^{\infty}\frac{\mathbf{Q}^n t^n}{n!}, t \ge 0\\ &= \mathbf{M}\left(\sum_{n=0}^{\infty}\frac{{\theta_{\mathbf{D}}}^n t^n}{n!}\right)\mathbf{M}^{-1}\\ &= \mathbf{M}e^{\theta_{\mathbf{D}}t}\mathbf{M}^{-1}\end{aligned}$$

   where $e^{\theta_{\mathbf{D}}t}$ is now the diagonal matrix having the $e^{\theta_j t}$ as diagonal elements.
   He has also given the transient solution in another form.
   For any $m, n \in \{0, 1, \ldots, N\}$ and $t > 0$

$$P_{mn}(t) = \delta_{mn}e^{-q_m t} + \int_0^t q_m e^{-q_m s}\sum_{k\neq m}P_{mk}(s)P_{kn}(t-s)\mathrm{d}s \tag{5.42}$$

   where $q_m = -q_{mm} = \sum_{n\neq m} q_{mn}$ and the q_{mn}'s are elements of $\mathbf{Q}$.

An interpretation of the above result is as follows: Starting from state m, the Markov chain is in state n at time t if it either remains in state m, in which case $m = n$, or moves at some time $s < t$ to a state k from which it moves to state n by time t. It should be noted that, due to the Markov property and homogeneity, the probability of the latter conditional event is equal to the conditional probability that, starting from state k at time 0, the Markov chain is in state n at time $t - s$.

5.2.2 Steady-State Solution

The steady-state probabilities for the presented BDP are provided as follows:

$$\pi_n = \frac{\lambda_{n-1}\lambda_{n-2}\dots\lambda_0}{\mu_n\mu_{n-1}\dots\mu_1}\pi_0, \quad n = 1, 2, \dots, N,$$

where

$$\pi_0 = \frac{1}{1 + \frac{\lambda_0}{\mu_1} + \frac{\lambda_0\lambda_1}{\mu_1\mu_2} + \dots + \frac{\lambda_{N-1}\lambda_{N-2}\dots\lambda_0}{\mu_N\mu_{N-1}\dots\mu_1}}.$$

A finite BDP with $\mu_0 = 0 = \lambda_N$ is ergodic, and the associated tridiagonal rate matrix is conservative such that one of its eigenvalue is zero, say $s_0 = 0$ (Kijima [9]). Steady-state probabilities exist when one of the roots of $B_{N+1}(s)$ is zero. Using

$$A_{N+1}(0) = \prod_{k=1}^{N}\mu_k \text{ and } B_k(0) = \prod_{j=0}^{k-1}\lambda_j, \quad k = 0, 1, \dots, N, \tag{5.43}$$

$$\pi_n = \frac{\left(\prod_{k=n+1}^{N}\mu_k\right)\left(\prod_{k=0}^{n-1}\lambda_k\right)}{\prod_{k=1}^{N}s_k}, \quad n = 0, 1, \dots, N. \tag{5.44}$$

Example 5.6 In cellular networks, a given geographical area is divided into cells. It has been assumed that the cells are statistically homogeneous. Consider only one type of traffic, i.e., voice. The arrival of incoming requests follow Poisson process with exponential distributed inter-arrival time with parameter λ and service follows exponential distribution with rate μ. Consider that the number of available channels is M. Let $\{X(t), t \geq 0\}$ be the stochastic process with a state space $\{j : 0 \leq j \leq M\}$ where j denotes the number of available channels. The generator matrix of the underlying system can be represented as a tri-diagonal matrix $\mathbf{Q}$. The elements of $\mathbf{Q}$ are provided as follows

- The elements of the upper diagonal $\mathbf{Q_{i0}}$ are represented by $A_{i1} = \lambda, \quad 0 \leq i \leq M - 1$,
- The elements of the lower diagonal are given by $\mathbf{Q_{i2}} = i\mu, 0 \leq i \leq M$.

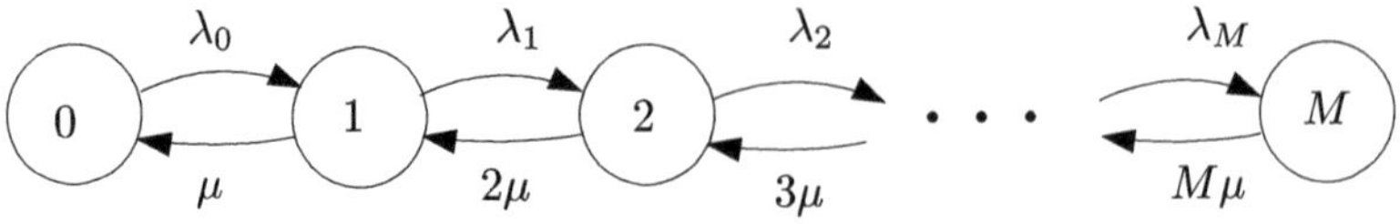

Fig. 5.6 State transition diagram of loss system in wireless networks

- The elements of the main diagonal, denoted as $\mathbf{Q_{i1}}$ represented as $B_{i1} = -(\lambda + i\mu), 0 \le i \le M$; $B_{i0} = -\lambda, \quad 0 \le i \le M$,

Example 5.7 *(Loss System with Application in Cellular Networks)* Consider the performance model of a single cell in a cellular wireless network as provided in [7]. The arrival of handoff and new calls follow Poisson process with arrival rates λ_H and λ_N, respectively. The service times of handoff and new calls are exponential distributed with rates μ_H and μ_N, respectively. The handoff calls are provided priority over new calls by employing guard channel policy. As per this policy, among the total number of channels M, a fixed number of channels (say, g) are exclusively reserved for the use of handoff calls only. If an arriving new call finds more than $M - g$ channels busy, the new call will be blocked and if an arriving handoff call finds all the channels occupied, the handoff call will be dropped and lost from the system. Find the handoff dropping probability and new call blocking probability for this system.

Solution: Let $\lambda = \lambda_H + \lambda_N$ and $\mu = \mu_H + \mu_N$. The state dependent arrival rates are given as follow

$$\lambda_n = \begin{cases} \lambda, & 0 \le n < M - g - 1, \\ \lambda_N, & M - g \le n \le M. \end{cases} \tag{5.45}$$

The service rates are given by $n\mu$; $1 \le n \le M$. The expression of the steady-state probabilities of the presented system can be calculated and provided by

$$\pi_n = \pi_0 \begin{cases} \frac{A^n}{n!}, & 0 \le n \le M - g, \\ \frac{A^{M-g} A_1^{n-M+g}}{n!}, & n \ge M - g \end{cases} \tag{5.46}$$

where $A = \frac{\lambda}{\mu}$, $A_1 = \frac{\lambda_N}{\mu_H + \mu_N}$ and

$$\pi_0 = \frac{1}{\sum_{n=0}^{M-g-1} \frac{A^n}{n!} + \sum_{n=M-g}^{M} \frac{A^{M-g} A_1^{n-M+g}}{n!}}.$$

The blocking probability of new call and dropping probability of handoff calls are considered the performance determining factors. The expression of dropping probability of handoff calls (P_d) is given by

$$P_d = \frac{\frac{A^{M-g}A_1^g}{M!}}{\sum_{n=0}^{M-g-1}\frac{A^n}{n!} + \sum_{n=M-g}^{M}\frac{A^{M-g}A_1^{n-M+g}}{n!}}. \quad (5.47)$$

The blocking probability of new call (P_b) is provided by

$$P_b = A^{M-g}\frac{\sum_{k=0}^{g}\frac{A_1^k}{(k+M-g)!}}{\sum_{n=0}^{M-g-1}\frac{A^n}{n!} + \sum_{n=M-g}^{M}\frac{A^{M-g}A_1^{n-M+g}}{n!}}. \quad (5.48)$$

Example 5.8 Stockmayer et al. [18] discussed a freely - jointed chain of two regularly alternating kinds of beads, with the structure . . . ABABAB . . .. These two kinds of beads have alternating jump rates λ and μ, respectively, i.e., the forward and backward jump rates for even labeled beads are λ and μ, respectively, and these rates are reversed for odd labeled beads. Let $X(t)$ be the position of the beads at time t. Assume that, the number of beads is finite, say N. Let $\pi_n(t) = P(X(t) = n), n = 0, 1, \ldots N$. The forward Kolmogorov equation for this system is given by

$$\pi_0'(t) = -\lambda\pi_0(t) + \lambda\pi_1(t), \quad (5.49)$$

$$\pi_j'(t) = \begin{cases} \lambda\pi_{j-1}(t) - (\lambda+\mu)\pi_j(t) + \mu\pi_{j+1}(t), & j = 1, 3, \ldots, N-1 \\ \mu\pi_{j-1}(t) - (\lambda+\mu)\pi_j(t) + \lambda\pi_{j+1}(t), & j = 2, 4, \ldots, N-1 \end{cases} \quad (5.50)$$

$$\pi_N'(t) = \begin{cases} \lambda\pi_{N-1}(t) - \lambda\pi_N(t), & N - \text{ odd} \\ \mu\pi_{N-1}(t) - \mu\pi_N(t), & N - \text{ even} \end{cases} \quad (5.51)$$

with

$$\pi_i(0) = a_i, \quad \forall\ i.$$

The state transition diagrams for even and odd numbers of bonds are shown in Figs. 5.7 and Fig. 5.8, respectively.

Let $\widehat{\psi}_n(s)$ be the Laplace transform of $\pi_n(t)$.

(a) When N is odd, then from Eqs. (5.49), (5.50) and (5.51)

Fig. 5.7 State-transition for Local-jump process with even number of bonds

$$
\begin{aligned}
(s+\lambda)\widehat{\psi}_0(s) - \lambda\widehat{\psi}_1(s) &= a_0,\\
-\lambda\widehat{\psi}_{n-1}(s) + (s+\lambda+\mu)\widehat{\psi}_n(s) - \mu\widehat{\psi}_{n+1}(s) &= a_n,\\
n = 1, 3, \ldots, N-2,&\\
-\mu\widehat{\psi}_{n-1}(s) + (s+\lambda+\mu)\widehat{\psi}_n(s) - \lambda\widehat{\psi}_{n+1}(s) &= a_n,\ n = 2, 4, \ldots, N-1,\\
-\lambda\widehat{\psi}_{N-1}(s) + (s+\lambda)\widehat{\psi}_N(s) &= a_N.
\end{aligned}
\tag{5.52}
$$

The above system of equations can be written as

$$
\mathbf{A}\widehat{\Psi} = [a_0,\ a_1,\ \ldots\ , a_N]', \tag{5.53}
$$

where

$$
\mathbf{A} = \begin{pmatrix}
s+\lambda & -\lambda & \cdot & \cdot\ \cdot\ \cdot & \cdot & \cdot \\
-\lambda & s+\lambda+\mu & -\mu & \cdot\ \cdot\ \cdot & \cdot & \cdot \\
\cdot & -\mu & s+\lambda+\mu & -\lambda\ \cdot\ \cdot & \cdot & \cdot \\
\cdot & \cdot & \cdot & \cdot\ \cdot\ \cdot & \cdot & \cdot \\
\cdot & \cdot & \cdot & \cdot\ \cdot\ -\mu & s+\lambda+\mu & -\lambda \\
\cdot & \cdot & \cdot & \cdot\ \cdot\ \cdot & -\lambda & s+\lambda
\end{pmatrix}_{N+1},
$$

and $\widehat{\Psi} = \left[\widehat{\psi}_0(s),\ \widehat{\psi}_1(s),\ \ldots\ , \widehat{\psi}_N(s)\right]'$. Let $D_r(s)$ be the determinant formed by taking the first r rows and r columns of $\mathbf{A}$. Define this subdeterminant recursively as follows:

$$
\begin{aligned}
D_0(s) &= 1,\\
D_1(s) &= s+\lambda,\\
D_r(s) &= \begin{cases} (s+\lambda+\mu)D_{r-1}(s) - \lambda^2 D_{r-2}(s), & r - \text{ even} \\ (s+\lambda+\mu)D_{r-1}(s) - \mu^2 D_{r-2}(s), & r - \text{ odd} \end{cases} \quad r = 2, 3, \ldots, N.
\end{aligned}
\tag{5.54}
$$

When N is odd, $j = 0, 1, \ldots, N$, the transient solution is:

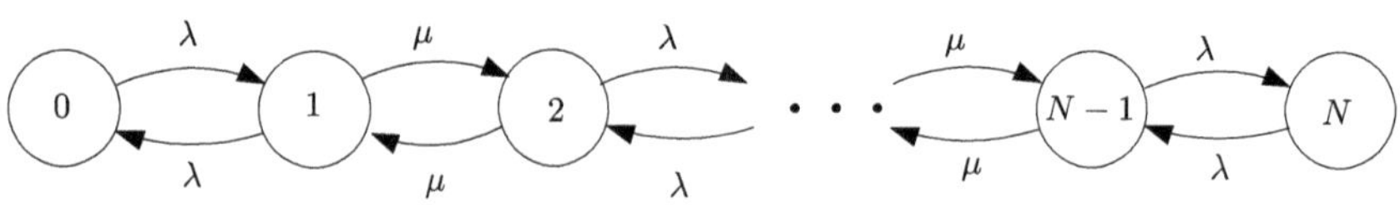

Fig. 5.8 State-transition for Local-jump process with odd number of bonds

$$q_j(t) = \frac{1}{N+1} + \sum_{k=1}^{N} \left\{ \sum_{i=0}^{j} a_i (\lambda\mu)^{(j-i)/2} C_{ij} \frac{D_i(s_k) D_{N-j}(s_k) e^{s_k t}}{b_k} \right.$$
$$\left. + \sum_{i=j+1}^{N} a_i (\lambda\mu)^{(i-j)/2} C_{ij}^{-1} \frac{D_j(s_k) D_{N-i}(s_k) e^{s_k t}}{b_k} \right\},$$

where

$$s_1 = -2\lambda,$$
$$s_k = \begin{cases} -(\lambda+\mu) - \sqrt{(\lambda+\mu)^2 - 4\lambda\mu \sin^2 \frac{k\pi}{N+1}}, & 2 \le k \le \frac{N+1}{2}, \\ -(\lambda+\mu) + \sqrt{(\lambda+\mu)^2 - 4\lambda\mu \sin^2 \frac{k\pi}{N+1}}, & \frac{N+3}{2} \le k \le N, \end{cases} \tag{5.55}$$

$$C_{ij} = \begin{cases} \sqrt{\frac{\lambda}{\mu}}, & i - \text{ even}, \ j - \text{ odd}, \\ \sqrt{\frac{\mu}{\lambda}}, & i - \text{ odd}, \ j - \text{ even}, \\ 1, & \text{otherwise}, \end{cases} \tag{5.56}$$

and

$$b_k = s_k \prod_{r=1,\ r\ne k}^{N} (s_k - s_r), \quad 1 \le k \le N. \tag{5.57}$$

(b) When N is even, taking Laplace transform of Eqs. (5.49), (5.50) and (5.51),

$$\begin{gathered} (s+\lambda)\widehat{\psi}_0(s) - \lambda\widehat{\psi}_0(s) = a_0, \\ -\lambda\widehat{\psi}_{n-1}(s) + (s+\lambda+\mu)\widehat{\psi}_n(s) - \mu\widehat{\psi}_{n+1}(s) = a_n, \\ n = 1, 3, \ldots, N-1, \\ -\mu\widehat{\psi}_{n-1}(s) + (s+\lambda+\mu)\widehat{\psi}_n(s) - \lambda\widehat{\psi}_{n+1}(s) = a_n, \ n = 2, 4, \ldots, N-2, \\ -\mu\widehat{\psi}_{N-1}(s) + (s+\mu)\widehat{\psi}_N(s) = a_N. \end{gathered} \tag{5.58}$$

Observe the subtle change between the set of Eqs. (5.58) and (5.52). The above set of equations (5.58) can be written as

$$\mathbf{B}\widehat{\Psi} = [a_0\, a_1 \ldots a_N]', \tag{5.59}$$

where

$$\mathbf{B} = \begin{pmatrix} s+\lambda & -\lambda & \cdot & \cdot\;\cdot\;\cdot & \cdot & \cdot \\ -\lambda & s+\lambda+\mu & -\mu & \cdot\;\cdot\;\cdot & \cdot & \cdot \\ \cdot & -\mu & s+\lambda+\mu & -\lambda\;\cdot\;\cdot & \cdot & \cdot \\ \cdot & \cdot & \cdot & \cdot\;\cdot\;\cdot & \cdot & \cdot \\ \cdot & \cdot & \cdot & \cdot\;\cdot\;-\lambda & s+\lambda+\mu & -\mu \\ \cdot & \cdot & \cdot & \cdot\;\cdot\;\cdot & -\mu & s+\mu \end{pmatrix}_{N+1} .$$

Let $\Delta_r(s)$ be the determinant formed by taking the last r rows and r columns of $\mathbf{B}$. Define this subdeterminant recursively as follows:

$$\begin{aligned} \Delta_0(s) &= 1, \\ \Delta_1(s) &= s+\mu, \\ \Delta_r(s) &= \begin{cases} (s+\lambda+\mu)\Delta_{r-1}(s) - \mu^2\Delta_{r-2}(s), & \text{if r is even ,} \\ (s+\lambda+\mu)\Delta_{r-1}(s) - \lambda^2\Delta_{r-2}(s), & \text{if r is odd ,} \end{cases} \end{aligned} \tag{5.60}$$

$$r = 2, 3, \ldots, N.$$

When N is even, for $j = 0, 1, \ldots, N$ the transient solution is given by

$$\pi_j(t) = \frac{1}{N+1} + \sum_{k=1}^{N} \left\{ \sum_{i=0}^{j} a_i C_{ij} \frac{D_i(x_k)\Delta_{N-j}(x_k)e^{x_k t}}{V_k} + \sum_{i=j+1}^{N} a_i C_{ij}^{-1} \frac{D_j(x_k)\Delta_{N-i}(x_k)e^{x_k t}}{V_k} \right\},$$

where

$$x_k = \begin{cases} -(\lambda+\mu) - \sqrt{(\lambda+\mu)^2 - 4\lambda\mu\sin^2\frac{k\pi}{N+1}}, & 1 \le k \le \frac{N}{2}, \\ -(\lambda+\mu) + \sqrt{(\lambda+\mu)^2 - 4\lambda\mu\sin^2\frac{k\pi}{N+1}}, & \frac{N}{2} < k \le N, \end{cases} \tag{5.61}$$

and

$$V_k = x_k \prod_{r=1,\ r\neq k}^{N} (x_k - x_r), \quad 1 \le k \le N. \tag{5.62}$$

5.3 Pure Birth Processes

Birth-death process (BDP) is said to be a pure birth process if the death rates, $\mu_i = 0$ for $i = 1, 2, \ldots$. The state transition diagram is shown in Fig. 5.9.

From the definition of a pure birth process, the Poisson process is a pure birth process with the same constant birth rates for all states.

Fig. 5.9 State transition diagram for the pure birth process

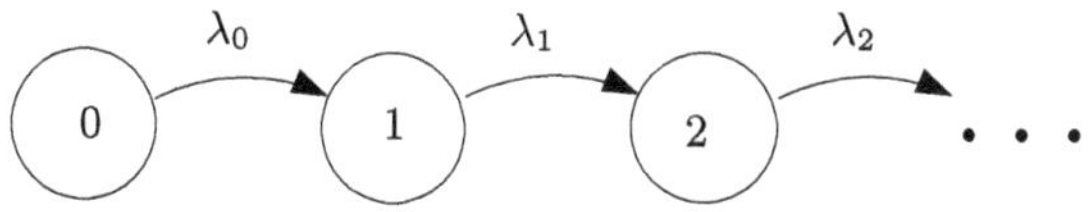

Consider this special case of pure birth process $\{X(t), t \geq 0\}$, with birth rates, $\lambda_i = \lambda$ for $i = 0, 1, \ldots$. The Kolmogorov forward equations (by setting $\lambda_i = \lambda$ for $i = 0, 1, \ldots$ and $\mu_i = 0$ for $i = 1, 2, \ldots$ in BDP) are,

$$\pi_0'(t) = -\lambda\pi_0(t)$$
$$\pi_n'(t) = \lambda\pi_{n-1}(t) - \lambda\pi_n(t), \quad n = 1, 2, \ldots$$

with initial conditions $\pi_0(0) = 1$ and $\pi_i(0) = 0,\ i = 1, 2, \ldots$.
Solving the above system of difference-differential equations, we get,

$$\pi_n(t) = \frac{e^{-\lambda t}(\lambda t)^n}{n!}, \quad n = 0, 1, \ldots.$$

Thus, $\{X(t), t \geq 0\}$, is a Poisson process with rate λ.

Example 5.9 Consider a pure birth process $\{X(t), t \geq 0\}$, with $\lambda_i = 2^i, i = 0, 1, \ldots$. Discuss the behavior of $X(t)$ as $t \to \infty$.
Solution: Let H_i denote the time in which the Markov chain spends in state i. Note that, the chain spends, on average, $E(H_i) = \frac{1}{\lambda_i} = 2^{-i}$ units of time in state i and then transits to state $i + 1$. As a result, it stays in each state for a shorter period of time and moves on to the next state more quickly.

Note that the chain will visit state i at time $H_0 + H_1 + \cdots + H_{i-1}$, the sum of the first i states holding times. Thus, the chain will visit all of the states by time

$$Y = \sum_{i=0}^{\infty} H_i.$$

The expectation of Y is given by $E(Y) = \sum_{i=0}^{\infty} E(H_i) = \sum_{i=0}^{\infty} 2^{-i} = 2 < \infty$, and it is concluded that on average all states $i = 0, 1, \ldots$ have been visited by time $t = 2$, a finite amount of time. The number of transitions in a finite interval of time is infinite. But, this implies that, all states will be visited in a finite amount of time, with probability 1, $P(Y < \infty) = 1$. Consequently, $X(Y + t) = \infty, t \geq 0$ with probability 1. Hence, $X(t) \to \infty$ as $t \to \infty$ with probability 1.

This is an example of an explosive Markov chain. A CTMC is called explosive if there is a positive probability that it will perform infinitely many transitions in a finite time; otherwise, it is called conservative. All CTMCs with finite state space are conservative Markov chains and most well-known birth-death processes are conservative Markov chains.

5.4 Pure Death Processes

A BDP is said to be a pure death process if the birth rates, $\lambda_i = 0$ for $i = 0, 1, \ldots, N$. Here state space is $\{0, 1, \ldots, N\}$. Initial condition is

$$\pi_N(0) = 1 \quad \text{and} \quad \pi_i(0) = 0, \ i \neq N$$

The state transition diagram is shown in Fig. 5.10

Consider the special case of pure death process $\{X(t), t \geq 0\}$, with death rates, $\mu_i = \mu$ for $i = 1, 2, \ldots, N$. The Kolmogorov forward equations (by setting $\lambda_i = 0$ for $i = 0, 1, \ldots, N-1$ and $\mu_i = \mu$ for $i = 1, 2, \ldots, N$ in BDP) are,

$$\begin{aligned}
\pi_0'(t) &= \mu\pi_1(t) \\
\pi_n'(t) &= -\mu\pi_n(t) + \mu\pi_{n+1}(t), \quad n = 1, 2, \ldots, N-1 \\
\pi_N'(t) &= -\mu\pi_N(t),
\end{aligned}$$

with these condition $\pi_N(0) = 1$.

Solving the above system of difference-differential equations,

$$\begin{aligned}
\pi_n(t) &= \frac{e^{-\mu t}(\mu t)^{N-n}}{(N-n)!}, \quad n = 1, 2, \ldots, N. \\
\pi_0(t) &= \left(\sum_{i=0}^{N-1} \frac{e^{-\mu t}(\mu t)^i}{i!}\right)^{-1}.
\end{aligned}$$

5.5 Poisson Processes

In this section, Poisson processes, which are a type of continuous-time stochastic process are described. This is a process that only changes value by means of jumps, and even then, the jumps are nicely spaced. It is the prototype of a pure jump process, and later it will be seen that it is the building block for an important class of stochastic processes known as Lévy processes. The Poisson process can be used to mimic real-life phenomena in which the interest is in counting the number of times a random event occurs in a certain time interval.

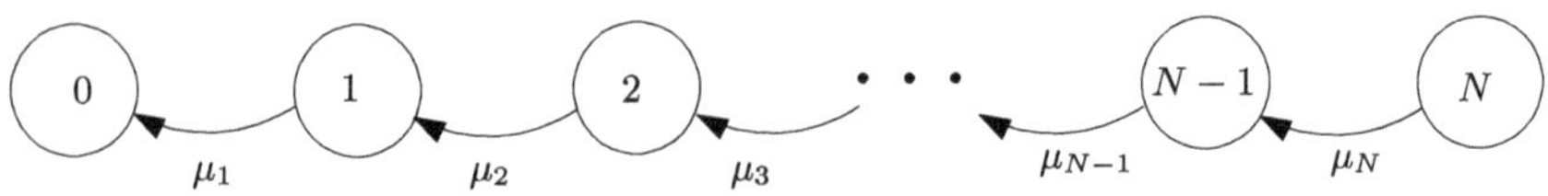

Fig. 5.10 State transition diagram for the pure death process

For example, the number of births in a hospital, the number of bank customers in a day, the number of phone calls made at a phone booth, the number of accidents on a specific stretch of highway, and so on. These processes are defined in terms of "exponential RVs," which are RVs having exponential distributions. Although there are some underlying assumptions, they appear to hold roughly in place.

5.5.1 *Homogeneous Poisson Processes*

In this section, two equivalent definitions of a Poisson process are provided with its basic properties.

Definition 5.2 (Poisson Process) A stochastic process $\{N(t), t \geq 0\}$ is said to be a Poisson process, with intensity or rate $\lambda > 0$, if the following conditions are satisfied:

1. $N(0) = 0$, a.s..
2. $N(t)$ has stationary and independent increments.
3. For every t, $N(t)$ has a Poisson distribution with parameter λt.

Remark 5.1 Poisson process is also referred as a counting process as $N(t)$ counts the number of arrivals upto time t.

Alternative Characterization of the Poisson Process
Consider an event E which may or may not occur at any time instant t. $\{N(t), t \geq 0\}$ is a non-negative integer-valued stochastic process where, $N(t)$ is the number of times E has occurred in the time period $(0, t]$. Define

$$\pi_n(t) = P(N(t) = n), \quad n = 0, 1, \ldots.$$

Furthermore, consider the following postulates:

1. $N(0) = 0$, i.e.,
$$\pi_0(0) = 1, \quad \pi_n(0) = 0 \text{ for } n \neq 0. \tag{5.63}$$
2. The number of events E that occurred in non-overlapping time intervals is independent.
3. $\pi_n(t)$ depends only on the length t of time interval and not on where this interval is situated. That is, $N(t_1 + t) - N(t_1)$ and $N(t)$ have same probability distribution for any $t \geq 0$.
4. In a time interval of infinitesimal length h, the probability of exactly one occurrence of E is $\lambda h + o(h)$ and that of more than one occurrence is $o(h)$.
   That is, if the interval between t and $t + h$ is of infinitesimal short duration h, then
$$\pi_1(h) = \lambda h + o(h),$$

$$\sum_{k=2}^{\infty} \pi_k(h) = o(h)$$

$$\Rightarrow \quad \pi_0(h) \;=\; 1 - \lambda h + o(h).$$

Theorem 5.1 *Under the above postulates (2), (3) and (4), $N(t)$ follows Poisson distribution with mean λt.*

Proof Consider $\pi_n(t+h)$ for $n \geq 0$ where n occurrences of event by epoch $t+h$ can happen in the following mutually exclusive ways

$$A_1, A_2, \ldots, A_n, A_{n+1}.$$

For $n \geq 1$
$A_1 : n$ occurrences of event by epoch t and no event occurred between t and $t+h$. Then

$$\begin{aligned} P(A_1) &= P(N(t)=n)P(N(t+h)=n) \\ &= \pi_n(t)\pi_0(h) \\ &= \pi_n(t)(1-\lambda h + o(h)). \end{aligned}$$

$A_2 : n-1$ occurrences of events by epoch t and 1 occurrence between t and $t+h$. Then

$$\begin{aligned} P(A_2) &= P(N(t)=n-1)P(N(t+h)=n) \\ &= \pi_{n-1}(t)\pi_1(h) \\ &= \pi_{n-1}(t)[(\lambda h) + o(h)]. \end{aligned}$$

For $n \geq 2$
$A_3 : n-2$ occurrence of events by epoch t and 2 occurrences between t and $t+h$. Then

$$P(A_3) = \pi_{n-2}(t)\pi_2(h) \leq \pi_2(h) \leq o(h)$$

and so on for $P(A_4), P(A_5), \ldots$. Thus,

$$\sum_{k=2}^{n} P(A_{k+1}) \;\leq\; \sum_{k=2}^{n} \pi_k(h) \;=\; o(h),$$

and therefore, for $n \geq 1$,

$$\begin{aligned}\pi_n(t+h) &= \sum_{i=1}^{n+1} P(A_i) = \pi_n(t)(1-\lambda h) + \pi_{n-1}(t)(\lambda h) + o(h),\\ \Longrightarrow \quad \frac{\pi_n(t+h)-\pi_n(t)}{h} &= -\lambda\pi_n(t) + \lambda\pi_{n-1}(t) + \frac{o(h)}{h}\\ \Longrightarrow \quad \lim_{h\to 0}\frac{\pi_n(t+h)-\pi_n(t)}{h} &= -\lambda\pi_n(t) + \lambda\pi_{n-1}(t) + \lim_{h\to 0}\frac{o(h)}{h}.\end{aligned}$$

Get a difference differential equation for $\pi_n(t)$ as follows

$$\pi_n'(t) = -\lambda(\pi_n(t) - \pi_{n-1}(t)),\ for\ \ n \geq 1. \tag{5.64}$$

For $n = 0$

$$\begin{aligned}\pi_0(t+h) &= \pi_0(t)\pi_0(h)\\ &= \pi_0(t)(1-\lambda h) + o(h)\\ \Longrightarrow \quad \frac{\pi_0(t+h)-\pi_0(t)}{h} &= -\lambda\pi_0(t) + \frac{o(h)}{h}\\ \Longrightarrow \quad \lim_{h\to 0}\frac{\pi_0(t+h)-\pi_0(t)}{h} &= -\lambda\pi_0(t) + \lim_{h\to 0}\frac{o(h)}{h}\end{aligned}$$

or

$$\pi_0'(t) = -\lambda\pi_0(t). \tag{5.65}$$

Using (5.63) and (5.65) and solving (5.64) for state probabilities $\pi_n(t)$, we get,

$$\pi_n(t) = \frac{e^{-\lambda t}(\lambda t)^n}{n!}, n = 0, 1, \ldots.$$

Thus, $\{N(t), t \geq 0\}$ follows a Poisson distribution with parameter or mean λt. □

Theorem 5.2 *Let $\{N(t), t \geq 0\}$ be a Poisson process with intensity $\lambda > 0$. Let $0 = t_0 < t_1 < \ldots < t_n$ are given. Prove that the increments $N(t_1) - N(t_0), N(t_2) - N(t_1), \ldots$ are independent and stationary and*

$$P(N(t_{j+1}) - N(t_j) = k) = \frac{\lambda^k (t_{j+1} - t_j)^k e^{-\lambda(t_{j+1}-t_j)}}{k!},\ k = 0, 1, \ldots.$$

Proof Left as an exercise.

5.5.1.1 Some Important Properties

Investigating the Poisson process unveils several crucial properties shaping its probabilistic nature. In this subsection, the essential properties are discussed.

1. **Auto-Covariance Function**

Theorem 5.3 *The auto-covariance function, namely, $cov(N(s), N(t))$, of the Poisson process $\{N(t),\ t \geq 0\}$ for $t, s \geq 0$ is given by*

$$cov(N(s), N(t)) = \lambda \min(s, t).$$

Proof

$$\begin{aligned}
E(N(s)N(t+s)) &= E(N(s)(N(s+t) - N(s) + N(s))) \\
&= E(N(s)(N(s+t) - N(s)) + E(N(s)^2) \\
&= E(N(s))E(N(s+t) - N(s)) + E(N(s)^2) \\
&\quad \text{(by independent increments property)} \\
&= \lambda s \lambda t + (\lambda s)^2 + \lambda s.
\end{aligned}$$

Therefore,

$$\begin{aligned}
cov(N(s), N(t+s)) &= E(N(s)N(s+t)) - E(N(s))E(N(s+t)) \\
&= \lambda s \lambda t + (\lambda s)^2 + \lambda s - \lambda s(\lambda(s+t)) \\
&= \lambda s.
\end{aligned}$$

Observe that the auto-covariance function is independent of time t, which is 1 because of the stationary increments property. In general,

$$cov(N(s), N(t)) = \lambda \min(s, t).$$

□

2. **Inter-arrival follows exponential distribution**

Theorem 5.4 *The inter-arrival time between the ith and $(i+1)$th occurrence of event has exponential distribution, i.e., Exp(λ), for $i = 0, 1, \ldots$. Also, they are independent RVs.*

Proof Let T_0 denote the time of first arrival. Then, for any $t \geq 0$,

$$\begin{aligned}
P(T_0 > t) &= P(N(t) = 0) \\
&= \frac{e^{-\lambda t}(\lambda t)^0}{0!} \\
&= e^{-\lambda t}.
\end{aligned}$$

Therefore T_0 follows Exp(λ). Suppose, for $i = 0, 1, \ldots$, t_i and t_{i+1} are instants of ith and $(i+1)$th arrival respectively, with $t_0 = 0$, and $T_i = t_{i+1} - t_i$ is the inter-arrival time. Due to the stationary increment property, for $i = 1, 2, \ldots$, T_i and T_0 have the same distribution. Hence, the inter-arrival time between ith arrival and $(i+1)$th arrival is also Exp(λ), i.e.,

$$P(T_i > t) = P\{N(t_{i+1}) - N(t_i) = 0|N(t_i) = i\} = P(N(t) = 0) = e^{-\lambda t}.$$

The proof of $T_i, i = 0, 1, \ldots$ are mutually independent RVs is left as an exercise problem.

□

Example 5.10 Suppose that, in an auto-insurance company, claims happen or arrive at times denoted by T_i satisfying $0 < T_1 < T_2 < \cdots$. Assume that, the inter-arrival times are independently exponential distributed with mean two days. Take $T_0 = 0$. Define, the total number of claims occurring till time t is

$$N(t) = \max\{i \geq 0|T_i \leq t\} \quad t \geq 0.$$

Find the probability that there is 10 claims for three days.
Solution: Given, that $N(t)$ follows Poisson with rate 0.5. Required probability is given by

$$P(N(3) = 10) = \frac{e^{-1.5}(1.5)^{10}}{10!} = 3.5457 \times 10^{-6}.$$

3. **Joint distribution of event times**

Now, the joint distribution of the event times $S_1, S_2, \ldots, S_n$ of the Poisson process $\{N(t), t \geq 0\}$, given that $N(t) = n$ is studied.

Theorem 5.5 *Given that $N(t) = n$, the n arrival times $S_1, S_2, \ldots, S_n$ have the same distribution as the order statistics corresponding to n independent RV which are uniformly distributed on the interval* $(0, t)$*, i.e.,*

$$f\{s_1, s_2, \ldots, s_n|N(t) = n\} = \begin{cases} \frac{n!}{t^n}, & 0 < s_1 < s_2 < \ldots < s_n < t \\ 0, & \textit{otherwise} \end{cases}.$$

Proof To obtain the conditional density of $S_1, \ldots, S_n$ given that $N(t) = n$, note that for $0 < S_1 < \ldots < S_n < t$, the event that $\{S_1 = s_1,\ S_2 = s_2, \ldots, S_n = s_n|N(t) = n\}$ is equivalent to the event that the first $n + 1$ inter-arrival times satisfies $\{T_1 = s_1,\ T_2 = s_2 - s_1, \ldots, T_n = s_n - s_{n-1}, T_{n+1} > t - s_n\}$.

Now, by using the property that $T_1, T_2, \ldots, T_n$ are IID exponential RVs having mean $\frac{1}{\lambda}$, for $0 < s_1 < s_2 < \cdots < s_n < t$

$$f\{s_1, s_2, \ldots, s_n|N(t) = n\} = \frac{\lambda e^{-\lambda s_1} \lambda e^{-\lambda(s_2 - s_1)} \ldots \lambda e^{-\lambda(s_n - s_{n-1})} e^{-\lambda(t - s_n)}}{e^{-\lambda t} \frac{(\lambda t)^n}{n!}}.$$

Hence,

$$f\{s_1, s_2, \ldots, s_n|N(t) = n\} = \begin{cases} \frac{n!}{t^n}, & 0 < s_1 < s_2 < \cdots < s_n < t \\ 0 & \text{otherwise} \end{cases}.$$

which proves the result.

4. **Superposition**
   When two Poisson processes merge to form a combined process, superposition occurs spontaneously. Customers that come into a bank, for example, can be either male or female, generating a Poisson process. As a result, the process that counts a customer's arrival is a superposition of the two processes. If there are n independent Poisson processes $\{N_i(t), t \geq 0\}$, $i = 1, 2, \ldots, n$ with respective rates, $\lambda_i, i = 1, 2, \ldots, n$, then the combined process that is characterized by

$$N(t) = N_1(t) + N_2(t) + \cdots + N_n(t),$$

   is the superposition of the n processes. The superposed process is determined in the next theorem.
   For instance, suppose there are two independent Poisson arrival processes, with the rates λ_1 and λ_2 respectively. Then the superposition of the two processes will again be a Poisson process with rate $(\lambda_1 + \lambda_2)$.

Theorem 5.6 *Suppose there are n independent Poisson processes* $\{N_i(t), t \geq 0\}$, $i = 1, 2, \ldots, n$ *with respective rates,* λ_i, $i = 1, 2, \ldots, n$, *then the superposed process defined by*

$$N(t) = N_1(t) + N_2(t) + \cdots + N_n(t),$$

is again a Poisson process with rate λ *where*

$$\lambda = \lambda_1 + \lambda_2 + \cdots + \lambda_n.$$

Proof We restrict the proof for two Poisson processes. However, the proof can be extended to n independent Poisson processes.

Let $N_1(t)$ and $N_2(t)$ be two independent Poisson arrival processes, with the rates λ_1 and λ_2, respectively. The arrival in the superimposed process will refer to either of two types of arrivals in two Poisson processes. Thus the combined process is

$$N(t) = N_1(t) + N_2(t).$$

Since both the independent processes $\{N_i(t), t \geq 0\}, i = 1, 2$, have stationary and independent increments, it follows that $\{N(t), t \geq 0\}$ inherits this property. It is quite obvious that $N(t) \in \{0, 1, \ldots\}$ and

$$\begin{aligned}
P(N(t) = n) &= \sum_{r=0}^{n} P(N_1(t) = r)P(N_2(t) = n - r) \\
&= \sum_{r=0}^{n} \frac{e^{-\lambda_1 t}(\lambda_1 t)^r}{r!} \frac{e^{-\lambda_2 t}(\lambda_2 t)^{n-r}}{(n-r)!} \\
&= \frac{e^{-(\lambda_1+\lambda_2)t}\{(\lambda_1 + \lambda_2)t\}^n}{n!}, \quad n = 0, 1, \ldots.
\end{aligned}$$

Thus $N(t)$ follows Poisson distribution with parameter $(\lambda_1+\lambda_2)t$ or the rate $\lambda_1+\lambda_2$. This proves the theorem for $n=2$. By mathematical induction principle one can prove the required result.

5. **Decomposition**

   Decomposition is the opposite of superposition in the sense that start with a single Poisson process and split it to create two or more counting processes. For example, the original process counts the number of calls received at a customer care center while the two decomposed processes might count the number of domestic and international calls separately. The next theorem describes the nature of these decomposed counting processes.

Theorem 5.7 *A Poisson process is produced by a random substream of a Poisson process. Assume that the number of times an event E occurs in a time interval of length t, $N(t)$, is a Poisson process with rate λ. Assume that each instance of E has a fixed chance of being recorded (p), and that the recording of one instance is unrelated to any other instances or to $N(t)$. $M(t)$ is also a Poisson process with the rate λp if $M(t)$ is the number of occurrences recorded in an interval of length t.*

Proof The event $\{M(t)=n\}$ can occur in mutually exclusive ways as follows:
A_r : E occurs $(n+r)$ times by epoch t and exactly n out of $(n+r)$ occurrences are recorded, probability of each occurrence to be recorded being p, $(r=0,1,\ldots)$.

$$\begin{aligned}P(A_r) &= P(E \text{ occurs } (n+r) \text{ times by epoch } t)\\ &\quad \times P(n \text{ occurrences are recordedgiven that } n+r \text{ number of occurrences})\\ &= \frac{e^{-\lambda t}(\lambda t)^{n+r}}{(n+r)!}\times\binom{n+r}{n}p^n(1-p)^r.\end{aligned}$$

Hence

$$P(M(t)=n)=\sum_{r=0}^{\infty}P(A_r)=\frac{e^{-\lambda pt}(\lambda pt)^n}{n!}.$$

Remark 5.2 Let $M_1(t)$ be the number of events not recorded in an interval of length t, then $M_1(t)$ is also a Poisson process with rate $\lambda(1-p)$ and $M(t)$ and $M_1(t)$ are independent.
Therefore, a Poisson process $\{N(t),\ t\geq 0\}$ with rate λ is decomposed into two independent Poisson process $\{M_1(t),\ t\geq 0\}$ and $\{M_2(t),\ t\geq 0\}$ with rates $\lambda(1-p)$ and λp respectively.

6. **Conditional Distribution**
   Here, we discuss an important result regarding the conditional distributions in the Poisson process.

Theorem 5.8 *Let $\{N(t),\ t\geq 0\}$ be a Poisson process with rate λ and $s<t$, then the conditional distribution of $\{N(s)|N(t)=n\}$ is Binomial$(n,\frac{s}{t})$ distribution.*

Proof For $k = 0, 1, \ldots, n$

$$
\begin{aligned}
P\{N(s) = k | N(t) = n\} &= \frac{P(N(s) = k,\ N(t) = n)}{P(N(t) = n)} \\
&= \frac{P(N(s) = k,\ N(t) - N(s) = n - k)}{P(N(t) = n)} \\
&= \frac{P(N(s) = k) P(N(t) - N(s) = n - k)}{P(N(t) = n)} \\
&= \frac{\frac{e^{-\lambda s}(\lambda s)^k}{k!} \times \frac{e^{-\lambda(t-s)}(\lambda(t-s))^{n-k}}{(n-k)!}}{\frac{e^{-\lambda t}(\lambda t)^n}{n!}} \\
&= \binom{n}{k} \left(\frac{s}{t}\right)^k \left(1 - \frac{s}{t}\right)^{n-k}.
\end{aligned}
$$

This proves the result.

7. **PASTA Property**

   The Poisson process has the so-called PASTA property (Poisson Arrivals See Time Averages): for instance, customers with Poisson arrivals see the system as if they came into the system at a random instant of time. This property will be discussed in detail in Chap. 6.

In this section, it is discussed some of the examples associated with time-homogeneous Poisson process.

Example 5.11 Consider about the vehicle insurance claims that the insurer has received. Assume that there are 10 claims on average filed each day. Assume as well that this pace is consistent all year long and throughout the day. Moreover assume that there can only be one claim in a suitably little period of time. What is the probability that a particular day will only have one or two claims reported? What is the probability that it will be less than two hours before the next reported claim?
Solution: The given situation can be modeled as Poisson process. The number of arrivals of car insurance claims to the insurer in time intervals of length t is a Poisson process,$\{N(t);\ t \geq 0\}$, with rate λ as 10 per day.

Required probability that there are less than 2 claims reported on a given day

$$
\begin{aligned}
P(N(t+1) - N(t) \leq 2) &= P(N(1) \leq 2) \\
&= P(N(1) = 1) + P(N(1) = 2) \\
&= \frac{e^{-\lambda \times 1}(\lambda \times 1)^0}{0!} + \frac{e^{-\lambda \times 1}(\lambda \times 1)^1}{1!} \\
&= e^{-\lambda}(1 + \lambda) \\
&= 11e^{-10}.
\end{aligned}
$$

Required probability that time until the next reported claim is less than 2 h, given at most 2 claims till now (2 hours $= \frac{2}{24}$ days) is $P(T < \frac{2}{24})$, where T is inter-arrival time which follows an exponential distribution with parameter λ. Thus, $P(T < \frac{2}{24}) = 1 - e^{-\lambda\frac{2}{24}} = 1 - e^{-\frac{5}{6}}$.

Example 5.12 Consider the situation of waiting for a bus on the bus stand. Assume that the bus arrivals follow a Poisson process with rate 5 per minute. That is, the mean inter-arrival time between the buses is 5 minutes. Suppose a person come to the bus stand at some time. What is the average waiting time to get the bus?

Solution: Suppose one arrives the bus stand at time t. Let W be the waiting time. Let $X_1, X_2, \ldots$ be the inter-arrival times. These are IID RV each follow exponential distribution with parameter λ. Since exponential distribution has memoryless property, hence the waiting time W will also follow exponential distribution with parameter λ. Hence, average waiting time is given by

$$E[W] = \frac{1}{\lambda} = 5.$$

Example 5.13 Consider X_t be the random telegraph signal, which jumps between two states, 0 and 1, according to the following rules. At time $t = 0$, the signal X_t has equal probability for the two states, i.e., $P(X_0 = 0) = P(X_0 = 1) = \frac{1}{2}$. Let the switching times be decided by a Poisson process $\{Y_t, t \geq 0\}$ with parameter λ which is independent of X_0. At time t, the signal

$$X_t = \frac{1}{2}\left(1 - (-1)^{X_0+Y_t}\right), \quad t > 0.$$

Check whether $\{X_t,\ t \geq 0\}$ is covariance/wide sense stationary or not?

Solution: The mean of X_t is given by

$$\begin{aligned} E[X_t] &= E\left[\frac{1}{2}\left(1 - (-1)^{X_0+Y_t}\right)\right] \\ &= \frac{1}{2} - \frac{1}{2}E[(-1)^{X_0}]E[(-1)^{Y_t}] \\ &= \frac{1}{2}. \end{aligned}$$

Now,

$$\begin{aligned} E[X_s X_t] &= \frac{1}{4}E\left[\left(1 - (-1)^{X_0+Y_s}\right)\left(1 - (-1)^{X_0+Y_t}\right)\right] \\ &= \frac{1}{4}\left(1 - E\left((-1)^{X_0+Y_s}\right) - E\left((-1)^{X_0+Y_t}\right) + E[(-1)^{2X_0+Y_s+Y_t}]\right) \\ &= \frac{1}{4}. \end{aligned}$$

The covariance function is given by

$$cov(X_s, X_t) = E[X_s X_t] - E[X_s]E[X_t] = 0.$$

Further, $E[X_t^2] = \frac{1}{4} < \infty$. Hence $\{X_t,\ t \geq 0\}$ is wide-sense stationary.

5.5.2 Non-homogeneous Poisson Processes

In this section, a generalization of the Poisson process by relaxing the requirement that the increments be stationary are studied. Let $\lambda : [0, \infty) \rightarrow [0, \infty)$ be a given function which is integrable over finite intervals and define

$$\Lambda(t) = \int_0^t \lambda(x)dx. \tag{5.66}$$

Definition 5.3 (Non-Homogeneous Poisson Process) A counting process $\{N(t), t \geq 0\}$, with $N(0) = 0$ is called a non-homogeneous Poisson process (NPP) if it has independent but not necessarily stationary increments and $N(t)$ follows Poisson distribution with parameter $\Lambda(t), t \geq 0$.

Then for $s < t$, $N(t) - N(s)$ has a Poisson distribution with parameter $\int_s^t \lambda(x)dx$. Let $N(t)$ denote the number of customers arriving in $(0, t)$. The arrival process $\{N(t); t \geq 0\}$ has a Poisson distribution, i.e.,

$$P(N(t) = i) = \frac{(\Lambda(t))^i e^{-\Lambda(t)}}{i!}. \quad i = 0, 1, \ldots$$

The next theorem without proof provides a necessary and sufficient condition for a counting process to be an NPP.

Theorem 5.9 *A counting process $\{N(t), t \geq 0\}$ is a $NPP(\lambda(\cdot))$ if and only if a.s..*

1. $N(0) = 0$, *a.s..*
2. $\{N(t), t \geq 0\}$ *has independent increments and*

$$P(N(t+h) - N(t) = 0) = 1 - \lambda(t)h + o(h),$$
$$P(N(t+h) - N(t) = 1) = \lambda(t)h + o(h),$$
$$P(N(t+h) - N(t) = k) = o(h),\ k = 2, 3, \ldots.$$

Remark 5.3 A non-homogeneous Poisson process is a Markov process.

The non-homogeneous Poisson process can be shown to have the following properties:

1. Conditioning on the number, given the total number of arrivals $N(t) = n$ in the interval $(0, t)$ from a non-homogeneous Poisson process, the arrival instants of these n arrivals are distributed independently in the interval $(0, t)$ with the density function as defined in (5.68).
2. Superposition. The superposition of two non-homogeneous Poisson processes with intensities $\lambda_1(t)$ and $\lambda_2(t)$ is a non-homogeneous Poisson process with intensity $\lambda(t) = \lambda_1(t) + \lambda_2(t)$.
3. Random selection. A random selection from a non-homogeneous Poisson process with intensity $\lambda(t)$ such that each arrival is selected, independent of the others, with the probability $p(t)$ (note, may depend on time) results in a non-homogeneous Poisson process with intensity $p(t)\lambda(t)$.
4. Random split. If a non-homogeneous Poisson process with intensity $\lambda(t)$ is randomly split into two subprocesses with the probabilities $p_1(t)$ and $p_2(t)$, where $p_1(t) + p_2(t) = 1$, then the resulting subprocesses are independent non-homogeneous Poisson processes with intensities $p_1(t)\lambda(t)$ and $p_2(t)\lambda(t)$.

Example 5.14 Let S_n be the time of the n-th occurrence. Find the distribution of S_n. Further, show that $T_n = S_n - S_{n-1}$, the nth inter-arrival time is not independent RV.
Solution:

$$P(S_n > t) = P(N(t) \le n-1) = \sum_{j=0}^{n-1} \frac{(\Lambda(t))^j e^{-\Lambda(t)}}{j!} = P(G_n > \Lambda(t)),$$

where G_n is a Gamma distributed RV with parameters n and 1.
When $\lim_{t\to\infty} \Lambda(t) = \infty$,

$$P(S_n = \infty) = \lim_{t\to\infty} P(S_n > t) = P(G_n > \lim_{t\to\infty} \Lambda(t)) = 0.$$

S_n is a continuous type RV with the probability density function

$$f_{S_n}(t) = \frac{(\Lambda(t))^{n-1} e^{-\Lambda(t)} \Lambda'(t)}{(n-1)!} = \frac{(\Lambda(t))^{n-1} e^{-\Lambda(t)} \lambda(t)}{(n-1)!}, \quad t \ge 0.$$

When $\lim_{t\to\infty} \Lambda(t) < \infty$, S_n is a mixed type RV with

$$P(S_n = \infty) = P(G_n > \lim_{t\to\infty} \Lambda(t)) > 0,$$

and density function of its continuous part is given by

$$f_{S_n}(t) = \frac{(\Lambda(t))^{n-1} e^{-\Lambda(t)} \lambda(t)}{(n-1)!}, \quad t \ge 0.$$

Let $T_n = S_n - S_{n-1}$ be the n-th inter-arrival time. For a non-homogeneous Poisson process $\{T_n, n = 1, 2, \ldots\}$ are not necessarily independent RVs. For $0 \le s \le t$,

$$\begin{aligned}
P\{T_{n+1} > t \mid S_n = s\} &= P\{S_{n+1} > s + t \mid S_n = s\} \\
&= P\{N(s + t) = n \mid S_n = s\} \\
&= P\{N(s + t) - N(s) = 1 \mid S_n = s\} \\
&= P(N(s + t) - N(s) = 1) = e^{-(\Lambda(s+t)-\Lambda(s))}.
\end{aligned}$$

Notice $S_n = t$ depends on the non-homogeneous Poisson process until time t. Hence, $\{N(s + t) - N(s) = 1\}$ and $\{S_n = t\}$ are independent RVs.

Example 5.15 Let $\{N(t), t \geq 0\}$ be a non-homogeneous Poisson process with intensity function $\lambda(t) = t$. Then the "time-changed" process $\{N(t) = N(\sqrt{2t}), t \geq 0\}$ can easily be seen to be a homogeneous Poisson process with rate 1. Further, show that $N(t) = N(t^2/2), t \geq 0$.

Solution: Firstly, we will show the result that if $\{N(t), t \geq 0\}$ is a non-homogeneous Poisson process with mean $\Lambda(t) = \int_0^t \lambda(s)\mathrm{d}s$. Then, $\{N(\Lambda^{-1}(t)), t \geq 0\}$ is a homogeneous Poisson process with intensity 1.

Since, $\Lambda(t) = \int_0^t \lambda(s)\mathrm{d}s$, and $\lambda : [0, \infty) \to [0, \infty)$, $\Lambda(\cdot)$ is a monotone increasing function which implies $\Lambda^{-1}(t)$ exists and $t = \Lambda(\Lambda^{-1}(t)) = \int_0^{\Lambda^{-1}(t)} \lambda(s)\mathrm{d}s$.

Consider

$$\begin{aligned}
P\left(N(\Lambda^{-1}(t)) = n\right) &= \frac{\left(\int_0^{\Lambda^{-1}(t)} \lambda(s)\mathrm{d}s\right)^n}{n!} e^{-\int_0^{\Lambda^{-1}(t)} \lambda(s)\mathrm{d}s} \\
&= \frac{t^n}{n!} e^{-t}.
\end{aligned}$$

In this case, $\Lambda^{-1}(t) = \sqrt{2t}$ which implies,

$$\begin{aligned}
P\left(N(\sqrt{2t}) = n\right) &= P\left(N(\Lambda^{-1}(t)) = n\right) \\
&= \frac{t^n}{n!} e^{-t},
\end{aligned}$$

and hence, the "time-changed" process $\{N(t) = N(\sqrt{2t}), t \geq 0\}$ is a homogeneous Poisson process with rate 1.

Further, consider $N(t^2/2)$, with mean $\int_0^{t^2/2} 1\mathrm{d}s = t^2/2$ which implies

$$\begin{aligned}
P\left(N(t^2/2) = n\right) &= \frac{(t^2/2)}{n!} e^{-(t^2/2)} \\
&= \frac{\left(\int_0^t s\mathrm{d}s\right)^n}{n!} e^{-\int_0^t s\mathrm{d}s} \\
&= \frac{(\Lambda(t))^n e^{-\Lambda(t)}}{n!},
\end{aligned}$$

which implies $N(t) = N(t^2/2), t \geq 0$.

Example 5.16 For a non-homogeneous Poisson process, the intensity function is given by

$$\lambda(t) = \begin{cases} 5 \text{ if } & t \in (1,2], (3,4], \ldots \\ 3 \text{ if } & t \in (0,1], (2,3], \ldots \end{cases}.$$

Find the probability that the number of observed occurrences in the time period $(1.25, 3]$ is more than two.

Solution: $N(3) - N(1.25)$ has a Poisson distribution with mean

$$m(3) - m(1.25) = \int_{1.25}^{3} \lambda(t)\mathrm{d}t = \int_{1.25}^{2} 5\mathrm{d}t + \int_{2}^{3} 3\mathrm{d}t = 6.75.$$

Hence,

$$\begin{aligned} P(N(3) - N(1.25) > 2) &= -e^{-6.75}\left(1 + 6.75 + \frac{(6.75)^2}{2}\right) + 1 \\ &= 0.9642. \end{aligned}$$

Example 5.17 A flexible model provides the expected number of failures $M(t)$ in the first t hours by

$$M(t) = at^b, \text{ for } a, b > 0.$$

The repair rate for this model is given by

$$m(t) = abt^{b-1} = \alpha t^{-\beta}, \text{ for } \alpha > 0, \beta < 1.$$

For the homogeneous Poisson process the repair rate will be $m(t) = \lambda$. For the non-homogeneous Poisson process the intensity becomes $\lambda(t) = m(t) = \alpha t^{-\beta}$. When $b = 1$ or $\beta = 0$, the model reduces to homogeneous one. The probabilities of a given number of failures for this model is given by

$$P(N(T) = k) = \frac{M(T)^k}{k!} e^{-M(T)}$$

$$P(N(T) = k) = \frac{(aT^b)^k}{k!} e^{-aT^b} = \frac{a^k T^{bk}}{k!} e^{-aT^b}.$$

For more discussions, readers may refer to [1], [2], [4], [5], [6], [7], [8], [9], [10], [11], [12], [13], [15], [16], [17], [18], [19], [20], [21], [22].

5.6 Exercise Problems

1. The birth-death process is called a pure death process if $\lambda_i = 0$ for all i. Suppose $\mu_i = i\mu, i = 1, 2, \ldots$ and initially $X_0 = n$. Show that X_t has $B(n, p)$ distribution with $p = e^{-\mu t}$.
2. The birth-death process $\{X_t, t \geq 0\}$ is called a pure birth process if $\mu_i = 0$ for every i and suppose that $\lambda_i = i\lambda$ for $i = 1, 2, \ldots$. For $X_0 = n$, show that

$$P\{X_t = i|X_0 = n\} = P_{ni}(t) = \binom{i-1}{i-n} e^{-n\lambda t} \left(1 - e^{-\lambda t}\right)^{i-n}, i = n, n+1, \ldots.$$

3. Consider a birth-death process with birth rate $\lambda_i = (i+1)\lambda, i \geq 0$ and $\mu_i = i\mu$, $i \geq 0$. Find the expected time to go from state 0 to 1.
4. Suppose that the probability of an accident happening in time t to $t + \Delta t$ is $(\nu + j\lambda)\Delta t + o(\Delta t)$ given that $j \geq 0$ accidents have occurred previous to time t, for $\nu, \lambda > 0$. Let $P_{ij}(t) = P\{X(t) = j|X(0) = i\}$, $i, j = 0, 1, \ldots$.

   (a) Write the forward Kolmogorov equations for the Markov process $\{X(t), t \geq 0\}$.
   (b) Prove that

$$P_{0n+1}(t) = \frac{\nu(\nu+\lambda)\ldots(\nu+n\lambda)}{(n+1)!\lambda^{n+1}} e^{-\nu t} \left(1 - e^{-\lambda t}\right)^{n+1}, \quad n = 0, 1, \ldots$$

5. In Example 5.3 by taking Laplace transforms of the differential-difference equations, prove that

$$P_0(t) = \sum_{m \geq 0} \left(e^{-\mu t} \frac{(\mu t)^m}{m!}\right)^2 = e^{-2\mu t} I_0(2\mu t),$$

   which is the form taken by (5.15) when $\lambda = \mu$.
6. Let τ be a finite stopping time, i.e., $P(\tau < \infty) = 1$. It is well known that the Radom-Nikodym derivative of the measure generated by the trajectories of the process $\{X(t), t \geq 0\}$ up to time τ, with respect to a σ-finite measure γ_τ equals (Stefanov (1991)):

$$\exp\left\{\sum_{i,j=0, i\neq j}^{\infty} N_{i,j}(\tau) \ln p_{i,j} - \sum_{i=1}^{N} p_{ii} T_i(\tau)\right\},$$

   where $\{p_{ij}\}_{i,j=0}^{N}$, $p_{ii} = \sum_{j=0, i\neq j} p_{ij}$, is the intensity matrix, $N_{i,j}(\tau)$ is the number of one-step transitions from state i to state j in the time interval $[0, \tau]$ and $T_i(\tau)$ is the sojourn time in state i during the time interval $[0, \tau]$.
   Prove that the mean passage times is given by

$$\sum_{i=0}^{n} E\left[T_i(\tau_{n,n+1})\right] = \frac{n+1}{\lambda_n},$$

$$\sum_{i=0}^{n} E\left[T_i(\tau_{n,n-1})\right] = \frac{m-n+1}{\mu_n},$$

$$\text{and} \quad \sum_{i=0}^{n} E\left[T_i(\tau_{n,n})\right] = \frac{1+m}{\lambda+\mu}.$$

7. Consider, the irreducible birth-death process, then the
   (a) state is transient if and only if

$$\sum_{n=1}^{\infty} \frac{\mu_1 \ldots \mu_n}{\lambda_1 \ldots \lambda_n} < \infty, \tag{5.67}$$

   (b) state is positive recurrent if and only if

$$\sum_{n=1}^{\infty} \frac{\lambda_1 \ldots \lambda_n}{\mu_1 \ldots \mu_n} < \infty, \tag{5.68}$$

   (c) state is null recurrent if and only if neither (5.67) nor (5.68) hold.
8. Consider a birth-death process with state space $S = \{0, 1, 2\}$ and rates $\lambda_0 = \mu_2 = \lambda$ and $\lambda_1 = \mu_1 = 0$. Find $P\{X(t) = n|X(0) = 0\}, \quad n = 0, 1, 2.$
9. Consider a birth-death process $\{X(t), t \geq 0\}$ with state space $S = \{0, 1, \ldots\}$ and birth rates are of the form $\lambda_n = n\lambda, \quad n = 0, 1, \ldots$ and death rates are of the form $\mu_n = n\mu,\ n = 0, 1, \ldots$ where λ and μ are non-negative constants. Assume $P_{ii}(j) = 1$ for fixed i. Take $E(X(t)) = \sum_{k=0}^{\infty} k P_{ik}(t)$. Prove that $E(X(t)) = ie^{(\lambda-\mu)t}$.
10. For a non-homogeneous Poisson process, the intensity function is given by

$$\lambda(t) = \begin{cases} t, & 0 < t \leq 5, \\ 4, & 5 < t < \infty \end{cases}.$$

   If $S_4 = 3$, calculate the probability that $S_6 > 5$.
11. Let $\{\tilde{N}(t), t \geq 0\}$ be a non-homogeneous Poisson process with intensity function $\lambda(\cdot)$. Suppose $\lambda(t) > 0$ for all $t > 0$ and $\int_0^{\infty} \lambda(t)\mathrm{d}t = \infty$, such that, the function $\Lambda(t) = \int_0^t \lambda(s)ds$ is a strictly increasing, continuous function on $(0, \infty)$ onto $(0, \infty)$ and therefore has an inverse function Λ^{-1}.
   Prove that, $\{N(t) = \tilde{N}(\Lambda^{-1}), t \geq 0\}$ can be seen to be a homogeneous Poisson process with rate 1 and $\tilde{N}(t) = N(\Lambda(t))$.
12. Patients see doctors according to a Poisson process at a rate of 8 per hour, and the average time it takes a doctor to examine a patient is 6 minutes. The doctor

sees each new patient as they arrive. Calculate the likelihood that patients will have to wait when they arrive.

13. Suppose there are n identical machines operating independently and serviced by a single repair crew. If a machine breaks down while another is being repaired it must wait its turn before repairs can start. Assume each machine has an operating time exponential with mean $\frac{1}{\mu}$ and a repair time exponential with mean $\frac{1}{\lambda}$. Let $X(t)$ = no. of machines in operating condition at time t. Model $X(t)$ as a Markov chain with state space $E = \{0, 1, \ldots, N\}$. Determine rate matrix Λ and the forward Kolmogorov equations. Determine the limiting equilibrium probability distribution of the process.
14. Assume that the passengers arrive at a bus station is a Poisson process with rate λ. The bus only departs after a deterministic time T. Let W be the combined waiting time for all the passengers. Compute the expected value of W.
15. Let $\{N(t), t \geq 0\}$ be a Poisson process with rate λ. Let X_1 be the epoch of the first arrival and X_n be the inter-arrival time between the $n-1$th and nth arrival. Let S_{n-1} be the epoch of the $n-1$th arrival.

    (a) Show that $P(X_1 > x) = e^{-\lambda x}$.
    (b) Show that $P\{X_n > x | S_{n-1} = \tau\} = e^{-\lambda x}$.

16. Suppose the arrival at a counter form a time-homogeneous Poisson process with parameter λ and suppose each arrival is of type A or of type B with respective probabilities p and $1-p$. Let $X(t)$ be the type of the last arrival before time t. Write down the forward Kolmogorov equations for the stochastic process $\{X(t), t \geq 0\}$. Find the time-dependent system state probabilities.
17. Let $Y_1, Y_2, Y_3, \ldots$ be a sequence of IID RVs with mean $E(Y_i) = 0$ and $Var(Y_i) = 4$. Define the discrete-time random process $\{X_n; n \in N\}$, where $X_n = Y_1 + Y_2 + \ldots + Y_n$ for all $n \in N$. Find the auto correlation $R_X(m, n)$ for all $m, n \in N$.
18. Let $\{N(t), t \geq 0\}$ be a Poisson process with parameter λ. Let T_0 denote the time of the first event and T_n denote the inter-arrival time between nth and $(n+1)$th arrivals.

    (a) Prove that $T_n, n = 0, 1, \ldots$ are mutually independent RVs.
    (b) Find $P\{T_1 > t | T_0 = s\}$.

19. Let $\{X(t), t \geq 0\}$ be a time-homogeneous Poisson process with $X(0) = j$. Consider the RV

$$T_j = \inf\{t : X(t) = j + 1\}.$$

    (a) Find the distribution of T_j.
    (b) Find the joint distribution of $(T_{2023}, T_{2024}, T_{2025})$.

20. Assume that a line of taxis is waiting for passengers to arrive. According to a Poisson process, these taxis pick up an average of 60 people every hour. Once two passengers have been gathered or three minutes have passed since the first passenger boarded the cab, the taxi will depart. Let's say you are the first person to enter the taxi. How long does it often take you to wait for the departure?

21. Let $N(t)$ be the RV denoting the number of events that occur upto and including time t. Let $\{N(t), t \geq 0\}$ be a Poisson process with parameter λ. Let T_i be the inter-arrival times of the ith event $i = 1, 2, \ldots$. Which of the following statements are TRUE?

    (a) $\sum_{i=1}^{n} T_i$ follows $B(n, \lambda)$.
    (b) $P(T_2 \leq t) = 1 - e^{-\lambda t}, 0 \leq t < \infty$.
    (c) T_i's are independent.
    (d) $cov(T_i, T_j) > 0$, for $i \neq j$.

22. Let us assume that cars arrive according to a Poisson process at rate 5 per hour. Assume each car will pick up a hitchhiker (usually stranger, for a ride in their car) with the probability $\frac{1}{5}$. You are third in line.

    (a) What is the probability that no car arrives in first 15 minutes?
    (b) What is the probability that you will have to wait for more than 2 h?

23. Let $\{N(t), t \geq 0\}$ be a Poisson process with rate λ. Consider, the compound Poisson process $Z(t) = Y_1 + Y_2 + \cdots + Y_{N(t)}$ where $\{Y_n, n = 1, 2, \ldots\}$ is a sequence of IID RVs. Prove that, the increment $Z(t) - Z(s)$ has characteristic function $e^{(\varphi_{Y_1}(x)-1)\lambda(t-s)}$, where $\varphi_{Y_1}(x)$ denotes the characteristic function of Y_1.
24. Let $\{N(t), t \geq 0\}$ be a Poisson process with rate 4. Find

    (a) Find $P(N(2) = 3, N(6) = 3)$.
    (b) Find the expected number of arrivals between time 2 and time 3 given that there was already 1 arrival by time 2.

25. Let $\{N(t), t \geq 0\}$ be a Poisson process.

    (a) Find the distribution of current life, δ_t and total life β_t.
    (b) Prove that the joint distribution of ν_t and δ_t is given by
    $$P(\nu_t > x, \delta_t > y) = \begin{cases} e^{-\lambda(x+y)}, & \text{if } x > 0, 0 < y < t \\ 0, & \text{if } y \geq t \end{cases}.$$

26. Consider a Poisson process with parameter λ. Let T_1 be the time of occurrence of the first event and let $N(T_1)$ denote the number of events occurred in the next T_1 units of time. Find the mean and variance of $N(T_1)T_1$.
27. Suppose $\{N(t), t \geq 0\}$ is a Poisson process and write $X(t) = N(t-)$. Is $N(1) - X(1-t)$ a Poisson process on $[0, 1]$? Why or why not?
28. Assume that passengers arrive at a bus station as a Poisson process with mean 6 minutes. A particular bus departs after a deterministic time $T = 5$ minutes. Let W be the combined waiting time for all passengers. Prove that, $E(W) = 75$ minutes. Suppose, assume T, the bus arrival time follows exponential distribution with mean 5 minutes, independent of the passengers' arrivals. Prove that, $E(W) = 150$ minutes.
    Hint: Find, conditional expectation of W given N, then find the unconditional expectation of W.

29. Items arrive to a service station according to a Poisson process $\{N(t), t \geq 0\}$ with rate λ. The items are stored until m have accumulated. Then the m items are served in a batch. The service time is exponential distributed with rate μ. During the service, items continue to arrive. There is a cost h_i per unit time of holding i items in the system. Assume the station is empty at time 0. Find the expected cost C_1 of holding the customers until m have arrived. Find the expected cost C_2 for holding the added arrivals in the system during the service.
30. In the Yule process, if instead of splitting into two particles each particle splits into exactly k particles ($k > 2$), then prove that the probability generating function is given by

$$\phi(t; s) = se^{-at}\left[1 - (1 - e^{-a(k-a)t}s^{k-1}\right]^{-1/(k-1)}.$$

References

1. Brown, M.: Spectral analysis, without eigenvectors, for Markov chains. Probab. Eng. Inform. Sci. **5**(2), 131–144 (1991)
2. Castaneda, L.B., Arunachalam, V., Dharmaraja, S.: Introduction to Probability and Stochastic Processes with Applications. Wiley, New Jersey (2012)
3. Chiang, C.L.: An introduction to stochastic processes and their applications. John Wiley, New York (1980)
4. Conolly, B.W., Parthasarathy, P.R., Dharmaraja, S.: A chemical queue. Math. Sci. **21**, 83–91 (1997)
5. Dharmaraja, S., Trivedi, K.S., Logothetis, D.: Performance modeling of wireless networks with generally distributed handoff interarrival times. Comp. Commun. **26**(15), 1747–1755 (2003)
6. Gaver, D.P., Jacobs, P.A., Latouche, G.: Finite birth-and-death models in randomly changing environments. Adv Appl Prob **16**(4), 715–731 (1984)
7. Haring, G., Marie, R., Puigjaner, R., Trivedi, K.S.: Loss formulas and their applications to optimization for cellular network. IEEE Trans. Vehic. Technol. **50**(3), 664–673 (2001)
8. Karlin, S., McGregor, J.: The classification of birth and death processes. Trans. Amer. Math. Soc. **86**, 366–400 (1957)
9. Kijima, M.: in *Markov Processes for Stochastic Modeling* (Vol. 6) (CRC Press, Boca Raton, 1997)
10. Ledermann, W.: G.E.H. Reuter, Spectral theory for the differential equations of simple birth and death processes. Philos. Trans. R. Soc. Londn. Ser. A Math. Phys. Sci. **246**(914), 321–369 (1954)
11. Muir, T.: The Theory of Determinants in the Historical Order of Development. Dover, New York (1960)
12. Murhy, J.A., O'donohoe, M.R.: Some properties of continued fractions with applications in Markov processes. IMA J. Appl. Math. **16**(1), 57–71 (1975)
13. Nikiforov, A.F., Uvarov, V. B., S. K.: Suslov, Classical orthogonal polynomials of a discrete variable (pp. 18-54). Springer, Berlin Heidelberg (1991)
14. Peng, N.F.: Spectral representations of the transition probability matrices for continuous time finite Markov chains. J. Appl. Prob. **33**, 28–33 (1996)
15. Parthasarathy, P.R., Dharmaraja, S.: Exact transient solutions of kinetics of first-order reactions with end effects. J. Math. Chem. **25**(2–3), 281–294 (1999)
16. Rosenlund, S.I.: Transition probabilities for a truncated birth-death process. Scandinav. J. Statis. 119–122 (1978)

17. Ross, S. M.: Introduction to Probability Models, Elsevier, 12/e, (2019)
18. Stockmayer, W.H., Gobush, W., Norvich, R.: Local-jump models for chain dynamics. Pure Appl. Chem. **26**(3–4), 537–544 (1971)
19. Szegö, G.: in *Orthogonal Polynomials*, vol. 23, 4th edn. (American Mathematical Society Colloquium Publications, Providence, RI, 1975)
20. Trivedi, K.S.: *Probability and Statistics with Reliability, Queuing, and Computer Science Applications*, 2nd edn. (John Wiley & Sons, 2001)
21. Van Doorn, E.: in *Stochastic Monotonicity and Queueing Applications of Birth-Death Processes* (Vol. 4). (Springer Science & Business Media, Berlin, 2012)
22. Wilkinson, J.H.: The Algebraic Eigenvalue Problem. Clarendon Press, Oxford, New York (1965)

Chapter 6
Simple Markov Queueing Models

The fundamental notions of CTMCs are discussed in detail in Chap. 4. A BDP is one such CTMC, which is discussed in Chap. 5, the structure of which leads to a simple solution for the system's steady-state probability. A broad class of simple queueing models can be developed using the theory of CTMCs and BDP. Using the theory of BDP, several queueing models can be constructed. A queueing model can be used to investigate queues or waiting lines. It provides a good mathematical formulation for several real-world queueing systems, and its analysis reveals properties such as the mean number of customers in a queue and the mean waiting time in a queue, which can be used in service system design and planning.

In this chapter, the above-mentioned models and variations of these models based on the arrival rate, service rate, and customer behavior are analyzed thoroughly. Further, the analysis of these models lead to the computation of effective performance measures of a system.

6.1 Introduction to Queueing Systems

Queueing theory is one of the main subjects in applied probability. It sometimes requires the most sophisticated tools of stochastic processes, and it often leads to simple and explicit answers. The first problem in queueing theory was raised by describing mathematical models for telephone calls. A Danish mathematician, statistician and engineer, Agner Krarup Erlang (1 January 1878–3 February 1929) was the first one who treated congestion problems in the beginning of 20th century. Consider a service facility where customers (need not be human beings) arrive, wait for their turn to receive the service and, after being served, leave. The underlying system in such a situation is called a queueing system. Some examples of queueing systems are as follows:

- Supermarket checkouts,

D. Selvamuthu, *Introduction to Stochastic Processes*, University Texts in the Mathematical Sciences, https://doi.org/10.1007/978-981-97-6152-4_6

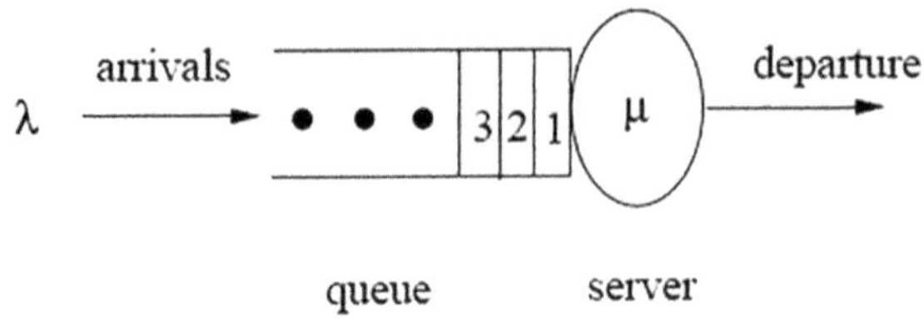

Fig. 6.1 A schematic representation of queueing structure

- Railway reservation counters,
- Computer service center,
- Calls allocation in communication system.

6.1.1 *Characterization*

A queueing system can be completely defined by the following characteristics:

- Arrival pattern of customers/jobs,
- Service pattern of customers/jobs,
- Number of service channels,
- System capacity,
- Queue discipline,
- Calling population.

Figure 6.1 shows a schematic representation of queueing structure. For example, consider a room where a walk-in-interview for a certain job is being held. Rule is that, on the interview day, the first 100 candidates will be interviewed in the order of their arrivals. The interview committee of 4 members take interview of one candidate at a time. The inter-arrival times between candidates to attend the interview, as well as the time taken by the interviewers to finish an interview, are exponentially distributed with some fixed parameters. Thus, for this example, the characteristics of the queueing system are as follows.

- Arrival pattern of customers—inter-arrival time is random with exponential distribution.
- Service pattern of customers—service time is random with exponential distribution.
- Number of servers—1 (although there are 4 interviewers).
- Population—100.
- Queue discipline—first come first service (FCFS).

6.1.2 Kendall's Notation $(A/B/m/K/n/D)$

Before starting the investigations of elementary queueing systems, let us introduce a notation proposed by British statistician and mathematician David George Kendall (15 January 1918–23 October 2007) to describe a queueing system. Kendall's notation is the standard notation used to represent different types of queueing models as per the characteristics described above. Here,

- A: Distribution of inter-arrival times,
- B: Distribution of service times,
- m: Number of servers in the system,
- K: Maximum number of customers in the system,
- n: Population size, number of sources of customers,
- D: Service Discipline.

Exponentially distributed RVs are notated by M (Markov or memoryless). If the population size and capacity are infinite, and the service discipline is first in first out (FIFO), then they are departed. Popular queue disciplines are first come first served (FCFS), last come first served (LCFS), shortest processing time (SPT), and random order.
The components A and B are generally chosen from the following set of different distributions:

- M: Exponential,
- D: Deterministic,
- E_k: Erlang type k $(k = 1, 2, \ldots)$,
- H_k: Hyper Exponential type k,
- G: General service,
- GI: General independent arrival.

Therefore, the above discussed example can be modeled by the $M/M/1/100/FCFS$ queueing system. Further, queues can be classified into two categories; one is Markov queues such as $M/M/1$, $M/M/c$, $M/M/c/k$ and the another one is non-Markov queues like $M/G/1$, $GI/M/1$, $GI/M/c$, $GI/G/1$, $GI/G/c$, $M/D/1$.

Example 6.1 ($M/M/c$ model) Calls to a technical support center are routed using a Poisson process at a pace of 10 per hour. The amount of time it takes a support representative to handle one customer is exponentially distributed with a mean of 7 minutes. Callers are assisted by 5 staff members at the support centre. The $M/M/5$ queueing structure can be used to model this type of system.

Example 6.2 ($M/M/c/k$ model) Consider a vehicle emission inspection station with five inspection stalls, each of which can only accommodate one vehicle. It is acceptable to imagine that the automobiles wait such that when a stall becomes available, the car at the front of the line pulls up to it. The station can only hold nine automobiles at a time (there are five in the station). During peak periods, the arrival pattern follows Poisson process, with a mean of one car every minute. The service time is exponential, with a mean of one minute. This scenario can be modeled by $M/M/5/9$ queueing model.

Example 6.3 (Retrial queueing model) A retrial model is a type of queueing system where customers who find the server busy upon arrival do not leave the system but instead wait in some orbit from which they can retry for the server after some arbitrary distributed time. In communication networks, due to the unavailability of the server, if a packet is lost, it may be retransmitted later by a retransmission mechanism such as the TCP (Transmission Control Protocol). This system can be analyzed by retrial queueing model.

Example 6.4 (Batch arrival and batch service model) A particular television station in a big city, desires to provide its services to thousands of viewers at a time on a Friday evening prime-time programme. Here, the service is provided in bulk at a time and the arrival of requests from the customers is also in bulk at one point in time. Therefore, this system can be studied by batch arrival and batch service model.

Example 6.5 (Vacation queueing model) A long distance calling card company's customer support hotline may not always be highly busy. The primary responsibility of customer service representatives (CSR) is to respond to consumer calls for assistance. During downtime, the CSR might call potential consumers and market the company's services and products. In this case, inbound calls represent queued clients, while outgoing calls represent extra jobs that can be described as server vacations. Such call center with many CSRs can be modeled as a multi-server vacation queueing model.

6.1.3 Markov Queueing Models

The dynamics of Markov queues can be effectively modeled using a BDP. As a result, the general theory of BDP will be presented first, and these findings will then be used to derive useful performance indicators for queueing systems. A BDP is a particular class of CTMC that comprises of the states $\{0, 1, \ldots\}$. State transitions take place as one unit up or down from the current state. The transition rates of the BDP are given as follows:

$$\begin{aligned} q_{ii+1} &= \lambda_i;\ i \geq 0, \\ q_{ii-1} &= \mu_i;\ i \geq 1, \\ q_{00} &= -\lambda_0, \\ q_{ii} &= -(\lambda_i + \mu_i);\ i \geq 1, \\ q_{ij} &= 0 \ \text{for} \ |i - j| > 1. \end{aligned}$$

The state transition diagram for this CTMC is shown in Figure 6.2. The states in the transition diagram represent the number of customers present in the system.

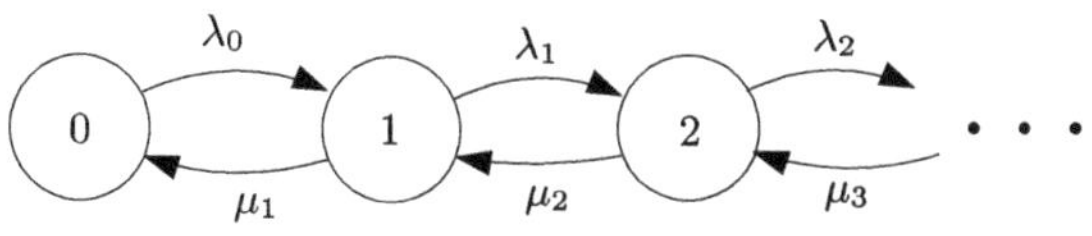

Fig. 6.2 State transition diagram for BDP

Customers' arrivals are represented by "births," and customers' departures are represented by "deaths". After applying the theory of CTMCs, the long-term fraction of time the system remains in state i (π_i) can be computed from $\pi \mathbf{Q} = 0$, subject to certain limitations over λ_i and μ_i. The steady-state probabilities of the system can be provided as follows:

$$\begin{aligned} \lambda_0 \pi_0 &= \mu_1 \pi_1, \\ (\lambda_i + \mu_i)\pi_i &= \lambda_{i-1}\pi_{i-1} + \mu_{i+1}\pi_{i+1}, \quad i = 1, 2, \ldots. \end{aligned}$$

For the explicit expression of the steady-state probabilities, one can refer to Chapter 5, Section 5.1.4.

These obtained results of steady-state probabilities of the system can be further utilized for analysis of the simple Markovian queueing models.

6.2 Single Server Queueing Models

In this section, the queueing system with both infinite capacity and finite capacity is presented.

6.2.1 *M/M/1 Queueing Model*

Here, the analysis of $M/M/1$ queueing model is provided in the steady-state as well as in transient states and the performance measures are discussed.

Definition 6.1 The $M/M/1$ queueing system is modeled by a queueing system with exponentially distributed inter-arrival and service time with parameter λ and μ, respectively, one server, infinite queue size, FIFO queueing discipline and infinite customer population.

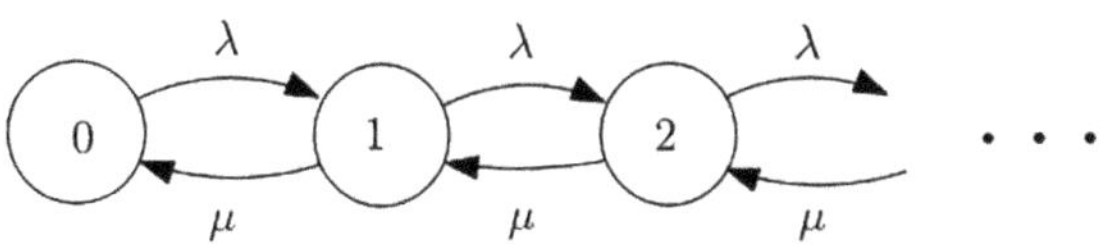

Fig. 6.3 $M/M/1$ queueing system

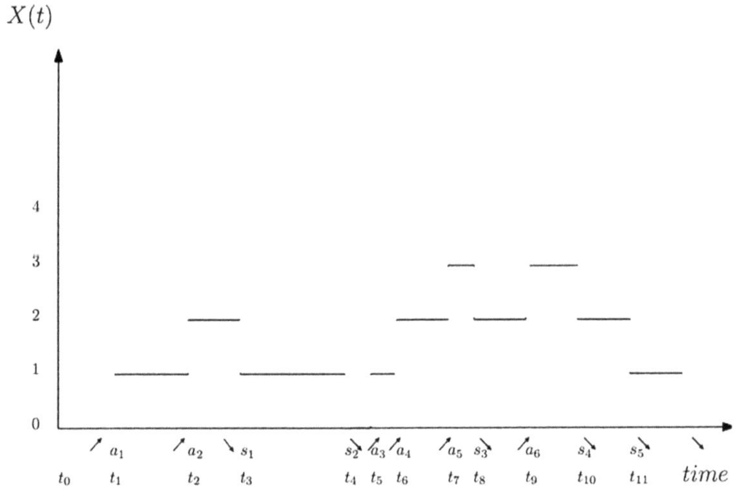

Fig. 6.4 Sample path for $M/M/1$ queueing system

Let $\{X(t), t \geq 0\}$ be the stochastic process defining the number of customers in a system at time t. The underlying stochastic process is a BDP with either the arrival of a new customer or the departure of the customer receiving service. The state transition rates of the process are given by

$$\begin{aligned} q_{ii+1} &= \lambda; \;\; i = 0, 1, \ldots \\ q_{ii-1} &= \mu; \;\; i = 1, 2, \ldots, \\ q_{00} &= -\lambda, \\ q_{ii} &= -(\lambda + \mu); \;\; i = 1, 2, \ldots, \\ q_{ij} &= 0 \text{ for } |i - j| > 1. \end{aligned}$$

Figure 6.4 represents the sample path ofan $M/M/1$ queueing system. Here, a_i denotes the arrivals in the system, s_i denotes the departure in the system and t_i represents the time when the system transits from one state to another state. The time between two successive arrivals is exponentially distributed and similarly, the time for service follows exponential distribution. Further, the stationary distribution, transient distributions and the performance measure of the systems are described.

6.2.1.1 Stationary Distribution

Following the methodology of the BDP, the stationary distribution of $M/M/1$ queueing model can be calculated. Let $\pi = (\pi_0, \pi_1, \ldots)$ denote the stationary distribution vector satisfying

$$\pi_i \geq 0, \quad \sum_i \pi_i = 1.$$

The following steady-state equations can be obtained from $\pi \mathbf{Q} = 0$,

$$0 = -\lambda \pi_0 + \mu \pi_1,$$
$$0 = \lambda \pi_{i-1} - (\lambda + \mu)\pi_i + \mu \pi_{i+1}, \quad i = 1, 2, \ldots.$$

Using the backward substitution to solve the above set of equations,

$$\pi_1 = \frac{\lambda}{\mu}\pi_0,$$
$$\pi_{i-1} = \frac{\lambda}{\mu}\pi_{i-2} = \ldots = \frac{\lambda^{i-1}}{\mu^{i-1}}\pi_0, \quad i = 3, 4, \ldots,$$

or

$$\pi_0 = 1 - \rho,$$
$$\pi_i = (1 - \rho)\rho^i, \quad i = 1, 2, \ldots,$$

where $\rho = \frac{\lambda}{\mu}$. If $\rho < 1$, the stationary distribution of the system exists, otherwise the stationary distribution does not exist. For $\rho \geq 1$, queue length will increase to infinity with probability one in the long run. Here ρ is called the offered load or traffic intensity or server utilization. Also, $\pi_0 = 1 - \rho$ is the probability that the queueing system $M/M/1$ has no customer in the system, and hence, the server is idle in the steady-state. Note that the stationary distribution is the modified geometric distribution with parameter ρ.

6.2.1.2 Mean Number of Customers in the System

Let X be the number of customers in the system in steady-state. Further, let $E(X)(= \lim_{t \to \infty} E(X(t)))$ denote the mean number of customers in the system, and $Var(X)$ denotes the variance of customers in the system in steady-state which can be computed as follows:

$$E(X) = \sum_n n\rho^n (1 - \rho) = \frac{\rho}{1 - \rho},$$
$$Var(X) = E[(X - E(X))^2] = \frac{\rho}{(1 - \rho)^2}.$$

6.2.1.3 Mean Number of Customers in the Queue

Let Q denote the number of customers in the queue in steady-state. Also, let $E(Q)$ and $var(Q)$ be the mean number and variance of customers in the queue in steady-state, respectively that can be obtained by

$$E(Q) = \sum_{n-1}^{\infty} n\pi_{n+1} = \frac{\rho^2}{1-\rho},$$

$$Var(Q) = E((Q - E(Q))^2) = \frac{\rho(1+\rho-\rho^2)}{(1-\rho)^2}.$$

6.2.1.4 Distribution of Response Time

Let R represents the RV for the response time/total time spent by a customer in the system. The distribution of response time of a customer in the system is given as follows. If a customer, on arrival, finds n customers in the system then

$$\begin{aligned} P\{0 < R \le t\} &= \sum_{n=0}^{\infty} P\{R \le t | X = n+1\} P\{X = n\} \\ &= \sum_{n=0}^{\infty} \int_0^t \frac{\mu^{n+1} x^n e^{-\mu x}}{n!} \left(1 - \frac{\lambda}{\mu}\right) \left(\frac{\lambda}{\mu}\right)^n dx \\ &= 1 - e^{-(\mu-\lambda)t}, \end{aligned}$$

which is a CDF of exponential distribution with parameter $\mu - \lambda$.

6.2.1.5 Mean Response Time of a Customer

The mean response time of a customer in the system $E(R)$ is obtained as

$$E(R) = \int_0^{\infty} e^{-(\mu-\lambda)t} dt = \frac{1}{\mu - \lambda}.$$

6.2.1.6 Distribution of Waiting Time

Let W represent the RV for the waiting time of a customer in the queue. For $t = 0$,

$$P\{W \le t\} = 1 - \rho.$$

For $t > 0$, if there are n customers in the system when a new customer arrives, the waiting time W for this customer before his/her service starts is

$$W = \begin{cases} 0, & \text{if } n = 0 \\ \tilde{S_1} + S_2 + \cdots + S_n, & \text{if } n = 1, 2, \ldots, \end{cases}$$

where $\tilde{S_1}$ is the remaining service time of the customer being served and $S_2, S_3, \ldots, S_n$ are the service times of the customers in the queue who arrived before the arrival of the new customer. Thus, $W|\{N = n\} \sim Gamma(n, \mu)$. For $t > 0$,

$$\begin{aligned} P\{0 < W \leq t\} &= \sum_{n=1}^{\infty} P\{W \leq t | X = n\} P\{X = n\} \\ &= \sum_{n=1}^{\infty} \int_0^t \frac{\mu^n x^{n-1} e^{-\mu x}}{(n-1)!} (1-\rho)\rho^n dx. \end{aligned}$$

Hence,

$$P\{W \leq t\} = \begin{cases} 0, & t < 0 \\ 1-\rho, & t = 0 \\ 1 - \rho + \sum_{n=1}^{\infty} \int_0^t \frac{\mu^n x^{n-1} e^{-\mu x}}{(n-1)!} dx (1-\rho)\rho^n, & 0 < t < \infty \end{cases}.$$

Further, after simplification it can be seen that, for $t > 0$,

$$P\{W \leq t\} = 1 - \rho e^{-\mu(1-\rho)t}, \; 0 < t < \infty.$$

Hence,

$$P\{W \leq t\} = \begin{cases} 0, & t < 0 \\ 1-\rho, & t = 0 \\ 1 - \rho e^{-\mu(1-\rho)t}, & 0 < t < \infty \end{cases}.$$

The distribution of W is given by

$$P\{W = 0\} = 1 - \rho,$$

$$f_W(t) = \rho(\mu - \lambda) e^{-(\mu - \lambda)t}, \; t > 0.$$

6.2.1.7 Mean Waiting Time of a Customer

The mean waiting time $E(W)$ is given by

$$E(W) = \int_0^\infty \rho e^{-(\mu-\lambda)t} dt = \frac{\rho}{\mu - \lambda}.$$

6.2.1.8 Little's Law

John Dutton Conant Little (February 1, 1928–September 27, 2024) was an American operations researcher who gave Little's law. The Little's law states that the long term mean number of customers in a stable queueing system. $E(X)$ is equal to the long term mean effective arrival rate (λ), multiplied by the mean time a customer spends in the system $E(R)$, i.e.,

$$E(X) = \lambda E(R). \tag{6.1}$$

The proof of the above result can be provided as follows. Consider a queueing system containing objects which arrive to the system with some rate. The sample path of the queueing system is shown in Fig. 6.5. In this Figure, a_i denotes the arrival epoch of the items, s_i represents the service epoch of items, and t_i denotes the time of transition from one state to another state by the system. If $E(X(t))$ is the mean value of $E(X(\tau))$, τ taking values over the interval $[0, t]$, then it is clear that:

$$E(X(t)) = \frac{1}{t} \int_0^t E(X(\tau)) d\tau.$$

The mean value of items over t is the integral of $A(t)$ over t i.e., the area covered by $A(t)$ over t divided by t. At time t, each of the items is waiting and hence is accumulating waiting time. Therefore,

$$\lambda(t) = \frac{X(t)}{t};\ E(X(t)) = \frac{A(t)}{t};\ E(R(t)) = \frac{A(t)}{X(t)}.$$

After simple calculation, it can be obtained that $E(X(t)) = \lambda(t)E(R(t))$. Due to the stochastic nature of the queueing model and end effects, the values of $E(X(t))$ and $E(R(t))$ change with respect to time t. End effects account for two situations: the waiting time of items that entered the system before time t, and the reduced waiting time of items that arrived during time t but have not yet departed. $E(X(t))$ and $\lambda(t)$ fluctuate slightly as t rises as items come and go. The end effects at the beginning and end of t become insignificant in comparison to the major area under the curve with the proper mathematical assumptions about the stationarity of the underlying stochastic process. As a result, when t increases, the changes in $E(X(t))$, $\lambda(t)$, and $E(R(t))$ become smaller and smaller percentages of their real values. As a result, as t increases to infinity, $E(X(t))$, $\lambda(t)$, and $E(R(t))$ each tend to a limit. Then, using the obvious symbols for the limits,

$$E(X) = \lim_{t\to\infty} E(X(t)), \lambda = \lim_{t\to\infty} \lambda(t), E(R) = \lim_{t\to\infty} E(R(t)).$$

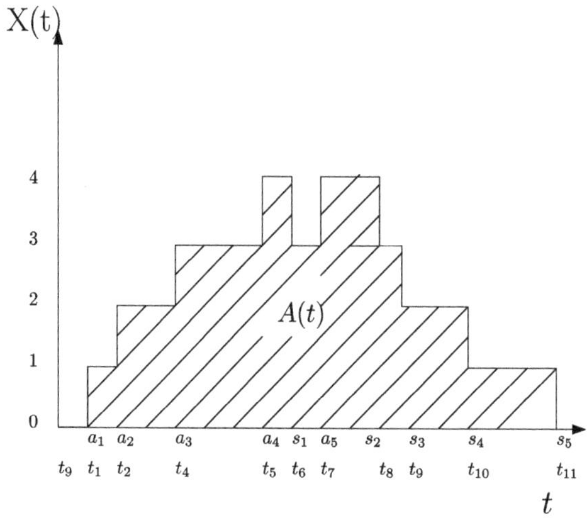

Fig. 6.5 Number of items in a queueing system versus time

Thus, it follows $E(X) = \lambda E(R)$.

Limitations of Little's Law

When the system is in steady-state mode, Little's law applies. On mean over the long term, the arrival rate of work for the system equals the departure rate of finished work from that system, the steady state has been attained. The throughput, arrival rate, and exit rate are all equal as a result. Little's law is applicable for $M/M/1$, $M/M/1/N$, $M/M/c$, $M/M/c/c$, $M/M/c/k$, $M/M/\infty$ models. There are two simple ways to establish a steady state in a queueing system subject to Little's Law.

- The best method of control is to reduce the amount of work needed such that it will never exceed the maximum amount of work in progress.
- Improve the system's processing speed to meet high demand and achieve optimal performance.

Measures for $M/M/1$ Queue Using Little's Law
Let R and W be the time spent by a customer in the system and the waiting time of a customer in the queue, respectively.
By Little's law,

$$E(R) = \frac{E(X)}{\lambda} = \frac{\rho}{\lambda(1-\rho)} = \frac{1}{\mu - \lambda},$$
$$E(W) = E(R) - \frac{1}{\mu} = \frac{1}{\mu-\lambda} - \frac{1}{\mu} = \frac{\lambda}{\mu(\mu-\lambda)}.$$

Theorem 6.1 *(Burke's Theorem) The output of a Poisson input queue with a single channel having exponential service time and in the steady state must be Poisson with the same rate as the input. This is valid for $M/M/1$, $M/M/c$, $M/M/\infty$ queues. The number of customers in the queue is independent of the departure process prior to t.*

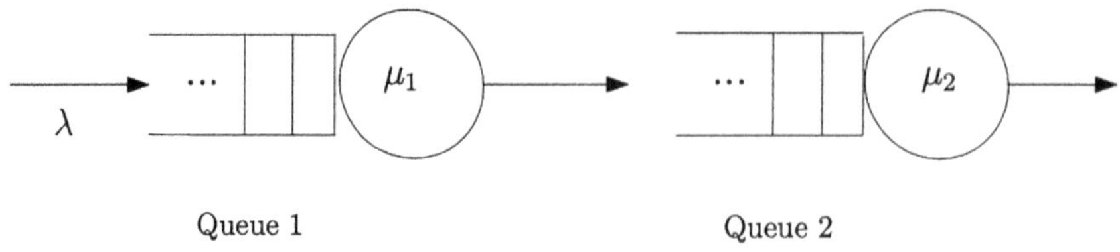

Fig. 6.6 Network of queueing system

Proof Since the arrival process is also a Poisson process with rate λ, the first half of the theorem is proved by showing that the time reverse of the arrival process is a Poisson process with rate λ. Given that the arrival process after t in the reversed process is the same as the departure process before t in the reversed process. As a result, in a Poisson input system, the number of customers in the system queue is independent of the arrivals after the point t. □

In many networking scenarios, a packet or customer must receive service from many servers before its final task is completed. Hence, departure from one queue might become arrival at another queue. Hence, the arrival processes are independent of each other.

Example 6.6 Consider a queueing system depicted in Fig. 6.6, where queue 1 is driven by a Poisson process with rate λ and the departures are routed to queue 2. Let $N_1(t)$ and $N_2(t)$ be the underlying stochastic processes for queue 1 and queue 2. As per the Burke's theorem for the proposed queueing system, processes $N_1(t)$ and $N_2(t)$ are independent. Therefore, the number of customers $N(t)$ in each queue is independent of the departure times prior to t.

6.2.1.9 PASTA Property

The Poisson Arrivals See Time Averages (PASTA) property is one of the central tools in queueing theory. Sometimes this property is referred to as ROP (Random Observer Property). For queueing systems with arrivals following Poisson process, i.e., for $M/\cdot/\cdot$ systems, this special property holds that arriving customers perceive the queueing system similarly to an external observer looking at the system at an arbitrary point in time. More precisely, the fraction of time the system is in a certain state is also the probability that a customer arriving at that moment sees the system in that state. This property is only true for arrivals with the Poisson process.

In equilibrium, each state j can be associated with two different probabilities:

1. The probability of the state as seen by an outside random observer π_j = probability that the system is in the state j at a random instant.
2. The probability of the state seen by an arriving customer π_j^* = probability that the system is in the state j just before (a randomly chosen) i^{th} customer arrival.

In general,

$$\pi_j \neq \pi_j^*.$$

That is,

$$\lim_{t\to\infty} P\{X(t) = j\} \neq \lim_{t\to\infty} P\{X(t_i^-) = j\}.$$

For instance, in a $D/D/1$ system which is empty at time 0, and with arrivals at $1, 3, \ldots$ and service times of unit 1, every arriving customer finds an empty system, whereas the fraction of time the system is empty is 1/2. However, in the case of a Poisson arrival process, it holds

$$\pi_j = \pi_j^*.$$

This property of Poisson arrivals is called PASTA property, which is the acronym for Poisson Arrivals See Time Averages. Intuitively, this property can be explained by the fact that Poisson arrivals occur completely random in time.

The arrival history before the instant of consideration, irrespective of whether a random instant or an arrival instant are considered is stochastically the same: a sequence of arrivals with exponentially distributed inter-arrival times. This follows from the memoryless property of the exponential distribution. The remaining time to the next arrival has the same exponential distribution irrespective of the time that has already elapsed since the previous arrival (the same also holds in reversed time, i.e., looking backwards). Since the stochastic characterization of the arrival process before the chosen instant is the same, the way the instant is selected does not matter. Therefore, the state distributions of the system, influenced by past arrival processes, must be identical in both cases.

A rigorous proof of the PASTA property can be found in Wolff [1] and Wolff [2].

Theorem 6.2 *The output process of an $M/M/1$ queue in steady state is again a Poisson process with the same mean rate as the input process.*

Proof It can be noted that if X and Y are two independent exponentially distributed variables with mean λ and μ, respectively. Then the PDF for the RV $Z = X + Y$ is

$$f(z) = \frac{\lambda\mu}{\mu - \lambda} e^{-\lambda z} - \frac{\lambda\mu}{\mu - \lambda} e^{-\mu z}.$$

Let the arrival rate of $M/M/1$ queue be λ and departure rate be μ. There can occur one of the following two cases: the server is idle (the steady-state probability will be $1 - \lambda/\mu$) and the server is not idle (the steady-state probability will be λ/μ).

For the first case, the departure time follows $f(z)$ (a waiting time for an arrival plus a service time). For the second case, the departure time follows $\mu e^{-\mu z}$. Therefore, the PDF $g(z)$ for the departure time is given by

$$g(z) = \left(1 - \frac{\lambda}{\mu}\right) f(z) + \frac{\lambda}{\mu} \left(\mu e^{-\mu z}\right)$$
$$= \frac{\lambda\mu}{\mu - \lambda} e^{-\lambda z} - \frac{\lambda\mu}{\mu - \lambda} e^{-\mu z} - \frac{\lambda^2}{\mu - \lambda} e^{-\lambda z} + \frac{\lambda^2}{\mu - \lambda} e^{-\mu z} + \lambda e^{-\mu z}.$$

Hence,

$$g(z) = \lambda e^{-\lambda z}, z \geq 0.$$

Therefore, the departure time follows the exponential distribution. This implies that the departure process is a Poisson process with mean λ. □

6.2.1.10 Transient Analysis

Considerable attention has been paid in obtaining the transient solution for the system size in an $M/M/1$ queueing system. A number of methods have been put forward to solve the following system of equations

$$\frac{d\pi_0(t)}{dt} = \mu\pi_1(t) - \lambda\pi_0(t), \tag{6.2}$$
$$\frac{d\pi_n(t)}{dt} = \mu\pi_{n+1}(t) - (\lambda + \mu)\pi_n(t) + \lambda\pi_{n-1}(t), \quad n = 1, 2, \ldots, \tag{6.3}$$

where $\pi_n(t)$ denotes the probability that the system size is n at time t. Here, a simple and direct approach has been proposed by Parthasarathy [3]. It has been assumed that initially there are i customers. Define

$$q_n(t) = \begin{cases} \exp((\lambda + \mu)t) \left[\mu\pi_n(t) - \lambda\pi_{n-1}(t)\right], & n = 1, 2, \cdots, \\ 0, & n = 0, -1, \cdots. \end{cases} \tag{6.4}$$

Let

$$H(s, t) = \sum_{n=-\infty}^{\infty} q_n(t) s^n. \tag{6.5}$$

Then Eqs. (6.2) and (6.3) become

$$\frac{\partial H(s, t)}{\partial t} = \left(\lambda s + \tfrac{\mu}{s}\right) H(s, t) - \mu q_1(t), \tag{6.6}$$
$$H(s, 0) = s^i \left[\mu\left(1 - \delta_{0i}\right) - \lambda s\right] \tag{6.7}$$

where δ is the Kronecker delta. Solving (6.6),

$$H(s,t) = H(s,0)\exp\left\{\left(\lambda s + \frac{\mu}{s}\right)t\right\} - \mu\int_0^t q_1(y)\exp\left\{\left(\lambda s + \frac{\mu}{s}\right)(t-y)\right\}dy. \tag{6.8}$$

If $\alpha = 2\sqrt{(\lambda\mu)}$ and $\beta = \sqrt{(\lambda/\mu)}$, then

$$\exp\left\{\left(\lambda s + \frac{\mu}{s}\right)t\right\} = \sum_{n=-\infty}^{\infty} (\beta s)^n I_n(\alpha t), \tag{6.9}$$

where $I_n(t)$ is a modified Bessel function of the first kind of order n (refer Appendix A.6). Comparing the coefficients of s^n on both sides of (6.8) for $n = 1, 2, \cdots,$

$$\frac{q_n(t)}{\beta^n} = \mu\left(1-\delta_{0i}\right)\beta^{-i}I_{n-i}(\alpha t) - \lambda\beta^{-i-1}I_{n-i-1}(\alpha t) - \mu\int_0^t q_1(y)I_n(\alpha(t-y))dy. \tag{6.10}$$

The above holds for $n = -1, -2, \cdots,$ with the left-hand side replaced by zero. Using $I_{-r} = I_r$, for $n = 1, 2, \cdots$

$$\mu\int_0^t q_1(y)I_n(\alpha(t-y))dy = \mu\beta^{-i}\left(1-\delta_{0i}\right)I_{n+i}(\alpha t) - \lambda\beta^{-i-1}I_{n+i+1}(\alpha t). \tag{6.11}$$

From Equation (6.10) and Equation (6.11), for $n = 1, 2, \cdots,$

$$q_n(t) = \mu\beta^{n-i}\left(1-\delta_{0i}\right)\left[I_{n-i}(\alpha t) - I_{n+i}(\alpha t)\right] + \lambda\beta^{n-i-1}\left[I_{n+i+1}(\alpha t) - I_{n-i-1}(\alpha t)\right]$$

such that

$$\pi_n(t) = \frac{\exp\{-(\lambda+\mu)t\}}{\mu}\sum_{k=1}^{n} q_k(t)\left(\frac{\lambda}{\mu}\right)^{n-k} + \left(\frac{\lambda}{\mu}\right)^n \pi_0(t)$$

and

$$\pi_0(t) = \int_0^t q_1(y)\exp\{-(\lambda+\mu)y\}dy + \delta_{0i}.$$

Example 6.7 Consider a mixed hair salon where clients are dealt with in the order they arrive. According to the data, there are 5 consumers who typically arrive per hour according to a Poisson process. Customers are always patience due to the business's stellar reputation. The data also demonstrate that client processing times are exponential spread, taking 10 minutes on average for each customer. $M/M/1$ queueing system can be used to model this system. Then, in the long run

- The mean number of customers in the system is

$$E(X) = \frac{\rho}{1-\rho} = \frac{5/6}{1-5/6} = 5.$$

- The mean number of customers waiting for haircut

$$E(Q) = \frac{\rho^2}{1-\rho} = \frac{25/36}{1-5/6} = 25/6.$$

Example 6.8 Consider the brand-new Chennai International Airport as an example. Assume it has a single runway that is only used for arrivals. There are 10 flights that come per hour, according to data. It is anticipated that the time (in minutes) needed for a plane to land will follow an exponential distribution with a mean of three minutes. Assume the arrivals process is Poisson. Assume that the system is characterized as a $M/M/1$ queueing system without losing generality.

- The probability that there is no waiting time to land is

$$\pi_0 = 1 - \rho = 0.5.$$

- The expected number of planes waiting to land is

$$E(Q) = \frac{\rho^2}{1-\rho} = 0.5.$$

- The expected waiting time is

$$E(R) = \frac{1}{\lambda}E(X) = 0.05 \text{ hours}.$$

6.2.2 $M/M/1/N$ Queueing Model

In this section, the $M/M/1$ queueing model with finite capacity will be described and the steady state as well as the transient solutions will be derived.

Definition 6.2 Consider an $M/M/1$ queueing model with the capacity of the system being finite N. Customers arriving after the system is full of capacity are rejected. This is a BDP with finite state space $S = \{0, 1, \ldots, N\}$. From state N, only transition to $N-1$ is possible.

The state transition diagram is shown in Fig. 6.7.

6.2.2.1 Steady-State Distributions

Equations for $M/M/1/N$ queueing model for determining the steady-state distribution are given by

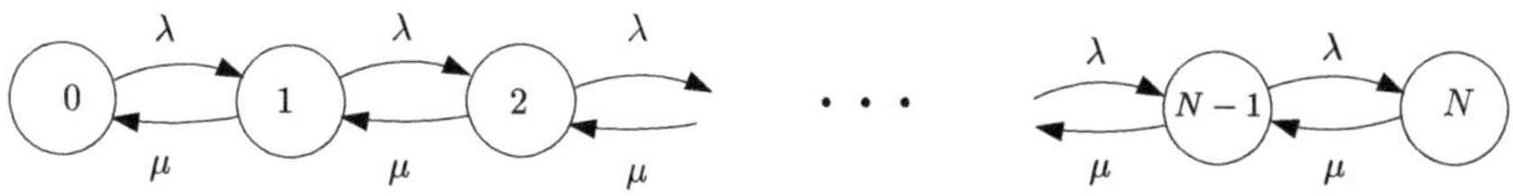

Fig. 6.7 State transition diagram of $M/M/1/N$ queueing model

$$\lambda\pi_0 = \mu\pi_1,$$
$$\lambda\pi_{i-1} - (\lambda + \mu)\pi_i + \mu\pi_{i+1} = 0, \qquad i = 1, 2, \ldots, N-1$$
$$\lambda\pi_{N-1} = \mu\pi_N.$$

Solving above system of equations and using $\sum_{i=0}^{N} \pi_i = 1$, for $\rho \neq 1$,

$$\pi_i = \left(\frac{1-\rho}{1-\rho^{N+1}}\right)\rho^i, \quad i = 0, 1, \ldots, N,$$

which is a truncated geometric distribution with truncation at N. When $\rho = 1, \pi_i = \frac{1}{N+1}, i = 0, 1, \ldots, N$.

6.2.2.2 Performance Measures of the System

1. As not every arriving customer gets entry into the system, the effective arrival rate is not λ. An arriving customer gets entry if and only if the system is not full. Hence, the effective arrival rate is

$$\lambda_{eff} = \lambda(1 - \pi_N).$$

2. The throughput of the system is given by

$$TP = \mu(1 - \pi_0).$$

3. Blocking probability is defined by the probability by which an arriving customer is not allowed to enter the system, i.e., when the system is full. Thus, blocking probability is

$$P_b = \pi_N.$$

By using Little's Law

$$E(R) = \frac{E(X)}{\lambda_{eff}}.$$

$E(X)$ is the expected value of number of customers in the system.

6.2.2.3 Transient Solution

The BDP $\{X(t), t \geq 0\}$ with the state space $\{0, 1, \ldots, N\}$ satisfies a finite system of equations

$$\begin{aligned} \pi_0'(t) &= -\lambda\pi_0(t) + \mu\pi_1(t), \\ \pi_n'(t) &= \lambda\pi_{n-1}(t) - (\lambda+\mu)\pi_n(t) + \mu\pi_{n+1}(t), \quad n = 1, 2, \ldots, N-1 \\ \pi_N'(t) &= \lambda\pi_{N-1}(t) - \mu\pi_N(t). \end{aligned} \tag{6.12}$$

Let for $s > 0$,

$$\widehat{\pi}_n(s) = \int_0^\infty e^{-st}\pi_n(t)dt. \tag{6.13}$$

Take $\pi_n(0) = \delta_{mn}$, where m is the initial state of the BDP, the Kronecker delta. Laplace transformation of (6.12) gives

$$\begin{aligned} s\widehat{\pi}_0(s) - \delta_{m0} &= -\lambda\widehat{\pi}_0(s) + \mu\widehat{\pi}_1(s), \\ s\widehat{\pi}_n(s) - \delta_{mn} &= \lambda\widehat{\pi}_{n-1}(s) - (\lambda+\mu)\widehat{\pi}_n(s) + \mu\widehat{\pi}_{n+1}(s), 1 \leq n \leq N-1, \\ s\widehat{\pi}_N(s) - \delta_{mN} &= \lambda\widehat{\pi}_{N-1}(s) - \mu\widehat{\pi}_N(s). \end{aligned} \tag{6.14}$$

This defines a linear system with $N+1$ equations. Define its $N+1$ by $N+1$ matrix as

$$\mathbf{E(s)} = \begin{pmatrix} s+\lambda & -\mu & & & & \\ -\lambda & s+\lambda+\mu & -\mu & & & \\ & -\lambda & s+\lambda+\mu & -\mu & & \\ & & & \ddots & & \\ & & & -\lambda & s+\lambda+\mu & -\mu \\ & & & & -\lambda & s+\mu \end{pmatrix}_{(N+1)\times(N+1)}.$$

Then Eq. (6.14) can be written in the matrix form

$$\mathbf{E(s)}[\widehat{\pi}_0(s), \widehat{\pi}_1(s), \ldots, \widehat{\pi}_N(s)]^T = [\delta_0, \delta_1, \ldots, \delta_N]^T. \tag{6.15}$$

Define $\mathbf{Q_{N+1}}(s) = |E(s)|$, i.e.,

$$\mathbf{Q_{N+1}}(s) = \begin{vmatrix} s+\lambda & -\mu & & & & \\ -\lambda & s+\lambda+\mu & -\mu & & & \\ & -\lambda & s+\lambda+\mu & -\mu & & \\ & & & \ddots & & \\ & & & -\lambda & s+\lambda+\mu & -\mu \\ & & & & -\lambda & s+\mu \end{vmatrix}_{(N+1)\times(N+1)}.$$

$$\mathbf{Q_{N+1}}(s) = s\begin{vmatrix} s+\lambda+\mu & \mu & & & & \\ \lambda & s+\lambda+\mu & \mu & & & \\ & \lambda & s+\lambda+\mu & \mu & & \\ & & & \ddots & & \\ & & & \lambda & s+\lambda+\mu & \mu \\ & & & & \lambda & s+\lambda+\mu \end{vmatrix}_{N\times N}.$$

It is well known that $\mathrm{Q_{N+1}}(s)$ has $N+1$ distinct real zeros, say, $\beta_0, \beta_1, \ldots, \beta_n$

$$0 = \beta_0 < \beta_1 < \cdots < \beta_N \tag{6.16}$$

and

$$\mathbf{Q_{N+1}}(s) = s\prod_{r=1}^{N}(s+\beta_r) \tag{6.17}$$

where $\beta_r = \lambda + \mu - 2\sqrt{\lambda\mu}\cos\left(\frac{r\pi}{N+1}\right)$, $r = 1, 2, \ldots, N$. By Equation (6.16), $|E(s)| > 0$ and (6.15) has a unique solution for $s > 0$. The solution of equation (6.15) by Cramer's rule is written as

$$\widehat{\pi}_n(s) = \frac{\widehat{B}_n(s)}{|E(s)|}. \tag{6.18}$$

$\widehat{B}_n(s)$ is obtained from $|E(s)|$ by replacing the nth column with $[\delta_{m0}, \delta_{m1}, \ldots, \delta_{mN}]^T$. Substituting the values of $\widehat{B}_n(s)$ and taking inversion of $\widehat{\pi}_n(s)$, the exact transient solution is given as [4]

$$\begin{aligned}\pi_n(t) = {} & \frac{(1-\rho)\rho^n}{(1-\rho^{N+1})} + \frac{2\rho^{1+(n-m)/2}}{N+1}\sum_{r=1}^{N}\frac{e^{-(\lambda+\mu)t+2\sqrt{\lambda\mu}t\cos(r\pi/(N+1))}}{1-2\sqrt{\rho}\cos\frac{r\pi}{N+1}+\rho} \\ & \times\left\{\sin\frac{(n+1)r\pi}{N+1} - \rho^{-1/2}\sin\frac{nr\pi}{N+1}\right\} \\ & \times\left\{\sin\frac{(m+1)r\pi}{N+1} - \rho^{-1/2}\sin\frac{mr\pi}{N+1}\right\}, \quad n = 0, 1, \ldots, N \end{aligned} \tag{6.19}$$

where $\rho = \frac{\lambda}{\mu}$.

Special Case: Consider $N = 2$ (refer Example 4.14), assuming $m = 0$, the differential-difference equation will be converted to

$$\begin{aligned}
\pi_0'(t) &= -\lambda\pi_0(t) + \mu\pi_1(t),\\
\pi_1'(t) &= \lambda\pi_0(t) - (\lambda+\mu)\pi_1(t) + \mu\pi_2(t),\\
\pi_2'(t) &= \lambda\pi_1(t) - \mu\pi_2(t).
\end{aligned} \tag{6.20}$$

and taking the Laplace transform of Equation (6.21)

$$\begin{aligned}
s\widehat{\pi}_0'(s) - 1 &= -\lambda\widehat{\pi}_0(s) + \mu\widehat{\pi}_1(s),\\
s\widehat{\pi}_1'(s) &= \lambda\widehat{\pi}_0(s) - (\lambda+\mu)\widehat{\pi}_1(s) + \mu\widehat{\pi}_2(s),\\
s\widehat{\pi}_2'(s) &= \lambda\widehat{\pi}_1(s) - \mu\widehat{\pi}_2(s).
\end{aligned} \tag{6.21}$$

Following the same computation procedure as described in Eq. (6.19),

$$\begin{aligned}
\pi_0(t) &= \frac{1-\rho}{1-\rho^3} + \frac{2\rho}{3}\sum_{r=1}^{2} \frac{e^{-(\lambda+\mu)t+2t\sqrt{\lambda\mu}\cos(r\pi/3)}}{1-2\sqrt{\rho}\cos\frac{r\pi}{3}+\rho} \times \left\{\sin\frac{r\pi}{3}\right\} \times \left\{\sin\frac{r\pi}{3}\right\}\\
&= \frac{1-\rho}{1-\rho^3} + \frac{2\rho}{3}\left[\frac{e^{-(\lambda+\mu)t+t\sqrt{\lambda\mu}}}{1+\rho-\sqrt{\rho}} \times \frac{\sqrt{3}}{2} \times \frac{\sqrt{3}}{2} + \frac{e^{-(\lambda+\mu)t-t\sqrt{\lambda\mu}}}{1+\rho+\sqrt{\rho}} \times \frac{\sqrt{3}}{2} \times \frac{\sqrt{3}}{2}\right],
\end{aligned}$$

$$\begin{aligned}
\pi_1(t) &= \frac{(1-\rho)\rho}{1-\rho^3} + \frac{2\rho^{3/2}}{3}\sum_{r=1}^{2} \frac{e^{-(\lambda+\mu)t+2t\sqrt{\lambda\mu}\cos(r\pi/3)}}{1-2\sqrt{\rho}\cos\frac{r\pi}{3}+\rho}\\
&\quad \times\left\{\sin\frac{2r\pi}{3}\right\} \times \left\{\sin\frac{r\pi}{3} - \rho^{-1/2}\sin\frac{r\pi}{3}\right\}\\
&= \frac{\rho}{1+\rho+\rho^2} + \frac{2\rho\sqrt{\rho}}{3}\left[\frac{e^{-(\lambda+\mu-\sqrt{\lambda\mu})t}}{1+\rho-\sqrt{\rho}} \times \frac{\sqrt{3}}{2}\left(\frac{-1+\sqrt{\rho}}{\sqrt{\rho}}\right) \times \frac{\sqrt{3}}{2}\right.\\
&\quad \left. + \frac{e^{-(\lambda+\mu+\sqrt{\lambda\mu})t}}{1+\rho-\sqrt{\rho}} \times \frac{\sqrt{3}}{2}\left(\frac{-1-\sqrt{\rho}}{\sqrt{\rho}}\right) \times \frac{\sqrt{3}}{2}\right],
\end{aligned}$$

$$\pi_2(t) = \frac{(1-\rho)\rho^2}{1-\rho^3} + \frac{2\rho^2}{3}\sum_{r=1}^{2}\frac{e^{-(\lambda+\mu)t+2t\sqrt{\lambda\mu}\cos(r\pi/3)}}{1-2\sqrt{\rho}\cos\frac{r\pi}{3}+\rho}$$
$$\times\left\{\sin\frac{r3\pi}{3}\right\}\times\left\{\sin\frac{r\pi}{3}-\rho^{-1/2}\sin\frac{r\pi}{3}\right\}$$
$$= \frac{\rho^2}{1+\rho+\rho^2} + \frac{2\rho^2}{3}\left[\frac{e^{-(\lambda+\mu-\sqrt{\lambda\mu})t}}{1-\rho+\sqrt{\rho}}\times\frac{-\sqrt{3}}{2\sqrt{\rho}}\times\frac{\sqrt{3}}{2}\right.$$
$$\left.+\frac{e^{-(\lambda+\mu+\sqrt{\lambda\mu})t}}{1-\rho+\sqrt{\rho}}\times\frac{\sqrt{3}}{2\sqrt{\rho}}\times\frac{\sqrt{3}}{2}\right].$$

The above time-dependent probabilities can be simplified and are given by

$$\pi_0(t) = \frac{1}{1+\rho+\rho^2} - \frac{\rho\sqrt{\rho}e^{-(\lambda+\mu-\sqrt{\lambda\mu})t}}{2\rho-2(1+\rho)\sqrt{\rho}} + \frac{\rho\sqrt{\rho}e^{-(\lambda+\mu+\sqrt{\lambda\mu})t}}{2\rho+2(1+\rho)\sqrt{\rho}},$$
$$\pi_1(t) = \frac{\rho}{1+\rho+\rho^2} + \frac{\rho(-\rho+\sqrt{\rho})e^{-(\lambda+\mu-\sqrt{\lambda\mu})t}}{2\rho-2(1+\rho)\sqrt{\rho}} - \frac{\rho(\rho+\sqrt{\rho})e^{-(\lambda+\mu+\sqrt{\lambda\mu})t}}{2\rho+2(1+\rho)\sqrt{\rho}},$$
$$\pi_2(t) = \frac{\rho^2}{1+\rho+\rho^2} + \frac{\rho^2 e^{-(\lambda+\mu-\sqrt{\lambda\mu})t}}{2\rho-2(1+\rho)\sqrt{\rho}} + \frac{\rho^2 e^{-(\lambda+\mu+\sqrt{\lambda\mu})t}}{2\rho+2(1+\rho)\sqrt{\rho}}.$$

Example 6.9 A mechanic discovers that the time required to fix an automobile follows an exponential distribution, with a mean of 40 min. The car being repaired is one of a maximum of 7 vehicles the facility can hold. The motor vehicles are repaired according to arrival order, and a Poisson process with a mean rate of 15 arrivals per 10-hour day can be used to estimate the arrivals. Find

1. the probability of the repair person being idle in the long run.
2. the proportion of the time the shop is not full.

Solution: This repair system can be modeled as a $M/M/1/N$ queueing model where each state of the system represents the number of motor cars being repaired. Here $\lambda = \frac{1}{40}$, $N = 7$, $\mu = \frac{1}{40}$. Since, $\rho = 1$, $\pi_0 + \ldots + \pi_7 = 1$. Hence, the probability of the repair person being idle in the long run will be

$$\pi_0 = \frac{1}{N+1} = \frac{1}{8}.$$

The proportion of the time the shop is not full is

$$1 - \pi_7 = 1 - \frac{1}{N+1} = \frac{7}{8}.$$

Example 6.10 Consider a mobile service provider's message switching center. Assume that the flow of traffic is a Poisson process, with a mean message rate of 180 per minute. A 600-character-per-second transmission rate is available on the line. Assume that messages have a mean length of 176 characters and are exponentially

Fig. 6.8 State transition diagram for $M/M/c$ queueing model

distributed. When the buffer capacity is N, the incoming messages are buffered for transmission. What is the minimum N such that $\pi_N < 0.005$.
Solution: Given $\lambda = 3$ and $\mu = \frac{600}{176}$. The minimum N to guarantee that $\pi_N < 0.005$ can be computed with $\rho = 0.88$ and $N = 26$.

$$\pi_N = \frac{(1-\rho)\rho^N}{1-\rho^{N+1}} < 0.005.$$

6.3 Multi-server Queueing Models

The multi-server queueing system describes a system where arrivals form a single queue and are governed by a Poisson process. There are a finite number of servers, and job service times are exponentially distributed.

6.3.1 *M/M/c Queueing Model*

This section will outline the $M/M/c$ queueing model, providing a description of its characteristics, and deriving both steady-state and transient solutions.

Definition 6.3 The $M/M/c$ queueing model is another Markov queueing system with arrivals according to a Poisson process with rate λ and service times follow an exponential distribution with parameter μ, c identical servers, and an infinite system capacity. In this queueing system, if the arriving customer finds n customers in the system and $n < c$, the customer is routed to any idle server otherwise he joins the waiting queue. Let $\{X(t), t \geq 0\}$ be a BDP where $X(t)$ denotes the number of customers in the system at time t. The birth rates are $\lambda_n = \lambda$, $n = 0, 1, \ldots$ and state-dependent death rates are

$$\mu_n = \begin{cases} n\mu, & 1 \leq n < c \\ c\mu, & n \geq c. \end{cases}$$

The state transition diagram is shown in Fig. 6.8.

6.3.1.1 Steady-State Distribution

The underlying BDP $\{X(t), t \geq 0\}$ having all the states are positive recurrent if and only if $\rho = \frac{\lambda}{c\mu} < 1$. If $\rho < 1$, the equilibrium distribution exists and is given by

$$\pi_n = \begin{cases} \frac{\lambda^n}{n!\mu^n}\pi_0, & 1 \leq n \leq c \\ \frac{\lambda^n}{c^{n-c}c!\mu^n}\pi_0, & n > c. \end{cases}$$

Using the normalizing condition, $\sum_{n=0}^{\infty} \pi_n = 1$,

$$\pi_0 = \left[\sum_{n=0}^{c-1} \frac{1}{n!}\left(\frac{\lambda}{\mu}\right)^n + \frac{1}{c!}\left(\frac{\lambda}{\mu}\right)^c \left(\frac{c\mu}{c\mu - \lambda}\right)\right]^{-1}.$$

In particular, for the single server case (i.e., $c = 1$), the equilibrium distribution follows geometric distribution with parameter ρ i.e.,

$$\pi_n = (1 - \rho)\rho^n, \quad n = 0, 1, \ldots .$$

Remark 6.1 It is known that the irreducible BDP is transient if and only if

$$\sum_{n=1}^{\infty} \frac{\mu_1 \ldots \mu_n}{\lambda_1 \ldots \lambda_n} < \infty, \tag{6.22}$$

and positive recurrent if and only if

$$\sum_{n=1}^{\infty} \frac{\lambda_1 \ldots \lambda_n}{\mu_1 \ldots \mu_n} < \infty, \tag{6.23}$$

and null recurrent if neither Eq. (6.22) nor Eq. (6.23) holds. For the above process, condition for transient reduces to

$$\sum_{n=1}^{\infty} \left(\frac{c\mu}{\lambda}\right)^n < \infty.$$

Thus, a state in $M/M/c$ queuing model is transient if and only if $c\mu < \lambda$. Condition for positive recurrence reduces to

$$\sum_{n=1}^{\infty} \left(\frac{\lambda}{c\mu}\right)^n < \infty.$$

Therefore, a state in $M/M/c$ queuing model is positive recurrent if and only if $\lambda < c\mu$. Consequently, a state in M/M/c queuing model is null recurrent if and only if $\lambda = c\mu$.

6.3.1.2 Performance Measures of the System

1. The probability that an arriving customer has to wait on arrival in steady-state

$$C\left(c, \frac{\lambda}{\mu}\right) = \sum_{n=c}^{\infty} \pi_n = \frac{\pi_c}{1-\rho} \quad \left(say, C\left(c, \frac{\lambda}{\mu}\right)\right).$$

   This is known as Erlang's C-formula. Here, $C\left(c, \frac{\lambda}{\mu}\right)$ is the proportion of time all the c servers are busy in a steady-state.
2. Let N_q denote the number of customers in the queue

$$P\{N_q = j, W > 0\} = P\{N = c + j\} = \rho^j \pi_c = \rho^j (1-\rho) C\left(c, \frac{\lambda}{\mu}\right).$$

$$P\{N_q = j | W > 0\} = \frac{P\{N_q = j, W > 0\}}{P\{W > 0\}}$$
$$= (1-\rho)\rho^j, \; j = 0, 1, \ldots$$

   The conditional distribution of queue length given a new customer has to wait for service is a modified geometric distribution with parameter $1 - \rho = 1 - \frac{\lambda}{c\mu}$.
3. Let B be the number of busy servers in the system. Then, the expected number of busy servers

$$E(B) = \sum_{n=0}^{c-1} n\pi_n + \sum_{n=c}^{\infty} c\pi_n = c\rho.$$

4. Let I be the number of idle servers in the system. Then, the expected number of idle servers

$$E(I) = E(c - B) = c - c\rho = c(1-\rho).$$

   From above,

$$E(X) = \sum_{n=0}^{\infty} n\pi_n = E(B) + E(Q).$$

   Now

$$E(Q) = \sum_{n=c}^{\infty}(n-c)\pi_n = \sum_{n=c}^{\infty}(n-c)\frac{(\frac{\lambda}{\mu})^n}{c!(c^{n-c})}\pi_0$$
$$= \frac{\rho}{1-\rho}C\left(c, \frac{\lambda}{\mu}\right).$$

Hence,

$$E(X) = c\rho + \frac{\rho}{1-\rho}C\left(c, \frac{\lambda}{\mu}\right).$$

Using Little's law

$$E(X) = \lambda E(R),$$

$$E(R) = \frac{E(X)}{\lambda} = \frac{1}{\mu} + \frac{\pi_c}{c\mu(1-\rho)^2}.$$

Using

$$\lambda E(W) = E(Q),$$

we obtain

$$E(W) = \frac{E(Q)}{\lambda} = \frac{1}{\mu} + \frac{\pi_c}{c\mu(1-\rho)^2}.$$

6.3.1.3 Transient Analysis

The approach to obtain the transient solution for the proposed model has been provided by Parthasarathy and Sharafali [5].

Let $\pi_n(t) = P\{X(t) = n\}, n = 0, 1, \ldots$ and let there be i units in the system at $t = 0$. The Kolmogorov forward equations for the system are

$$\frac{d}{dt}\pi_0(t) = \mu\pi_1(t) - \lambda\pi_0(t), \tag{6.24}$$
$$\frac{d}{dt}\pi_n(t) = (n+1)\mu\pi_{n+1}(t) - (\lambda + n\mu)\pi_n(t) + \lambda\pi_{n-1}(t), \ 1 \leqq n < c, \tag{6.25}$$
$$\frac{d}{dt}\pi_n(t) = c\mu\pi_{n+1}(t) - (\lambda + c\mu)\pi_n(t) + \lambda\pi_{n-1}(t), \quad n \geqq c, \tag{6.26}$$

with $\pi_n(0) = \delta_{in}$, the Kronecker delta, λ and μ are the arrival and service rates, respectively. Define

$$P(z,t) = q_{c-1}(t) + \sum_{n=0}^{\infty}\pi_{c+n}(t)z^{n+1}; \quad P(z,0) = z^{\tau(i)}, \tag{6.27}$$

with

$$q_{c-1}(t) = \sum_{k=0}^{c-1} \pi_k(t), \quad \tau(i) = (i+1-c)\left[1 - \sum_{k=0}^{c-1} \delta_{ik}\right]. \tag{6.28}$$

The system of equations (6.24), (6.25), and (6.26) yields

$$\frac{\partial}{\partial t}P(z,t) = \left[\frac{c\mu}{z} + \lambda z - (\lambda + c\mu)\right]\left[P(z,t) - q_{c-1}(t)\right] + \lambda(z-1)\pi_{c-1}(t). \tag{6.29}$$

The solution of this differential equation is obtained as

$$\begin{aligned} P(z,t) &= P(z,0)\exp\left\{-\left[\lambda + c\mu - \frac{c\mu}{z} - \lambda z\right]t\right\} \\ &+ \int_0^t \left[\lambda(z-1)\pi_{c-1}(u) + \left\{(\lambda + c\mu) - \frac{c\mu}{z} - \lambda z\right\} q_{c-1}(u)\right] \\ &\times \exp\left\{-\left[\lambda + c\mu - \frac{c\mu}{z} - \lambda z\right](t-u)\right\} du. \end{aligned} \tag{6.30}$$

It is well known that if $\alpha = 2\sqrt{c\lambda\mu}$ and $\beta = \sqrt{\lambda/c\mu}$,

$$\exp\left\{\left(\lambda z + \frac{c\mu}{z}\right)t\right\} = \sum_{n=-\infty}^{\infty} (\beta z)^n I_n(\alpha t),$$

where $I_n(\cdot)$ is the modified Bessel function of the first kind of order n. Using this and comparing the coefficients of z^n on either side, for $n = 1, 2, \ldots$

$$\begin{aligned} \beta^{-n+1}\pi_{n+c-1}(t) &= \exp(-(\lambda + c\mu)t)\beta^{1-\tau(i)} I_{n-\tau(i)}(\alpha t) \\ &+ \int_0^t \exp(-(\lambda + c\mu)(t-u))[\lambda I_{n-1}(\alpha(t-u)) \\ &-\beta I_n(\alpha(t-u))\pi_{c-1}(u) + (\lambda + c\mu)\beta I_n(\alpha(t-u)) \\ &-\lambda I_{n+1}(\alpha(t-u)) + I_{n-1}(\alpha(t-u))q_{c-1}(u)]du, \end{aligned} \tag{6.31}$$

and for $n = 0$,

$$\begin{aligned} \beta q_{c-1}(t) = \exp(-(\lambda + c\mu)t)\beta^{1-\tau(i)} I_{\tau(i)}(\alpha t) + \int_0^t \exp(-(\lambda + c\mu)(t-u)) \\ \left[\lambda \left\{I_1(\alpha(t-u)) - \beta I_0(\alpha(t-u))\right\} \pi_{c-1}(u)\right. \\ \left. + \left\{\beta(\lambda + c\mu)I_0(\alpha(t-u)) - 2\lambda I_1(\alpha(t-u))\right\} q_{c-1}(u)\right] du. \end{aligned} \tag{6.32}$$

As $P(z, t)$ does not contain terms with negative powers of z, the right-hand side of (6.31) with n replaced by $-n$, must be zero. Thus,

$$\exp(-(\lambda + c\mu)t)\beta^{1-\tau(i)} I_{n+\tau(i)}(\alpha t)$$
$$+ \int_0^t \exp(-(\lambda + c\mu)(t-u))\Big[\lambda \{I_{n+1}(\alpha(t-u)) - \beta I_n(\alpha(t-u))\}\pi_{c-1}(u)$$
$$+ \{(\lambda + c\mu)\beta I_n(\alpha(t-u)) - \lambda(I_{n-1}(\alpha(t-u)) + I_{n+1}(\alpha(t-u)))\}$$
$$q_{c-1}(u)\Big]du = 0, \tag{6.33}$$

where $I_{-k}(\cdot) = I_k(\cdot)$. Usage of (6.33) in (6.31) considerably simplifies the working and results in elegant expressions for $\pi_n(t)$. This yields, for $n = 1, 2, \ldots,$

$$\begin{aligned}\pi_{n+c-1}(t) &= \exp(-(\lambda + c\mu)t)\beta^{n-t(i)} \left[I_{n-\tau(i)}(\alpha t) - I_{n+\tau(i)}(\alpha t)\right] \\ &\quad + n\beta^n \int_0^t \frac{I_n(\alpha(t-u))}{t-u} \exp(-(\lambda + c\mu)(t-u)) p_{c-1}(u)du.\end{aligned} \tag{6.34}$$

Now, the probabilities $\pi_n(t)$, $n = 0, 1, \ldots, c-1$, remain to be solved. For this the system of equations (6.24) and (6.25) are considered subject to (6.32). Leaving out the last of the equation from the system in (6.25), the remaining equations together with (6.24) can be expressed in the form

$$\frac{d}{dt}\pi(t) = A\pi(t) + (c-1)\mu\pi_{c-1}(t)e_{c-1}, \tag{6.35}$$

where

$$\pi(t) = (\pi_0(t), \pi_1(t), \ldots, \pi_{c-2}(t))^{\mathrm{T}}, \quad A = \left(a_{kj}\right)_{(c-1)\times(c-1)},$$

with

$$a_{kj} = \begin{cases} \lambda, & j = k-1, k = 1, 2, \cdots, c-2 \\ -(\lambda + k\mu), & j = k, k = 0, 1, \cdots, c-2 \\ (k+1)\mu, & j = k+1, k = 0, 1, \cdots, c-3 \end{cases} \tag{6.36}$$

and $\boldsymbol{e}_{c-1}$ is a column vector of order $(c-1)$ with 1 in the last place and zero in the remaining places.

In the sequel, for any function $f(\cdot)$, let $\widehat{f}(s)$ denote its Laplace transform. Now, by taking the Laplace transform, the solution of (6.33) is obtained as

$$\widehat{\pi}(s) = [sI - A]^{-1}\left[(c-1)\mu\widehat{\pi}_{c-1}(s)\boldsymbol{e}_{c-1} + \boldsymbol{\pi}(0)\right], \tag{6.37}$$

with

$$\pi(0) = (\delta_{i0}, \delta_{i1}, \cdots, \delta_{ic-2})^{\mathrm{T}}. \tag{6.38}$$

Thus, only $\widehat{\pi}_{c-1}(s)$ remains to be found. It is observed that if

$$\mathbf{e} = (1, 1, \cdots, 1)^{\mathrm{T}}, \quad e^{\mathrm{T}}\widehat{\pi}(s) + \widehat{\pi}_{c-1}(s) = \widehat{q}_{c-1}(s). \tag{6.39}$$

Using (6.32) in this and simplifying,

$$\widehat{\pi}_{c-1}(s) = \frac{\left\{\frac{p-\sqrt{p^2-\alpha^2}}{\alpha\beta}\right\}^{\tau(i)} - se^{\mathrm{T}}(sI - A)^{-1}\pi(0)}{s + \lambda - \left(\frac{p-\sqrt{p^2-\alpha^2}}{2}\right) + (c-1)\mu\, se^{\mathrm{T}}(sI - A)^{-1}e_{c-1}}, \tag{6.40}$$

where $p = s + \lambda + c\mu$. Let

$$(sI - A)^{-1} = \left(\widehat{a}_{kj}(s)\right)_{(c-1)\times(c-1)}. \tag{6.41}$$

It is noted that $(sI - A)$ is almost lower triangular. It can be obtained, for $k = 0, 1, \ldots, c-2$,

$$\widehat{a}_{kj}(s) = \begin{cases} \dfrac{u_{c-1,j+1}(s)u_{k,0}(s) - u_{k,j+1}(s)u_{c-1,0}(s)}{(j+1)\mu\, u_{c-1,0}(s)}, & j = 0, 1, \ldots, c-3 \\ \dfrac{u_{k,0}(s)}{u_{c-1,0}(s)}, & j = c-2 \end{cases} \tag{6.42}$$

where $u_{k,j}(s)$ are recursively given as,

$$\begin{aligned} u_{k,k} &= 1, \quad k = 0, 1, \ldots, c-2 \\ u_{k+1,k} &= \frac{s + \lambda + k\mu}{(k+1)\mu}, \quad k = 0, 1, \ldots, c-3 \\ u_{k+1,k-j} &= \frac{(s + \lambda + k\mu)u_{k,k-j} - \lambda u_{k-1,k-j}}{(k+1)\mu}, \quad j \le k;\, k = 1, 2, \ldots, c-3 \\ u_{c-1,j} &= \begin{cases} [s + \lambda + (c-2)\mu]u_{c-2,j} - \lambda u_{c-3,j}, & j = 0, 1, \ldots, c-3 \\ s + \lambda + (c-2)\mu, & j = c-2 \end{cases} \\ u_{kj} &= 0, \end{aligned}$$

for other k and j. The advantage in using these relations is that no determinant is evaluated. However, for smaller c, one can use the usual method to find $(sI - A)^{-1}$. Using these in (6.40), implies

$$\widehat{\pi}_{c-1}(s) = \frac{\left\{\frac{p-\sqrt{p^2-\alpha^2}}{\alpha\beta}\right\}^{\tau(i)} - \sum_{j=0}^{c-2}\sum_{k=0}^{c-2}\delta_{ij}s\widehat{a}_{kj}(s)}{s + \lambda - \frac{p-\sqrt{p^2-\alpha^2}}{2} + (c-1)\mu s\sum_{k=0}^{c-2}\widehat{a}_{kc-2}(s)} \tag{6.43}$$

and for $k = 0, 1, \ldots, c-2$

$$\widehat{\pi}_k(s) = \sum_{j=0}^{c-2} \delta_{ij}\widehat{a}_{kj}(s) + (c-1)\mu\widehat{a}_{kc-2}(s)\widehat{\pi}_{c-1}(s). \tag{6.44}$$

It is observed that $\widehat{a}_{kj}(s)$ are all rational algebraic functions in s. The cofactor of the (i, j)th element of $(sI - A)$ is a polynomial of degree $c - 2 - |i - j|$. In particular, the cofactors of the diagonal elements are polynomials in s of degree $c - 2$ with the leading coefficient equal to 1. Also, $u_{c-1,0}(s) = 0$ is the characteristic equation of A. Since $a_{00} = -\lambda \neq 0$, it is also known that the characteristic roots of A are all distinct and negative [6]. Hence, the inverse transform $\widehat{a}_{kj}(t)$ of $\widehat{a}_{kj}(s)$ can be obtained by partial fraction decomposition. Let s_k, $k = 0, 1, \cdots, c-2$ be the characteristic roots of the matrix A. Then

$$\widehat{a}_{kj}(s) = \sum_{m=0}^{c-2} \frac{\rho_{kj}^{(m)}}{s - s_m}, \tag{6.45}$$

where the constants $\rho_{kj}^{(m)}$ are given by

$$\rho_{kj}^{(m)} = \lim_{s \to s_m} \left[(s - s_m)\,\widehat{a}_{kj}(s)\right]. \tag{6.46}$$

Thus,

$$a_{kj}(t) = \sum_{m=0}^{c-2} \rho_{kj}^{(m)} \exp(s_m t). \tag{6.47}$$

Similarly,

$$\sum_{k=0}^{c-2} (c-1)s\widehat{a}_{kj}(s) = c - 1 + \widehat{b}_j(s), \quad j = 0, 1, \ldots, c-2, \tag{6.48}$$

where $\widehat{b}_j(s)$ can be resolved as

$$\widehat{b}_j(s) = \sum_{m=0}^{c-2} \frac{B^{(m)}}{s - s_m}, \tag{6.49}$$

with

$$B_j^{(m)} = \lim_{s \to s_m} \left[(s - s_m)(c-1)\sum_{k=0}^{c-2} s\widehat{a}_{kj}(s)\right]. \tag{6.50}$$

Then, $b_j(t)$, the inverse transform of $\widehat{b}_j(s)$ is

$$b_j(t) = \sum_{m-0}^{c-2} B_j^{(m)} \exp(s_m t), \quad j = 0, 1, \cdots, c-2. \tag{6.51}$$

Using (6.50) in (6.43) yields

$$\widehat{\pi}_{c-1}(s) = \frac{\left\{\frac{p-\sqrt{p^2-\alpha^2}}{\alpha\beta}\right\}^{\tau(i)} - (c-1)^{-1}\sum_{j=0}^{c-2}\delta_{ij}[c-1+\widehat{b}_j(s)]}{\frac{p+\sqrt{p^2-\alpha^2}}{2} - \mu + \mu\widehat{b}_{c-2}(s)}$$

$$= \left(\frac{2}{\alpha}\right)\frac{p-\sqrt{p^2-\alpha^2}}{\alpha}\left[\left\{\frac{p-\sqrt{p^2-\alpha^2}}{\alpha\beta}\right\}^{\tau(i)}\right.$$

$$\left. -(c-1)^{-1}\sum_{j=0}^{c-2}\delta_{ij}\left\{c-1+\widehat{b}_j(s)\right\}\right]$$

$$\cdot\left[1-\left(\frac{\mu}{c\lambda}\right)^{1/2}\frac{p-\sqrt{p^2-\alpha^2}}{\alpha}\left(1-\widehat{b}_{c-2}(s)\right)\right]^{-1}, \tag{6.52}$$

which on inversion yields

$$\pi_{c-1}(t) = \sum_{m=0}^{\infty}\left(\frac{\mu}{c\lambda}\right)^{m/2}\sum_{k=0}^{m}(-1)^k\binom{m}{k}\int_0^t b_{c-2}^{*k}(t-u)$$

$$\cdot\left[\beta^{-\tau(i)}\exp(-(\lambda+c\mu)u)\{I_{m+\tau(i)}(\alpha u) - I_{m+\tau(i)+2}(\alpha u)\}.\right.$$

$$-\sum_{j=0}^{c-2}\delta_{ij}\int_0^u\{\delta(u-v)+(c-1)^{-1}b_j(u-v)\}\exp(-(\lambda+c\mu)v)$$

$$\left.\cdot\{I_m(\alpha v) - I_{m+2}(\alpha v)\}dv\right]du, \tag{6.53}$$

where $b_{c-2}^{*k}(t)$ is the k-fold convolution of $b_{c-2}(t)$ with itself. It is noted that $b_{c-2}^{*0}(t) = \delta(t)$. Also for $k = 0, 1, \ldots, c-2$,

$$\pi_k(t) = \sum_{j=0}^{c-2}\delta_{ij}\tilde{a}_{kj}(t) + (c-1)\mu\int_0^t \tilde{a}_{kc-2}(u)\pi_{c-1}(t-u)du. \tag{6.54}$$

Thus (6.34) (6.53) and (6.54) completely determine all the transient state probabilities.

Example 6.11 In a particular supermarket, there are seven counters. Assume that each cashier at the counter can service a mean of three clients every five minutes and that there are nine customers who arrive on the mean every five minutes. Assume that service times follow an exponential distribution and that arrivals follow a Poisson process. In the long run, find the following measures.

1. Mean number of customers in the queue.
2. Mean time a customer spends in the system.

Solution: For given $\lambda = 9/5$, $\mu = 3/5$. In the long run:

1. Mean number of customers in the queue

$$E(Q) = \frac{\rho\pi_c}{(1-\rho)^2} = 0.02825.$$

2. Mean time a customer spends in the system

$$E(R) = \frac{E(Q)}{\lambda} + \frac{1}{\mu} = 1.6823 \text{ min}.$$

Example 6.12 (Heterogeneous two processor system) Consider a system with two processors where the service rates of the processors differ. The queueing structure is demonstrated in Fig. 6.9. Assume without loss of generality that $\mu_1 > \mu_2$. The states of the system are denoted by (n_1, n_2), where $n_1 \geq 0$ is the number of jobs in the queue including any at the faster server, and $n_2 \in \{0, 1\}$ is the number of jobs at the slower server. Jobs wait in line corresponding to the order of their arrival. When both servers are idle, the faster server is scheduled for service before the slower one [7]. The state transition diagram of the system is given in Fig. 6.10. It is assumed that there are 0 units in the system at $t = 0$. The forward equations for the system are as follows:

$$\begin{aligned}
P'_{00}(t) &= -\lambda P_{00}(t) + \mu_1 P_{10}(t) + \mu_2 P_{01}(t),\\
P'_{10}(t) &= \lambda P_{00}(t) - (\lambda + \mu_1) P_{10}(t) + \mu_2 P_{11}(t),\\
P'_{01}(t) &= -(\lambda + \mu_2) P_{01}(t) + \mu_1 P_{11}(t),\\
P'_{11}(t) &= \lambda P_{10}(t) + \lambda P_{01}(t) - (\lambda + \mu_1 + \mu_2) P_{11}(t) + (\mu_1 + \mu_2) P_{21}(t),\\
P'_{n1}(t) &= \lambda P_{n-11}(t) - (\lambda + \mu_1 + \mu_2) P_{n1}(t) + (\mu_1 + \mu_2) P_{n+11}(t), n = 2, 3, \ldots,
\end{aligned}$$

with $P_{00}(0) = 1$. Here, $P_{nm}(t)$ is the probability that the system is in state (n, m) at time t.

The detailed derivation of the above time-dependent probabilities is presented in [7]. The time-dependent probabilities for this system are

$$P_{11}(t) = \sum_{m=0}^{\infty} \left(\frac{\mu}{\lambda}\right)^{m/2} \sum_{k=0}^{m} (-1)^k \binom{m}{k} \tag{6.55}$$
$$\times \int_0^t \frac{1}{\mu} b_2^{*k}(t-u) \Big[\exp(-(\lambda+\mu)u)\{I_m(\alpha u) - I_{m+2}(\alpha u)\}$$
$$- \int_0^u b_1(u-v)\exp(-(\lambda+\mu)v)\{I_m(\alpha v) - I_{m+2}(\alpha v)\}dv \Big] du,$$

where b_2^{*k} is the $k-$ fold convolution of $b_2(t)$ with itself. Further,

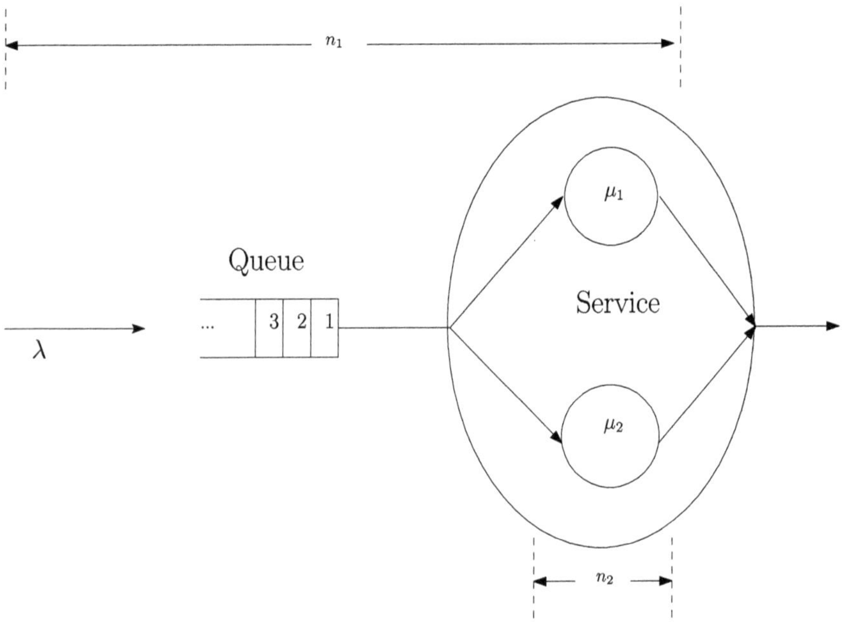

Fig. 6.9 Queueing structure for the heterogeneous two processors system

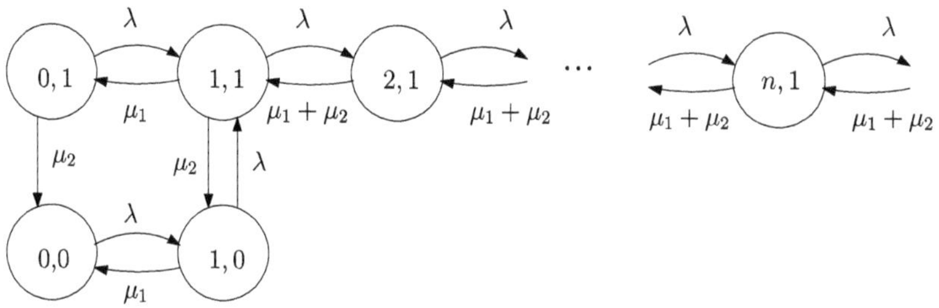

Fig. 6.10 State transition diagram for the heterogeneous two processors system

$$P_{00}(t) = a_{11}(t) + \int_0^t (\mu_2 a_{12}(u) + \mu_1 a_{13}(u))\, P_{11}(t-u)du,$$

$$P_{10}(t) = a_{21}(t) + \int_0^t (\mu_2 a_{22}(u) + \mu_1 a_{23}(u))\, P_{11}(t-u)du,$$

$$P_{01}(t) = a_{31}(t) + \int_0^t (\mu_2 a_{32}(u) + \mu_1 a_{33}(u))\, P_{11}(t-u)du.$$

The steady-state probabilities for this system are given by [8]

$$\pi_{01} = \frac{\rho}{1+2\rho}\frac{\lambda}{\mu_2}\pi_{00}, \tag{6.56}$$

$$\pi_{10} = \frac{1+\rho}{1+2\rho}\frac{\lambda}{\mu_1}\pi_{00}, \tag{6.57}$$

$$\pi_{11} = \frac{\rho}{1+2\rho}\frac{\lambda(\lambda+\mu_2)}{\mu_1\mu_2}\pi_{00}, \tag{6.58}$$

$$\pi_{n1} = \rho\pi_{n-11} = \rho^{n-1}\pi_{11}, \quad n = 2, 3, \ldots, \tag{6.59}$$

where $\rho = \frac{\lambda}{\mu_1+\mu_2}$ and,

$$\pi_{00} = \left[1 + \frac{\lambda(\lambda+\mu_2)}{\mu_1\mu_2(1+2\rho)(1-\rho)}\right]^{-1}.$$

Example 6.13 (Multi-server with catastrophes) Consider a multi-server heterogeneous queueing system with catastrophes as mentioned by Dharmaraja and Rakesh [9]. The arrivals occur in accordance with a Poisson process with mean arrival rate λ. There are c servers with different service rates μ_i $(i = 1, 2, \ldots, c)$ for each of the c servers and the service times for each server follow exponential distributions. Since in most the real life situations the inter-arrival times and service times are found to be continuous RVs following Markov property, they may be better described by an exponential distribution. Moreover, the Markov assumption helps us to obtain closed

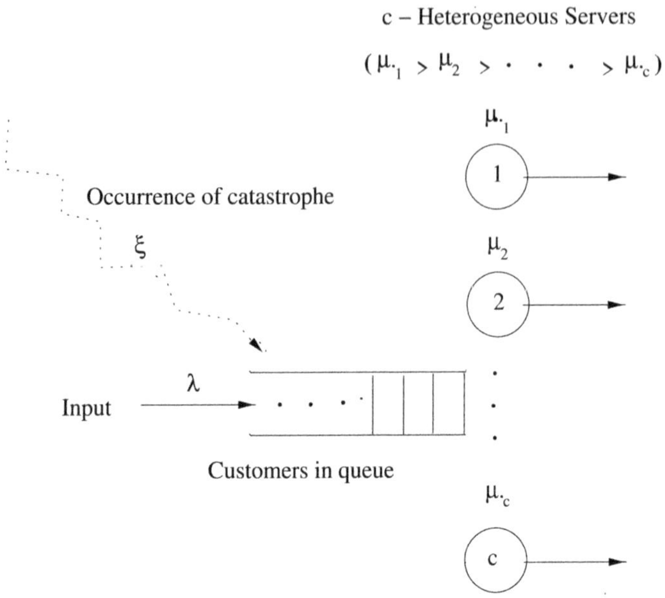

Fig. 6.11 A Markovian queueing model with heterogeneous servers and catastrophes

form results. One server is required for each customer, and FCFS queue discipline is used. Without losing generality, it is assumed that the servers are arranged in decreasing order of service speed. This implies that the customers are always served by the fastest servers, i.e., when $k < c$ customers are present, servers $1, 2, \ldots, k$ are used. This implies, if a faster server becomes available and there is no other customers waiting, a client may switch to it immediately within a service time. When the system is not empty, catastrophes can also happen at the service facilities as a Poisson process with a rate of ξ, in addition to arrival and service processes. Every time a catastrophe happens in the system, all the users are instantly killed, all the c servers are momentarily idle, and the servers are ready to serve again.

The schematic representation of the proposed system is provided in Fig. 6.11. Let $\{X(t), t \geq 0\}$ be the number of customers present in the system at time t. Let $\pi_n(t) = P\{X(t) = n\}, n = 0, 1, \ldots$ be the probability that there are n customers in the system at time t and $P(z, t)$ the corresponding probability generating function. It is considered that there is no customer in the system at $t = 0$.

The queueing model under investigation is governed by the following differential-difference equations:

$$\frac{d\pi_0(t)}{dt} = -(\lambda + \xi)\pi_0(t) + \mu_1\pi_1(t) + \xi \tag{6.60}$$

$$\frac{d\pi_n(t)}{dt} = -\left(\lambda + \xi + \sum_{i=1}^{n} \mu_i\right)\pi_n(t) + \lambda\pi_{n-1}(t) + \sum_{i=1}^{n+1} \mu_i \pi_{n+1}(t),$$
$$n = 1, 2, \ldots, c-1 \qquad (6.61)$$

$$\frac{d\pi_c(t)}{dt} = -\left(\lambda + \xi + \sum_{i=1}^{c} \mu_i\right)\pi_c(t) + \lambda\pi_{c-1}(t) + \sum_{i=1}^{c} \mu_i \pi_{c+1}(t) \qquad (6.62)$$

$$\frac{d\pi_n(t)}{dt} = -\left(\lambda + \xi + \sum_{i=1}^{c} \mu_i\right)\pi_n(t) + \lambda\pi_{n-1}(t) + \sum_{i=1}^{c} \mu_i \pi_{n+1}(t),$$
$$n = c+1, c+2, \ldots \qquad (6.63)$$

with $\pi_n(0) = \delta_{0n}$, the Kronecker delta symbol.

Theorem 6.3 *The time-dependent probabilities of the system size of a Markov queueing model with heterogeneous servers and catastrophes which is governed by the differential-difference Equations (6.60–6.63) are given by*

$$\pi_k(t) = b_{k,0}(t) + \xi \int_0^t b_{k,0}(u)du$$
$$+\gamma \int_0^t b_{k,c-1}(u)\pi_c(t-u)du, k = 0, 1, \ldots, c-1$$

$$\pi_c(t) = \sum_{n=0}^{\infty} \sum_{m=0}^{n} \frac{(-1)^m}{\gamma} \left(\frac{\alpha}{2\lambda}\right)^{n+1} \binom{n}{k} \left[\int_0^t E(t-u)\right.$$
$$\int_0^u F^{C(m)}(u-v) \exp\{-(\lambda+\gamma+\xi)v\} \frac{I_{n+1}(\alpha v)}{v} dudv$$
$$+\xi \int_0^t G(t-u) \int_0^u F^{C(m)}(u-v) \exp\{-(\lambda+\gamma+\xi)v\}$$
$$\left.\frac{I_{n+1}(\alpha v)}{v} dudv\right]$$

for $n = 1, 2, \ldots,$

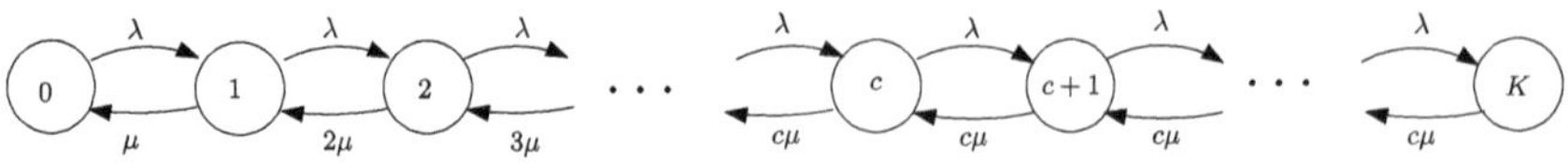

Fig. 6.12 State transition diagram of $M/M/c/K$ queueing model

$$\pi_{n+c}(t) = n\beta^n \int_0^t \exp\{-(\lambda + \xi + \gamma)(t-u)\} \frac{I_n(\alpha(t-u))}{(t-u)} \pi_c(u)du$$

where $G(t) = \int_0^t E(u)du$ *and* $F^{C(m)}(t)$ *is* $m -$ *fold convolution of* $F(t)$ *with itself with* $F^{C(0)}(t) = \delta(t)$*, the Dirac delta function.*

Proof The proof of the theorem is left to the reader as an exercise. □

Special Cases
Case (1): When $c = 2$, which is the case of two heterogeneous servers system (refer Example 6.12).
Case(2): When all the c servers are homogeneous and there is no catastrophe, the model results agree with that of $M/M/c$ model (refer Sect. 6.3.1.3).

6.3.2 M/M/c/K Queueing Model

In this section, we will delineate the $M/M/c$ queueing model incorporating a finite capacity K. The discussion will encompass a description of its characteristics, focusing on the derivation of the steady-state solution.

Definition 6.4 In this queueing system, the arrivals follow the Poisson process with rate λ and service time is exponentially distributed with parameter μ. The number of servers is c with finite system capacity K $(c < K)$. If an arriving customer finds n customers already in the system, where if

- $n < c$, he is routed to an idle server,
- $n \geq c$, he joins the waiting queue.

Also, since the system capacity is finite K, therefore customers are forced to leave the system if already K are present in the system.

The state transition diagram is shown in Fig. 6.12.

6.3.2.1 Steady-State Distribution

This queueing model is a birth-death process with state dependent rates

$$\mu_n = \begin{cases} n\mu, & \text{if } 1 \leq n < c, \\ c\mu, & \text{if } c \leq n < K. \end{cases}$$

As evaluated before for other queueing systems, the steady-state or equilibrium solution of the system is given by

$$\pi_n = \begin{cases} \frac{1}{n!}\left(\frac{\lambda}{\mu}\right)^n \pi_0, & \text{if } 0 \leq n < c, \\ \frac{1}{c^{n-c}c!}\left(\frac{\lambda}{\mu}\right)^n \pi_0, & \text{if } c \leq n < K. \end{cases}$$

6.3.2.2 Performance Measures of the System

For this queueing system, the performance measures are defined as follows:

1. Expected number of customers in the system in the steady-state is

$$E(X) = \sum_{n=1}^{K} n\pi_n.$$

2. Expected queue length is given by

$$E(Q) = \sum_{n=c}^{K} (n-c)\pi_n.$$

3. Expected time spent by a customer in the system is given using Little's law,

$$E(R) = \frac{E(X)}{\lambda_{eff}}, \text{ where } \lambda_{eff} = \lambda(1 - \pi_K).$$

4. Expected time spent by a customer waiting is given by

$$E(W) = \frac{E(Q)}{\lambda_{eff}}.$$

Example 6.14 Consider a vehicle emission inspection facility with three examination booths that can fit a single vehicle each. It makes sense to suppose that vehicles queue up such that the first vehicle to move up to an open stall will do so. At most four automobiles can wait at once in the station (there are totally seven stations). During busy times, there is typically one automobile arriving every minute according to a Poisson distribution. The service time is exponentially distributed with a mean 6 min.

Find

1. The mean number of customers in the system during the peak periods.

2. The mean waiting time during the peak period.

Solution: This problem can be modeled as $M/M/3/7$ system with $\lambda = 1$ and $\mu = \frac{1}{6}$.

1. The mean number of customers in the system during the peak periods is

$$E(X) = \sum_{n=1}^{7} n\pi_n = 6.06 \text{ cars}.$$

2. The mean waiting time during the peak period is

$$E(R) = \frac{E(X)}{\lambda_{eff}} = 12.3 \text{ min}.$$

Example 6.15 Consider a service center with two servers. Each server performs with rate of two services every hour with an exponential distribution. Assuming the system can accommodate no more than three consumers, assume that customers arrive at a Poisson process with a rate of three per hour. Model this problem as a $M/M/2/3$ system and find the fraction of potential customers entering the system.
Solution: The proportion of potential customers who enter the system will be

$$1 - \pi_3 = \frac{1}{22!}\left(\frac{3}{2}\right)^3 = 0.8111.$$

6.3.3 M/M/c/c Loss System

In this section, we will outline the $M/M/c/c$ loss queueing model, providing a comprehensive description of its characteristics. The emphasis will be on deriving the steady-state solution within this context.

Definition 6.5 Here, the assumption of infinite capacity queue is relaxed. Here, it has been considered that the capacity of the system is c and there are c servers providing service. This type of system is also known as *Erlang loss system*. The customer arrives at the system following the Poisson process with arrival rate λ and will be served with the exponentially distributed rate μ.

After deducing from the balance equations of the system, the steady-state probabilities are calculated as

$$\pi_n = \frac{\left(\frac{\lambda}{\mu}\right)^n}{n!}\left[\sum_{k=0}^{c} \frac{\left(\frac{\lambda}{\mu}\right)^k}{k!}\right]^{-1}, \quad n = 0, 1, \ldots, c.$$

When $n = c$, the resultant formula for p_c is called *Erlang B formula*.

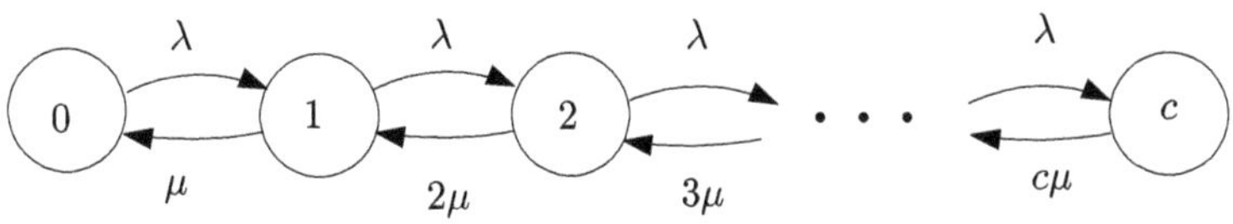

Fig. 6.13 $M/M/c/c$ queueing model

$$B(c,r) = \pi_c = \frac{\left(\frac{\lambda}{\mu}\right)^c}{c!}\left[\sum_{k=0}^{c}\frac{\left(\frac{\lambda}{\mu}\right)^k}{k!}\right]^{-1}, \quad n = 0, 1, \ldots, c.$$

The state transition diagram is shown in Fig. 6.13.

The probability that an arriving customer does not have to wait in the queue is given as

$$C(c,r) = \frac{\left(\frac{\lambda}{\mu}\right)^c}{c!\left(1-\frac{\lambda}{\mu}\right)}\left[\sum_{k=0}^{c-1}\frac{\left(\frac{\lambda}{\mu}\right)^k}{k!} + \frac{\left(\frac{\lambda}{\mu}\right)^c}{c\left(1-\frac{\lambda}{\mu}\right)}\right]^{-1}, \quad n = 0, 1, \ldots, c.$$

This expression is known as *Erlang C formula*. From the closed form expression of steady-state probabilities it is clear that the system state follows a truncated Poisson distribution with parameter ρ.

Example 6.16 Calls to a particular mobile service provider's technical center are received with a Poisson process at a rate of 30 per hour. An exponential distribution with a mean of 5 minutes shows how long it takes a staff member to complete one customer request on average. Three technicians are available to help callers at the support center. Find the number of servers required such that the call center is to reach a 90% likelihood of calls arriving on time.
Solution: The goal of the call center is to reach a 90% likelihood of calls arriving on time. The number of servers required for this can be computed by $1 - C(c,r) \geq 0.9$. It is found that $c = 6$ servers is the minimum number of servers that satisfies this requirement.

6.3.3.1 Performance Measures of the System

Some of the important performance measures for $M/M/c/c$ system are derived as follows.

1. The probability that all servers are busy

$$P_{busy} = \pi_c.$$

2. The mean number of customers in the system is given by

$$E(X) = \rho(1 - \pi_c).$$

3. The effective arrival rate

$$\lambda_{eff} = \lambda(1 - \pi_c).$$

6.3.3.2 Transient Solution

The Kolmogorov forward equation of the system can be written as

$$\begin{aligned}
\pi_0'(t) &= -\lambda\pi_0(t) + \mu\pi_1(t),\\
\pi_j'(t) &= \lambda\pi_{j-1}(t) - (\lambda + j\mu)\pi_j(t) + (j+1)\mu\pi_{j+1}(t), \quad j = 1, 2, \ldots, c-1,\\
\pi_c'(t) &= \lambda\pi_{c-1}(t) - c\mu\pi_c(t).
\end{aligned}$$

Assume that $\pi_j(0) = \delta_{mj}$. The transient solution of $M/M/c/c$ system involves the roots of Charlier polynomials which cannot be expressed in a closed form. Take $\rho = \frac{\lambda}{\mu}$. These are also the roots of $|A(s)| = D_{c+1}(s) = 0$, and they are evaluated numerically where

$$|A(s)| = \begin{vmatrix} s+\rho & -1 & & \\ -\rho & s+\rho+1 & -2 & \\ & & \ddots & \\ & & -\rho & s+c \end{vmatrix}_{c+1\times c+1}.$$

The subdeterminant $D_n(s)$ is obtained from $|A(s)|$ by taking the first n rows and n columns with $D_0(s) = 1$. Therefore, $|A(s)| = sD_c(s+1)$. It is well known that the roots of $|A(s)| = 0$ are real, negative, and distinct.

The transient solution of system size probabilities is given as follows:

$$\pi_j(t) = \frac{\rho^j/j!}{\sum_{k=0}^{c}\frac{\rho^k}{k!}} + c!\rho^{c-m}\sum_{r=1}^{c}\frac{D_j(x_r)e^{(x_r\mu t)}}{x_r D_c'(x_r+1)D_c(x_r)}, \quad j = 0, 1, \ldots, c$$

where x_r are the roots of Charlier polynomial $D_c(s+1) = 0$.

Special Cases

1. For $M/M/1/1$, the time-dependent probabilities are given by

$$\pi_0(t) = \frac{1}{\sum_{k=0}^{1} \frac{\rho^k}{k!}} + \rho^{1-m} \frac{D_0(x_1)e^{(x_1 \mu t)}}{x_1 D_1'(x_1+1)D_c(x_1)}$$
$$= \frac{1}{1+\rho} + \frac{\rho}{1+\rho} e^{-(\mu+\lambda)t},$$
$$\pi_1(t) = \frac{\rho}{1+\rho} - \frac{1}{1+\rho} e^{-(\mu+\lambda)t}.$$

2. For $M/M/2/2$, the time-dependent probabilities of the system are provided by.

$$\pi_j(t) = \frac{\rho^j/j!}{1+\rho+\rho^2/2} + \rho^{j-m} \sum_{r=1}^{2} \frac{D_m(x_r)D_j(x_r)e^{(x_r \mu t)}}{x_r(x_r+1-\rho)}, \; j = 0, 1, 2$$

where $x_1, x_2 = \frac{-1}{2}\left(2\rho + 3 \pm (1+4\rho)^{1/2}\right)$ and

$$D_0(s) = 1; \quad D_1(s) = s + \rho; \quad D_2(s) = s^2 + s(1+2\rho) + \rho^2.$$

Example 6.17 There are c circuits available on a rural telephone switch to transport c calls. If all circuits are occupied, a new call is blocked. Suppose calls have duration (in minutes) which arrives following the Poisson process with rate $\lambda = 2$ and the service is exponentially distributed with rate $\mu = 3$. Find out the minimum number of circuits for $g = 1$ such that the blocking probability denoted by $P_b < 0.05$ and the dropping probability $P_d < 0.02$.
Solution: Using (5.47) and (5.48), the minimum value of c is found to be 8 to satisfy the given threshold values (Fig. 6.14).

6.4 $M/M/\infty$ self-service System

In this section, the $M/M/\infty$ self-service system(which is also called infinite server system) will be outilned, offering a thorough description of its characteristics. The focus will be on deriving the steady-state solution within this particular framework.

Definition 6.6 There is no queue in this Markov queueing model. Assume that the clients arrive following the Poisson process with the parameter λ, for service. Every arriving client finds a server who is idle right away because there are an infinite number of servers. Let $X(t)$ denote the number of customers in the system at time t. The underlying stochastic process $\{X(t), t \geq 0\}$ of the infinite server system is a BDP with birth rate as

$$\lambda_n = \lambda, \quad n = 0, 1, \ldots \tag{6.64}$$

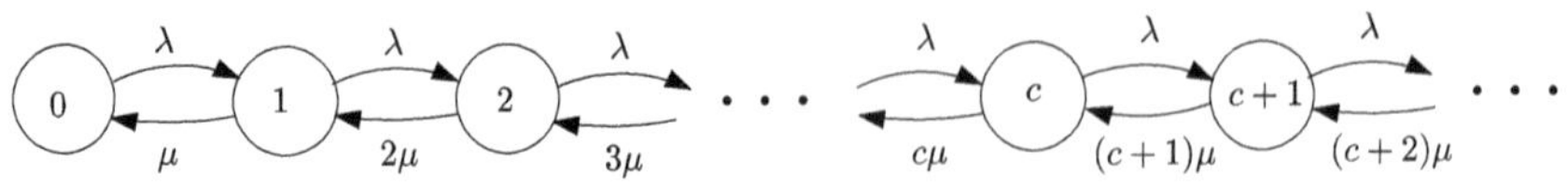

Fig. 6.14 $M/M/\infty$ self-service system

and with state dependent death rates:

$$\mu_n = n\mu, \quad n = 1, 2, \ldots. \tag{6.65}$$

This BDP is a special case of the branching process with immigration corresponding to $\alpha = \mu$ and $p = 0$.

6.4.1 Steady-State Distribution

The balance equations of the $M/M/\infty$ self-service system takes the form

$$\lambda\pi_j = (j+1)\mu\pi_{j+1}, \quad j = 0, 1, \ldots.$$

Irreducibility follows from the fact that the embedded chain is an irreducible simple random walk, and positive recurrence will follow if the above equations can be solved. The stochastic process $\{X(t), t \geq 0\}$ is always positive recurrent for any $\lambda > 0$ and $\mu > 0$. Let $\rho = \lambda/\mu$ and $\pi_j = \frac{\rho^j}{j!}\pi_0$, $j = 1, 2, \ldots$. Forcing these to sum to one (via using the Taylor's series expansion for e^x), $1 = e^{\rho}\pi_0$, or $\pi_0 = e^{-\rho}$ is obtained. Thus, the steady-state distribution π_n is given by:

$$\pi_n = \frac{\rho^n e^{-\rho}}{n!}, \quad n = 0, 1, \ldots. \tag{6.66}$$

This follows a Poisson distribution with parameter ρ. Hence, the expected number of customers in the steady state in the system is ρ. Also, since there is no queue in this queueing model, all waiting times are zero and the mean time spent in the system is the same as its mean service time and hence is equal to $\frac{1}{\mu}$. This means that all customers passing through such a system independently spend an exponentially distributed time in the system.

Remark 6.2 The above stationary distribution result should not be surprising, for the more general $M/G/\infty$ queue, and obtained the same stationary distribution (refer (9.28) in Section 9.5.1). Since exponential service times are now assumed, the results are derived using CTMC methods. For a general service time distribution it can not be done because the underlying stochastic process $\{X(t), t \geq 0\}$ does not form a

CTMC; hence more general methods will be used which will be discussed in Chap. 9.

6.4.2 Transient Solution

Let the joint distribution be given by

$$\pi_k(t) = P\{X(t) = k\},\ k = 0, 1 \ldots.$$

It satisfies

$$\frac{d\pi_0(t)}{dt} = \lim_{\Delta t \to 0} \frac{\pi_0(t + \Delta t) - \pi_0(t)}{\Delta t} = -\lambda\pi_0(t) + \mu\pi_1(t),$$

for $k = 1, 2, \ldots$

$$\frac{d\pi_k(t)}{dt} = \lambda\pi_{k-1}(t) - (\lambda + k\mu)\pi_k(t) + (k+1)\mu\pi_{k+1}(t).$$

Solving the above system of difference-differential equations with the initial conditions, the distribution of $X(t)$ is obtained as

$$\pi_k(t) = \frac{1}{k!}\left(\rho(1 - e^{-\mu t})\right)^k \exp\left\{-\rho(1 - e^{-\mu t})\right\},\ k = 0, 1, \ldots.$$

The time-dependent distribution of $X(t)$ follows non-homogeneous Poisson distribution with parameter $\rho(1 - e^{-\mu t})$. Furthermore, note that $\pi_k(t)$, for $t \to \infty$, turns out to be

$$\pi_k = \frac{e^{-\rho}\rho^k}{k!},\ k = 0, 1, \ldots. \tag{6.67}$$

The limiting distribution of $X(t)$ follows Poisson distribution with parameter ρ. These results of time-dependent distribution of $X(t)$ and of limiting distribution coincide with the results presented in [11].

The above time-dependent result also can be obtained using probabilistic argument as follows:

Let $Y(t)$ denote the number of customers who arrives at the time period $(s, t]$. If a customer arrives at time $s \in (s, t]$, the probability that he is still in the process of being served at time t is $e^{-\mu(t-s)}$. Thus, if a customer arrives at time chosen uniformly $(s, t]$, the probability that he is still in the process of being served at time t denoted by q_t is:

$$q_t = \frac{1}{t}\int_0^t e^{-\mu(t-s)} ds = \frac{1 - e^{-\mu t}}{\mu t}.$$

Let $X(t)$ denote the number of customers arriving in $(s, t]$, who are still in the process of being served at time t. It follows that the conditional distribution of $X(t)$ given $Y(t)$ follows binomial distribution with parameters k and q_t, i.e., (left as an exercise)

$$P\{X(t) = n | Y(t) = k\} = \binom{k}{n} (q_t)^n (1 - q_t)^{n-k}, \quad n = 0, 1, \dots k.$$

Since $Y(t)$ follows Poisson distribution with parameter λt, after some simple calculation, we get,

$$P\{X(t) = n\} = \frac{e^{-\lambda t q_t} (\lambda t q_t)^n}{n!}, \quad n = 0, 1, \dots.$$

Example 6.18 Imagine a dinner buffet held in a large hall where diners serve themselves. The arrival of diners follows a Poisson process, with one diner entering every 10 seconds. Each diner typically takes 1/8 of a minute to serve themselves. Calculate the following performance measures.

1. Mean number of customers serving themselves.
2. Mean time spent in the system.

Solution: The number of patrons who serve themselves as well as the wait time will be computed as follows. The steady-state probability of the system is

$$\pi_n = \frac{\rho^n e^{-\rho}}{n!} = \frac{0.75^n e^{-0.75}}{n!}, n = 0, 1, \dots.$$

1. The Mean number of customers serving themselves is

$$E(X) = 0.75.$$

2. The Mean time spent in the system is

$$E(R) = 0.1250 \text{ min}.$$

Example 6.19 A television station in a major metropolitan area is interested in estimating the typical audience size for a Friday night prime-time program. Previous studies indicate that the number of people turning on their televisions during prime time follows a Poisson distribution with a mean of 100,000 per hour. Each person is assumed to randomly select one of the five major TV stations. Additionally, studies show that the typical viewer watches for 90 minutes, and their viewing times are approximately exponentially distributed. Determine the mean number of viewers during prime time.

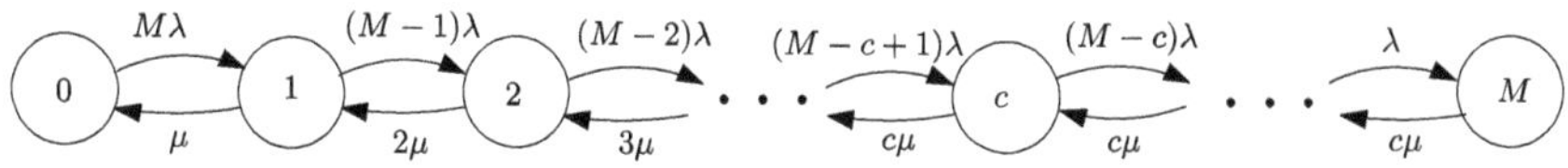

Fig. 6.15 Finite source queueing system

Solution: The mean arrival rate (people tuning in the television station during prime time on Friday evening) is 100,000/5 = 20,000 /hour and the mean service time is 90 min or 1.5 hours. The mean number of viewers during prime time is

$$E(X) = \frac{\lambda}{\mu} = \frac{20000}{2/3} = 30,000 \text{ people.}$$

6.5 Finite Source Queueing Systems

In the previously mentioned models, it has been assumed that the source of arriving population is infinite. Now consider a finite calling size population of size M, and future event occurrence probabilities are functions of the state of the system. This model is frequently used in the field of machine maintenance, where the calling population consists of a finite number of machines, an arrival represents a breakdown, and the repair personnel serve as the servers.

Definition 6.7 In this model, the calling population model is finite, with c servers, exponentially distributed service time with rate μ. If a calling unit is not in the system at time t, the time a calling unit spends outside the system is exponentially distributed with parameter λ. The finite source queueing system can be represented by Figure 6.16. The arrival and service rates of the proposed system are given by

$$\lambda_n = \begin{cases} (M-n)\lambda, & 0 \le n \le M \\ 0, & n \ge M \end{cases} ; \quad \mu_n = \begin{cases} n\mu, & 0 \le n < c \\ c\mu, & n \ge c \end{cases} .$$

6.5.1 *Steady-State Distribution*

As evaluated before for other queueing systems, the steady-state or equilibrium solution of the system is given by

$$\pi_n = \begin{cases} \frac{M!}{(M-n)!n!} \left(\frac{\lambda}{\mu}\right)^n \pi_0, & 1 \le n < c, \\ \frac{M!}{c^{n-c}c!(M-n)!} \left(\frac{\lambda}{\mu}\right)^n \pi_0, & c \le n \le M. \end{cases}$$

6.5.2 Performance Measures of the System

1. The mean number of customers in the system

$$E(X) = \sum_{n=1}^{M} n\pi_n.$$

2. The overall effective arrival rate of the system

$$\lambda_{eff} = \sum_{n=0}^{M} (M - n)\lambda\pi_n.$$

3. The mean waiting time of customers

$$E(R) = E(X) - \rho(M - E(X)).$$

Example 6.20 Suppose that five robots are used by a particular Semi Conductor Company to create its circuit boards. The robots malfunction, and the corporation has two repair personnel on hand to fix them. When one is fixed, it is believed that the time until the subsequent breakdown will have an exponential distribution with a mean of 30 hours. The company always maintains a sufficient backlog of work to guarantee that all robots in working order will be operational. The estimated exponential distribution of the repair times for each service has a mean of 3 hours. The company manager is interested in learning the typical number of robots in use at any given moment, the anticipated downtime of a robot needing repair, and the anticipated proportion of idle time for each repairer.
Solution: Here $M = 5,\ c = 2, \lambda = \frac{1}{30}$, and $\mu = \frac{1}{3}$. The mean number of customers in the system is $E(X) = 0.465$ and the expected waiting time is $E(R) = 3.075$ hour.

Example 6.21 (Double-ended with catastrophe) Consider a double-ended queueing system with catastrophes as mentioned by Dharmaraja et al. [12]. Let $\{N(t), t \geq 0\}$ be a time-homogeneous BDP subject to catastrophes, defined on the state space $S = \{-N, -(N-1), \ldots, -1, 0, 1, \ldots, N\}$, with N a positive integer, such that transitions occur according to the following scheme:

1. $n \to n+1$ with rate $\lambda_n = \lambda(N-n)$, for $n = -N, -N+1, \ldots, N-1$
2. $n \to n-1$ with rate $\mu_n = \mu(n+N)$, for $n = -N+1, -N+2, \ldots, N-1, N$
3. $n \to 0$ with rate ξ for $n \neq 0$.

Therefore, births occur with rates λ_n, deaths occur with rates μ_n and catastrophes with rate ξ, the effect of each catastrophe being the instantaneous transition to the state 0.

The transition probabilities, $P_{jn}(t) = P\{N(t) = n|N(t) = j\}, \quad j, n \in S$, of the system are given by,

$$\begin{aligned}
P_{j-N}(t) &= e^{-\xi t}\widetilde{P}_{j+N0}(t) + \xi\int_0^t e^{-\xi\tau}\widetilde{P}_{N0}(\tau)d\tau \\
&= e^{-\xi t}\frac{\mu^{j+N}}{(\lambda+\mu)^{2N}}\left[1-e^{-(\lambda+\mu)t}\right]^{j+N}\left[\mu+\lambda e^{-(\lambda+\mu)t}\right]^{N-j} + \\
&\quad \xi\frac{\mu^{2N}}{(\lambda+\mu)^{2N-1}}\sum_{i=0}^{N}\binom{N}{i}(-1)^i\sum_{k=0}^{N}\binom{N}{k}\rho^k\left(\frac{1-e^{-[\xi+(\lambda+\mu)(i+k)]t}}{\xi+(\lambda+\mu)(i+k)}\right).
\end{aligned} \tag{6.68}$$

Similarly, for $n = -N+1, -N+2, \ldots, -1$, $P_{jn}(t)$ is

$$\begin{aligned}
&= e^{-\xi t}\widetilde{P}_{j+Nn+N}(t) + \xi\int_0^t e^{-\xi\tau}\widetilde{p}_{Nn+N}(\tau)d\tau \\
&= e^{-\xi t}\frac{\mu^{j+N}}{(\lambda+\mu)^{2N}}\left[1-e^{-(\lambda+\mu)t}\right]^{j+N}\left[\mu+\lambda e^{-(\lambda+\mu)t}\right]^{N-j} \times \\
&\quad \sum_{i=0}^{\min\{j+N,N+n\}}\binom{j+N}{i}\binom{N-j}{N+n-i}\left[\frac{\lambda+\mu e^{-(\lambda+\mu)t}}{\mu\left(1-e^{-(\lambda+\mu)t}\right)}\right]^i\left[\frac{\lambda\left(1-e^{-(\lambda+\mu)t}\right)}{\mu+\lambda e^{-(\lambda+\mu)t}}\right]^{N+n-i} + \\
&\quad \xi\frac{\mu^{j+N-n}\lambda^{N+n}}{(\lambda+\mu)^{2N-1}}\sum_{i=0}^{N+n}\binom{N}{i}\binom{N}{n-i}\sum_{l=0}^{2N+n-2i}\binom{2N+n-2i}{l}(-1)^l + \\
&\quad \sum_{m=0}^{i}\binom{i}{m}\rho^{-m}\sum_{q=0}^{i-n}\binom{i-n}{q}\rho^q\left(\frac{1-e^{-[\xi+(\lambda+\mu)(2N+n-2i+l+m+q)]t}}{\xi+(\lambda+\mu)(2N+n-2i+l+m+q)}\right).
\end{aligned}$$

Finally, for $n = 0, 1, \ldots, N$,

$$\begin{aligned}
P_{jn}(t) &= e^{-\xi t}\widetilde{P}_{j+Nn+N}(t) + \xi\int_0^t e^{-\xi\tau}\widetilde{P}_{Nn+N}(\tau)d\tau \\
&= e^{-\xi t}\frac{\mu^{j+N}}{(\lambda+\mu)^{2N}}\left[1-e^{-(\lambda+\mu)t}\right]^{j+N}\left[\mu+\lambda e^{-(\lambda+\mu)t}\right]^{N-j} \times \\
&\quad \sum_{i=0}^{N}\binom{j+N}{i}\binom{N-j}{N+n-i}\left[\frac{\lambda+\mu e^{-(\lambda+\mu)t}}{\mu\left(1-e^{-(\lambda+\mu)t}\right)}\right]^i\left[\frac{\lambda\left(1-e^{-(\lambda+\mu)t}\right)}{\mu+\lambda e^{-(\lambda+\mu)t}}\right]^{N+n-i} + \\
&\quad \xi\frac{\mu^{j+N-n}\lambda^{N+n}}{(\lambda+\mu)^{2N-1}}\sum_{i=0}^{N}\binom{N}{i}\binom{N}{n-i}\sum_{l=0}^{2N+n-2i}\binom{2N+n-2i}{l}(-1)^l + \\
&\quad \sum_{m=0}^{i}\binom{i}{m}\rho^{-m}\sum_{q=0}^{i-n}\binom{i-n}{q}\rho^q\left(\frac{1-e^{-[\xi+(\lambda+\mu)(2N+n-2i+l+m+q)]t}}{\xi+(\lambda+\mu)(2N+n-2i+l+m+q)}\right).
\end{aligned} \tag{6.69}$$

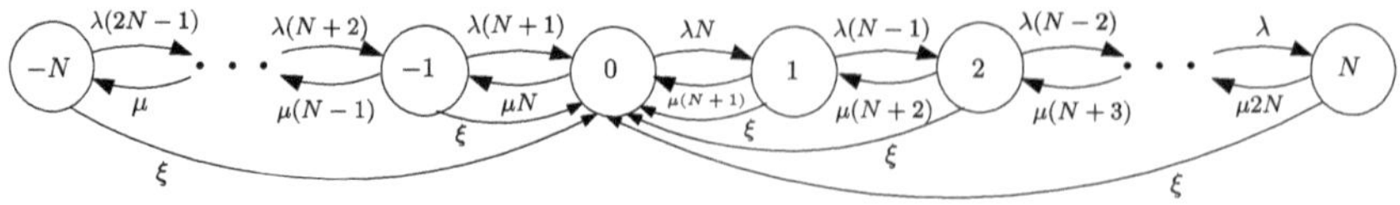

Fig. 6.16 State transition diagram for Example 6.21

A queueing theory homepage has been created where readers are informed about relevant sources, for example books, software, conferences, journals, etc. It is highly recommended to visit it at http://web2.uwindsor.ca/math/hlynka/queue.html

For the reader wishing to obtain more information on the topic of the queueing theory that is not covered in this text book, Kulkarni [13], Wolff [2] are highly recommended references. (Fig. 6.16).

6.6 Exercise Problems

1. Consider a system with two servers and unlimited space for the waiting room. The arrival rate is λ and the service rate is μ. Write down the Kolmogorov backward equation for the above process.
2. Consider the Queueing model where customers arrives as a Poisson process with a rate of 5 per minute and service time follows an exponential distribution with rate μ service is provided to one customer at a time. Given that 90% of time the system contains less than or equal to 4 customers, then find the value of μ.
3. In an airport, there is a separate runway for arrivals only. An arriving aircraft joins a single queue for the runway. Here, the arrival follow Poisson process with the rate 20 arrivals per hour and service time is exponential distribution with a rate 27 arrivals per hour. Find the waiting time in the queue.
4. The time spent on a job has an exponential distribution, with a mean of 30 minutes, according to a television repairman. If he fixes the sets in the order they arrive and if the arrivals are roughly Poisson, with a mean rate of 10 every eight-hour day. Find the expected number of units in the queue?
5. In an $M/M/1$ queueing system with service rate μ and customers arrival follows Poisson process with rate λ, where $\lambda < \mu$. Suppose one extra identical server can be added, then which of the following scenarios is better.

   (a) Separate the two operators. Therefore, there will be two $M/M/1/\infty$ queues. In this case, it is assumed that an arriving customer can wither join the first queue or the second with equal probability.
   (b) Combine the two operators together. Therefore, it will be considered as $M/M/2/\infty$ queue.

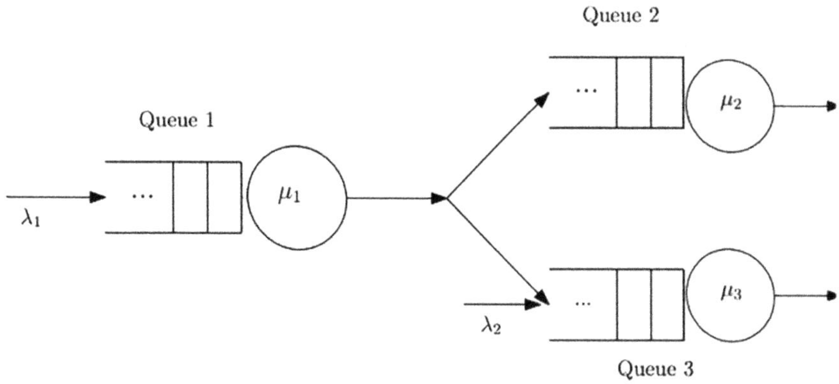

Fig. 6.17 Network of queueing system

6. A woman is the only proprietor of a unisex hair salon. She operates the salon on a first-come, first-served basis and does not take appointments. She discovers that Friday mornings are her busiest time, therefore she is thinking about hiring a part-time helper and perhaps even moving to a bigger space. She chooses to conduct a thorough analysis of the circumstance before making a decision because she has a master's degree in operations research (OR) before beginning her profession. As a result, she carefully records each Friday morning and discovers that there are 5 clients on average every hour, consistent with a Poisson process. Customers were always patient because of her stellar reputation. Consider the system as $M/M/1$ queue and calculate the following measures:

   (a) What is the mean number of customers in the shop?
   (b) What is the total number of customers waiting for a haircut?
   (c) What is the percentage of time an arrival can walk right in without having to wait at all?
   (d) If the waiting room has only four seats at present, what is the probability that a customer, upon arrival, will not be able to find a seat and have to stand?

7. Consider a queueing system depicted in Fig. 6.17, where queue 1 is driven by a Poisson process with rate λ_1 and the departures are randomly routed to queue 2 or queue 3. Queue 3 has additional arrival process with rate λ_2. Prove the Burke's theorem for the proposed queueing system.
8. Consider a single-server hair salon, where the owner serves customers alone. Customers arrive according to a Poisson process with a rate of 4 per hour, and the service time follows an exponential distribution. While customers are willing to wait for service, the waiting area can hold a maximum of 5 customers. Determine the mean time a customer spends in the system.
9. Consider a modified $M/M/1/N$ queueing system where the arrival and service rates are given by $\lambda_0 = \lambda + \sqrt{\lambda\mu}$, $\lambda_i = \lambda$ for $i = 1, 2, \ldots, N-1$, $\mu_i = \mu$ for

$i = 1, 2, \ldots, N-1$ and $\mu_N = \mu + \sqrt{\lambda\mu}$. Calculate the transient probabilities for this system.

10. Suppose M machines are subject to failures and repairs. The failure times are exponential distributed with parameter μ, and repair times are exponentially distributed with parameter λ. Let $X(t)$ denote the number of machines that are failed at time t. If there is only one repairman, then under appropriate reasonable assumptions, let $\{X(t), t \geq 0\}$ be a birth-death process on $S = \{0, 1, \ldots, M\}$ with birth rates $\lambda_n = (M-n)\lambda, \quad n = 0, 1, \ldots, M-1$ and death rate $\mu_n = \mu, \quad n = 1, 2, \ldots, M$. Find the stationary distribution for this process.
11. A Poisson process with a rate of two per hour is formed when jobs arrive at a server with single service at a time. Such jobs service times have an exponential distribution with a mean of 20 minutes. In the system, only maximum four jobs can be active at any time. Find the likelihood that a work will be rejected due to storage space, assuming that the percentage of processing power used by smaller jobs is insignificant. Find out how many mean number of jobs in the system at steady-state as well.
12. Let $Y(t)$ denote the number of customers who arrive in the time period $(s, t]$. If a customer arrive a time $s \in (0, t]$, the probability that he is still in the process of being served at time t is $e^{-\mu(t-s)}$. Thus if a customer arrives at time chosen uniformly $(s, t]$, the probability that he is still in the process of being served at time t is denoted by q_t is:

$$q_t = \frac{1}{t}\int_0^t e^{-\mu(t-s)} ds = \frac{1 - e^{-\mu t}}{\mu t}.$$

Let $X(t)$ denote the number of customers arriving in $(s, t]$, that are still in the process of being served at time t. Prove that the conditional distribution of $X(t)$ given $Y(t) = k$ is $B\left(k, \frac{1-e^{-\mu t}}{t}\right)$.

13. Prove that the compound Poison process $\{Z(t), t \geq 0\}$ is a time-homogeneous stochastic process with independent increments.
14. Every Wednesday night, free vision screenings are provided by the eye clinic at City Hospital. Three ophthalmologists are on call. The real-time is found to be spread with potential distribution around the average test time of 20 minutes. Patients are accepted on a first-come, first-served basis and clients arrive according to a Poisson process with a mean of 6/hr. Consider the system as $M/M/c$ queueing system and compute the following:

    (a) What is the mean number of people waiting?
    (b) What is the total amount of time a patient spends at the clinic?
    (c) What is the average percentage idle time of each of the doctors?

15. Two servers make up a service centre, and each wraps up two services each hour following exponential distribution. What percentage of potential consumers join the system if the customers arrive at the Poisson rate of three per hour and the

system can only accommodate a maximum of three clients? Jobs arrive in a Poisson process with rate λ at a computer center with two computers called A and B. It is splitted the arriving work evenly between the two computers using one of the following two splitting rules:

Rule 1: Each arriving job flips a fair coin to decide which computer queue to join; let's say heads it goes to computer A and tails it goes to computer B.

Rule 2: Assuming the jobs are numbered in sequence, the even numbered jobs are directed to computer A and the odd numbered jobs are directed to computer B.

Clearly, Rule 1 is randomized and Rule 2 is deterministic. Under a given rule, the arrival processes at the two computers are identical stochastically. Consider one such process. How would you describe it under Rule 1? under Rule 2? Is the process under either rule a Poisson process? under both rules? If the process differs under the two rules, describe the difference. Is there a sense in which one rule is preferable to the other? Explain.

16. Items arrive at a waiting station according to a Poisson process at a rate of λ. Buses arrive at the station according to a Poisson process with a rate of μ, promptly collecting items from the system. Upon arrival at an unoccupied system, a bus departs without delay. Assume that the capacity of each bus is a RV following a geometric distribution with parameter $1-p$. Furthermore, when the queue is empty and an item arrives, there is a probability p that a bus is available to accommodate the arrival without delay.

    (a) If a bus arrives and encounters i items awaiting, show that the actual quantity Y that departs in a batch follows a truncated geometric distribution,

    $$P(Y=n)=p^{n-1}(1-p)\mathbf{1}(n<i)+p^{n-1}\mathbf{1}(n=i).$$

    (b) Let $X(t)$ be the number of items in the system at time t.
    (i) a CTMC with specified transition rates,

    $$q_{ii+1}=\lambda(1-p)\mathbf{1}(i=0)+\lambda\mathbf{1}(i\geq 1),$$
    $$q_{ii-n}=\mu p^{n-1}(1-p)\mathbf{1}(1\leq n\leq i-1)+\mu p^{i-1}\mathbf{1}(n=i)(i\geq 1).$$

    (ii) Show that $\{X(t),t\geq 0\}$ is ergodic if and only if $\lambda<\mu+p\lambda$.
    (iii) In this scenario, show that its stationary distribution is given by

    $$\pi_i=\pi_0(1-p)\left(\frac{\lambda}{\mu+p\lambda}\right)^i,\quad i=1,2,\ldots.$$

17. Consider a $M/M/c/c$ system with two types of customers, preemptive priority policy and retrial phenomenon. When a type 2 customer arrives to the system and finds all the servers occupied, type 2 customer joins orbit of infinite capacity to avail the liberty of retrial. If the system is occupied with c type 2 customer and a type 1 customer arrives, it is given preemptive priority over a type 2 customer. As per this policy, the type 1 customer will preempt the service of a type 2 customer and starts service in its place. The preempted type 2 customer will be dropped from the system. Compute the probability when any type of customer is busy in the system and average number of customers in the system.
18. Consider a vehicle emission inspection facility with three examination booths that can fit a single vehicle each. It makes sense to suppose that vehicles queue up such that the first vehicle to move up to an open stall will do so. At most four automobiles can wait at once in the station (there are seven total). During the busiest times, there is typically one automobile arriving every minute according to a Poisson distribution. The average service time is exponential and is 6 minutes. The chief inspector is interested in knowing the typical number using the system at busy times, the typical wait time (including service), and the anticipated number of people per hour who are unable to enter the station due to capacity.
19. In a parking lot, the vehicles arrive with the arrival rate $\lambda = 5$ and the service time is exponentially distributed with rate $\mu = 7$. Obtain the optimal value of c such that $\pi_c < 0.05$.
20. Consider $M/M/c$ queueing model with $c < \infty$, traffic intensity $\rho = \lambda/c\mu < 1$, and stationary distribution p_i. Let $\{X(t), t \geq 0\}$ denotes the stochastic process for the underlying queueing system and $\{Y(t), t \geq 0\}$ denote the number of items in the queue waiting to be served (not in service) at time t. Show that $Y(t)$ is a stationary process and specify its distribution. Also, show that

$$E[Y(0)] = \frac{p_0 \rho (c\rho)^c}{(1-\rho)^2 c!}.$$

Consider the waiting time $W(t) = \min\{u \geq t : X(u) < c\}$ until a server is available at or after time t (sometimes called the virtual waiting time in the queue of an arrival at time t before it can enter service). Justify that $W(t)$ is a stationary process, and find $E[W(0)]$. Show that

$$P\{W(0) \leq t \mid X(0) = c\} = 1 - e^{-rt}$$

and specify the rate r. Show that $P\{W(0) > 0\} = P\{X(0) > c\}$ and

$$P\{W(0) > t\} = \sum_{i=c}^{\infty} p_i P_i\{M(t) < i - c\} = P\{X(0) > c\} e^{-(c\mu - \lambda)}$$

where $\{M(t), t \geq 0\}$ is a Poisson process with rate $c\mu$ denoting departure times when c servers are busy. Finally, prove $P\{X(0) \geq c+m\} = (\lambda/c\mu)^m$.

21. Consider Example 6.13, and obtain the transient probabilities of the system. Prove that for a without catastrophe system, the system presents a $M/M/c$ queueing system.
22. Prove that, the PDF of the length of busy period, denoted by B, for $c = 2$ in $M/M/c$ queue is given by

$$f_B(x) = \frac{e^{-(\lambda+2\mu)x}}{x} \sum_{r=0}^{\infty} (r+1) \left(\frac{\mu}{2\lambda}\right)^{\frac{r+1}{2}} I_{r+1}(2\sqrt{2\lambda\mu}), \quad x > 0.$$

Further, show that

$$E(B) = \frac{1}{\mu(1-\rho)}; \quad E(B^2) = \frac{2-\rho}{\mu^2(1-\rho)^3}.$$

23. Consider Example 6.21. Prove that,

$$E(N(t)|N(0)=0) = \frac{1 - N(\lambda-\mu)e^{-\xi t} - N(\lambda+\mu)e^{-(\xi+\lambda+\mu)t}}{\lambda+\mu} + \frac{\xi}{\xi+\lambda+\mu}\left(1 - N(\lambda+\mu)e^{-(\xi+\lambda+\mu)t}\right)$$

and

$$\begin{aligned} var(N(t)|N(0)=0) = & \frac{e^{-\xi t}}{(\lambda+\mu)^2}\{2N\lambda - e^{-2(\lambda+\mu)t}[2N\lambda^2 + N(\mu^2-\lambda^2)] \\ & - e^{-(\lambda+\mu)t}[2N\lambda(\mu-\lambda) - N(\mu^2-\lambda^2)]\} \\ & + \frac{e^{-\xi t}}{(\lambda+\mu)^2}\left[N(\lambda-\mu) - N(\lambda+\mu)e^{-(\lambda+\mu)t}\right]^2 \\ & + \xi \int_0^t e^{-\xi\tau}\left[Var(\widetilde{N}(\tau)) + (E(\widehat{N}(\tau)))^2\right] d\tau \\ & - (E\{N(t)|N(0)=0\})^2. \end{aligned}$$

24. Consider an $M/M/\infty$ queueing system with stochastic process $\{X(t), t \geq 0\}$ where the arrivals follow Poisson process with rate λ and service is exponentially distributed with rate μ. To derive its transition probabilities at time t, we will consider two independent populations of items: those present at time 0, and those arrivals in $(0, t)$ that are still in the system at time t. Now, each item present at time 0 is still in the system at time t with the probability $e^{-\mu t}$ that the residual service time exceeds t (the exponential distribution is memoryless). Then if there are i items present at time 0, the number of these still present at time t has the

Binomial distribution with parameters i and $1 - e^{-\mu t}$. Next, note that since this process is a special case of the $M/G/\infty$ process, it is known that the number of the arrivals in $(0, t)$ that are still present at time t has a Poisson distribution with mean

$$\eta(t) = \int_0^t e^{-\mu(t-s)}\lambda ds = (\lambda/\mu)\left(1 - e^{-\mu t}\right).$$

Then the number of items in the system at time t is the sum of these independent binomial and Poisson RVs. Consequently, the transition probabilities are

$$P_{ij}(t) = \sum_{k=0}^{i} \binom{i}{k} e^{-k\mu t}\left(1 - e^{-\mu t}\right)^{i-k} \frac{\eta(t)^{j-k}}{(j-k)!} e^{-\eta(t)}.$$

References

1. Wolff, R.W.: Poisson arrivals see time averages. Opns. Res. **30**, 223–231 (1982)
2. Wolff, R.W.: Stochastic Modeling and the Theory of Queues. Prentice Hall, London (1989)
3. Parthasarathy, P.R.: A transient solution to an M/M/1 queue - A simple approach. Adv. Appl. Prob. **19**, 997–998 (1987)
4. Parthasarathy, P.R., Dharmaraja, S.: Exact transient solutions of kinetics of first-order reactions with end effects. J. Math. Chem. **25**(2–3), 281–294 (1999)
5. Parthasarathy, P.R., Sharafali, M.: Transient solution to the many-server Poisson queue: a simple approach. J. Appl. Probab. **26**(3), 584–594 (1989)
6. W. Ledermann, G.E.H. Reuter, Spectral theory for the differential equations of simple birth and death processes. Philos. Trans. R. Soc. Londn. Ser. A Math. Phys. Sci. **246**(914), 321–369 (1954)
7. Dharmaraja, S.: Transient solution of a two-processor system. Math. Comput. Modelling **32**, 1117–1123 (2000)
8. K.S. Trivedi, *Probability and Statistics with Reliability, Queuing, and Computer Science Applications*, 2nd edn. (John Wiley & Sons, 2001)
9. Dharmaraja, S., Kumar, R.: Transient solution of a Markovian queuing model with heterogeneous servers and catastrophes. Opsearch **52**, 810–826 (2015)
10. Haring, G., Marie, R., Puigjaner, R., Trivedi, K.S.: Loss formulas and their applications to optimization for cellular network. IEEE Trans. Veh. Technol. **50**(3), 664–673 (2001)
11. J.F. Shortle , J.M. Thompson , D. Gross, C.M. Harris, *Fundamentals of Queueing Theory* (John Wiley & Sons, 2018)
12. Dharmaraja, S., Di Crescenzo, A., Giorno, V., Nobile, A.G.: A continuous-time Ehrenfest model with catastrophes and its jump-diffusion approximation. J. Stat. Phys. **161**, 326–345 (2015)
13. Kulkarni, V.G.: Modelling and Analysis of Stochastic Systems. Chapman & Hall, London (1995)

Chapter 7
Advanced Markov Queueing Models

In Chapter 6, the simple Markov queueing models were described in detail. The simple Markov queueing models mean that the underlying stochastic process is a BDP. In this chapter, some assorted additional results for the advanced queueing models are provided. The advanced Markov queueing models imply that the underlying stochastic process is a non-BDP or a general CTMC. Most of this new material follows in a logical way from previous chapters in the sense that it ties up some loose theoretical ends and should also help to provide a more complete picture of the kinds of queueing models that may occur in real life. This chapter continues the development of Markov queueing models that are amenable to analytic methods and is concerned especially with Markov models where the underlying stochastic process is a non-birth-death process. That is, changes of more than one are allowed over infinitesimal time intervals but insists on retaining the memory less or Markov property. The Chapman-Kolmogorov backward and forward equations, with the resultant balance equations, are all valid, and together form the essence of the approach to solution for these non-birth-death Markov models.

This chapter provides a detailed analysis of the advanced Markov queueing models. Section 7.1 describes Quasi-Birth-Death processes which are the simplest and most well-studied class of multidimensional Markov chains. A solution methodology known as the matrix geometric method has been provided to handle these queueing models. Section 7.3 studies retrial queueing models. In section 7.4 the balking and reneging scenarios in queues have been discussed. Section 7.5 analyzes bulk arrival and bulk service queueing models. Section 7.2 discusses queueing models in which the server goes on some vacation. In Section 7.6, time-dependent probabilities for the fluid queueing models are derived.

In this chapter, the objective is to explore generalizations that extend beyond the limitations of the exponential distribution and/or the Poisson process. Versatile modelling approaches have been sought that retain the ability to analyze queues through the lens of Markov chains. Phase-type distribution (PH), Markovian arrival processes (MAPs), and correlation play crucial roles in the realm of stochastic mod-

D. Selvamuthu, *Introduction to Stochastic Processes*, University Texts in the Mathematical Sciences, https://doi.org/10.1007/978-981-97-6152-4_7

eling and queueing theory. A PH distribution is a probability distribution constructed through a convolution or mixture of exponential distributions, particularly applicable in scenarios where interconnected Poisson processes unfold sequentially in distinct phases. Markov arrival processes, on the other hand, describe the arrival patterns of customers in a system, emphasizing the memoryless property of the time between arrivals, a characteristic inherent in Markov processes. These processes are valuable for modeling various real-world phenomena, such as network traffic and customer arrivals. In the context of queueing theory, understanding and incorporating correlation is essential, as it captures the interdependence between different events or entities in a system. Whether exploring the intricate relationships between service times, queue lengths, or other system variables, accounting for correlation enhances the accuracy and applicability of models, providing a more realistic representation of complex dynamic systems. Together, these concepts contribute to a comprehensive toolkit for analyzing and modeling diverse stochastic processes, enabling a nuanced understanding of dynamic systems and their inherent complexities.

Definition 7.1 (Stochastic process whose sojourn time follows exponential distribution) A RV X is said to be exponential distributed with PDF

$$f(x) = \alpha e^{-\alpha x},\ \alpha > 0, x > 0.$$

It is used to model the sojourn time spent in any state of irreducible CTMC.

Definition 7.2 (Stochastic process whose sojourn time follows Erlang-k distribution) A RV X has an Erlang-k distribution if X is the sum of k independent RV$X_1, X_2, \ldots, X_k$ having a common exponential distribution with mean $\frac{1}{\alpha}$, where $k = 2, 3, \ldots$. The common notation for Erlang-k distribution is $E_k(\alpha)$ or E_k. The probability density function of an $E_k(\alpha)$ is given by

$$f(x) = \frac{\alpha^k}{(k-1)!} x^{k-1} e^{-\alpha x}, x \geq 0.$$

A stochastic process $\{X(t), t \geq 0\}$ whose sojourn time follows Erlang-k distribution is defined on the state space $\{0, 1, 2, \ldots, k\}$ with an infinitesimal generator matrix

$$\mathbf{Q} = \begin{pmatrix} \mathbf{0} & \mathbf{0} \\ \mathbf{h} & \mathbf{H} \end{pmatrix},$$

where

$$\mathbf{H} = \begin{pmatrix} -\alpha & \alpha & \cdots & 0 \\ 0 & -\alpha & \cdots & 0 \\ \vdots & \vdots & \ddots & \vdots \\ 0 & 0 & \cdots & -\alpha \end{pmatrix}, \quad \mathbf{h} = \begin{pmatrix} 0 \\ 0 \\ \vdots \\ \alpha \end{pmatrix}.$$

Here, the order of the matrix $\mathbf{H}$ is $k \times k$ and the order of the matrix $\mathbf{h}$ is $k \times 1$, Where 0 denotes absorbing state, and $(0, 1, 0, 0, \ldots, 0)$ is the initial probability vector.

Definition 7.3 (Stochastic process whose sojourn time follows hyperexponential distribution) A RV X has hyperexponential distribution if X is a mixture of k independent exponential RVs X_i with probability β_i, $i = 1, \ldots, k$ and mean $\frac{1}{\rho_i}$. The notation for hyperexponential distribution is $H_k(\beta_1, \ldots, \beta_k; \rho_1, \ldots, \rho_k)$, or simply H_k. The PDF is given by

$$f(x) = \sum_{i=1}^{k} \beta_i \rho_i e^{-\rho_i x}; \quad 0 \le \beta_i \le 1, \text{for all } i, \quad \sum_{i=1}^{k} \beta_i = 1.$$

$\{X(t),\ t \ge 0\}$ whose sojourn time follows hyper-exponential distribution, the infinitesimal generator matrix with state space $\{0, 1, \ldots, k\}$ is given by

$$\mathbf{Q} = \begin{pmatrix} \mathbf{0} & \mathbf{0} \\ \mathbf{h} & \mathbf{H} \end{pmatrix},$$

where

$$\mathbf{H} = \begin{pmatrix} -\rho_1 & 0 & \cdots & 0 \\ 0 & -\rho_2 & \cdots & 0 \\ \vdots & \vdots & \ddots & \vdots \\ 0 & 0 & \cdots & -\rho_k \end{pmatrix}, \quad \mathbf{h} = \begin{pmatrix} \rho_1 \\ \rho_2 \\ \vdots \\ \rho_k \end{pmatrix}.$$

Here, the order of the matrix $\mathbf{H}$ is $k \times k$ and the order of the matrix $\mathbf{h}$ is $k \times 1$. State 0 is an absorbing state and $(0, \beta_1, \beta_2, \ldots, \beta_k)$ is the initial probability vector.

Example 7.1 Let's consider a single-server queueing system at a service center, where customers arrive to receive assistance (like a help desk or customer service center).

1. **M/M/1 Queueing System**
   Description: This is a classic single-server queue where arrivals follow a Poisson process (inter-arrival times are independent and exponential distributed), and service times are exponential distributed.
2. **Heterogeneous Customers with Erlang-*k* Arrival Process**
   Description: Customers arrive according to an Erlang distribution of order k, which is the sum of k independent exponential distributed RVs.
3. **Hyper-Exponential Arrival Process**
   Description: Arrivals follow a hyper-exponential distribution, which is a mixture of exponential distributions with different rates.

Definition 7.4 (Phase-type distribution) A non-negative RV X has a phase-type distribution (PH) if its distribution function and PDF are given by

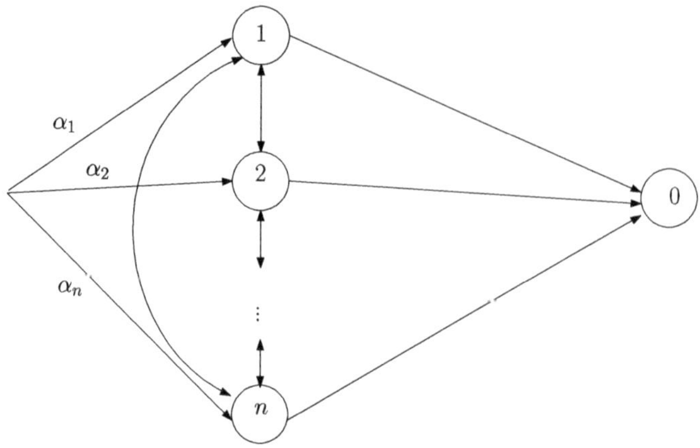

Fig. 7.1 State transition diagram for PH distribution

$$F(x) = P\{X \leq x\} = 1 - \alpha e^{-\mathbf{H}x}\mathbf{e} = 1 - \alpha\Big(\sum_{n=0}^{\infty} \frac{(-x)^n}{n!}\mathbf{H}^n\Big)\mathbf{e}, x \geq 0,$$

and

$$f(x) = \alpha \mathbf{H} e^{-\mathbf{H}x}\mathbf{e}, \quad \text{for } x > 0,$$

where $\mathbf{e}$ denotes column vector of dimension n consisting of 1's, α be an initial probability vector of order n. $\mathbf{H}$ is an $n \times n$ matrix such that all the diagonal elements are negative, and non-diagonal elements are non-negative. Figure 7.1 shows the state transition diagram for the PH distribution.

Consider the CTMC $\{X(t), t \geq 0\}$ with the state space $\{0, 1, \ldots, n\}$ and an infinitesimal generator matrix

$$\mathbf{Q} = \begin{pmatrix} \mathbf{0} & \mathbf{0} \\ \mathbf{h} & \mathbf{H} \end{pmatrix}.$$

Since $\mathbf{Q}$ is the generator matrix of the CTMC,

$$\mathbf{H}_{ii} < 0, \quad \mathbf{H}_{ij} \geq 0, \quad h_i \geq 0 \quad \text{for} 1 \leq i, j \leq n,$$
$$\text{and} \quad \mathbf{He} + \mathbf{h} = 0, \quad \alpha_0 + \alpha\mathbf{e} = 1.$$

The PH distribution is denoted by $(\alpha, \mathbf{H})$ with dimension n.

Example 7.2 Consider a $M/PH/1$ queueing system where the service is completed in two phases. The phase one is exponential distributed with parameter μ_1 and the phase two with parameter μ_2. After the completion of phase one, if the work is found satisfactory, the customer can depart with probability p otherwise with complementary probability $1 - p$, the customer enters phase two of the system. At the end of phase two, the customer departs automatically. Figure 7.2 shows the state transition

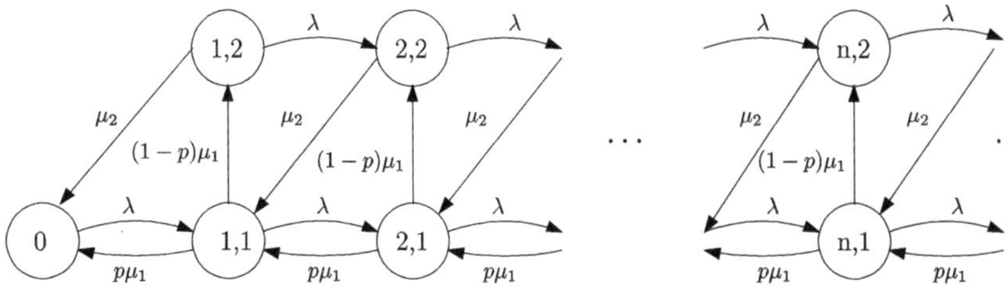

Fig. 7.2 State transition diagram for $M/PH/1$ queue

diagram for the presented system. The arrival of customers follows a Poisson process with rate λ. With this description, the queue may be represented as a continuous-time Markov process on the state space $S = \{(i, j); i \geq 0, \ j = 1, \ 2\}$. Here, i denotes the number of customers in the system and j denotes the phase of service. The generator matrix for this Quasi-birth-death (QBD) process is given as follows:

$$\mathbf{Q} = \begin{bmatrix} -\lambda & \lambda\alpha & 0 & 0 & 0 & \cdots \\ \mathbf{h} & \mathbf{H} - \lambda I & \lambda I & 0 & 0 & \cdots \\ 0 & \mathbf{h}\alpha & \mathbf{H} - \lambda I & \lambda I & 0 & \cdots \\ 0 & 0 & \mathbf{h}\alpha & \mathbf{H} - \lambda I & \lambda I & \cdots \\ \vdots & \vdots & \vdots & \ddots & \ddots & \ddots \end{bmatrix},$$

$$\alpha = [1, 0], \mathbf{h} = \begin{pmatrix} \mu_1 p \\ \mu_2 \end{pmatrix}, \mathbf{H} = \begin{pmatrix} -\mu_1 & \mu_1(1-p) \\ 0 & -\mu_2 \end{pmatrix}.$$

Transitioning from a phase-type distribution to a Markovian arrival process involves establishing a connection between the probabilistic characteristics of arrival times and the underlying structure of a Markov process. A PH distribution provides a flexible framework for modeling the time until an event happens, constructed by a convolution or mixture of exponential distributions. To create a Markovian arrival process, this distribution is integrated into the broader context of a Markov chain. Each phase in the distribution corresponds to a distinct state in the Markov process, and the transitions between states align with the evolution of the arrival process. The memoryless property inherent in both the PH distribution and the Markov arrival process ensures that the timing of arrivals is not influenced by past events, simplifying the modeling of dynamic systems.

Definition 7.5 (Markovian Arrival Process) Markovian arrival process (MAP) is a CTMC $\{N(t), I(t), t \geq 0\}$ defined over the two dimensional state space $\{1, 2, ...\} \times \{1, 2, ..., m\}$. It keeps track of phases associated with an arrival. Consider matrices $\{\mathbf{D_0}, \mathbf{D_1}\}$ in such a way that all the elements of these two matrices are non-negative except the diagonal elements of $\mathbf{D_0}$. The generator matrix $\mathbf{Q}^*$ for the underlying Markov chain has the entries as $\mathbf{D_0}$ and $\mathbf{D_1}$.

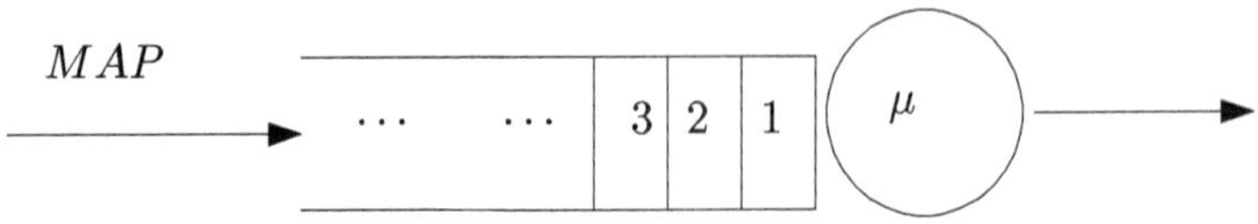

Fig. 7.3 Queueing diagram for $MAP/M/1$ queue

$$\mathbf{Q} = \begin{bmatrix} \mathbf{D_0} & \mathbf{D_1} & & \\ & \mathbf{D_1} & \mathbf{D_0} & \\ & & \ddots & \ddots \end{bmatrix}.$$

Here, $\mathbf{D_0}$ denotes the transitions without an arrival and $\mathbf{D_1}$ denotes the transitions with an arrival. Matrices $\mathbf{D_0} = (d_{ij}^0)$ and $\mathbf{D_1} = (d_{ij}^1)$ are defined as follows:

$$\begin{aligned} d_{ii}^0 &= -\rho_i, \quad 1 \le i \le m, \\ d_{ij}^0 &= \rho_i p_{ij}(0), \quad j \neq i, \\ d_{ij}^1 &= \rho_i p_{ij}(1), \quad 1 \le i, j \le m. \end{aligned}$$

Here $(\mathbf{D_0}, \mathbf{D_1})$ is called a matrix representation of the underlying MAP.

Example 7.3 Consider Example 7.11 where the arrival of incoming traffic follows MAP with matrices C_H, C_N of order K [7]. Let the state space for the stochastic process $\{X(t), t \ge 0\}$ is given with the state space $\{(i, u), \ 0 \le i \le M, \ 1 \le u \le K\}$.

- The lower diagonal elements $\mathbf{Q_{i2}}$ represent the decrement in the number of ongoing traffic. For $i = 1, 2, \dots, M$, $\mathbf{Q_{i2}} = i\mu\mathbf{I}$, where $\mathbf{I}$ is the matrix of order $K \times K$.
- The upper diagonal element $\mathbf{Q_{i0}}$ represent increment in the number of ongoing traffic due to handoff or new ve traffic arrival as

$$\mathbf{Q_{i0}} = \begin{cases} C_H + C_N, & 0 \le i < M - g \\ C_H, & M - g \le i < M \\ 0, & \text{otherwise} \end{cases}.$$

- The main diagonal elements Q_{i1} corresponds to no change in the number of ongoing traffic given as follows.

$$\mathbf{Q_{i1}} = \begin{cases} C_0 - i\mu\mathbf{I}, & 0 \le i < M - g \\ C_0 + C_N - i\mu\mathbf{I}, & M - g \le i < M \\ -i\mu\mathbf{I}, & i = M \end{cases}.$$

Example 7.4 Consider a service station in which the customer arrival rate depends on the environment whether it is raining or not. The two underlying environments are represented by 1 and 2. The transitions between these two environments may happen with or without the arrival of customers to the system. The matrix $\mathbf{D_0}$ contains the transition rates without arrival and $\mathbf{D_1}$ contains those with arrival.

$$\mathbf{D_0} = \begin{bmatrix} -5 & 3 \\ 2 & -10 \end{bmatrix}$$

$$\mathbf{D_1} = \begin{bmatrix} 0 & 2 \\ 8 & 0 \end{bmatrix}.$$

All four rates $\{3, 2, 2, 8\}$ correspond to changes of state in the underlying Markov chain $\{I(t) : t \geq 0\}$ with state space $\{1, 2\}$. Let $N(t)$ denotes the number of arrivals in $[0, t]$, then the CTMC $\{N(t), I(t), t \geq 0\}$ is a Markov arrival process with state space $\{1, 2, ...\} \times \{1, 2\}$ and infinitesimal generator of this process will be

$$\mathbf{Q} = \begin{bmatrix} \mathbf{D_0} & \mathbf{D_1} & & \\ & \mathbf{D_1} & \mathbf{D_0} & \\ & & \ddots & \ddots \end{bmatrix}.$$

Instead of the input flow of calls with the Poisson process having "memoryless property", the calls in the wireless technology have burstiness and correlation properties. Incoming traffic can be correlated, homogeneous, heterogeneous, bursty or any combination of these. MAP, a generalization of the stationary Poisson process, is an appropriate mathematical representation for the arrival process in the case of homogeneous traffic (where traffic is of the same type). PH distribution forms a versatile family of probability distributions in which a process is completed in multiple phases. It is useful when there is a need to use more generalized processes and distributions such that the performance of the system is not over/under estimated.

7.1 Quasi-Birth-Death Processes

There are various CTMCs that can be used to represent queueing systems with Poisson arrivals and exponential distributed service times. The birth-death process is one such example. In the birth-death process, the process moves up and down between levels one at a time, but the time between these transitions might have a more intricate distribution as described in the blocks. Although this CTMC will not be a birth-death process, the resulting infinitesimal generator matrix will have birth-death like structure and, hence, it is called Quasi-Birth-Death (QBD) process.

Definition 7.6 (Quasi-Birth-Death Processes) A QBD process is a continuous-time Markov chain $\{(X(t), Y(t)), t \geq 0\}$ which has state space defined as $\{(i, j) : i \geq 0,\ 1 \leq j \leq n\}$. The first coordinate i in the state space is defined as the level, and the second coordinate j is defined as the phase. The number n, a positive integer, can be either finite or infinite. The CTMC is defined as a QBD process if the one-step transition from a state is restricted to either the same level or in the two adjacent levels, i.e., the Markov chain can have transition from state (i, j) to $(i - 1, j)$, (i, j), $(i +$

1, j). The infinitesimal generator matrix $\mathbf{Q}$ of the QBD process is of the following block matrix form:

$$\mathbf{Q} = \begin{bmatrix} \mathbf{B_1} & \mathbf{B_0} & \mathbf{0} & \mathbf{0} & \mathbf{0} & \cdots \\ \mathbf{C} & \mathbf{A_1} & \mathbf{A_0} & \mathbf{0} & \mathbf{0} & \cdots \\ \mathbf{0} & \mathbf{A_2} & \mathbf{A_1} & \mathbf{A_0} & \mathbf{0} & \cdots \\ \mathbf{0} & \mathbf{0} & \mathbf{A_2} & \mathbf{A_1} & \mathbf{A_0} & \cdots \\ \vdots & \vdots & \vdots & \ddots & \ddots & \ddots \end{bmatrix},$$

where $\mathbf{A_0}, \mathbf{A_1}, \mathbf{A_2}, \mathbf{B_0}, \mathbf{B_1}$, and $\mathbf{C}$ are matrices of finite orders. Hence, this process is named as quasi-birth-death process. The zero entries in the $\mathbf{Q}$ matrix are zero matrices of appropriate sizes. It should be noted that in the case of one-dimensional birth-death process the entries of the generator matrix are scalars.

Example 7.5 Consider a $MAP/M/1$ queueing system where the arrival of customers follows MAP with matrices $(\mathbf{D_0}, \mathbf{D_1})$ of dimension m and the service is provided with exponential distribution with rate μ. The arriving customers wait in a queue of infinite size for their service. Figure 7.3 represents the schematic diagram of the described system. With this description, the queue may be represented as a CTMC on the state space $\Omega = \{(i, j); i \geq 0,\ 1 \leq j \leq m\}$, where, i denotes the number of customers in the system and j denotes the phase of MAP. The generator matrix for this QBD process is given as follows.

$$\mathbf{Q} = \begin{bmatrix} \mathbf{B_1} & \mathbf{A_0} & \mathbf{0} & \mathbf{0} & \mathbf{0} & \cdots \\ \mathbf{A_2} & \mathbf{A_1} & \mathbf{A_0} & \mathbf{0} & \mathbf{0} & \cdots \\ \mathbf{0} & \mathbf{A_2} & \mathbf{A_1} & \mathbf{A_0} & \mathbf{0} & \cdots \\ \vdots & \vdots & \vdots & \ddots & \ddots & \ddots \end{bmatrix},$$

where $\mathbf{B_1} = \mathbf{D_0}$, $\mathbf{A_0} = \mathbf{D_1}$, $\mathbf{A_2} = \mu\mathbf{I}$ and $\mathbf{A_1} = \mathbf{D_0} - \mu\mathbf{I}$.

Example 7.6 Consider two heterogeneous servers with parallel queues. The arrival of customers follows an independent Poisson process with arrival rate λ and each arriving customer joins a queue which has less number of customers. If both the queues have an equal number of customers, the customer can join any queue with equal probability. The service times are exponential distributed with rate μ_1 and μ_2. The customers can switch queues while waiting for their service, and join the queue which is having less number of customers. The stochastic process $\{(X(t), Y(t)), t \geq 0\}$ can be modeled as a QBD by obtaining its component matrices. The state space of the underlying process $\{(X(t), Y(t)), t \geq 0\}$ is defined as follows: $S = \{(i, j); i \geq 0,\ j \geq 0\}$, where, i denotes the total number of customers in queue 1 and j denotes total number of customers in queue 2. The generator matrix has QBD structure with the components as

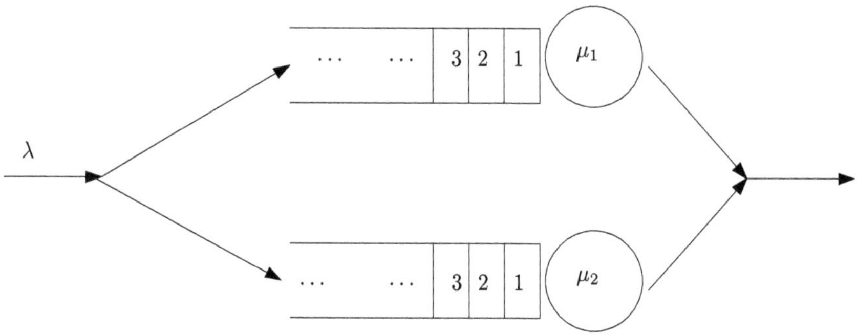

Fig. 7.4 Queueing structure for quasi-birth-death process

$$\mathbf{A_1} = \begin{bmatrix} -\lambda-\mu_1-\mu_2 & \lambda/2 & \lambda/2 \\ \mu_1+\mu_2 & -\lambda-\mu_1-\mu_2 & 0 \\ \mu_1+\mu_2 & 0 & -\lambda-\mu_1-\mu_2 \end{bmatrix}, \mathbf{A_0} = \mathbf{B_0} = \begin{bmatrix} 0\,0\,0 \\ \lambda\,0\,0 \\ \lambda\,0\,0 \end{bmatrix},$$

$$\mathbf{C} = \mathbf{A_2} = \begin{bmatrix} 0 & \mu_2 & \mu_1 \\ 0 & 0 & 0 \\ 0 & 0 & 0 \end{bmatrix}, \mathbf{B_1} = \begin{bmatrix} -\lambda & \lambda/2 & \lambda/2 \\ \mu_2 & -\lambda-\mu_2 & 0 \\ \mu_1 & 0 & -\lambda-\mu_1 \end{bmatrix}.$$

The QBD process can be classified into two categories: level independent QBD and level dependent QBD.

Definition 7.7 (Level Independent Quasi-Birth-Death Processes) If the transition rates in the generator matrix of a QBD process does not depend on the specific value of level i, the QBD process is defined as level independent QBD (LIQBD) process.

Example 7.7 Consider a single server queueing system with an associated inventory. The customers arrive at the system following a Poisson process with rate λ and the service time follows an exponential distribution with rate μ. When a customer leaves the system an item is also removed. The inventory level can vary from 0 to S. Assume that the inventory is replenished to the level S as soon as it reaches the level 0. If $N(t)$ denotes the number of customers at time t and $I(t)$ denotes the inventory level at time t, the CTMC $\{(N(t), I(t)), t \geq 0\}$ is a QBD process where the block matrices in the infinitesimal generator matrix $\mathbf{B_1}, \mathbf{A_0}, \mathbf{A_1}, \mathbf{A_2}$ are of order $S+1$ and described as follows,

$$\mathbf{B_1} = \begin{bmatrix} -\lambda & & & \\ & -\lambda & & \\ & & \ddots & \\ & & & -\lambda \end{bmatrix}, \mathbf{B_0} = \mathbf{A_0} = \begin{bmatrix} \lambda & & & \\ & \lambda & & \\ & & \ddots & \\ & & & \lambda \end{bmatrix},$$

$$\mathbf{C} = \mathbf{A_2} = \begin{bmatrix} \mu & & & \\ & \mu & & \\ & & \ddots & \\ & & & \mu \end{bmatrix}, \mathbf{A_1} = \begin{bmatrix} -\lambda-\mu & & & \\ & -\lambda-\mu & & \\ & & \ddots & \\ & & & -\lambda-\mu \end{bmatrix}.$$

7.1.1 Matrix Geometric Method

The matrix geometric method (MGM) is widely utilized for accurate analysis of commonly encountered types of queueing systems that can be described as $QBDs$. It is worth noting that similar techniques, known as the more generic matrix analytic methods, can be utilized in contexts other than $QBDs$. However, the MGM can only be used if the system can be divided into two parts: the initial and repeating portions, which are common in all $QBDs$. To do MGM-based analysis, it is assumed that the CTMC with generator matrix $\mathbf{Q}$ is irreducible. The generator matrix will have a structure as follows:

$$\mathbf{Q} = \begin{bmatrix} \mathbf{B_1} & \mathbf{B_0} & \mathbf{0} & \mathbf{0} & \mathbf{0} & \cdots \\ \mathbf{B_2} & \mathbf{A_1} & \mathbf{A_0} & \mathbf{0} & \mathbf{0} & \cdots \\ \mathbf{0} & \mathbf{A_2} & \mathbf{A_1} & \mathbf{A_0} & \mathbf{0} & \cdots \\ \mathbf{0} & \mathbf{0} & \mathbf{A_2} & \mathbf{A_1} & \mathbf{A_0} & \cdots \\ \vdots & \vdots & \vdots & \ddots & \ddots & \ddots \end{bmatrix}.$$

Furthermore, it has been assumed that $\mathbf{A_0} + \mathbf{A_1} + \mathbf{A_2}$ is an irreducible infinitesimal generator with stationary probability $(\pi_i)_{i \in S}$ such that $\pi(\mathbf{A_0} + \mathbf{A_1} + \mathbf{A_2}) = \mathbf{0}$ and $\sum_i \pi_i = 1$. The elements of π denote the probabilities of being in the respective phase for any level greater than zero. While drawing the rate diagram and performing arc cuts between levels i and $i+1$ for any $i \geq 0$, then the net flow from left to right is $\pi \mathbf{A_0 e}$ and likewise the net flow from right to left is $\pi \mathbf{A_2 e}$ (note that the transitions in $\mathbf{A_1}$ correspond to those within a level; however, $\mathbf{A_0}$ are those that go up a level and $\mathbf{A_2}$ those that go down a level). For the system to be stable (much like the $M/M/1$ queue result), the net flow to the right ought to be smaller than the net flow to the left. Thus, once π is obtained, the condition that the CTMCs is stable can be stated as $\pi \mathbf{A_2 e} > \pi \mathbf{A_0 e}$, where $\mathbf{e}$ is a column vector with all entries 1.

If the CTMC is stable, the steady-state probabilities exist; let us call them π. Let π = $[\pi_0, \pi_1, \pi_2, \ldots]$, where $\pi_0, \pi_1, \pi_2, \ldots$ are row vectors of appropriate order. Since $\pi \mathbf{Q} = 0$, the following system of equations are obtained:

$$\begin{aligned}
\pi_0 \mathbf{B_1} + \pi_1 \mathbf{B_2} &= \mathbf{0}, \\
\pi_0 \mathbf{B_0} + \pi_1 \mathbf{A_1} + \pi_2 \mathbf{A_2} &= \mathbf{0}, \\
\pi_1 \mathbf{A_0} + \pi_2 \mathbf{A_1} + \pi_3 \mathbf{A_2} &= \mathbf{0}, \\
\pi_2 \mathbf{A_0} + \pi_3 \mathbf{A_1} + \pi_4 \mathbf{A_2} &= \mathbf{0}, \\
&\vdots
\end{aligned} \tag{7.1}$$

For these system of equations, the solution for all $i \geq 1$, $\pi_\mathbf{i} = \pi_\mathbf{1}\mathbf{R}^{\mathbf{i-1}}$ is computed with the normalization condition $\sum_\mathbf{i} \pi_\mathbf{i} = \mathbf{1}$, where $\mathbf{R}$ is the solution of $\mathbf{A_0} + \mathbf{R}\mathbf{A_1} + \mathbf{R}^2\mathbf{A_2} = 0$. This is known as the matrix geometric solution due to the matrix geometric relation between the stationary probabilities.

In fact, the crux of the MGM is in computing the $\mathbf{R}$ that satisfies quadratic form $\mathbf{A_0} + \mathbf{R}\mathbf{A_1} + \mathbf{R}^2\mathbf{A_2} = \mathbf{0}$. The matrix $\mathbf{R}$ is known as the auxiliary matrix. Thus, once $\mathbf{R}$ is known, π_0 is obtained in terms of π_1 such that it satisfies both $\pi_0\mathbf{B_1} + \pi_1\mathbf{B_2} = \mathbf{0}$ and $\pi_0\mathbf{B_0} + \pi_1\mathbf{A_1} + \pi_1\mathbf{R}\mathbf{A_2} = \mathbf{0}$. However, as with any CTMC, the set of equations are not linearly independent. Thus, drop one of them and use the normalizing condition that the elements of π sum to one. Although it is worthwhile to point out that there have been several articles in the literature that focus on efficiently computing $\mathbf{R}$ recursively, once $\mathbf{R}$ is computed, π can be obtained and used to determine steady-state performance measures such as queue length, waiting times, etc.

Example 7.8 Consider an $M/M/1$ queueing system where customers arrive at the system with arrival rate λ and the server provides service with rate μ. The system has an infinite size queue length for the waiting of the customers (refer, Fig. 6.1). The customers are served in a first come first served queueing discipline.

The stochastic process $\{X(t), t \geq 0\}$ for the underlying system denotes the number of customers in the system at time t. The state space of this process is defined as $S = \{0, 1, \ldots\}$. The generator matrix of the underlying stochastic process is given by

$$\mathbf{Q} = \begin{pmatrix} \mathbf{B_0} & \mathbf{A_0} & \mathbf{0} & \mathbf{0} & \mathbf{0} & \cdots \\ \mathbf{B_2} & \mathbf{A_1} & \mathbf{A_0} & \mathbf{0} & \mathbf{0} & \cdots \\ \mathbf{0} & \mathbf{A_2} & \mathbf{A_1} & \mathbf{A_0} & \mathbf{0} & \cdots \\ \mathbf{0} & \mathbf{0} & \mathbf{A_2} & \mathbf{A_1} & \mathbf{A_0} & \cdots \\ \vdots & \vdots & \vdots & \ddots & \ddots & \ddots \end{pmatrix}.$$

Here, the matrices $\mathbf{B_0}, \mathbf{A_0}, \mathbf{A_2}, \mathbf{B_2}, \mathbf{A_1}$ are reduced to single elements, i.e., $\mathbf{B_0} = -[\lambda]$, $\mathbf{A_0} = [\lambda]$, $\mathbf{A_1} = [-(\lambda + \mu)]$, $\mathbf{B_2} = \mathbf{A_2} = [\mu]$. The steady-state equations of the proposed model can be written by using the system of equations (7.1).

$$
\begin{aligned}
-\pi_0\lambda + \pi_1\mu &= 0,\\
\pi_0\lambda + \pi_1[-(\lambda+\mu)] + \pi_2\mu &= 0,\\
\pi_1\lambda + \pi_2[-(\lambda+\mu)] + \pi_3\mu &= 0,\\
\pi_2\lambda + \pi_3[-(\lambda+\mu)] + \pi_4\mu &= 0,\\
&\vdots
\end{aligned}
$$

Using the matrix-geometric methodology, the stationary distribution of the model is

$$\pi_n = (1-\rho)\rho^n, \quad n = 0, 1, \ldots .$$

where $\rho = \lambda/\mu < 1$.

Definition 7.8 (Level Dependent Quasi-Birth-Death Processes) Let $\{(X(t), J(t)) : t \geq 0\}$ be a continuous-time Markov process on the state space $S = \{(i, j) : i \geq 0, j = 1, 2, \ldots, m\}$ where i denotes the level of the process and j denotes the phase. The infinitesimal generator matrix of the continuous-time LDQBD process is of the form

$$
\mathbf{Q} = \begin{bmatrix}
\mathbf{A}_1^{(0)} & \mathbf{A}_0^{(0)} & \mathbf{0} & \mathbf{0} & \cdots \\
\mathbf{A}_2^{(1)} & \mathbf{A}_1^{(1)} & \mathbf{A}_0^{(1)} & \mathbf{0} & \cdots \\
\mathbf{0} & \mathbf{A}_2^{(2)} & \mathbf{A}_1^{(2)} & \mathbf{A}_0^{(2)} & \cdots \\
\mathbf{0} & \mathbf{0} & \mathbf{A}_2^{(3)} & \mathbf{A}_1^{(3)} & \cdots \\
\vdots & \vdots & & \ddots & \ddots
\end{bmatrix}
$$

The structure of the generator matrix $\mathbf{Q}$ reveals that its transitions and hence the block matrices $\mathbf{A}_0^{(i)}$, $\mathbf{A}_1^{(i)}$ and $\mathbf{A}_2^{(i)}$, for each level $i \geq 0$, may be different or in fact they are level dependent.

The LDQBD process finds wide applicability as a model for systems with complex dynamics. It is most prevalent in the modeling of queuing systems with MAP input, or for those with phase-type distributions. These models can help assess the impact of retrial customers who are initially denied access to service but who retry the system after some random amount of time. Such models are particularly important in large-scale service systems, such as large customer contact centers. In addition to queuing models, LDQBD processes are now emerging in reliability modeling and analysis and inventory systems modeling.

Although the structure, and many of the attributes of the LDQBD, mirror those of the level-independent version, their analysis and computational aspects are significantly more complicated. To compute the limiting probabilities for such QBD processes is a challenging problem. A key issue is to determine the truncation point. Many researchers have analyzed different LDQBD using some efficient methodologies.

7.1.2 *Matrix Analytic Method*

Matrix analytic method is a complicated version of the matrix-geometric method which is used to solve more general systems like $M/G/1$, $G/M/1$, LDQBD. Consider a system with the generator matrix given as

$$\mathbf{Q} = \begin{pmatrix} \mathbf{B_1} & \mathbf{A_{0,0}} & \mathbf{0} & \mathbf{0} & \mathbf{0} & \cdots \\ \mathbf{A_{2,1}} & \mathbf{A_{1,1}} & \mathbf{A_{0,1}} & \mathbf{0} & \mathbf{0} & \cdots \\ \mathbf{0} & \mathbf{A_{2,2}} & \mathbf{A_{1,2}} & \mathbf{A_{0,2}} & \mathbf{0} & \cdots \\ \mathbf{0} & \mathbf{0} & \mathbf{A_{2,3}} & \mathbf{A_{1,3}} & \mathbf{A_{0,3}} & \cdots \\ \vdots & \vdots & \vdots & \ddots & \ddots & \ddots \end{pmatrix},$$

If the CTMC is stable, then the steady-state probabilities exist; let us call them π. Write π as $[\pi_0, \pi_1, \pi_2, \ldots]$, where π_0 is a row vector and $\pi_1, \pi_2, \ldots$ are row vectors of appropriate order. Since $\pi \mathbf{Q} = 0$, the following set of equations are obtained:

$$\begin{aligned} \pi_0 \mathbf{B_1} + \pi_1 \mathbf{A_{2,1}} &= \mathbf{0}, \\ \pi_0 \mathbf{A_{0,0}} + \pi_1 \mathbf{A_{1,1}} + \pi_2 \mathbf{A_{2,2}} &= \mathbf{0}, \\ \pi_1 \mathbf{A_{0,1}} + \pi_2 \mathbf{A_{1,2}} + \pi_3 \mathbf{A_{2,3}} &= \mathbf{0}, \\ \pi_2 \mathbf{A_{0,2}} + \pi_3 \mathbf{A_{1,3}} + \pi_4 \mathbf{A_{2,4}} &= \mathbf{0}, \\ &\vdots \end{aligned}$$

For these system of equations, the solution for all $i \geq 1$, $\pi_i = \pi_{i-1} \mathbf{R^i}$ is computed with the normalization condition $\sum_i \pi_i = 1$. This is known as the matrix geometric solution due to the matrix geometric relation between the stationary probabilities. Clearly, for that to be a solution, $\mathbf{A_{0,i-1}} + \mathbf{R^i A_{1,i}} + \mathbf{R^i R^{i+1} A_{2,i+1}} = 0$. Thus, once $\mathbf{R^i}$ is known, π_i is obtained using the normalization condition.

7.2 Retrial Queueing Models

In classical queueing theory, it is frequently believed that a customer who cannot be serviced immediately upon arrival either enters the waiting line and is then served according to some queueing discipline, or exits the system permanently. When frustrated clients abandon the queue, it is thought that they are quitting the system permanently. However, the assumption concerning the loss of customers who chose to exit the system is only a first-order approximation to the actual scenario. Typically, such a consumer returns to the system after a random period of time and attempts to obtain service again. Queues of this type can be seen in a variety of everyday situations, such as wireless networks, healthcare systems, airport systems, cellular networks, etc. Therefore, the concept of retrial queues (or queues with returning customers,

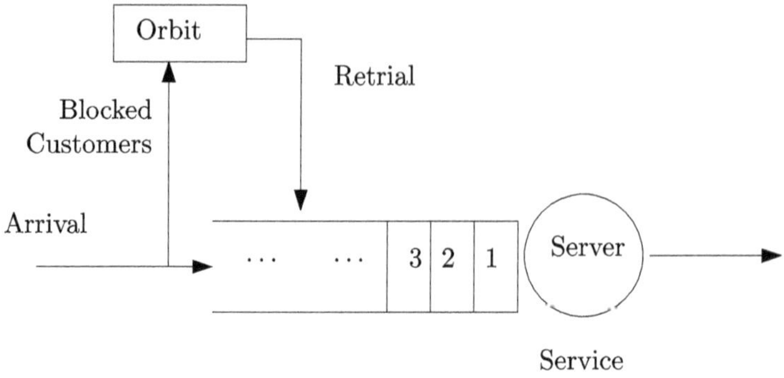

Fig. 7.5 Retrial queue general structure

repeated attempts, etc.) have been introduced. Retrial queues can alternatively be thought of as a subset of queueing networks. These networks, in their most basic form, have two nodes: a main node where blocking is possible and a delay node for retrial attempts. More nodes can be provided to represent specific retrial queues with a specified structure and queueing discipline. The waiting space for the retrial process is referred to as the orbit, and the size of the orbit can be either finite or infinite.

The main aim of this section is to construct retrial queues in various scenarios as well as to discuss algorithmic solution for these retrial queueing systems. For a few simple models, compact form expressions of stationary distributions is considered difficult. Therefore, some algorithmic methods and techniques are introduced for the analytically intractable problems. Retrial queues can be analyzed for several types of queueing models such as $M/M/1$, $M/M/c$, $MAP/PH/1$, $MAP/PH/c$, etc. Here, two examples are provided for $M/M/c$ retrial queueing system and $MAP/PH/c$ retrial queueing system with guard channel policy.

Example 7.9 Consider an $M/M/1/1$ retrial queue with setup time. Customers arrive at the server according to a Poisson process with rate λ.. The service time of customers follows an exponential distributed time with mean ν. Customers that see the server busy upon arrival join the orbit and retry for service after some exponential distributed time with mean $1/\mu$. When the system becomes empty, the server is turned off immediately. Customers who see the empty server wait at the server and the server is turned on. However, the server needs some setup time to be active to serve the waiting customer. It has been assumed that the setup time is exponential distributed with mean $1/\alpha$. Customers that see the server in setup state join the orbit and behave the same way as other customers in the orbit. Figure 7.6 shows $M/M/1/1$ retrial queueing system.

Let $X(t)$ and $Y(t)$ denote the state of the server and the number of customers in the orbit, respectively.

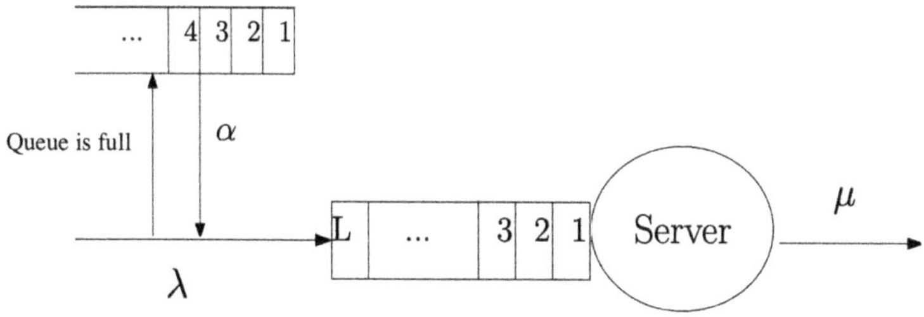

Fig. 7.6 Queueing structure for retrial system which is an LDQBD

$$X(t) = \begin{cases} 0, & \text{server is empty,} \\ 1, & \text{server is busy,} \\ 2, & \text{server is in setup process.} \end{cases}$$

The state space of the underlying stochastic process $\{(X(t), Y(t)), t \geq 0\}$ is given by

$$S = \{(i, j); i = 0, 1, 2, \ j \geq 0\}.$$

Suppose, there is no setup time, the limiting distribution of the system state for the $M/M/1$ retrial queue is given by

$$\pi_{0j} = \frac{\rho^j}{j!\mu^j}(1-\rho)^{1+\lambda/\mu}\prod_{k=0}^{j-1}(\lambda + k\mu), \quad j = 0, 1, \ldots,$$

$$\pi_{1j} = \frac{\rho^{j+1}}{j!\mu^j}(1-\rho)^{1+\lambda/\mu}\prod_{k=1}^{j}(\lambda + k\mu), \quad j = 0, 1, \ldots,$$

where $\rho = \frac{\lambda}{\nu}$. The derivation of the above results is left as an exercise.

Example 7.10 Consider a single server system with a finite buffer of capacity L. The arrival of customers follows the Poisson process with arrival rate λ and the service time follows an exponential distribution with parameter μ. When the buffer is full, an arriving customer joins a waiting space of infinite size and waits for an exponential distributed period with rate α. Let $X(t)$ be the number of customers who have arrived at the system but have not entered the system yet (i.e., in the waiting space) at time t and $Y(t)$ be the number of customers in the system including one who is waiting for service at time t. The state space of the underlying stochastic process $(X(t), Y(t)), t \geq 0$ is defined as follows: $S = \{(i, j); i \geq 0, 0 \leq j \leq L\}$ The generator matrix and the component matrices are given as follows

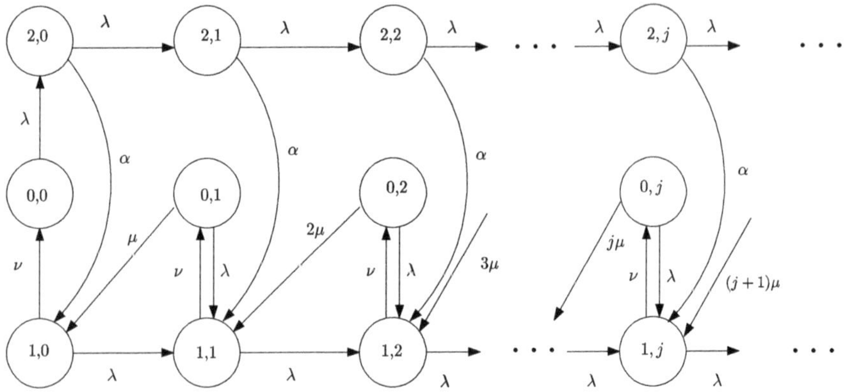

Fig. 7.7 State transition diagram for the balking queueing system

$$\mathbf{Q} = \begin{pmatrix} \mathbf{B_1} & \mathbf{A_0} & \mathbf{0} & \mathbf{0} & \mathbf{0} & \cdots \\ \mathbf{A_{2,1}} & \mathbf{A_{1,1}} & \mathbf{A_0} & \mathbf{0} & \mathbf{0} & \cdots \\ \mathbf{0} & \mathbf{A_{2,2}} & \mathbf{A_{1,2}} & \mathbf{A_0} & \mathbf{0} & \cdots \\ \mathbf{0} & \mathbf{0} & \mathbf{A_{2,3}} & \mathbf{A_{1,3}} & \mathbf{A_0} & \cdots \\ \vdots & \vdots & \vdots & \ddots & \ddots & \ddots \end{pmatrix},$$

where

$$\mathbf{B_1} = \begin{pmatrix} -\lambda & \lambda & 0 & \cdots \\ 0 & -\lambda & \lambda & \cdots \\ \vdots & \ddots & \ddots & \vdots \\ 0 & 0 & 0 & -\lambda \end{pmatrix}, \mathbf{A_0} = \begin{pmatrix} 0 & 0 & \cdots & 0 \\ 0 & 0 & \cdots & 0 \\ \vdots & \vdots & \cdots & \vdots \\ 0 & 0 & \cdots & \lambda \end{pmatrix}, \mathbf{A_{2,n}} = \begin{pmatrix} 0 & n\alpha & 0 & \cdots & 0 \\ 0 & 0 & n\alpha & \cdots & 0 \\ \vdots & \vdots & \vdots & \cdots & \vdots \\ 0 & 0 & 0 & \cdots & n\alpha \\ 0 & 0 & 0 & \cdots & 0 \end{pmatrix},$$

$$\mathbf{A_{1,n}} = \begin{pmatrix} -\lambda - n\alpha & \lambda & 0 & \cdots & 0 & 0 \\ \mu & -\lambda - n\alpha - \mu & \lambda & \cdots & 0 & 0 \\ \vdots & \ddots & \ddots & \cdots & \vdots & \vdots \\ 0 & 0 & 0 & \cdots & -\lambda - n\alpha - \mu & \lambda \\ 0 & 0 & 0 & \cdots & \mu & -\lambda - \mu \end{pmatrix}$$

for $n \geq 1$. The order of the component matrices is $(L + 1 \times L + 1)$.

Example 7.11 Consider the Example 5.7 with retrial phenomenon. In this example, consider that the quality of service (QoS) requirements for voice traffic is considered as both soft and hard, depending on the particular application. The arrival of hard/handoff and soft/new types of traffic will follow the Poisson process with exponential distributed arrival rates λ_H and λ_N, respectively. The service follows an exponential distribution with service rates μ. In this scenario, the handoff traffic is given priority over the new traffic using the guard channel policy where g channels are reserved exclusively for the use of handoff traffic. The blocked new traffic which is not able to receive the service will wait in an orbit of finite capacity J. The blocked calls may retry for their service from the orbit after some random amount of time which follows exponential distribution with rate μ_r and probability p or may exit the system with complimentary probability [8].

Let $\{X(t), t \geq 0\}$ be the stochastic process with a state space $\{(j, k) : j = 0, 1, \ldots, M;\ k = 0, 1, \ldots, J\}$.

- The elements of the upper diagonal $\mathbf{Q_{j0}}$ are represented by $A_{j2} = jp\mu_r, j = 0, 1, \ldots, J;$
$$A_{j1} = \begin{cases} \lambda = \lambda_H + \lambda_N, & 0 \leq j \leq M - g - 1 \\ \lambda_H, & j = M - g, M - g + 1, \ldots, \leq M - 1 \end{cases}.$$
- The elements of the lower diagonal are given by $\mathbf{Q_{j2}} = j\mu$.
- The elements of the main diagonal, denoted as $\mathbf{Q_{j1}}$ are represented as $B_{j2} = k(1-p)\mu_r, k = 0, 1, \ldots, J;\ B_{j1} = -(\lambda + j\mu + k\mu_r), k = 0, 1, \ldots, J;$
$$B_{j0} = \begin{cases} 0, & j = 0, 1, \ldots, M - g - 1 \\ \lambda_N, & j = M - g, M - g + 1, \ldots, M \end{cases}.$$

Example 7.12 Consider Example 7.11 with PH distributed service times with order N [2]. Let the state space for the stochastic process $\{X(t), t \geq 0\}$ be given by $\{(i, l, S_i),\ i = 1, 2, \ldots, M,\ l = 0, 1, \ldots, J\}$.

- The expressions of A^i_{l1} and A^i_{l2}, for $0 \leq i \leq M - 1$, are $A^i_{l1} = \lambda I,\quad 0 \leq l \leq J,\ A^i_{l2} = l\mu_{rc} I,\ l = 1, 2, \ldots, J$.
- C^i_{l1} represents a decrement in the ongoing traffic due to service completion by one of the i, $i = 1, 2, \ldots, M$ traffic and is given by $V(i)$, for $l = 0, 1, \ldots, J$.
- B^i_{l1} represents no service completion whereas B^i_{l0} represents the increment in the number of traffic in orbit. The decrement in the number of traffic in orbit is possible when the traffic leaves the orbit and is given by B^i_{l2}. The expressions are given below.

$$B^i_{l1} = \begin{cases} -(\lambda + l\mu_{rq} + l\mu_{rc})I, & i = 0,\ 0 \leq l \leq J \\ W(i) - (\lambda + l\mu_{rq} + l\mu_{rc})I, & 1 \leq i \leq M - 1,\ 0 \leq l \leq J \\ W(i) - (\lambda - l\mu_{rq})I & i = M,\ 0 \leq l \leq J - 1 \\ W(i) - l\mu_{rq} I, & i = M,\ l = J \end{cases}.$$

$$B^i_{l0} = \begin{cases} 0, & 0 \leq i \leq M - 1 \\ \lambda I, & i = M \end{cases};\ .\ B^i_{l2} = l\mu_{rq}\ 0 \leq i \leq M,\ 1 \leq l \leq J\,.$$

Example 7.13 Consider Example 7.12 with MAP arrival of incoming traffic [3].

Let the state space of the underlying stochastic process be given as:

$$\{(i, l, u),\ 0 \leq i \leq M,\ 0 \leq l \leq J,\ 1 \leq u \leq K\}.$$

The block matrix entries are given as follows:

- $\mathbf{Q}_{i0}$ is a lower diagonal matrix with $\mathbf{A}^{\mathbf{i}}_{\mathbf{l1}}$ as main diagonal and $\mathbf{A}^{\mathbf{i}}_{\mathbf{l2}}$ as lower diagonal entries. The matrices $\mathbf{A}^{\mathbf{i}}_{\mathbf{l1}}$ and $\mathbf{A}^{\mathbf{i}}_{\mathbf{l2}}$ corresponds to the increment in the number of traffic in service due to fresh arrivals of traffic and due to successful retrials of traffic, respectively. The expressions are given as:

$$\mathbf{A}^{\mathbf{i}}_{\mathbf{l1}} = C_N + C_H, \ \ 0 \leq l \leq J; \ \ A^i_{l2} = l\mu_{rc}I, \ \ 1 \leq l \leq J.$$

- Consider the block $\mathbf{Q}_{\mathbf{i2}}$. It is a diagonal matrix with diagonal elements C^i_{l1}. The expression for $\mathbf{C}^{\mathbf{i}}_{\mathbf{l1}}$ is given as $\mathbf{C}^{\mathbf{i}}_{\mathbf{l1}} = i\mu\mathbf{I}$.
- Finally, consider the blocks $\mathbf{Q}_{\mathbf{i1}}$. It is a tri-diagonal matrix with main diagonal elements $\mathbf{B}^{\mathbf{i}}_{\mathbf{l1}}$, as

$$\mathbf{B}^{\mathbf{i}}_{\mathbf{l1}} = \begin{cases} C_0 - (i\mu + l\mu_{rq} + l\mu_{rc})I, \ i = 0, 1, \ldots, M-1, \ l = 0, 1, \ldots, J \\ C_0 + C_H - (i\mu + l\mu_{rq})I, \ \ i = M, \ l = 0, 1, \ldots, J-1 \\ -(i\mu + l\mu_{rq})I, \qquad\qquad i = M, \ l = J \end{cases}.$$

Similarly, B^i_{l0} represents the upper diagonal given as

$$B^i_{l0} = \begin{cases} 0 \quad 0 \leq i \leq M-1, \ 0 \leq l \leq J-1 \\ C_N, \ i = M, \ 0 \leq l \leq J-1 \end{cases}.$$

Lastly, B^i_{l2} is given by : $B^i_{l2} = l\mu_{rq}; \ i = 0, 1, \ldots, M, \ \ l = 1, 2, \ldots, J.$

Example 7.14 In cellular networks, a given geographical area is divided into cells. It has been assumed that the cells are statistically homogeneous. The incoming traffic are categorized as video (vo), voice (ve), and data queries (d). The quality of service (QoS) requirements for video traffic is considered as both soft and hard, depending on the particular application. Video traffic has soft QoS for a few specific applications (like, video on demand), whereas it can have hard QoS for some other ones (like, live video). The video traffic is therefore divided into video 1 (vo1) and video 2 (vo2) traffic. Each inbound traffic is classified as new (n_i, i = ve, vo1, vo2, d) and handoff (h_i, i = ve, vo1, vo2, d). A continuous-time MAP is used to represent each traffic's arrival since it enables correlation between the traffic's individual arrival times. Numerous prioritization policies have been put out in the literature for traffic prioritization. One of the most used techniques, fixed guard channel policy (GC), is a decent balance between performance and complexity. In this study, the GC policy is used to maintain connectivity and provide various types of traffic preference. According to this policy, a specific number of channels, let's say g, (g changes for each type of traffic), are set aside for each type of traffic. Some other detailed assumptions are as follows:

- **Arrival Process:** The arrival of each traffic is modeled by using MAP and represented by MAP_1 (ve), MAP_2 (vo1), MAP_3 (vo2) and MAP_4 (d). The arrival process for ve traffic is stated briefly. Let C^{ve} be the infinitesimal generator matrix with dimension K. Let C_0^{ve}, C_n^{ve}, C_h^{ve} ne the matrices that govern the transition of no call arrivals, new call arrivals and handoff call arrivals, respectively. Then, $C^{ve} = C_0^{ve} + C_n^{ve} + C_h^{ve}$. The arrival rates for new and handoff ve traffic are given as $\lambda_{nve} = \pi C_n^{ve} e$ and $\lambda_{hve} = \pi C_h^{ve} e$, respectively. Here, π is the stationary probability vector of the generator C^{ve} and e is column vector of all 1's. On the similar track, the arrival processes for vo1, vo2, and d traffic can be defined.

- **Cell Residence Time:** The CRT (R_s) is PH distributed with representation (α, T) and dimension r.
- **Call Holding Time:** The CHT (H_i, $i =$ ve, vo, d) of type i traffic has PH distribution with representation (β_i, S_i) and dimension h_i.
- **Service Process:** According to Neuts (1994), the residual CRT ($\bar{R}_s$) and residual CHT ($\bar{H}_i$) are also PH distributed. Let $S^{i,N}$ be the service times of new traffic of type i, defined as $S^{i,N} = min(H_i, \overline{R}_s)$. Since, the minimum of two PH distributions is also a PH distribution. Therefore, $S^{i,N}$ is also PH distributed with representation (δ_{N_i}, L_{N_i}) and dimension rh_i. Similarly, $S^{i,H}$ is PH distributed with representation (δ_{H_i}, L_{H_i}) and dimension rh_i. Here, $\delta_{N_i} = \beta_i \otimes \alpha$, $L_{N_i} = S_i \oplus T$, $\delta_{H_i} = \beta_i \otimes \alpha$, $L_{H_i} = S_i \oplus T$.
- **Retrial Process:** The retrial process R_t is PH distributed with representation (δ, L) and dimension n. The inter-retrial times or departure times are denoted by $R = min(R_t, \overline{R}_s)$, whichever happens first. The retrial process has two absorbing states either due to successful retrial or departure from the system.
- **Call Admission Control Scheme:** As per the proposed CAC scheme, the incoming traffic is classified in the decreasing order of priority as h_{ve}, n_{ve}, h_{vo1}, n_{vo1}, h_{vo2}, n_{vo2}, h_d, n_d. It has been assumed that ve and d traffic are assigned only one channel whereas vo are assigned four channels. Therefore, h_1, h_2, h_3 and h_4 are reserved for h_{ve}, h_{vo1}, h_{vo2}, h_d, whereas n_1, n_2 and n_3 channels are reserved for n_{ve}, n_{vo1}, n_{vo2}, respectively. Since, n_d is of the least priority and can tolerate the delay, no channel has been reserved for it. Let M be the total number of available channels. To perform the analysis, define the following notations:

$$M_1 = M - \sum_{i=1}^{3}(h_i + n_i) - h_4; \quad M_2 = M - \sum_{i=1}^{3}(h_i + n_i)$$

$$M_3 = M - \sum_{i=1}^{2}(h_i + n_i) - h_3; \quad M_4 = M - \sum_{i=1}^{2}(h_i + n_i)$$

$$M_5 = M - h_1 - n_1 - h_2; \qquad M_6 = M - h_1 - n_1; \qquad M_7 = M - h_1.$$

Analytical Model

Consider a cellular system with M channels and J orbit size. Let $\{X(t), t \geq 0\}$ be a stochastic process for a given system with the state space

$$\begin{aligned}\Omega =& \{i, j, k, l, u_{ve}, u_{vo1}, u_{vo2}, u_d, \mathbf{S}_{i-j-k}^{ve}, \mathbf{S}_j^{vo}, \mathbf{S}_k^{d}, \mathbf{r}_l;\ 0 \leq i \leq M,\ 0 \leq j \leq j_m,\\ & 0 \leq k \leq k_m,\ 0 \leq l \leq J,\ 1 \leq u_m \leq K_m,\ 1 \leq u_d \leq 3\},\end{aligned}$$

where

- i is the number of traffic receiving service,
- j is the number of vo traffic,
- k is the number of d traffic,
- l is the total number of new ve, handoff vo2, new vo2, handoff d and new d traffic in the orbit,
- $u_{ve}, u_{vo1}, u_{vo2}, u_d$ are the phases of MAPs for ve, vo1, vo2 and d traffic arrival process, respectively,
- $\mathbf{S}^{ve}_{i-j-k}$, $\mathbf{S}^{vo}_{j}$, $\mathbf{S}^{d}_{k}$ are the set of phase of service for ve, vo (1 and 2) and d traffic, respectively,
- $\mathbf{r}_l$ is the set of retrial phase of l traffic in the orbit,

and, $m =$ ve, vo1, vo2, $j_m = min\{i, \left[\frac{M_6}{4}\right]\}$ and $k_m = min\{i-j, M_2\}$. Here, $[x]$ means the greatest integer function of x. The infinitesimal generator matrix $\mathbf{Q}$ for this $LDQBD$ process and the blocks of the matrix $\mathbf{Q}$ are provided as follows.

$$\mathbf{Q} = \begin{pmatrix} \mathbf{Q_{01}} & \mathbf{Q_{00}} & & & \\ \ddots & \ddots & \ddots & & \\ & \mathbf{Q_{i2}} & \mathbf{Q_{i1}} & \mathbf{Q_{i0}} & \\ & & \ddots & \ddots & \ddots \\ & & & \mathbf{Q_{M2}} & \mathbf{Q_{M1}} \end{pmatrix}$$

$$\mathbf{Q_{i0}} = \begin{pmatrix} \mathbf{A^i_{01}} & \mathbf{A^i_{00}} & & & \\ \ddots & \ddots & \ddots & & \\ & \mathbf{0} & \mathbf{A^i_{j1}} & \mathbf{A^i_{j0}} & \\ & & \ddots & \ddots & \ddots \\ & & & \mathbf{0} & \mathbf{A^i_{j_m1}} \end{pmatrix} \quad \mathbf{Q_{i2}} = \begin{pmatrix} \mathbf{C^i_{01}} & \mathbf{0} & & & \\ \ddots & \ddots & \ddots & & \\ & \mathbf{C^i_{j2}} & \mathbf{C^i_{j1}} & \mathbf{0} & \\ & & \ddots & \ddots & \ddots \\ & & & \mathbf{C^i_{j_m2}} & \mathbf{C^i_{j_m1}} \end{pmatrix}$$

and $\mathbf{Q_{i1}} = diag(\mathbf{B^i_0}, \ldots, \mathbf{B^i_j}, \ldots, \mathbf{B^i_{j_m}})$.
Note that the notations $\mathbf{I_w}$, $\mathbf{W_r(i)}$, $\mathbf{V_r(i, 1)}$, $\mathbf{I(w, s)}$ and $\widehat{\mathbf{V_r}}\mathbf{(i)}$ as defined in [1] have been used to carry out the analysis. A detailed explanation for the blocks of the generator matrix is given as follows.

- Q_{i0} **block:** The blocks $\mathbf{A^i_{j1}}$ and $\mathbf{A^i_{j0}}$ represent the cases when due to successful retrial or arrival of fresh traffic, the number of ve or d traffic and vo traffic goes up by one. The main diagonal elements $\mathbf{A^{i,j,1}_{k1}}$ represent the case, when the number of ve traffic goes up by one. Similarly, the upper diagonal elements $\mathbf{A^{i,j,1}_{k0}}$ shows increment in the d traffic. Each of $\mathbf{A^{i,j,1}_{k1}}$ and $A^{i,j,1}_{k0}$ is a lower diagonal block matrices with main diagonal elements $\mathbf{A^{i,j,1,k,1}_{l,1}}$ and $\mathbf{A^{i,j,1,k,0}_{l,1}}$, respectively and lower diagonal elements as $\mathbf{A^{i,j,1,k,1}_{l,2}}$ and $\mathbf{A^{i,j,1,k,0}_{l,2}}$.

- $\mathbf{A_{l,1}^{i,j,1,k,1}}$ represent the increment due to the arrival of fresh ve traffic. Similarly, $\mathbf{A_{l,1}^{i,j,1,k,0}}$ denotes the increment in the fresh d traffic.
- $\mathbf{A_{l,2}^{i,j,1,k,1}}$ and $\mathbf{A_{l,2}^{i,j,1,k,0}}$ represents the increment due to the successful retrial of new ve and d traffic from the orbit.

The upper diagonal block $\mathbf{A_{j0}^{i}}$ is a diagonal matrix with $\mathbf{A_{k1}^{i,j,0}}$, $0 \leq k \leq k_m$ as diagonal entries. Each of $\mathbf{A_{k1}^{i,j,0}}$ is a lower diagonal matrix with main diagonal elements $\mathbf{A_{l,1}^{i,j,0,k,1}}$ and lower diagonal elements $\mathbf{A_{l,2}^{i,j,0,k,1}}$.

- $\mathbf{A_{l,1}^{i,j,0,k,1}}$ represents the increase in the number of vo traffic due to the fresh arrivals of vo (1 and 2) traffic.
- $\mathbf{A_{l,2}^{i,j,0,k,1}}$ shows the case when an increase in the vo traffic is due to successful retrials by the vo traffic from the orbit.

The expressions of the block entries are given as follows, for $0 \leq i \leq M-1$, $0 \leq j \leq j_m$, $0 \leq k \leq k_m$ and $0 \leq l \leq J$:

$$\begin{aligned}
\mathbf{A_{l,1}^{i,j,1,k,1}} &= \lambda_{VE}^{i,j} \oplus C_0^{vo1} \oplus C_0^{vo2} \oplus C_0^{d} \oplus W_{ve}(i-j-k) \oplus W_{vo}(j) \oplus W_d(k) \oplus W_r(l) \\
\mathbf{A_{l,1}^{i,j,1,k,0}} &= C_0^{ve} \oplus C_0^{vo1} \oplus C_0^{vo2} \oplus \lambda_D^{i,j} \oplus W_{ve}(i-j-k) \oplus W_{vo}(j) \oplus W_d(k) \oplus W_r(l) \\
\mathbf{A_{l,1}^{i,j,0,k,1}} &= C_0^{ve} \oplus \lambda_{VO}^{i,j} \oplus C_0^{d} \oplus W_{ve}(i-j-k) \oplus W_{vo}(j) \oplus W_d(k) \oplus W_r(l) \\
\mathbf{A_{l,2}^{i,j,1,k,1}} &= C_0^{ve} \oplus C_0^{vo1} \oplus C_0^{vo2} \oplus C_0^{d} \oplus W_{ve}(i-j-k) \oplus W_{vo}(j) \oplus W_d(k) \oplus g^{i,j}\widehat{V}_r(l) \\
\mathbf{A_{l,2}^{i,j,1,k,0}} &= C_0^{ve} \oplus C_0^{vo1} \oplus C_0^{vo2} \oplus C_0^{d} \oplus W_{ve}(i-j-k) \oplus W_{vo}(j) \oplus W_d(k) \oplus h^{i,j}\widehat{V}_r(l) \\
\mathbf{A_{l,2}^{i,j,0,k,1}} &= C_0^{ve} \oplus C_0^{vo1} \oplus C_0^{vo2} \oplus C_0^{d} \oplus W_{ve}(i-j-k) \oplus W_{vo}(j) \oplus W_d(k) \oplus m^{i,j}\widehat{V}_r(l)
\end{aligned}$$

where

$$\lambda_{VE}^{i,j} = \begin{cases} C_{hv}^{ve} + C_{nv}^{ve}, & 0 \leq i+3j < M_7 \\ C_{hv}^{ve}, & M_7 \leq i+3j < M \\ 0, & \text{otherwise} \end{cases}; \quad g^{i,j} = \begin{cases} p_{nve}, & 0 \leq i+3j < M_7 \\ 0, & \text{otherwise} \end{cases};$$

$$\lambda_{D}^{i,j} = \begin{cases} C_{hd}^{d} + C_{nd}^{d}, & 0 \leq i+3j < M_1 \\ C_{hd}^{d}, & M_1 \leq i+3j < M_2 \\ 0, & \text{otherwise} \end{cases}; \quad h^{i,j} = \begin{cases} p_{hd} + p_{nd}, & 0 \leq i+3j < M_1 \\ p_{hd}, & M_1 \leq i+3j < M_2 \\ 0, & \text{otherwise} \end{cases};$$

$$\lambda_{VO}^{i,j} = \begin{cases} (C_{hvo1}^{vo1} + C_{nvo1}^{vo1}) \oplus (C_{hvo2}^{vo2} + C_{nvo2}^{vo2}), & 0 \leq i+3j+4 < M_3 \\ (C_{hvo1}^{vo1} + C_{nvo1}^{vo1}) \oplus C_{hvo2}^{vo2}, & M_3 \leq i+3j+4 < M_4 \\ (C_{hvo1}^{vo1} + C_{nvo1}^{vo1}) \oplus C_0^{vo2}, & M_4 \leq i+3j+4 < M_5 \\ C_{hvo1}^{vo1} \oplus C_0^{vo2}, & M_5 \leq i+3j+4 < M_6 \\ 0, & \text{otherwise} \end{cases};$$

$$m^{i,j} = \begin{cases} p_{hvo2} + p_{nvo2}, & 0 \leq i+3j+4 < M_3 \\ p_{hvo2}, & M_3 \leq i+3j+4 < M_4 \\ 0, & \text{otherwise} \end{cases}.$$

$\mathbf{Q_{i2}}$ block: For each $1 \le i \le M$, the blocks $\mathbf{C^{i}_{j1}}$ corresponds to the case of ve or d traffic completion whereas $\mathbf{C^{i}_{j2}}$ represents the traffic completion by the vo traffic. $\mathbf{C^{i}_{j1}}$ is a lower diagonal matrix with $\mathbf{C^{i,j,1}_{k1}}$ as main diagonal and $\mathbf{C^{i,j,1}_{k2}}$ as lower diagonal elements. Each of $\mathbf{C^{i,j,1}_{k1}}$ and $\mathbf{C^{i,j,1}_{k2}}$ is a diagonal matrix with diagonal elements $\mathbf{C^{i,j,1,k,1}_{l,1}}$ and $\mathbf{C^{i,j,1,k,2}_{l,1}}$, respectively.

- $\mathbf{C^{i,j,1,k,1}_{l,1}}$ shows reduction in the number of ongoing ve traffic.
- $\mathbf{C^{i,j,1,k,2}_{l,1}}$ represents the reduction in the d traffic due to service completion.

For $1 \le i \le M, 0 \le j \le j_m, 0 \le k \le k_m, 0 \le l \le J$,

$$\mathbf{C^{i,j,1,k,1}_{l,1}} = C^{ve}_0 \oplus C^{vo1}_0 \oplus C^{vo2}_0 \oplus C^{d}_0 \oplus V_{ve}(i-j-k) \oplus W_{vo}(j) \oplus W_d(k) \oplus W_r(l)$$
$$\mathbf{C^{i,j,1,k,2}_{l,1}} = C^{ve}_0 \oplus C^{vo1}_0 \oplus C^{vo2}_0 \oplus C^{d}_0 \oplus W_{ve}(i-j-k) \oplus W_{vo}(j) \oplus V_d(k) \oplus W_r(l).$$

$\mathbf{C^{i}_{j2}}$ is a diagonal matrix with diagonal entries as $\mathbf{C^{i,j,2}_{k,1}}$. Each of $\mathbf{C^{i,j,2}_{k,1}}$ is a diagonal matrix with its diagonal elements $\mathbf{C^{i,j,2,k,1}_{l,1}}$. $\mathbf{C^{i,j,2,k,1}_{l,1}}$ represents the decrement in the number of vo traffic.

For $1 \le i \le M, 1 \le j \le j_m, 0 \le k \le k_m$ and $0 \le l \le J$,

$$\mathbf{C^{i,j,2,k,1}_{l,1}} = C^{ve}_0 \oplus C^{vo1}_0 \oplus C^{vo2}_0 \oplus C^{d}_0 \oplus W_{ve}(i-j-k) \oplus V_{vo}(j) \oplus W_d(k) \oplus W_r(l).$$

- Q_{i1} **block:** $\mathbf{B^{i}_{j}}$ represents no change in the number of traffic receiving service where, $B^i_j = diag(B^{i,j}_0, \ldots, B^{i,j}_k, \ldots, B^{i,j}_{k_m})$. Each of $\mathbf{B}^{i,j}_k$ is a tri-diagonal matrix.
- The main diagonal element $\mathbf{B^{i,j,k}_{l,1}}$ exhibits no change in the traffic in service as well as no change in the number of traffic in the orbit.
  For $0 \le i \le M, 0 \le j \le j_m, 0 \le k \le k_m$, and $0 \le l \le J$,

$$\mathbf{B^{i,j,k}_{l,1}} = C^{ve}_0 \oplus C^{vo1}_0 \oplus C^{vo2}_0 \oplus C^{d}_0 \oplus W_{ve}(i-j-k) \oplus W_{vo}(j) \oplus W_d(k) \oplus W_r(l).$$

- The lower diagonal elements $\mathbf{B^{i,j,k}_{l,2}}$ shows departure of the traffic from the orbit without service.
  For $0 \le i \le M, 0 \le j \le j_m, 0 \le k \le k_m$, and $1 \le l \le J$,

$$\mathbf{B^{i,j,k}_{l,2}} = C^{ve}_0 \oplus C^{vo1}_0 \oplus C^{vo2}_0 \oplus C^{d}_0 \oplus W_{ve}(i-j-k) \oplus W_{vo}(j) \oplus W_d(k) \oplus V_r(l,1).$$

- The upper diagonal elements $\mathbf{B^{i,j,k}_{l,0}}$ represents increase in the number of traffic in the orbit.
  For $0 \le i \le M, 0 \le j \le j_m, 0 \le k \le k_m$, and $0 \le l \le J-1$,

$$\mathbf{B^{i,j,k}_{l,0}} = C^{vo1}_0 \oplus n^{i,j,k} \oplus W_{ve}(i-j-k) \oplus W_{vo}(j) \oplus W_d(k) \oplus W_r(l),$$

where

$$n^{i,j,k} = \begin{cases} C_0^{nd} \oplus C_0^{ve} \oplus C_0^{vo2}, & M_1 \le i+3j < M_2 \\ (C_{hd}^d + C_{nd}^d) \oplus C_0^{ve} \oplus C_0^{vo2}, & M_2 \le i+3j < M_3 \\ C_{nvo2}^{vo2} \oplus (C_{hd}^d + C_{nd}^d) \oplus C_0^{ve}, & M_3 \le i+3j < M_4 \\ (C_{nvo2}^{vo2} + C_{hvo2}^{vo2}) \oplus (C_{hd}^d + C_{nd}^d) \oplus C_0^{ve}, & M_4 \le i+3j < M_7 \\ C_{nv}^{ve} \oplus (C_{nvo2}^{vo2} + C_{hvo2}^{vo2}) \oplus (C_{hd}^d + C_{nd}^d) \oplus C_0^{ve}, & M_7 \le i+3j < M \\ 0, & \text{otherwise.} \end{cases}$$

Special Case: Poisson arrivals, exponential service and retrial times, four types of traffic [5]

Let the state space for the stochastic process $\{X(t), t \ge 0\}$ be defined as : $\Omega = \{(i, j, k, l);\ 0 \le i \le M;\ 0 \le j \le j_m;\ 0 \le k \le k_m;\ 0 \le l \le J\}$. The expression for the block matrices of the generator matrix $\mathbf{Q}$ for this particular case is given as follows.

- The expression of $\mathbf{A_{l,1}^{i,j,1,k,1}}$ and $\mathbf{A_{l,2}^{i,j,1,k,1}}$. For $0 \le i \le M-1, 0 \le j \le j_m, 0 \le k \le k_m$, $A_{l,1}^{i,j,1,k,1} = \lambda_{VE}^{i,j}$, $l = 0, 1, \ldots, J$; $A_{l,2}^{i,j,1,k,1} = g^{i,j} * l * \mu_{rc}$, $l = 1, \ldots, J$, where

$$\lambda_{VE}^{i,j} = \begin{cases} \lambda_{hv} + \lambda_{nv}, & 0 \le i+3j < M_7 \\ \lambda_{hv}, & M_7 \le i+3j < M; \\ 0, & \text{otherwise} \end{cases} \quad g^{i,j} = \begin{cases} p_{nv}, & 0 \le i+3j < M_7 \\ 0, & \text{otherwise.} \end{cases}$$

- For $0 \le i \le M-1, 0 \le j \le j_m, 0 \le k \le k_m - 1$, $A_{l,1}^{i,j,1,k,0} = \lambda_D^{i,j}$, $l = 0, 1, \ldots, J$; $A_{l,2}^{i,j,1,k,0} = h^{i,j} * l * \mu_{rc}$, $l = 1, 2, \ldots, J$ where,

$$\lambda_D^{i,j} = \begin{cases} \lambda_{hd} + \lambda_{nd}, & 0 \le i+3j < M_1 \\ \lambda_{hd}, & M_1 \le i+3j < M_2; \\ 0, & \text{otherwise,} \end{cases} \quad h^{i,j} = \begin{cases} p_{hd} + p_{nd}, & 0 \le i+3j < M_1 \\ p_{hd}, & M_1 \le i+3j < M_2 \\ 0, & \text{otherwise.} \end{cases}$$

- For $0 \le i \le M-1$, $0 \le j \le j_m - 1$, $0 \le k \le k_m$, the expressions for $A_{l,1}^{i,j,0,k,1}$ and $A_{l,2}^{i,j,0,k,1}$: $A_{l,1}^{i,j,0,k,1} = \lambda_{VO}^{i,j}$, $l = 0, 1, \ldots, J$; $A_{l,2}^{i,j,0,k,1} = m^{i,j} * l * \mu_{rc}$, $l = 1, \ldots, J$, where

$$\lambda_{VO}^{i,j} = \begin{cases} \lambda_{hvo1} + \lambda_{nvo1} + \lambda_{hvo2} + \lambda_{nvo2}, & 0 \le i+3j+4 < M_3 \\ \lambda_{hvo1} + \lambda_{nvo1} + \lambda_{hvo2}, & M_3 \le i+3j+4 < M_4 \\ \lambda_{hvo1} + \lambda_{nvo1}, & M_4 \le i+3j+4 < M_5 \\ \lambda_{hvo1}, & M_5 \le i+3j+4 < M_6 \\ 0, & \text{otherwise} \end{cases}$$

$$m^{i,j} = \begin{cases} p_{hvo2} + p_{nvo2}, & 0 \le i+3j+4 < M_3 \\ p_{hvo2}, & M_3 \le i+3j+4 < M_4 \\ 0, & \text{otherwise.} \end{cases}$$

– For $1 \le i \le M$, $0 \le j \le j_m$, $0 \le k \le k_m$ and $0 \le l \le J$, $C_{l,1}^{i,j,1,k,1} = (i - j - k) * \mu_{ve}$; $C_{l,1}^{i,j,2,k,1} = j * \mu_{vo}$; $C_{l,1}^{i,j,1,k,2} = k * \mu_d$. Finally, consider the blocks of Q_{i1}. For $0 \le i \le M, 0 \le j \le j_m, 0 \le k \le k_m$ and $0 \le l \le J$,

$$B_{l,1}^{i,j,k} = -(\lambda_{VE}^{i,j} + \lambda_{VO}^{i,j} + \lambda_{D}^{i,j}) - (i - j - k) * \mu_{ve} - j * \mu_{ve} - k * \mu_d - l * \mu_{rq} - l * \mu_{rc}$$

$$B_{l,0}^{i,j,k} = n^{i,j,k}; \quad B_{l,2}^{i,j,k} = l * \mu_{rq},$$

where

$$n^{i,j,k} = \begin{cases} \lambda_{nd}, & M_1 \le i + 3j < M_2 \\ \lambda_{hd} + \lambda_{nd}, & M_2 \le i + 3j < M_3 \\ \lambda_{nvo2} + \lambda_{hd} + \lambda_{nd}, & M_3 \le i + 3j < M_4 \\ \lambda_{hvo2} + \lambda_{nvo2} + \lambda_{hd} + \lambda_{nd}, & M_4 \le i + 3j < M_7 \\ \lambda_{nv} + \lambda_{hvo2} + \lambda_{nvo2} + \lambda_{hd} + \lambda_{nd}, & M_7 \le i + 3j < M \\ 0, & \text{otherwise.} \end{cases}$$

In the literature, many methodologies have been proposed for the efficient computation of steady-state distributions of the system such as matrix geometric method, matrix analytic method, block Jacobi method, etc. Next, a detailed description over matrix geometric method is provided for the steady-state analysis of the system.

7.3 Balking and Reneging Queueing Models

In real life, it often happens that arrivals become discouraged when the queue is long and do not wish to wait. If the customer refuses to join the queueing system, then this scenario is termed as balking. Whereas, some customers who have a tendency to be impatient might not always be deterred by long lines; instead, they might join the line to see how long it might take them to get service while still having the option to leave if they determine that the wait will be intolerably long. This scenario is referred to as reneging. The balking and reneging scenario will be described in detail with the help of $M/M/c$ queueing system. In the case of balking, consider that an arriving customer who finds that all the servers are busy may join the system with probability p or balk (does not join) with probability $1 - p$. In the case of reneging, it is assumed that the time spent by such a customer in the system follows an exponential distribution with parameter β. For both scenarios, the steady-state probabilities will be obtained.

Fig. 7.8 State transition diagram for the reneging queueing system

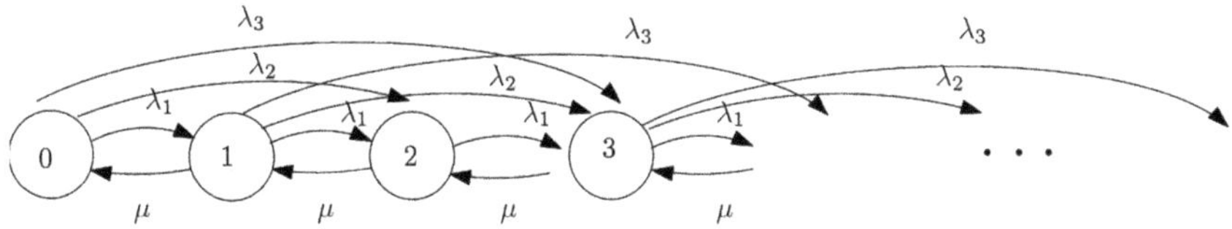

Fig. 7.9 State transition diagram for bulk arrival system

7.3.1 *Balking M/M/c Model*

Consider an $\mathbf{M/M}/c$ queue with balking. The arriving customer will enter the system if any one server is not busy. This model is considered as BDP with the arrival rate for the system being $\lambda_n = \lambda,\ n = 0, 1, \ldots, c-1, \lambda_n = p\lambda,\ n = c, c+1, \ldots$. The service rate is $\mu_n = n\mu,\ n = 1, 2, \ldots, c,\ \mu_n = c\mu,\ n = c+1, c+2, \ldots$.
The steady-state equations are presented as follows

$$\begin{aligned}
\pi_0\lambda &= \pi_1\mu, \\
\pi_i(\lambda + i\mu) &= \pi_{i-1}\lambda + \pi_{i+1}(i+1)\mu, \quad i = 1, 2, \ldots, c-1, \\
\pi_c(p\lambda + c\mu) &= \pi_{c-1}\lambda + \pi_{c+1}c\mu, \\
\pi_i(p\lambda + c\mu) &= \pi_{i-1}\lambda + \pi_{i+1}c\mu, i > c.
\end{aligned}$$

The steady-state probabilities are given by

$$\pi_n = \begin{cases} \pi_0 \dfrac{1}{n!}\left(\dfrac{\lambda}{\mu}\right)^n, & n = 1, 2, \ldots, c \\ \pi_0 \dfrac{1}{c!}\left(\dfrac{p}{c}\right)^{n-c}\left(\dfrac{\lambda}{\mu}\right)^n, & n > c \end{cases}$$

where π_0 is determined using the condition $\sum\limits_{n=0}^{\infty} \pi_n = 1$.

7.3.2 *Reneging M/M/c Model*

Consider the $M/M/c$ queueing system with reneging. In this situation, the customer will not leave the system when the customer is in service. Therefore, the arrival rate is $\lambda_n = \lambda,\ n = 0, 1, \ldots$. and the service rate is $\mu_n = n\mu, n = 1, 2, \ldots, c\ \mu_n = (n-c)\beta + c\mu,\ n = c+1, c+2, \ldots$. The steady-state equations are presented as follows

$$\begin{aligned}\lambda\pi_0 &= \mu\pi_1,\\(\lambda + i\mu)\pi_i &= \lambda\pi_{i-1} + (i+1)\mu\pi_{i+1}, i = 1, 2, \ldots, c-1,\\(\lambda + c\mu)\pi_c &= \lambda\pi_{c-1} + (c\mu + \beta)\pi_{c+1},\\(\lambda + c\mu + (j-c)\beta)\pi_j &= \lambda\pi_{j-1} + [c\mu + (j-c+1)\beta]\pi_{j+1}, j > c.\end{aligned}$$

The steady-state probabilities are given by

$$\pi_n = \begin{cases} \pi_0 \frac{1}{n!}\left(\frac{\lambda}{\mu}\right)^n, & n = 1, 2, \ldots, c\\ \pi_0 \dfrac{\lambda^n}{c!\mu^c \prod_{k=1}^{n-c}(c\mu + k\beta)}, & n > c\end{cases}$$

The probability π_0 is obtained from $\sum_{n=0}^{\infty} \pi_n = 1$.

Example 7.15 Consider an $M/M/c$ queueing system with balking and reneging scenarios. Obtain the steady-state probabilities for the system. The arrival rate is given by $\lambda_n = \lambda,\ n = 0, 1, \ldots, c-1,\ \lambda_n = p\lambda,\ n = c, c+1, \ldots$. And the service rate is $\mu_n = n\mu, n = 1, 2, \ldots, \mu_n = (n-c)\beta + c\mu,\ n = c+1, c+2, \ldots$. The steady-state probabilities are given by

$$\pi_n = \begin{cases} \pi_0 \frac{1}{n!}\left(\frac{\lambda}{\mu}\right)^n, & n = 1, 2, \ldots, c\\ \pi_0 \dfrac{p^{n-c}\lambda^n}{c!\mu^c \prod_{k=1}^{n-c}(c\mu + k\beta)}, & n > c\end{cases}$$

The probability π_0 is obtained from $\sum_{n=0}^{\infty} \pi_n = 1$.

7.4 Bulk Queueing Models

In queueing theory, a bulk queue (also referred as a batch queue) is a general queueing model where jobs arrive in and/or are served in groups of random size. In Kendall's notation for single queueing models, the RV denoting bulk arrivals or service is denoted with a superscript, for example $M^N/M^Y/1$denotes an $M/M/1$ queue where the arrivals are in batches determined by the RV N and the services are in bulk determined by the RV Y.

7.4.1 Bulk Arrival Queueing Model

In the queueing system, it is necessary to know whether customers can arrive simultaneously (batch or bulk arrivals), and if so, the probability distribution describing the size of the batch. This type of process is Markov process in the sense that future behavior is a function only of the present and not the past.

Definition 7.9 Consider a $M/M/1$ queueing system with bulk arrival in the system. The arrival process follows a Poisson process and the actual number of customers in the arriving module is a RV N which takes the value n with probability c_n. If λ_n is the arrival rate of a Poisson process of batch size n, then clearly $c_n = \frac{\lambda_n}{\lambda}$, where λ is the composite arrival rate of all batches and is $\sum_{n=1}^{\infty} \lambda_n$. This overall arrival process is a compound Poisson process. Figure 7.10 shows the bulk arrival queueing system.

7.4.1.1 Stationary Distribution

The steady-state equations for the system can be given as follows:

$$\begin{aligned} -\lambda\pi_0 + \mu\pi_1 &= 0, \\ -(\lambda+\mu)\pi_n + \mu\pi_{n+1} + \lambda\sum_{k=1}^{n}\pi_{n-k}c_k &= 0, n = 1, 2, \ldots. \end{aligned} \tag{7.2}$$

To solve the system of equations, a generating function approach has been used. Define,

$$C(z) = \sum_{n=1}^{\infty} c_n z^n, \ (|z| \le 1); \quad P(z) = \sum_{n=0}^{\infty} \pi_n z^n, \ (|z| \le 1)$$

as the generating functions of the batch-size probabilities $\{c_n\}$ and the steady-state probabilities $\{\pi_n\}$, respectively. The main objective is to determine $P(z)$ from $C(z)$ and from this to find the unknown steady-state probabilities $\{\pi_n\}$. Multiplying Eq. (7.2) by z^n and summing the results together gives

$$0 = -\lambda\sum_{n=1}^{\infty}\pi_n z^n - \mu\sum_{n=1}^{\infty}\pi_n z^n + \frac{\mu}{z}\sum_{n=1}^{\infty}\pi_n z^n + \lambda\sum_{n=1}^{\infty}\sum_{k=1}^{n}\pi_{n-k}c_k z^n. \tag{7.3}$$

Since, the generating function of $\sum_{k=1}^{n}\pi_{n-k}c_k$ is the product of the respective generating functions given as

$$\sum_{n=1}^{\infty}\sum_{k=1}^{n}\pi_{n-k}c_k z^n = \sum_{k=1}^{\infty} c_k z^k \sum_{n=k}^{\infty}\pi_{n-k} z^{n-k} = C(z)P(z).$$

From Eq. (7.3), it can be obtained that

$$P(z) = \frac{\mu\pi_0(1-z)}{\mu(1-z) - \lambda z(1 - C(z))}. \tag{7.4}$$

Here, π_0 is determined from $P(1) = 1$ using the above mentioned equation.
The mean number of customers in the system will be given by

$$E(X) = \rho \frac{(E[N] + E[N^2])}{2(1 - \rho E[N])}$$

Special Case: Consider the case when $N = 1$ therefore, $E[N] = 1$. In this scenario, the mean number of customers in the system will be the same as $M/M/1$ system and is given by

$$E(X) = \rho \frac{(1 + 1)}{2(1 - \rho.1)} = \frac{\rho}{1 - \rho},$$

where $\rho = \frac{\lambda}{\mu}$.

Example 7.16 Consider that N is geometrically distributed RV. Suppose that

$$c_n = (1 - \alpha)\alpha^{n-1}, 0 < \alpha < 1, \quad n = 1, 2, \ldots.$$

It follows that $\rho E[N] = \frac{\rho}{1-\alpha}$,

$$C(z) = (1 - \alpha) \sum_{n=1}^{\infty} \alpha^{n-1} z^n = \frac{z(1 - \alpha)}{1 - \alpha z}.$$

The steady-state probabilities of the system are given by

$$P(z) = \left(1 - \frac{\rho}{1 - \alpha}\right) \left(\sum_{n=0}^{\infty} \{[\alpha + \rho]z\}^n - \sum_{n=0}^{\infty} \alpha\{[\alpha + \rho]z\}^n z^{n+1}\right).$$

Solving the above equation yields

$$\pi_n = \left(1 - \frac{\rho}{1 - \alpha}\right)[\alpha + \rho]^{n-1} \rho, \quad n = 1, 2, \ldots.$$

Example 7.17 Consider a multistage machine-line process that produces an assembly in quantity. After the first stage many items are found to have one or more defects, which must be repaired before they enter the second stage. One employee is responsible for making the necessary modifications to return the assembly to the process flow. The amount of flaws per item is routinely recorded, and only very seldom does it exceed two. The interarrival times for both units with one and two defective parts are found to be governed closely by exponential distributions, with parameters $\lambda_1 = 1$ per hour and $\lambda_2 = 2$ per hour, respectively. There are many different types of parts that are found defective that an exponential distribution does actually provide a good fit for the worker's service-time distribution, with a mean $1/10$ min. Find the average number of units in the system.

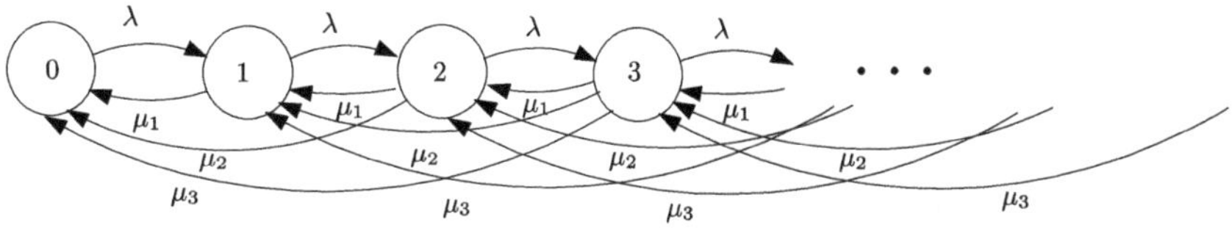

Fig. 7.10 State transition diagram for the bulk service system

Solution: The average number of units in the system is computed as follows: $\lambda = 3$, $c_1 = \frac{\lambda_1}{\lambda} = \frac{1}{3}$, and $c_2 = \frac{\lambda_2}{\lambda} = \frac{2}{3}$. The batch size mean and second moment are $E[N] = \frac{5}{3}$ and $E[N^2] = 3$. Therefore

$$E(X) = \frac{\frac{5}{6} + \frac{3}{2}}{2(\frac{1}{6})} = 7.$$

7.4.2 *Bulk Service Queueing Model*

Definition 7.10 Consider a single-server Markovian queueing model with bulk service. The customers are assumed to arrive according to a Poisson process, service times are exponential distributed with a single server, and the K customers are served at a time. On the basis of service facility, there can be two variations of the model, namely full-batch model and partial-batch model. Figure 7.11 shows the state transition diagram for the bulk service queueing system.

7.4.2.1 Stationary Distribution

– **Partial-Batch Model:** The steady-state balance equations are given by

$$-\lambda\pi_0 + \mu\pi_1 + \mu\pi_2 + \ldots + \mu\pi_K = 0,$$
$$-(\lambda + \mu)\pi_n + \mu\pi_{n+K} + \lambda\pi_{n-1} = 0, \quad n = 1, 2, \ldots.$$

The second equation can be written in the operator notation as

$$[\mu D^{K+1} - (\lambda + \mu)D + \lambda]\pi_n = 0, \quad n = 0, 1, \ldots.$$

Hence if $(r_1, r_2, \ldots, r_{K+1})$ are the roots of the operator or characteristic equation, then

$$\pi_n = \sum_{i=1}^{K+1} C_i r_i^n, \quad n \geq 0.$$

Since $\sum_{n=0}^{\infty} \pi_n = 1$, each r_i must be less than one. By Rouche's Theorem, it can be found that there is exactly one root r_0 in (0, 1). Therefore,

$$\pi_n = (1 - r_0)r_0^n; 0 < r_0 < 1, \ n = 0, 1, \ldots .$$

Special Case: Let X denote the number of customers in the system. Then the mean number of customers is the same as $M/M/1$ system and is given by

$$E(X) = \frac{r_0}{1 - r_0},$$

where $r_0 = \rho$.

- **Full-Batch Model**: Let's consider that the batch size is K. If there are less than K customers then the server waits until K customers are there. The steady-state balance equations are given by

$$\begin{aligned} -\lambda\pi_0 + \mu\pi_K &= 0, \\ -\lambda\pi_n + \mu\pi_{n+K} + \lambda\pi_{n-1} &= 0, \quad n = 1, 2, \ldots, K, \\ -(\lambda + \mu)\pi_n + \mu\pi_{n+K} + \lambda\pi_{n-1} &= 0, \quad n = K, K+1, \ldots . \end{aligned}$$

Following a similar procedure for the computation, the steady-state probabilities are obtained as

$$\pi_n = \begin{cases} \pi_0 \dfrac{(1 - r_0)^{n+1}}{1 - r_0}, & n = 1, 2, \ldots, K - 1 \\ \pi_0 \dfrac{\lambda r_0^{n-K}}{\mu}, & n \geq K - 1 \end{cases} \tag{7.5}$$

where

$$\pi_0 = \left(\frac{\mu r_0^{K+1} - (\lambda + \mu)r_0 + \lambda + \mu K(1 - r_0)}{\mu(1 - r_0)^2} \right)^{-1}.$$

Special Case: Let the batch size $K = 1$ and $r_0 = \rho$, the result coincide with $M/M/1$ system given as follows

$$\pi_0 = 1 - \rho, \ \ \pi_i = (1 - \rho)\rho^i, \ \ i = 0, 1, \ldots .$$

Example 7.18 Consider a particular self-propelled vehicle car wash that wants to make a decision to modify its operational process. It puts in new equipment that makes it possible to wash two cars simultaneously (and one if no other cars wait). A car that enters the wash while one car is already in joins the wash and finishes the wash. There is no cap on the number of people waiting. Arrivals are Poisson with an average of 20 per hour exponential distributed, with a mean of 5 minutes, which is the time needed to wash a car. Find the average number of cars in the system.

Solution: Here, $\lambda = 20,\ \mu = 12,\ K = 2$. The characteristic equation is given by

$$12r^3 - 32r + 20 = 0.$$

The roots are $r = 1, (-3+\sqrt{69})/6, (-3-\sqrt{69})/6$. The positive root with an absolute value less than one is selected as $r_0 = (-3+\sqrt{69})/6$. Therefore $E(X) = \frac{0.884}{0.116} = 7.6$ cars.

7.5 Vacation Queueing Models

Many studies have worked on queues with vacation over the last two decades. Vacation queues occur when the server stops serving customers entirely during the vacation time. Vacation queueing models have been extensively used in various real life problems such as modern telecommunication systems, computer communication networks, production, manufacturing systems, etc. In a multiple vacation policy, if the server discovers no customer waiting for service at the end of the server's vacation, it begins another vacation and continues until a customer is identified waiting for service. If there is no customer waiting in the line at the end of a single vacation policy, the server remains inactive. A busy period begins for both policies if there is at least one customer waiting in the queue at the vacation completion moment. Vacation queueing models can be applied to various queueing models namely, $M/M/1$ queue, $M/M/c$ queue, $M/M/1/N$ queue, $M/M/c/K$ queue, $M/M/c/c$ loss system, $M/M/\infty$ self service system. Figure 7.9 represents a vacation queueing system.

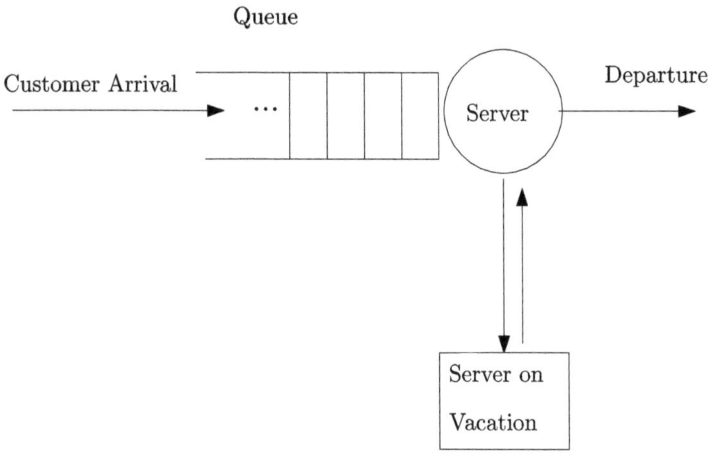

Fig. 7.11 Queueing diagram for vacation queues

Example 7.19 (Single Working Vacation Model) Consider an $M/M/1$ queueing model with a single working vacation, slow service and impatient customers as described by Sudhesh et al. [12]. The arrival process follows the Poisson process with rate λ and the service time is considered to be exponential distributed with rate μ_1. The server goes on vacation when the system is empty and the arriving customer is served at a lower service rate μ_0, following an exponential distribution. If there are no customers to serve when the server's vacation period ends, the server becomes dormant in the system. Otherwise, the server starts providing service with normal service rate μ_1. The vacation time of the server is exponential distributed with rate γ. Thus, under the single vacation model, the server may be idle for a period of time, whereas the multiple vacation model does not have any idle time for the server. Customers become impatient as a result of the server's delayed service during a single vacation. That is, each individual customer can renege in the queue by activating an independent and exponential distributed impatient timer with parameter ξ. If the customer's service is not completed before the timer ends, he or she abandons the system and never returns. The state transition diagram of this model is given in Figure 7.11.

Based on the model description, the stochastic process can be denoted as $\{(X(t), Y(t)), t \geq 0\}$, $X(t) \geq 0$; $Y(t) = 0, 1$. Here, $X(t)$ denotes the number of customers at time t and $Y(t)$ denotes the state of the server at time t, where $Y(t) = 0$ when the server is on vacation and serves the customer with slow service rate and $Y(t) = 1$ when the server is busy and serves the customer with normal service rate. Let

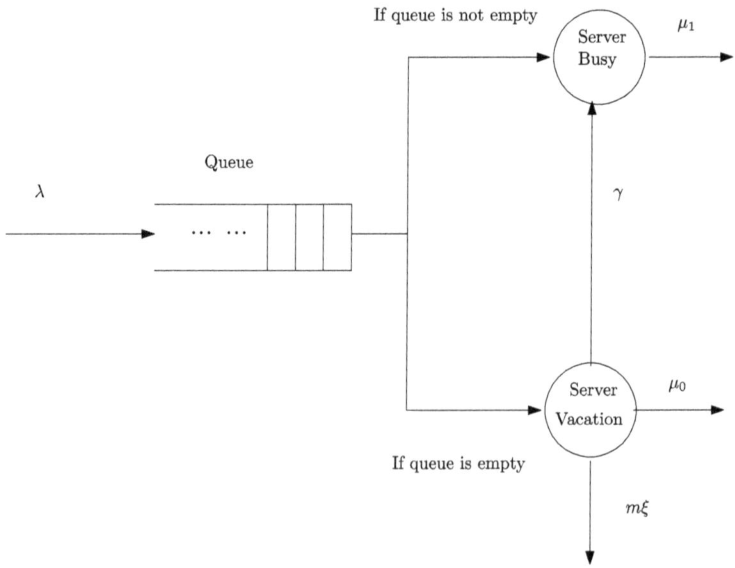

Fig. 7.12 Queueing diagram for vacation queues

$$P_{n0}(t) = P\{X(t) = n,\ Y(t) = 0\},\quad n = 0, 1, \ldots,$$

$$P_{n1}(t) = P\{X(t) = n,\ Y(t) = 1\},\quad n = 0, 1, \ldots$$

Then $P_{nk}(t), n = 0, 1, 2, \ldots,\ k = 0, 1$, satisfy the forward Kolmogrov equations as follows:

$$P'_{00}(t) = -(\lambda + \gamma)P_{00}(t) + \mu_1 P_{11}(t) + (\mu_0 + \xi)P_{10}(t), \tag{7.6}$$

$$P'_{10}(t) = -(\lambda + \mu_0 + \xi + \gamma)P_{10}(t) + \lambda P_{00}(t) + (\mu_0 + 2\xi)P_{20}(t), \tag{7.7}$$

$$P'_{n0}(t) = -(\lambda + \mu_0 + n\xi + \gamma)P_{n0}(t) + \lambda P_{n-10}(t) + (\mu_0 + (n+1)\xi)P_{n+10}(t), n = 2, 3, \ldots, \tag{7.8}$$

$$P'_{01}(t) = -\lambda P_{01}(t) + \gamma P_{00}(t), \tag{7.9}$$

$$P'_{11}(t) = -(\lambda + \mu_1)P_{11}(t) + \lambda P_{01}(t) + \mu_1 P_{21}(t) + \gamma P_{10}(t), \tag{7.10}$$

$$P'_{n1}(t) = -(\lambda + \mu_1)P_{n1}(t) + \lambda P_{n-11}(t) + \mu_1 P_{n+11}(t) + \gamma P_{n0}(t), n = 2, 3, \ldots, \tag{7.11}$$

with $P_{00}(0) = 1$, i.e., the system is empty at time $t = 0$.
The next theorem (without proof) provides the transient probabilities $P_{n1}(t)$, for $n \geq 1$, which are derived using probability generating function method in terms of the modified Bessel function (refer, Appendix A.7).

Theorem 7.1 *The probabilities $P_{n1}(t)$ are obtained, for $n = 1, 2, \ldots$, from Eqs. (7.10) and (7.11) in terms of the modified Bessel function as*

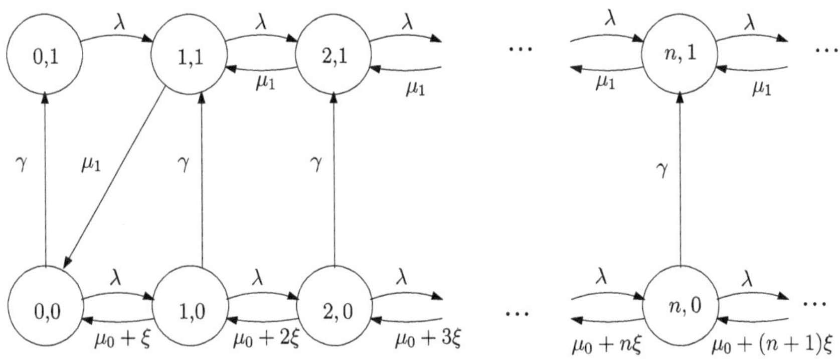

Fig. 7.13 Queueing diagram for $M/M/1$ queueing model with single working vacation

$$P_{n1}(t) = \gamma \int_0^t e^{-(\lambda+\mu_1)(t-u)} \sum_{m=0}^{\infty} P_{m0}(u)\beta_1^{n-m}[I_{n-m}(\alpha_1(t-u)) - I_{n+m}(\alpha_1(t-u))]du$$
$$+ \mu_1 \int_0^t e^{-(\lambda+\mu_1)(t-u)} P_{01}(u)\beta_1^{n+1}[I_{n-1}(\alpha_1(t-u)) - I_{n+1}(\alpha_1(t-u))]du, \tag{7.12}$$

where $P_{01}(t)$ is obtained from Eq. (7.9) as

$$P_{01}(t) = \gamma \int_0^t e^{-\lambda(t-u)} P_{00}(u)du \tag{7.13}$$

and $I_n(t)$ is the modified Bessel function of the first kind of order n, $\alpha_1 = 2\sqrt{\lambda\mu_1}$ and $\beta_1 = \sqrt{\frac{\lambda}{\mu_1}}$.

The following theorem (without proof) expresses $P_{n0}(t)$ in terms of $P_{00}(t)$, for $n = 1, 2, \dots$ and gives $P_{00}(t)$ explicitly.

Theorem 7.2 *The probabilities $P_{n0}(t)$ are obtained for $n = 1, 2, \dots$ from (7.6), (7.7),(7.8) using the continued fraction method and confluent hypergeometric functions as*

$$P_{n0}(t) = Q_n(t) * P_{00}(t), \tag{7.14}$$

where

$$P_{00}(t) = \sum_{k=0}^{\infty}\sum_{r=0}^{k}\sum_{i=0}^{r}\binom{k}{r}\binom{r}{i}\left(\frac{\mu_0+\xi}{\lambda\gamma}\right)^i \lambda^r\gamma^k e^{-(\lambda+\gamma)t}\frac{t^k}{k!} * e^{-\lambda t}\frac{t^{r-i-1}}{(r-i-1)!}$$
$$*Q_1^i(t) * \frac{\lambda}{\beta_1^{r-i+1}} e^{-(\lambda+\mu_1)t}[I_{r-i-1}(\alpha_1(t) - I_{r-i+1}(\alpha_1(t))]^{*(r-i)} \tag{7.15}$$
$$*\left[\sum_{m=0}^{\infty}\frac{\lambda}{\beta_1^{m+1}} e^{-(\lambda+\mu_1)t}[I_{m-1}(\alpha_1(t) - I_{m+1}(\alpha_1(t))] * Q_m(t)\right]^{*(k-r)}$$

$$Q_n(t) = \lambda^n \sum_{k=0}^{\infty}(-\lambda)^k \frac{\prod_{j=1}^{k}(\mu_0 + (n+j)\xi)}{\xi^k k!} a_{n+k}(t) * \sum_{i=1}^{\infty}\lambda^i b_i(t), \tag{7.16}$$

$$a_k(t) = \frac{1}{\xi^{2k-1}k!}\sum_{r=1}^{k}\prod_{j=1}^{k}\frac{(-1)^{r-1}(\mu_0 + j\xi)}{(r-1)!(k-r)!}e^{-(\gamma+\mu_0+r\xi)t}, k = 1, 2, \dots, \tag{7.17}$$

$$b_k(t) = \sum_{i=1}^{k}(-1)^{i-1}a_i(t) * b_{k-i}(t), k = 2, 3, \dots; b_1(t) = a_1(t), \tag{7.18}$$

where $*$ denotes the convolution and $*(k-r)$ denotes the $(k-r)$ fold convolution.

The performance measures, mean and variance of the system are computed as follows

1. Mean:

$$E(X(t)) = \lambda t - \mu_0 \sum_{n=1}^{\infty} \int_0^t P_{n0}(u)du - \xi \sum_{n=1}^{\infty} n \int_0^t P_{n0}(u)du$$
$$- \mu_1 \sum_{n=1}^{\infty} \int_0^t P_{n1}(u)du. \tag{7.19}$$

2. Variance:

$$\text{Var}(X) = E(X^2) - [E(X)]^2$$

$$E(X^2(t)) = \lambda t + 2\lambda \int_0^t E(X(u))du - \mu_0 \sum_{n=1}^{\infty} (2n-1) \int_0^t P_{n0}(u)du$$
$$- \xi \sum_{n=1}^{\infty} (2n^2 - n) \int_0^t P_{n0}(u)du - \mu_1 \sum_{n=1}^{\infty} (2n-1) \int_0^t P_{n1}(u)du, \tag{7.20}$$

where, $P_{n0}(t)$ and $P_{n1}(t)$, for $n = 1, 2, \ldots$, are given in Eq. (7.14) and (7.12), respectively.

Special Case: If the concept of vacation is removed in Example 7.19, i.e., $\xi = 0$, $\gamma = 0$, $\mu_0 = 0$, the Kolmogorov forward equations of the model will be modified as follows

$$P_{01}^{'}(t) = -\lambda P_{01}(t) + \mu_1 P_{11}(t),$$
$$P_{n1}^{'}(t) = -(\lambda + \mu_1) P_{n1}(t) + \lambda P_{n-11}(t) + \mu_1 P_{n+11}(t), \quad n = 1, 2, \ldots.$$

These equations are same as Eqs. (6.3) and (6.2) as described in Chap. 6. After applying the same approach to obtain the solution, the time-dependent probabilities are obtained which are equivalent to the $M/M/1$ queueing system. The same model with general distributed service time will be discussed in Chap. 9 for the non-Markov case.

7.6 Fluid Queueing Models

According to classical queueing theory, the main component in assessing a system is a server (or numerous servers), where customers line up to get services. A probability distribution describes the uncertainty surrounding the arrival and service timings of consumers. As a result, the essence of the model is discrete, with each individual consumer playing a crucial part. Queues have been extensively discussed in the

literature and have many practical uses. Because the incoming traffic on high-speed networks is typically in the form of data packets, it is impossible to identify specific users. Therefore, it is presumed that this flood of data packets is a continuous entity and is referred to as fluid. A fluid queue is defined by two random patterns namely, the amount of fluid with arrivals to the queue and the maximum amount of fluid that the servers can serve. The fluid that cannot be decided instantly accumulates in the reservoir. In these models, a stochastic mechanism decides how much fluid will flow into the reservoir at any given time. Some claim that the fluid leaves the reservoir at a rate that can either be continuous or random in nature. The fluid reservoir, whose function is comparable to that of a customer line, is referred to as a fluid queue in literature. The term "fluid buffer" is also used to describe the temporary storage of fluid in the reservoir (Fig. 7.14).
In queueing theory, there are numerous applications where individual customers are so small that they can hardly be distinguished. It is then easier to imagine a continuous stream of work that flows into the system instead of customers. The resulting queueing models are called fluid models. Fluid models are used to represent systems where some quantity accumulates and is depleted, gradually over time, subject to some random environment. For example, fluid models are now universally acknowledged as suitable representations of contemporary manufacturing and telecommunication networks. The queue derived from the fluid model is referred to as the fluid queue.

Definition 7.11 (Fluid Queue) Let us denote the content of the reservoir at time t by $C(t)$ and the state of the background process by $X(t)$ at time t with state space S. Then, the two dimensional process $\{(X(t), C(t)), t \geq 0\}$ is called fluid queue modulated by the stochastic process $X(t)$.

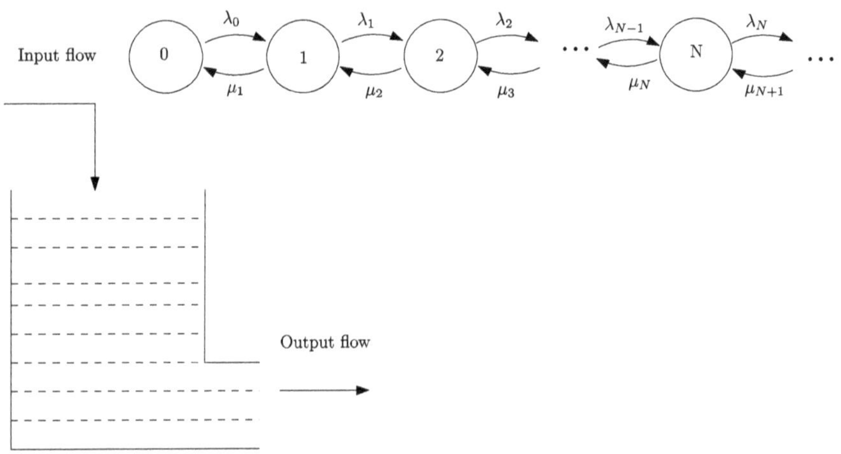

Fig. 7.14 Queueing structure of fluid queue

7.6.1 Modulated by a BDP

Consider an infinite capacity buffer which receives and releases fluid modulated by a BDP evolving in the background. The state space of the background BDP may be finite or infinite. Let us denote the state of the background process at time t by $X(t)$ with state space S. Further, let λ_i denotes the birth rate and μ_i the death rate in state i, $i \in S$, of $\{X(t),\ t \geq 0\}$. The flow rate of the fluid into and out of an infinite capacity buffer are determined by the actual state of the background process. Let us denote the content of the reservoir at time t by $C(t)$ and let r_j denote the net flow rate of the fluid when the background process is in state j. The rate of change of content of the buffer $C(t)$ when $X(t) = j$ is given by

$$\frac{dC(t)}{dt} = \begin{cases} r_j \text{ if } C(t) > 0, \text{ or } (C(t) = 0, \text{ and } r_j \geq 0) \\ 0 \text{ if } C(t) = 0, r_j < 0 \end{cases} \tag{7.21}$$

Clearly, the two-dimensional process $\{(X(t), C(t)), t \geq 0\}$ constitutes a Markov process that possesses a unique stationary distribution under suitable stability conditions. The stationary probabilities $p_i,\ i \in S$ of the background BDP, are given by

$$p_i = \frac{\pi_i}{\sum_{j \in S} \pi_j}, i \in S,$$

where $\pi_i = (\lambda_0 \lambda_1 \ldots \lambda_{i-1})/(\mu_1 \mu_2 \ldots \mu_i), i = 1, 2, \ldots$ and $\pi_0 = 1$ are called the potential coefficient. For a limiting distribution $C(t)$ to exist as $t \to \infty$, the stationary net input rate should be negative, i.e.,

$$\sum_{i=0}^{\infty} \pi_i r_i < 0.$$

Letting

$$F_j(t, u) = P\{X(t) = j, C(t) \leq u\}, j \in S, t, u \geq 0, \tag{7.22}$$

the Kolmogorov forward equations for the Markov process $\{(X(t), C(t), t \geq 0)\}$ are given by [12]

$$\begin{aligned} \frac{\partial F_0(t, u)}{\partial t} &= -r_0 \frac{\partial F_0(t, u)}{\partial u} - \lambda_0 F_0(t, u) + \mu_1 F_1(t, u), \\ \frac{\partial F_j(t, u)}{\partial t} &= -r_j \frac{\partial F_j(t, u)}{\partial u} - (\lambda_j + \mu_j) F_j(t, u) + \lambda_{j-1} F_{j-1}(t, u) + \mu_{j+1} F_{j+1}(t, u), \\ & j \in S \backslash \{0\}, t, u \geq 0. \end{aligned}$$

When the process is in equilibrium, then $\frac{\partial F_j(t,u)}{\partial t} = 0$ and let $\lim_{t\to\infty} F_j(t,u) = F_j(u)$.

7.6.2 *Modulated by M/M/1 Model*

Consider a fluid model driven by an $M/M/1$ queue. Hence, $S = \{0, 1, \ldots\}$. Let λ and μ denote the mean arrival and service rates, respectively. When the background process is in state $i \in S \setminus \{0\}$, a fluid commodity gets accumulated in an infinite capacity buffer with a constant rate $r > 0$. Whenever the background process is in state 0, the fluid flows out of the buffer with a constant rate $r_0 < 0$. Let $C(t)$ denotes the buffer content at time t. The stochastic process $\{(X(t), C(t)), t \geq 0\}$ becomes a two-dimensional Markov process. Let

$$F_j(t,x) = P\{X(t) = j, C(t) \leq x\}, \quad t, x \geq 0; \quad j \in S$$

denote the distribution function of the Markov process $\{(X(t), C(t)), t \geq 0\}$. The Kolmogorov forward equations governing this process are given by:

$$\frac{\partial F_0(t,x)}{\partial t} + r_0 \frac{\partial F_0(t,x)}{\partial x} = -\lambda F_0(t,x) + \mu F_1(t,x),$$
$$\frac{\partial F_j(t,x)}{\partial t} + r \frac{\partial F_j(t,x)}{\partial x} = \lambda F_{j-1}(t,x) - (\lambda+\mu) F_j(t,x) + \mu F_{j+1}(t,x), \; j = 1, 2, \ldots.$$

The initial conditions are given by

$$F_0(0,x) = 1, \quad F_j(0,x) = 0 \quad \text{for } j = 1, 2, \ldots$$

and boundary conditions are given by

$$F_0(t,0) = q_0(t) \quad \text{and} \quad F_j(t,0) = 0 \quad \text{for } j = 0, 1, \ldots.$$

Note that, whenever the net input rate of fluid flow into the buffer is positive, the buffer cannot stay empty.
There can be two approaches to obtain the transient solution.

7.6.2.1 Transient Analysis: Recursive Approach

Consider a fluid queue driven by an $M/M/1$ queueing model $\{X(t), \geq 0\}$ taking values in $S = \{0, 1, \ldots\}$, where $X(t)$ denotes the number of customers in the system at time t. Let λ and μ denote the mean arrival and service rates, respectively. The arrivals follow the Poisson process and the service times are exponential distributed. During the busy periods of the server, a fluid commodity which is referred as 'credit' accumulates in an infinite capacity buffer at a constant rate $r > 0$. The credit buffer depletes the fluid during the idle periods of the server at a constant rate $r_0 < 0$

as long as the buffer is non-empty. If $C(t)$ denotes the content of buffer at time t, the two-dimensional process $\{(X(t), C(t)), t \geq 0\}$ constitutes a Markov process. If If $F_j(t,x) = P\{X(t) = j, C(t) \leq x\}$, $j \in S$, $t, x \geq 0$, the Kolmogorov forward equations for the Markov process $\{(X(t), C(t)), t \geq 0\}$ are given by

$$\frac{\partial F_0(t,x)}{\partial t} = -r_0 \frac{\partial F_0(t,x)}{\partial x} - \lambda F_0(t,x) + \mu F_1(t,x)$$
$$\frac{\partial F_j(t,x)}{\partial t} = -r \frac{\partial F_j(t,x)}{\partial x} + \lambda F_{j-1}(t,x) + \mu F_{j+1}(t,x) - (\lambda + \mu) F_j(t,x)$$
$$j \in S \backslash \{0\}, t, x \geq 0$$

subject to the initial conditions

$$F_0(0,x) = 1, \ \ F_j(0,x) = 0, \ \text{for } j = 1, 2, \ldots$$

and the boundary condition

$$F_j(t,0) = q_j(t), \ \text{for } j = 1, 2, \ldots.$$

Here $q_j(t)$ represents the probability that at time t the buffer is empty and the state of the background Markov process is j. The content of the buffer decreases and thus it becomes empty only when the net input rate of the fluid into the buffer is negative. Therefore, when the buffer becomes empty at any time t, the background process should necessarily be in state zero corresponding to which the effective input rate is $r_0 < 0$. Hence, $q_j(t) = 0$ for $j = 1, 2, \ldots$ as $r > 0$ for $j = 1, 2, \ldots$.
The transient distribution of the buffer content is given by

$$P\{C(t) > x\} = 1 - \sum_{j=0}^{\infty} F_j(t,x).$$

Let P denotes the transition probability matrix of the uniformized Markov chain associated with the process $[X(t), t \geq 0]$. If p and q are defined by

$$p = \frac{\lambda}{\lambda + \mu}, \text{ and } q = \frac{\mu}{\lambda + \mu}$$

the non-zero entries of the matrix P are

$$P_{00} = q, \ P_{01} = p, \ \text{and for } i \geq 1,$$

$$P_{ii-1} = q, \ P_{ii+1} = p.$$

For every $i \geq 0$, let $(\pi(0)P^n)_i$ denote the ith entry of the row vector $(\pi(0)P^n)$ with $\pi(0) = (1, 0, 0, \ldots)$ denoting the initial probability vector. The expression for the joint distribution of the buffer content of the fluid queue and the state of the $M/M/1$ queue is obtained and given as follows

$$F_i(t, x) = \sum_{n=0}^{\infty} e^{-(\lambda+\mu)t} \frac{[(\lambda + \mu)t]^n}{n!} \sum_{k=0}^{n} \binom{n}{k} \left(\frac{x}{rt}\right)^k \left(1 - \frac{x}{rt}\right)^{n-k} b_i(n, k),$$

where the coefficients $b_i(n, k)$ are given by the following recursive expressions. For $i = 0$,

$$\begin{aligned} b_0(n, n) &= (\pi(0)P^n)_0, \text{ for } n \geq 0, \\ b_0(n, k) &= \frac{rq}{r - r_0}[b_0(n-1, k) + b_1(n-1, k)] - \frac{r_0}{r - r_0} b_0(n, k+1) \\ &\quad \text{for } n \geq 1, k = 0, 1, \ldots, n-1. \end{aligned}$$

For $i = 1, 2, \ldots,$

$$\begin{aligned} b_i(n, 0) &= 0, \text{ for } n = 0, 1, \ldots \\ b_i(n, k) &= pb_{i-1}(n-1, k-1) + qb_{i+1}(n-1, k-1), \text{ for } n = 1, 2, \ldots, \end{aligned}$$

$k = 1, 2, \ldots, n$.

For all $n = 1, 2, \ldots, k = 0, 1, \ldots, n-1$,

$$b_0(n, k) - \beta^{n-k} b_0(n, n) = \alpha \sum_{i=k}^{n-1} [b_0(n-1, i) + b_1(n-1, i)]\beta^{i-k},$$

where $\alpha = rq/(r - r_0)$ and $\beta = r_0/(r_0 - r)$.
For $i \geq 1$, $b_i(n, k) = 0$ for $0 \neq n < i$ and

$$b_i(n, k) = \begin{cases} 0, & 0 \leq k < i, \\ p^i \sum_{l=0}^{(k-i)/2} s(i, l) p^l q^l b_0(n - 2l - i, k - 2l - i), & i \leq k \leq n, \\ 0, & k > n \end{cases}$$

where the numbers $s(i, l)$ are referred to as ballot numbers and is given by

$$s(i, l) = \frac{i(2l + i - 1)!}{l!(i + l)!}.$$

7.6.2.2 Transient Analysis: Probabilistic Approach

The interest lies in obtaining the probability that the content of buffer at time t is greater than x i.e., $P\{C(t) > x\}$. The transient solution is obtained by probabilistic approach given in [6]. Let $\mathbf{Q}$ be the infinitesimal generator matrix of CTMC $\{X(t), t \geq 0\}$. Let

$$Z = \{Z_n : n = 0, 1, \ldots\}$$

be a DTMC with state space S and transition probability matrix $\mathbf{P} = \mathbf{I} + \frac{\mathbf{Q}}{\lambda+\mu}$ and $\{N(t), t \geq 0\}$ be a Poisson process with rate $\lambda + \mu$, that is independent of Z. Then $X(t) = Z_{N(t)}$ for $t \geq 0$. Assume that the background process has been uniformized and thus have n transitions during the period $(0, t)$, i.e., $N(t) = n$, at times

$$0 < \tau_1 < \tau_2 < \ldots < \tau_n < t.$$

These events split $(0, t)$ into $n + 1$ intervals with lengths

$$Y_1 = \tau_1, Y_2 = \tau_2 - \tau_1, \ldots, Y_{n+1} = t - \tau_n.$$

Thus, each interval is associated with a net rate whose value is based on the state of the process during the interval.
The state space S is partitioned into two subsets B_0 and B_1 such that $B_0 = \{0\}$ and $B_1 = \{1, 2, \ldots\}$. Define $V_n^0 = \sum_{i=0}^{n} \mathbf{1}_{Z_i \in B_0}$ and $V_n^1 = \sum_{i=0}^{n} \mathbf{1}_{Z_i \in B_1}$ where $\mathbf{1}_A$ is the indicator function.
Define

$$V_n = (V_n^0, V_n^1).$$

Let k_1 and $k_2 = n + 1 - k_1$ denote the number of intervals associated with rate r_0 and r respectively. Note that, $k_1 \geq 1$ since the system starts in state 0 at time 0. Refer (k_1, k_2) as a partition of $n + 1$. Thus, conditioning on the number of transitions n and $V_n = (V_n^0, V_n^1) = (k_1, k_2) = \mathbf{k}$, obtain

$$\begin{aligned}
P\{C(t) > x\} &= \sum_{n=0}^{\infty} \sum_{k_1+k_2=n+1} P\{N(t) = n, V_n = \mathbf{k}\} P\{C(t) > x | N(t) = n, V_n = \mathbf{k}\} \\
&= \sum_{n=0}^{\infty} \sum_{k_1+k_2=n+1} P\{N(t) = n\} P\{V_n = \mathbf{k} | N(t) = n\} \qquad (7.23) \\
&\quad P\{C(t) > x | N(t) = n, V_n = \mathbf{k}\} \\
&= \sum_{n=0}^{\infty} e^{-(\lambda+\mu)t} \frac{((\lambda + \mu)t)^n}{n!} \sum_{k_2=0}^{n} G[n, (k_1, k_2)] M(t, x, n, k_2) \qquad (7.24)
\end{aligned}$$

where

$$M(t, x, n, k_2) = P\{C(t) > x | N(t) = n, V_n^1 = k_2\} \tag{7.25}$$

and

$$G[n, (k_1, k_2)] = P\{V_n = \mathbf{k} | N(t) = n\}.$$

Further,

$$G[n, (k_1, k_2)] = \sum_{i \in S} G_i[n, (k_1, k_2)]. \tag{7.26}$$

Here $G_i[n, (k_1, k_2)]$ is the probability of partition $\mathbf{k}$ given that the number of transitions are n and the state visited after the last transition is i. If i and j are the states visited after the last $(n-1)$th and nth transitions, then (k_1, k_2) is equal to the previous partition $+\ 1$ at the entry corresponding to the net rate associated with state j. Let $p = \frac{\lambda}{\lambda+\mu}$ and $q = 1 - p$. Since, for $i \geq 1$, the only possibilities for j are $i-1$ and $i+1$, conditioning on the state visited after the $(n-1)$th transition,

$$G_0[n, (k_1, k_2)] = G_1[n-1, (k_1 - 1, k_2)]q \tag{7.27}$$

and for $i \in S \setminus \{0\}$,

$$G_i[n, (k_1, k_2)] = G_{i-1}[n-1, (k_1, k_2 - 1)]p + G_{i+1}[n-1, (k_1, k_2 - 1)]q. \tag{7.28}$$

The recursive function $G_i[n, (k_1, k_2)]$ satisfies the initial conditions

$$G_0[0, (1, 0)] = \pi_0^{(0)} = 1,$$
$$G_i[0, (0, 1)] = \pi_i^{(0)} = 0 \text{ for } i \in S \setminus \{0\}.$$

Now, compute $M(t, x, n, k_2)$. Let $U_1, U_2, \ldots, U_n$ be uniformly distributed IID RVs in $(0, 1)$ and $U_{(1)}, U_{(2)}, \ldots, U_{(n)}$ be their order statistics with $U_{(0)} = 0$ and $U_{(n+1)} = 1$. Then, τ_i, the time of the ith transition has the same distribution as $tU_{(i)}$. Thus

$$Y_1 \equiv tU_{(1)}, Y_2 \equiv t(U_{(2)} - U_{(1)}), \ldots, Y_{n+1} \equiv t(1 - U_{(n)}).$$

Since Y_i's are exchangeable RV, and also the background process is an $M/M/1$, the time spent in any state follows exponential property. Thus, rearranging the intervals, letting the first k_1 intervals to be associated with the net rate r_0 and the next k_2 intervals associated with the net rate r. Now, the event $\{C(t) > x\}$ given $\{N(t) = n, V_n^1 = k_2\}$ can be written as

$$\begin{aligned}
\{C(t) > x | N(t) = n, V_n^1 = k_2\} &= \{C(t) > x | N(t) = n, V_n^1 = k_2\} \\
&= \{C(t) > x | N(t) = n, V_n^0 = k_1\} \\
&= \{r_0(Y_1 + \ldots + Y_{k_1}) + r(Y_{k_1+1} + \ldots + Y_{n+1}) > x\} \\
&= \{r_0(tU_{(1)} + \ldots + t(U_{(k_1)} - U_{(k_1-1)})) + \\
&\quad r(t(U_{(k_1+1)} - U_{k_1}) + \ldots + t(U_{(n+1)} - U_{(n)})) > x\} \\
&= \{t(r_0 - r)U_{(k_1)} + rt > x\}.
\end{aligned}$$

Thus, for $x \in [0, rt)$

$$P\{C(t) > x | N(t) = n, V_{n^1} = k_2\} = P\{(t(r_0 - r)U_{(k_1)} + rt) > x\} \\ = P\{U_{(k_1)} < \frac{x - rt}{t(r_{0-r})}\}. \quad (7.29)$$

It is known that if $X_1, X_2, \ldots, X_n$ are $U(0, 1)$ IID RVs, then the sth order statistics $X_{(s)}$ follows $Beta(s, n + 1 - s)$. Therefore, the RHS of Eq. (7.29) is the cumulative distribution function of $Beta(k_1, n + 1 - k_1)$ evaluated at $\frac{x-rt}{t(r_0-r)}$. Thus,

$$M[t, x, n, k_2] = \sum_{j=0}^{k_1} \binom{n}{j} \left(\frac{x - rt}{t(r_0 - r)}\right)^j \left(1 - \frac{x - rt}{t(r_0 - r)}\right)^{n-j}. \quad (7.30)$$

Thus, the solution is obtained as, for $t \geq 0$ and $x \in [0, rt)$,

$$P[C(t) > x] = \sum_{n=0}^{\infty} e^{-(\lambda+\mu)t} \frac{[(\lambda + \mu)t]^n}{n!} \sum_{k_2=0}^{n} G[n, (k_1, k_2)]M[t, x, n, k_2] \quad (7.31)$$

where $M(t, x, n, k_2)$ is given in Eq. (7.30) and $G[n, (k_1, k_2)]$ is given in Eqs. (7.26), (7.27) and (7.28). Further, the value of $q_0(t)$ can be obtained from (7.29) by letting $x = 0$ and state of the background process as $i = 0$ and using the recursive relation described in (7.27) as follows [6]:

$$q_0(t) = \sum_{n=0}^{\infty} \left[e^{-(\lambda+\mu)t} \frac{-[(\lambda + \mu)t]^n}{n!} \sum_{k_2=0}^{n} G[n, (k_1, k_2)]M[t, 0, n, k_2] \right].$$

7.6.2.3 Steady-State Analysis: Continued Fraction Approach

Let $\{X(t), t \geq 0\}$ be the birth-death single-server queueing model with the state space $S = \{0, 1, 2, \ldots\}$, where $X(t)$ denotes the number of customers in the system at time t. Let λ_n and μ_n denote the mean arrival and service rates, respectively, when there are n customers in the queue. During the busy periods of the server, a fluid commodity which is referred as credit accumulates in an infinite fluid buffer at a constant rate $r > 0$. The credit buffer depletes the fluid during the idle periods of the server at a constant rate $r_0 < 0$ as long as the buffer is non-empty. Let $C(t)$ be the content of the buffer at time t. Clearly, the two-dimensional process $\{(X(t), C(t)), t \geq 0\}$ constitutes a Markov process which possesses a unique stationary distribution under a suitable stability condition and is obtained in [10]. In order that a limiting distribution for $C(t)$ exists as $t \to \infty$, the stationary net input rate should be negative, i.e.,

$$r_0 + r\sum_{i=1}^{\infty} \pi_i < 0,$$

where $\pi_i = \lambda_0\lambda_1 \ldots \lambda_{i-1}/(\mu_1\mu_2 \ldots \mu_i)$, $i = 1, 2, \ldots$ and $\pi_0 = 1$ are called the potential coefficients.

Let

$$F_j(t, u) = P\{X(t) = j, C(t) \leq u\}, j \in S, t, u \geq 0.$$

The Kolmogorov forward equations for the Markov process $\{X(t), C(t)\}$ are given by

$$\begin{aligned}
\frac{\partial F_0(t, u)}{\partial t} &= -r_0\frac{\partial F_0(t, u)}{\partial u} - \lambda_0 F_0(t, u) + \mu_1 F_1(t, u),\\
\frac{\partial F_j(t, u)}{\partial t} &= -r\frac{\partial F_j(t, u)}{\partial u} - (\lambda_j + \mu_j)F_j(t, u) + \mu_{j+1}F_{j+1}(t, u) + \lambda_{j-1}F_{j-1}(t, u),\\
& \quad j \in S\backslash\{0\}, t, u \geq 0.
\end{aligned}$$

In the equilibrium state $\partial F_j(t, u)/\partial t = 0$ and $\lim_{t\to\infty} F_j(t, u) = F_j(u)$ and the above system reduces to a system of ordinary differential equations

$$\begin{aligned}
\frac{\mathrm{d}}{\mathrm{d}u}F_0(u) &= -\frac{\lambda_0}{r_0}F_0(u) + \frac{\mu_1}{r_0}F_1(u),\\
\frac{\mathrm{d}}{\mathrm{d}u}F_j(u) &= \frac{\lambda_{j-1}}{r}F_{j-1}(u) + \frac{\mu_{j+1}}{r}F_{j+1}(u) - \frac{(\lambda_j + \mu_j)}{r}F_j(u), u \geq 0, j = 1, 2, \ldots.
\end{aligned}$$

When the net input rate of fluid flow into the buffer is positive, the buffer content increases and buffer cannot stay empty. It follows that the solution of the above mentioned ODEs must satisfy the boundary conditions

$$\begin{aligned}
F_j(0) &= 0, \ j = 1, 2, \ldots\\
F_0(0) &= a, \ \text{ for some constant } 0 < a < 1.
\end{aligned}$$

Let $\widehat{f}(s)$ denote the Laplace transform of the function $f(t)$. On taking the Laplace transforms of the system of ODEs,

$$\left(s + \frac{\lambda_0}{r_0}\right)\widehat{F}_0(s) - \frac{\mu_1}{r_0}\widehat{F}_1(s) = a \tag{7.32}$$

$$\left(s + \frac{\lambda_j + \mu_j}{r}\right)\widehat{F}_j(s) - \frac{\lambda_{j-1}}{r}\widehat{F}_{j-1}(s) - \frac{\mu_{j+1}}{r}\widehat{F}_{j+1}(s) = 0, \ j = 1, 2, \ldots \tag{7.33}$$

Define

$$f_0(s) := \widehat{F}_0(s),$$
$$f_j(s) := (-1)^j \frac{\mu_1\mu_2\ldots\mu_j}{r_0 r^{j-1}} \widehat{F}_j(s),\ j = 1, 2, \ldots$$

In terms of the $f_j(s)$, the system of equations 7.32 and 7.33 can be written as

$$\left(s + \frac{\lambda_0}{r_0}\right) f_0(s) + f_1(s) = a,$$
$$\left(s + \frac{\lambda_1 + \mu_1}{r}\right) f_1(s) + \frac{\lambda_0\mu_1}{r_0 r} f_0(s) + f_2(s) = 0,$$
$$\left(s + \frac{\lambda_j + \mu_j}{r}\right) f_j(s) + \frac{\lambda_{j-1}\mu_j}{r^2} f_{j-1}(s) + f_{j+1}(s) = 0,\ j = 2, 3, \ldots.$$

The above system of equations can be written as

$$f_0(s) = \frac{a}{s + \lambda_0/r_0 + f_1(s)/f_0(s)},$$
$$\frac{f_1(s)}{f_0(s)} = \frac{-\lambda_0\mu_1/(r_0 r)}{s + (\lambda_1 + \mu_1)/r + f_2(s)/f_1(s)},$$
$$\frac{f_j(s)}{f_{j-1}(s)} = \frac{-\lambda_{j-1}\mu_j/(r^2)}{s + (\lambda_j + \mu_j)/r + f_{j+1}(s)/f_j(s)},\ j = 2, 3, \ldots.$$

Therefore, $f_0(s)$ can be represented by a continued fraction as follows (Refer Appendix A.8):

$$f_0(s) = \cfrac{a}{s + \frac{\lambda_0}{r_0} - \cfrac{\frac{\lambda_0\mu_1}{r_0 r}}{s + \frac{\lambda_1+\mu_1}{r} - \cfrac{\frac{\lambda_1\mu_2}{r^2}}{s + \frac{\lambda_2+\mu_2}{r} - \ldots}}}. \tag{7.34}$$

Now, consider the arrival and service rates to be λ and μ, respectively. Since, $\pi = \rho^i,\ i = 1, 2, \ldots$ with $\rho = \lambda/\mu$, it follows that

$$\rho < \frac{r_0}{r_0 - r}.$$

Substituting $\lambda_j = \lambda$ and $\mu_j = \mu$, in Eq. (7.34),

$$f_0(s) = \cfrac{a}{s + \frac{\lambda}{r_0} - \cfrac{\frac{\lambda\mu}{r_0 r}}{s + \frac{\lambda+\mu}{r} - \cfrac{\frac{\lambda\mu}{r^2}}{s + \frac{\lambda+\mu}{r} - \ldots}}} \tag{7.35}$$

which can be written as

$$f_0(s) = \frac{a}{s + \lambda/r_0 - \lambda\mu/(r_0 r) f(s)},$$

where

$$f(s) = \frac{1}{s + \frac{\lambda+\mu}{r} - \frac{\lambda\mu}{r^2} f(s)} \tag{7.36}$$

which results in a quadratic equation in $f(s)$ whose negative root is

$$f(s) = \frac{p - \sqrt{p^2 - \alpha^2}}{\alpha^2/2}, \tag{7.37}$$

where $\alpha = 2\sqrt{\lambda\mu}/r$ and $p = s + \lambda/r + \mu/r$. Substituting for $f(s)$, from Eq. (7.37) in Eq. (7.36), and pulling out the term $(s + \lambda/r_0)$ from the denominator,

$$f_0(s) = \frac{a}{s + \lambda/r_0} + a \sum_{k=1}^{\infty} \left(\frac{r}{2r_0}\right)^k \frac{[p - \sqrt{p^2 - \alpha^2}]^k}{(s + \lambda/r_0)^{k+1}}$$

On inversion,

$$F_0(u) = ae^{-\lambda u/r_0} + ae^{-(\lambda\mu+\mu^2)/r} \sum_{k=1}^{\infty} \left(\frac{r}{2r_0}\right)^k \tag{7.38}$$

$$\times \frac{\alpha^k}{\Gamma(k)} \int_0^u v^k \frac{I_k(\alpha(u-v))}{u-v} e^{((\lambda+\mu)/r - \lambda/r_0)v} dv, \tag{7.39}$$

where $I_k(.)$ is the modified Bessel function of the first kind of order k (refer, Appendix A.7). To obtain other probabilities $F_j(u)$, $j \geq 1$, write

$$f_1(s) = -\frac{\lambda\mu}{r_0 r} f_0(s) f(s),$$

$$f_j(s) = -\frac{\lambda\mu}{r^2} f_{j-1}(s) f(s), \; j = 2, 3, \ldots,$$

which can be put together as

$$f_j(s) = \frac{r}{r_0} \left(\frac{-\lambda\mu}{r^2}\right)^j (f(s))^j f_0(s), \; j = 1, 2, \ldots. \tag{7.40}$$

$$\widehat{F}_j(s) = \left(\frac{\lambda}{r}\right)^j (f(s))^j f_0(s), \; j = 0, 1, \ldots. \tag{7.41}$$

Substituting the values of $f(s)$ and $f_0(s)$, Eq. (7.41) can be written as

$$\widehat{F}_j(s) = a\left(\frac{r}{2\mu}\right)^j \sum_{k=0}^{\infty} \frac{[p - \sqrt{p^2 - \alpha^2}]^{j+k}}{(s + \lambda/r_0)^{k+1}}, \ j = 1, 2, \ldots \tag{7.42}$$

which on inverting yields, for $j = 1, 2, 3, \ldots, u \geq 0$,

$$F_j(u) = a\left(\frac{r}{2\mu}\right)^j e^{-((\lambda+\mu)/r)u} \sum_{k=0}^{\infty} \left(\frac{r}{2r_0}\right)^k \frac{(j+k)\alpha^{j+k}}{\Gamma(k+1)}, \tag{7.43}$$

$$\times \int_0^u v^k \frac{I_{j+k(\alpha(u-v))}}{(u-v)} e^{((\lambda+\mu)/r - \lambda/r_0)v} dv. \tag{7.44}$$

Interested readers may refer to [1–12].

7.7 Exercise Problems

1. Consider an $M/M/1$ queue with mean service time as $\frac{1}{\mu}$ and customers arrival following Poisson process with rate λ, where $\lambda < \mu$. An arrived customer may leave the system directly with a probability of $\frac{i}{i+1}$, $i = 0, 1, \ldots,$ when they find there are already i customers in the system.
   (a) Find the steady-state probability distribution of the queueing system in terms of λ and μ.
   (b) Assuming that the queueing system is in steady-state, find the loss probability (probability that an arrived customer chooses not to join the queueing system).
   (c) If the mean service time μ^{-1} can be assigned as fast as possible (i.e., $\lambda << \mu$), recommend the service rate μ if the loss probability has to be less than 1%.

2. Consider a $M/M/1$ queueing system with balking. Here, if a customer enters the queue when it reaches its maximum size, he exits the system without getting served. Consider b_n to be a series of monotonically decreasing function of the system size multiplying by the average arrival rate λ such that $\lambda_n = b_n\lambda$ and $0 \leq b_{n+1} \leq b_n \leq 1$. Prove that the steady-state probabilities of the system are given by

$$\pi_n = \pi_0\left(\frac{\lambda}{\mu}\right)^n \prod_{i=1}^{n} b_{i-1}, \quad n = 1, 2, \ldots.$$

3. Consider a $M/M/1$ queueing system with balking and reneging. Consider reneging function $r(n)$.

The updated death rates will be $\mu_n = \mu + r(n)$ and arrival rates will be $\lambda_n = b_n\lambda$, $0 \le b_{n+1} \le b_n \le 1$. Prove that the steady-state probabilities of the system will be

$$\pi_n = \pi_0 \lambda^n \prod_{i=1}^{n} \left(\frac{b_{i-1}}{\mu + r(i)} \right), \quad n = 1, 2, \ldots.$$

4. Let $\{X(t), t \ge 0\}$ be a CTMC indicating the customer count in a service system at time t, with transition rates defined as

$$q_{ij} = \lambda(j = i + 1) + \mu(j = \max\{0, i - K\}).$$

Here, λ, μ, and K are positive, with the condition that $\lambda < K\mu$. This denotes a system where customers come according to a Poisson process with rate λ and are served in batches as described. When there are $i \ge K$ customers in the system, batches of K customers leave at the rate μ; conversely, when $i < K$ customers are present, all customers depart at the rate μ. Justify that $\{X(t), t \ge 0\}$ is ergodic and that its stationary distribution is given by $\pi_{0i} = r^i(1 - r)$ for $i \ge 0$, where r is the unique root in the interval (0, 1) of the specified equation $\mu r^{K+1} - (\lambda + \mu)r + \bar{\lambda} = 0$.

5. Let $\{X(t), t \ge 0\}$ represent the number of objects in an input-output system that functions as described. Items arrive in batches according to a Poisson process with rate λ, and the sizes of the batches are independent and identically distributed, following a geometric distribution $(1 - \alpha)\alpha^{n-1}$. The items are delivered by a single server, and the service durations are independently and exponential distributed with a rate of $\mu > \lambda/(1 - \alpha)$. Show that $\{X(t), t \ge 0\}$ constitutes a CTMC and identify its transition rates. Prove that its stationary distribution is

$$\pi_n = \pi_0 \left(\frac{\lambda}{\mu} \right) \left(\frac{\lambda + \alpha\mu}{\mu} \right)^{n-1} \quad n = 1, 2, \ldots.$$

6. Examine the system discussed in the previous exercise, noting that at the end of a service period with i items in the system, $\min\{i, K\}$ items depart, where K denotes the batch service capacity. The items available at the beginning of service, along with all arrivals during the service period up to the quantity K, will depart at the end of the service. Here, $K\mu > \frac{\lambda}{1-\alpha}$. Show that the quantity of items in the system $\{X(t), t \ge 0\}$ constitutes a CTMC characterized by specific transition rates

$$\begin{aligned} q_{ij} &= \lambda(1 - \alpha)\alpha^{k-1} 1(j = i + k) \\ &\quad + \mu[1(j = i - B, i > K) + 1, \ \ (j = 0, i \le K)] \end{aligned}$$

Show that its stationary distribution is

$$\pi_n = \frac{\pi_0 \lambda r^{n-1}}{\left(\mu \sum_{k=0}^{K-1} r^k\right)}, \quad n = 1, 2, \ldots,$$

where r is the unique solution in (0, 1) of

$$\mu r^K + \mu(1-\alpha) \sum_{k=0}^{K-1} r^k = \lambda + \mu.$$

7. Consider a single capacity buffer which is modulated by a BDP evolving in the background. Let $\{X(t), t \geq 0\}$ denote the state of the process at time t. The flow rates of the fluid into and out of a single capacity buffer are determined by the actual state of the background process. Prove that the final solution is given by

$$\lim_{t\to\infty} P\{C(t) \leq u, X(t) = 0\} = F_0(u) = \pi_0 + \frac{r_1}{r_0}\pi_1 e^{-(\lambda/r_0+\mu/r_1)u}$$
$$\lim_{t\to\infty} P\{C(t) \leq u, X(t) = 1\} = F_1(u) = \pi_1 - \pi_1 e^{-(\lambda/r_0+\mu/r_1)u}$$

and

$$\lim_{t\to\infty} P\{C(t) \leq u\} = F_0(u) + F_1(u) = 1 - \left(1 - \frac{r_1}{r_0}\right)\left(\frac{\lambda}{\lambda+\mu}\right) e^{-(\lambda/r_0+\mu/r_1)u}$$

where $\pi_0 = \frac{\mu}{\lambda+\mu}$ and $\pi_1 = \frac{\lambda}{\lambda+\mu}$.
8. A ferry transports cars across a river, but it only departs once it has a full load of 10 cars. The loading process is considered instantaneous, and the round-trip travel time follows an exponential distribution with an average duration of 15 minutes. Cars arrive at the ferry dock following a Poisson process with an average rate of 30 cars per hour. The question is: on average, how many cars are waiting at the dock at any given moment, ready to be loaded onto the ferry?
9. For an $M/M/1$ retrial queue with reneging customers, calculate the proportion of customers who ultimately receive service.
10. A small store has a single phone line. Calls to the store are well modeled by an $M/M/1$ retrial queue with arrival rate $\lambda = 10$ per hour, service rate $\mu = 15$ per hour, and retrial rate $\gamma = 6$ per hour.

    (a) What is the average length of time for a customer to reach a clerk at the store?
    (b) What is the average rate that call attempts are made to the store? (Call attempts include calls that are answered and calls that receive a busy signal.)
    (c) What is the fraction of call attempts that receive a busy signal?
    (d) Suppose that you are measuring call attempts to the store and you are not aware that some customers are making redial attempts. That is, you assume each call attempt is from a distinct customer. You decide to model the system

as an $M/M/1/1$ queue, where the arrival rate is the rate of call attempts found in (b). Based on these assumptions, what is the fraction of calls that receive a busy signal? Compare this to the actual result found in (c).

11. Consider a machine which is turned off when there is no work. It is turned on and restarts work when enough orders, say N, arrived to the machine. The setup times are negligible. The processing times are exponential distributed with mean $1/\mu$ and the average number of orders arriving per unit time is $\lambda(< \mu)$.

References

1. Alfa, Attahiru Sule, and Wei Li. PCS networks with correlated arrival process and retrial phenomenon. IEEE Transactions on Wireless Communications 1, no. 4 : 630-637, (2002)
2. Alfa, Attahiru Sule, and K .P. Sapna Isotupa. An $M/PH/k$ retrial queue with finite number of sources. Computers & Operations Research 31, no. 9, 1455-1464, (2004)
3. Chakravarthy, S.R.: Introduction to Matrix-Analytic Methods in Queues 1: Analytical and Simulation Approach - Basics. ISTE Ltd, London and John Wiley and Sons, New York (2022)
4. Chakravarthy, S.R.: Introduction to Matrix-Analytic Methods in Queues 2: Analytical and Simulation Approach - Basics. ISTE Ltd, London and John Wiley and Sons, New York (2022)
5. Dharmaraja, S., Jindal, V., Alfa, A.S.: Phase-type models for cellular networks supporting voice, video and data traffic. Mathematical and computer modelling **47**(11–12), 1167–1180 (2008)
6. S. Dharmaraja, and S. Kapoor, (2023). On the time-dependent solution of fluid models driven by an M/M/1 queue using a probabilistic approach. International Journal of Operational Research, 46(1), 65–72.
7. D. P. Gaver, P. A. Jacobs, and Guy Latouche. Finite birth-and-death models in randomly changing environments. Advances in applied probability 16, no. 4, 715-731, (1984)
8. V. Jain and R. Raj and S. Dharmaraja, Numerical optimization of loss system with retrial phenomenon in cellular networks,International Journal of Operations Research, In Press, 2022
9. M. F. Neuts, Matrix Geometric Solutions in Stochastic Models: An Algorithmic Approach. John Hopkins University Press, Baltimore (1981).
10. P. R. Parthasarathy, K. V. Vijayashree, and R. B. Lenin. An $M/M/1$ driven fluid queue-continued fraction approach. Queueing Systems 42(2002): 189–199.
11. J. F. Shortle , James M. Thompson , Donald Gross, Carl M. Harris. Fundamentals of queueing theory. John Wiley & Sons; 2018 Apr 10
12. R. Sudhesh, A. Azhagappan, and S. Dharmaraja, (2017). Transient analysis of M/M/1 queue with working vacation, heterogeneous service and customers' impatience. RAIRO-Operations Research, 51(3), 591–606.

Chapter 8
Point Processes and Non-Markov Processes

In the earlier seven chapters, a detailed study on Markov process in the form of DTMC and CTMC is presented. This chapter sets out the fundamentals of the Point processes, Renewal processes, processes, and are discussed in Sects. 8.1, 8.2 and 8.3, respectively. In Sects. 8.4 and 8.5, the basic concepts of the semi-Markov process and Markov regenerative process are discussed. The steady-state and transient analysis for various stochastic models has been provided.

8.1 Point Processes

Point processes are stochastic processes that are used to model events that occur at random intervals relative to the time axis or the space axis, i.e., they are used to describe data that are localized at a finite set of time points.

A point process can be specified in terms of arrival times, or inter-arrival times. Let $T_1, T_2, \ldots,$ be RVs describing the arrival times of the events that occur, and thus $\{T_i, i = 1, 2, \ldots\}$ forms a point process. Let the realization of a point process be the event $T_1 = t_1, T_2 = t_2, \ldots,$ for times $0 < t_1 < t_2, \ldots$. Let $X_1, X_2, \ldots$ be a set of RVs describing the inter-arrival times. Clearly, there is a one-to-one correspondence between any set of arrival times and inter-arrival times as,

$$X_1 = T_1 \text{ and } X_i = T_i - T_{i-1} \; \forall \, i = 2, 3, \ldots \quad \Longrightarrow \; T_n = \sum_{i=1}^{n} X_i \; \forall \, n = 1, 2, \ldots.$$

Here, t is considered as time and t_n as the nth arrival time or nth arrival epoch. When T_i are RVs, $\{T_i, i = 1, 2, \ldots\}$ is called a random point process. A random point process $\{T_i, i = 1, 2, \ldots\}$ for which the inter-arrival times $\{X_n, n = 1, 2, \ldots\}$ forms an IID sequence RVs is called a renewal process.

When more than one arrival is not allowed to occur at the same time, the point process

D. Selvamuthu, *Introduction to Stochastic Processes*, University Texts in the Mathematical Sciences, https://doi.org/10.1007/978-981-97-6152-4_8

is referred to as simple point process.
Another way to describe a set of arriving observations is in terms of the number of arrivals that occurred over any time interval, and such process is called a **Counting Process**. A counting process is a non-negative, integer-valued, increasing stochastic process.
Let $\{N(t), t \geq 0\}$ be a counting process, then the number of arrivals observed in the interval $(t_1, t_2]$ is defined as

$$\Delta N_{(t_1,t_2)} = N(t_2) - N(t_1)$$

and sometimes $\Delta N_{(t_1,t_2)}$ is called as increment of the point process between t_1 and t_2.
A sequence of RVs that arises naturally in a counting process is the "sojourn time" T_i between the ith and $(i+1)$th event for $i = 0, 1, \ldots$, and the quantity $N(t)$ can be expressed in terms of T_i by using the standard technique of "indicator functions". The indicator function 1_A of an event A equals 1 if the event occurs and 0 otherwise. Letting $S_n = T_0 + T_1 + \ldots + T_n$, the time of the nth event, it holds that

$$N(t) = \sum_{n=0}^{\infty} 1_{S_n \leq t},$$

and the number of events in a given time interval $(s, t]$ can be defined as $N(s, t) := N(t) - N(s)$.
Depending on the assumptions, a counting process may be known under other names and fit into other general classes of processes.

1. When the sojourn times are independent with T_i following an exponential distribution with mean $1/\mu_i$, it leads to a "pure birth process".
2. When the sojourn times T_i are IID, the resulting process is known as a "renewal process".
3. In addition, the T_i has an exponential distribution with mean $1/\mu$ (so that a process that is both a pure birth process and a renewal process), the result is a homogeneous "Poisson process" with rate $\lambda := 1/\mu$ (the mean number of events per time unit).

Example 8.1 Consider a non-homogeneous Poisson process $\{N(t), t \geq 0\}$ with rate function $\lambda(t)$ which is a point process. Compute the distribution of the time of the next arrival given the previous arrival time.
Solution: Let $\{T_i, i = 1, 2, \ldots\}$ be the RV which denotes the inter-arrival time to find $P\{T_i \leq t_i | T_{i-1} = t_{i-1}\}$. Since a non-homogeneous Poisson process has independent increment property, thus

$$P\{T_i > t_i | T_{i-1} = t_{i-1}\} = P\{\Delta N(t_{i-1}, t_i] = 0\} = exp\left[-\int_{t_{i-1}}^{t_i} \lambda(t)dt\right],$$

which implies that the CDF is $P\{T_i \le t_i | T_{i-1} = t_{i-1}\} = 1 - exp\left[-\int_{t_{i-1}}^{t_i} \lambda(t)dt\right]$. Thus, the probability density function for the next arrival time is given by

$$f_{T_i}(t_i | T_{i-1} = t_{i-1}) = \lambda(t_i) exp\left[-\int_{t_{i-1}}^{t_i} \lambda(t)dt\right].$$

Remark 8.1 1. A counting process for which the inter-arrival times are IID RV with an arbitrary distribution is said to be a renewal process.

2. A Poisson process can be defined as a counting process for which the inter-arrival times are IID with an exponential distribution. A renewal process is a more general counting process than Poisson process.
3. A continuous-time stochastic process in which the embedded jump chain (the discrete process registering what values the process takes) is a Markov chain, and where the holding times (time between jumps) are RVs with any distribution, whose distribution function may depend on the two states between which the move is made, is called a semi-Markov process or Markov renewal process.
4. When $P\{X_i = k\} = 1$ for some $k > 0$, for every i, $\{S_n = nk, n = 1, 2, \ldots\}$ is called a deterministic renewal process.

8.2 Renewal Processes

Renewal theory is the branch of probability theory that generalizes Poisson processes for arbitrary interarrival (holding) times. Let X_1 be the time to the first renewal, renewal, and let $X_n, n = 2, 3, \ldots$ be the time between $(n-1)$th renewal and nth renewal. Assume that $X_n, n = 1, 2, \ldots$ are IID RVs with distribution function F. Let

$$\mu = E(X_n) = \int_0^\infty x dF(x)$$

which will be positive and finite.

Definition 8.1 Define the time of the nth renewal by

$$S_n = \sum_{i=1}^n X_i.$$

Let $N(t)$ be the number of renewals by time t such that

$$N(t) = \max\{n : S_n \le t\}.$$

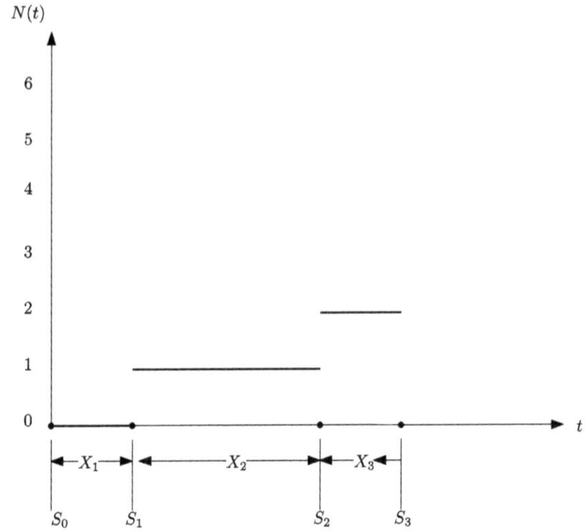

Fig. 8.1 Sample path of renewal process

Then, the counting process $\{N(t), t \geq 0\}$ is called a renewal counting process. If for some n, $S_n = t$, then a renewal is said to occur at time t; S_n gives the time of the nth renewal and is called the nth renewal time (Fig. 8.1).

Distribution

Consider a renewal process $\{N(t), t \geq 0\}$ with interarrival time distribution F. Note that the event $\{N(t) \geq n\}$ is equivalent to the event $\{S_n \leq t\}$. Hence,

$$P\{N(t) \geq n\} = P\{S_n \leq t\} = F^{(n)}(t)$$

where $F^{(n)}$ is n-fold convolution of F with $F^{(0)} = 1$.

$$\begin{aligned} P\{N(t) = n\} &= P\{N(t) \geq n\} - P\{N(t) \geq n+1\} \\ &= F^{(n)}(t) - F^{(n+1)}(t). \end{aligned}$$

For example, when inter-arrival time is uniformly distributed between 0 and 1, the n-fold convolution of F is given by

$$F^{(n)}(t) = \frac{t^n}{n!}, \quad n = 1, 2, \ldots; \quad 0 \leq t \leq 1$$

which can be shown from the mathematical induction.

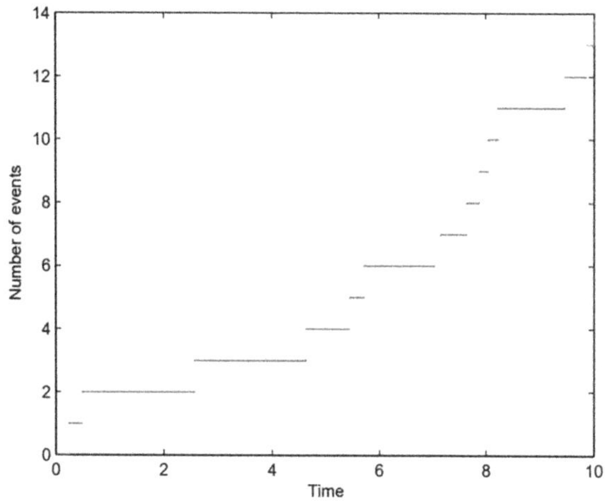

Fig. 8.2 Sample path of Poisson process with $\lambda = 2$

$$\begin{aligned}F^{(n+1)}(t) &= P\{S_{n+1} \le t\}\\ &= \int_0^t P\{S_{n+1} \le t \mid S_n = x\}\frac{x^{n-1}}{(n-1)!}dx\\ &= \int_0^t P\{X_{n+1} \le t - x\}\frac{x^{n-1}}{(n-1)!}dx\\ &= \frac{t^{n+1}}{(n+1)!}, \quad 0 \le t \le 1.\end{aligned}$$

Example 8.2 Prove that Poisson process $\{N(t), t \ge 0\}$ is a counting renewal process whose inter-arrival time has an exponential distribution with parameter λ, i.e., $X_n, n = 1, 2, \ldots$ follows an exponential distribution with parameter λ.
Solution: Distribution of $N(t)$ is given by

$$P\{N(t) = n\} = \frac{e^{-\lambda t}\,(\lambda t)^n}{n!}, n = 0, 1, \ldots$$

For $n = 1, 2, \ldots,$

$$\begin{aligned}P\{X_n > x\} &= P\{N(S_{n-1} + x) = n - 1 | N(S_{n-1}) = n - 1\}\\ &= P\{N(x) = 0 | N(0) = 0\} = e^{-\lambda x}, x \ge 0.\end{aligned}$$

Hence, the distribution of X_n is $F(x) = 1 - e^{-\lambda x}, x \ge 0$.

Sample path of the Poisson process, generated using MATLAB software, with $\lambda = 2$ is shown in Figure 8.2.

Example 8.3 Suppose that in a system, a unit fails according to a Poisson process with rate $\lambda = 3$ per day. Suppose that there are 6 spare units in an inventory and the next supply is not due in 4 days. Find the probability that the system will be out of order in the next 4 days.
Solution: The probability that the system will be out of order in the next 4 days is

$$\sum_{n=7}^{\infty} P(N(4) = n) = 1 - \sum_{n=0}^{6} P(N(4) = n) = 0.954\,.$$

8.2.1 Renewal Function and Key Renewal Theorem

The renewal function serves as a fundamental concept in stochastic processes, specifically addressing the renewal theory's exploration of recurring events over time. It provides a mathematical framework to analyze the cumulative occurrence of events and is closely linked to the key renewal theorem, a pivotal result in renewal theory. The key renewal theorem establishes insightful connections between the renewal function and the asymptotic behavior of the underlying stochastic processes, shedding light on the long-term properties of recurring events.
Renewal Function
Let

$$M(t) = E[N(t)] = \sum_{n=1}^{\infty} P(N(t) \geq n) = \sum_{n=1}^{\infty} F^{(n)}(t).$$

Then $M(t)$ is called the renewal function. For example, the renewal process whose inter-arrival times are IID with uniform distribution between 0 and 1, has the renewal function for $0 \leq t \leq 1$ which is given by

$$M(t) = \sum_{n=1}^{\infty} F^{(n)}(t) = \sum_{n=1}^{\infty} \frac{t^n}{n!} = e^t - 1.$$

The function $m(t) = M'(t)$ is called the renewal density function of the renewal process.

$$\begin{aligned} m(t) &= \lim_{h \to 0^+} \frac{P\{\text{one or more renewals in}\}\ (t, t+h)\}}{h} \\ &= \lim_{h \to 0^+} \frac{1}{h} \sum_{n=1}^{\infty} P\{n \text{ renewals in}\}\ (t, t+h)\} \\ &= \sum_{n=1}^{\infty} \lim_{h \to 0^+} \frac{1}{h} \left[F^{(n)}(t+h) - F^{(n)}(t)\right] = M'(t). \end{aligned}$$

Consider

$$M(t) = \sum_{n=1}^{\infty} F^{(n)}(t).$$

In equivalent form using the Laplace transform,

$$M^*(s) = \sum_{n=1}^{\infty} F^{(n)*}(s) = \sum_{n=1}^{\infty} \frac{f^{(n)*}(s)}{s},$$

where $f^{(n)}(t) = F^{(n)'}(t)$.

$$M^*(s) = \frac{1}{s}\sum_{n=1}^{\infty} \left[f^*(s)\right]^n = \frac{f^*(s)}{s\left[1 - f^*(s)\right]}.$$

Hence,

$$m^*(s) = \frac{f^*(s)}{1 - f^*(s)} \text{ and } f^*(s) = \frac{m^*(s)}{1 + m^*(s)}.$$

Renewal Equation

The renewal function $M(t), 0 \leq t < \infty$, uniquely determines the inter-arrival time distribution F. For example, $M(t) = \lambda t$ corresponds to the exponential distribution with parameter λ. Renewal equations are useful in deriving the quantity of interest associated with a renewal process as a function of time. A renewal equation is expressed as a recursive form through an integral equation.

$$M(t) = E[N(t)].$$

This can be evaluated by conditioning on X_1, the time of first renewal, i.e.,

$$M(t) = \int_0^{\infty} E[N(t) \mid X_1 = x] dF(x).$$

This integral can be evaluated by dividing it into two cases, whether the first renewal occurs before t or not and the other is the case where the first renewal occurs before or after t. In the former case, since there are no renewals observed by time t,

$$E[N(t)|X_1 = x > t] = 0.$$

In the latter case, the first renewal occurs before time t and the expected number of renewals between x and t will be $M(t - x)$ from the definition of the renewal function. Hence,

$$E[N(t)|X_1 = x < t] = 1 + M(t - x).$$

And therefore,

$$M(t) = \int_0^t (1 + M(t-x))dF(x) = F(t) + \int_0^t M(t-x)dF(x).$$

This integral equation is called a renewal equation if $M(\cdot)$ is considered as unknown. The above equation can be written as

$$M = F + M * f.$$

Now, taking Laplace transform on both sides,

$$M^* = F^* + (M * f)^* = F^* + M^* f^*.$$

Substituting $F^*(s) = \frac{f^*(s)}{s}$ and simplifying, implies

$$f^*(s) = \frac{sM^*(s)}{1 + sM^*(s)}.$$

The renewal equation can be generalized for $Z(t)$, the unknown function associated with a renewal process with distribution function F,

$$Z(t) = Q(t) + \int_0^t Z(t-x)dF(x),$$

where $Q(t)$ is a known function.

Renewal Times
Let $X_1, X_2, \ldots$ be the time between its successive occurrences. Then

$$S_0 = 0; \; S_{n+1} = S_n + X_{n+1}, \quad n = 0, 1, \ldots$$

define the times of occurrence assuming that the time origin is taken to be an instant of such an occurrence. The sequence $\{S_n, n = 0, 1, \ldots\}$ is called a renewal process provided that $X_1, X_2, \ldots$ be IID non-negative RVs. Then S_n are called renewal times. Note that, the renewal process $\{S_n, n = 0, 1, \ldots\}$ is said to be recurrent if $X_n < \infty$ almost surely for every n; otherwise it is called transient.

Age, Excess, and Spread at Time t
Suppose $\{N(t), t \geq 0\}$ is a renewal process.

1. **Age**: The age at time t of the renewal process is defined by

$$A(t) = t - S_{N(t)}.$$

2. **Excess**: The excess at time t of the renewal process is defined by

$$Y(t) = S_{N(t)+1} - t.$$

3. **Spread**: The spread at time t of the renewal process is defined by

$$X_{N(t)+1} = A(t) + Y(t).$$

For example, let $Y(t)$ be the excess at time t. Let $g(t) = E[Y(t)]$. Using the above renewal equation, $g(t)$ can be found. Conditioning on X_1 gives

$$g(t) = \int_0^\infty E[Y(t)|X_1 = x]dF(x).$$

Now,

$$E[Y(t)|X_1 = x > t] = x - t,$$

while for $X_1 = x < t$

$$E[Y(t)|X_1 = x < t] = g(t - x).$$

Hence,

$$g(t) = \int_0^t g(t - x)dF(x) + \int_t^\infty (x - t)dF(x),$$

is the renewal equation for $g(t)$.

Long-Run Properties

One can study the long-run properties of the renewal process of which there are two types. One is to obtain the long-run (time) average of the quantity of interest and the other is to obtain the pointwise limit. For example, the long-run average of age is given by

$$\lim_{t\to\infty} \frac{\int_0^t A(s)ds}{t},$$

while the pointwise limit of the expected age is given by

$$\lim_{t\to\infty} E[A(t)].$$

Long-Run Renewal Rate

One can study the average number of renewals (per unit time) in the long-run. It is called the long-run renewal rate. For a renewal process $\{N(t), t \geq 0\}$ having distribution function F for interarrival times, the long-run renewal rate is given by

$$\lim_{t\to\infty}\frac{N(t)}{t}=\frac{1}{\mu}\ \text{with probability } 1,$$

where

$$\mu=\int_0^\infty xdF(x).$$

Since $S_{N(t)}$ is the last renewal time prior to t and $S_{N(t)+1}$ is the first renewal time after t. It is known that

$$S_{N(t)}\le t\le S_{N(t)+1}\ \text{ or }\ \frac{S_{N(t)}}{N(t)}\le\frac{t}{N(t)}\le\frac{S_{N(t)+1}}{N(t)}.$$

But,

$$\begin{aligned}\lim_{t\to\infty}\frac{S_{N(t)}}{N(t)}&=\lim_{t\to\infty}\frac{X_1+X_2+\ldots+X_{N(t)}}{N(t)}\\&=E[X]=\mu\ \text{ with probability } 1\end{aligned}$$

and

$$\lim_{n\to\infty}\frac{S_{N(t)+1}}{N(t)}=\lim_{t\to\infty}\frac{S_{N(t)+1}}{N(t)+1}\frac{N(t)+1}{N(t)}=\mu\ \text{ with probability } 1.$$

Hence,

$$\lim_{t\to\infty}\frac{N(t)}{t}=\frac{1}{\mu}\ \text{ with probability } 1.$$

Central Limit Theorem

As $n\to\infty$,

$$Z_n=\frac{S_n-n\mu}{\sigma\sqrt{n}}\Rightarrow Z\sim N(0,1).$$

As $t\to\infty$, $N(t)$ becomes normal distributed with $E(N(t))\sim\frac{t}{\mu}$ and $Var(N(t))\sim\frac{\sigma^2 t}{\mu^3}$. As $t\to\infty$, using central limit theorem,

$$Z(t)=\frac{N(t)-\frac{t}{\mu}}{\sigma\sqrt{\frac{t}{\mu^3}}}\Rightarrow Z\sim N(0,1).$$

Renewal Limit Theorems

From the previous consideration, we are to deduce that the arrival time process and the renewal or counting process are linked. The arrival time process, in other words, is the partial sum process for a set of independent, identically distributed RVs of inter-arrival times. As a result, the fundamental limit theorems for partial sum processes (such as the law of large numbers and the central limit theorem) should have analogous results for the renewal process. This section will look at the limiting behavior of the renewal process. The major findings known as renewal limit theorems, are useful for a variety of stochastic systems.

Elementary Renewal Theorem

It states the basic limit in the law of large numbers in first order moment as well as with probability 1. For a renewal process $\{N(t), t \geq 0\}$ having mean interarrival time $0 < \mu < \infty$,

$$\lim_{t\to\infty} \frac{E[N(t)]}{t} = \lim_{t\to\infty} \frac{M(t)}{t} = \frac{1}{\mu}.$$

That is, the limiting mean rate of arrivals is $\frac{1}{\mu}$.

Note that the limit value of a sequence of RVs is not same as the limit value of a sequence of mean of RVs.

Consider the reward renewal processes. Let

$$R(t) = \sum_{n=1}^{N(t)} R_n$$

be reward earned by time t. If $E[R] < \infty$, $E[X] < \infty$, then

$$\frac{E[R(t)]}{t} \to \frac{E[R]}{E[X]} \text{ as } t \to \infty.$$

Asymptotic property:

$$\frac{R(t)}{t} \to \frac{E[R]}{E[X]} \text{ as } t \to \infty, \text{ with probability 1.}$$

Lattice

A non-negative RV X is said to be lattice if there exists $d \geq 0$ such that

$$\sum_{n=0}^{\infty} P(X = nd) = 1.$$

The largest d having this property is said to be the period of X. If X is a lattice and X has a distribution function of F, then we say that F is lattice. For instance, if X follows Poisson distribution with mean λ, then X is a lattice with period 1. If $P\{X = \sqrt{2}\} = P\{X = \sqrt{3}\} = 0.5$, then X is not a lattice.

Key Renewal Theorem

Let $\{N(t), t \geq 0\}$ be a renewal process with continuous interarrival time, non-lattice distribution function F with mean $\mu < \infty$. If $h : [0, \infty) \to [0, \infty)$ is integrable and non-increasing, as $t \to \infty$

$$\int_0^t h(t-x)dM(x) \to \frac{1}{\mu}\int_0^\infty h(x)dx,$$

where $M(t)$ is the renewal function.

Example 8.4 Let $Y(t)$ be the excess at time t. It is known that

$$E[Y(t)] = h(t) + \int_0^t h(t-x)dM(x),$$

where

$$h(t) = \int_t^\infty (x-t)dF(x),$$

$$\lim_{t\to\infty} E[Y(t)] = \lim_{t\to\infty} \int_0^t h(t-x)dM(x).$$

Assume that the second moment of an interarrival time is finite, then $h(t)$ is integrable and non-increasing. Assume that F is not lattice.
Now, key renewal theorem can be applied to obtain the limiting mean excess. After simplifications,

$$\lim_{t\to\infty} E[Y(t)] = \frac{E[X^2]}{2E[X]}.$$

Blackwell's Theorem
Let $M(t) = E[N(t)]$ be the renewal function. The theorem states that, asymptotically, the expected number of renewals in an interval is proportional to the length of the interval; the proportionally constant is $\frac{1}{\mu}$. If F is not lattice, then

$$\lim_{t\to\infty} [M(t+h) - M(t)] = \frac{h}{\mu}.$$

If F is lattice with period d, then

$$\lim_{n\to\infty} E[\text{number of renewals at } nd] = \frac{d}{\mu},$$

since the number of renewals at time nd will be 0 or 1.

8.2.2 *Some Other Renewal Processes*

In this section, certain variants of the renewal process, namely alternating renewal process, delayed renewal process, and reward renewal process are discussed.

8.2.2.1 Alternating Renewal Process

Let $X_1, X_2, \ldots$ be IID RVs which constitutes *on times*. Let $Y_1, Y_2, \ldots$ be IID RVs which constitute *off times*. Assume that $E(X+Y) < \infty$ and $X+Y$ has distribution F. Suppose that renewals occur at the end of every X_i, whereas no renewals at the end of every Y_i. Assume that X_i and Y_i are independent RVs. Then $\{X_i + Y_i, i = 1, 2, \ldots\}$ is called an alternating renewal process.

Example 8.5 For example, consider the following situation. A machine works for time X_1, then breaks down and has to be repaired (which takes time Y_1), then works for a time X_2, then is down for a time Y_2, and so on. Suppose that the machine is as good as new after each repair, then this constitutes an alternating renewal process.

8.2.2.2 Delayed Renewal Process

It's not always appropriate to demand that the first renewal takes place at $S_1 = X_1$. For example, in situations when the occurrences are irregular, when a system component needs to be replaced, it is possible that one will not be found. Consider the scenarios when a working component already exists in position at time 0. For this reason, a delayed renewal process is defined to be a sequence $S_0, S_1, S_2, \ldots$ where

$$S_n = S_0 + \sum_{i=1}^{n} X_i \text{ and}$$

(i) the inter-arrival times $X_1, X_2, \ldots$ are positive and IID RV, as in an ordinary renewal process, and
(ii) the initial delay $S_0 \geq 0$ is independent of the interarrival times X_i.

The initial delay RV S_0 does not have to have the same distribution as the inter-arrival time RVs X_i. As a result, a delayed renewal process is one in which the initial arrival time, $X_1 = t_1$, is allowed to have a different distribution $P\{X_1 \leq x\} = F_1(x); x \geq 0$ than F, the distribution of all the remaining IID inter-arrival times $\{X_n, n \geq 2\}$. The renewal process is considered to be a non-delayed variant when there is no such delay, that is, when $X_1 \sim F$ as usual.

8.2.2.3 Reward Renewal Process

Let X_1 be the time to the first renewal and let $\{X_n, n = 2, 3, \ldots\}$ be the time between $(n-1)$th renewal and nth renewal. Assume that $\{X_n, n = 1, 2, \ldots\}$ are IID RVs with distribution function F. Let $\mu = E(X_n) = \int_0^\infty x dF(x)$ which will be positive. Let R_n be nth reward at the time of the nth renewal. Usually, R_n may depend on X_n. Let

$$R(t) = \sum_{n=1}^{N(t)} R_n,$$

be the reward earned by time t and $N(t)$ counts the number of renewals by time t. Note that unlike the X_n, each R_n may take negative values as well as positive values.

Example 8.6 Consider an age replacement model. Let X be the lifetime of the component, which is random with distribution function F. The component is replaced by a new one upon failure or at a fixed time period T, whichever comes first. This replacement policy is called an age replacement policy. The cost of a new component is c_1 and the additional cost incurred by a failure is c_2. Find the long-run average cost.

Solution: Let $N(t)$ be the number of replacements of components by time t. Let $R(t)$ be the amount of cost incurred by time t, then $\{R(t), t \geq 0\}$ becomes a reward renewal process. The long-run average cost can be expressed as

$$\text{Long-run average cost} = \frac{E[R]}{E[X]} = \frac{E[\text{cost during a cycle}]}{E[\text{cycle length}]}.$$

Now,

$$E[\text{cycle length}] = E[min(X, T)] = \int_0^T (1 - F(x))dx.$$

Given that, the reward or cost during a cycle, R, as:

$$R = \begin{cases} c_1, & X > T \\ c_1 + c_2, & X \leq T \end{cases}.$$

Hence, the expected value is

$$E[R] = c_1 P\{X > T\} + (c_1 + c_2)P\{X \leq T\} = c_1 + c_2 F(T).$$

The required long-run average cost is given by

$$\frac{E[R]}{E[X]} = \frac{E[\text{cost during a cycle}]}{E[\text{cycle length}]} = \frac{c_1 + c_2 F(T)}{\int_0^T (1 - F(x))dx}.$$

8.3 Hawkes Process

Hawkes processes are a non-Markov extension of homogeneous Poisson processes whose intensity is dependent on both time and history, implying that an arrival has an impact on the probability of next arrivals.

The British Statistician, Alan G. Hawkes (14 July 1938-9 November 2023), first proposed these processes in 1971 with the goal of modelling earthquakes. His plan was to simulate the replica effects after an earthquake. However, these processes have a wide range of application in various fields including finance, criminology, neurology, genome analysis, and sociology.

Definition 8.2 (Hawkes Process) A counting process $\{N(t), t \geq 0\}$ is a Hawkes process if $N(0) = 0$ and

$$P\{N(t+h) = n|N(t) = n\} = 1 - \lambda(t)h + o(h), \; \forall\, n = 0, 1, \ldots$$
$$P\{N(t+h) = n|N(t) = n-1\} = \lambda(t)h + o(h), \; \forall\, n = 1, 2, \ldots$$
$$P\{N(t+h) = n|N(t) = n-k\} = o(h), \; \forall\, n : n-k = 0, 1, \ldots \; k = 2, 3, \ldots.$$

The intensity function λ is given by

$$\lambda(t) = \mu + \int_0^t h(t-u)dN(u)$$

where $h : [0, \infty) \to [0, \infty)$ is called excitation function (or memory kernel) and μ is the baseline intensity, h reflects the influence of past events on the conditional intensity. The general form of this definition is given in Definition 11.11 in Sect. 11.

A typical sample path of the Hawkes process and the corresponding intensity is shown in Fig. 8.3.

Hawkes process is a self-exciting process as the arrival of an event causes the conditional intensity function to increase. Empirical observations in several fields in periods like earthquakes, finance distress, crime and riots and social media trends, motivate to develop models act to capture feedback effects and contagion effects. A natural choice in this context is to model these phenomena using Hawkes processes. These are a type of point processes suitable for modeling phenomena of self-excitation and mutual excitation, i.e., situations in which the occurrence of an event makes future events more likely to happen. To capture self-exciting properties as well as clustering effects, Hawkes processes have wide applications in cosmology, ecology, epidemiology, seismology, neuroscience and DNA modeling.

On the contrary, the simplest case of point processes, the so-called time-homogeneous Poisson processes, assumes that the arrival of an event follows a constant rate, which is the intensity function, and that the occurrence of a new event is independent of the occurrence of previous events. This is the reason why Hawkes process is a non-Markov extension of a Poisson process.

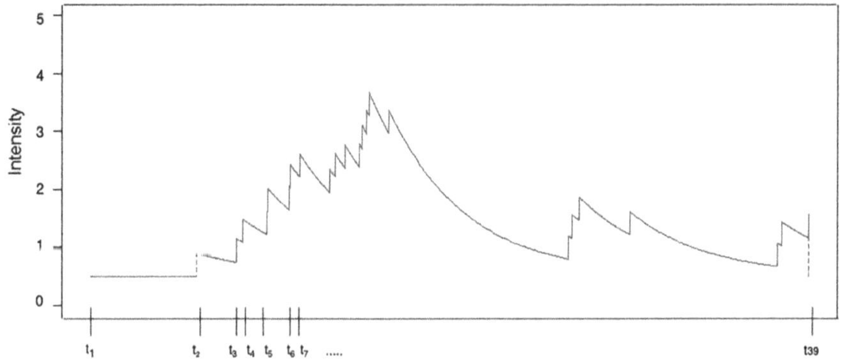

(a) Sample path of Intensity process

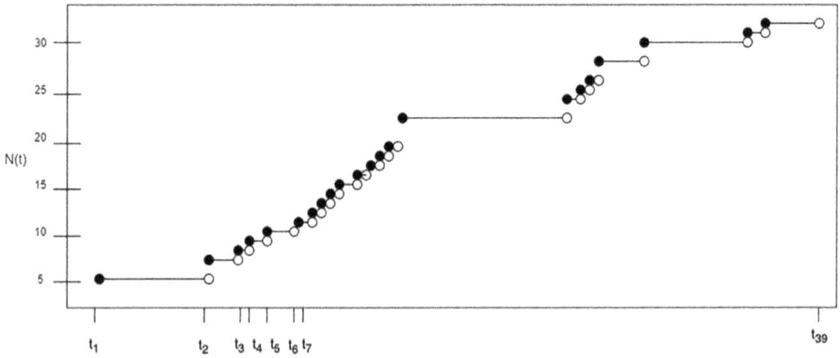

(b) Sample path of Hawkes process

Fig. 8.3 Sample path of Hawkes process and the corresponding intensity

Example 8.7 Let $\{N(t), t \geq 0\}$ be a Hawkes process with conditional intensity $\lambda(t)$ given by

$$\lambda(t) = \mu + \int_0^t h(t-u)dN(u).$$

Let $h(t) = \alpha e^{-\beta t}$ where $\alpha, \beta > 0$ and $\frac{\alpha}{\beta} < 1$. Find $E[\lambda(t)]$ for large t.
Solution: Consider

$$f(t) = E[\lambda(t)] = \mu + \int_0^t \alpha e^{-\beta(t-u)} E[dN(u)]. \tag{8.1}$$

By using $E[dN(t)] = E[\lambda(t)]dt$, (8.1) becomes

$$f(t) = \mu + \int_0^t \alpha e^{-\beta(t-u)} f(u)du.$$

Using Laplace transform, it is obtained that

$$f(t) = E[\lambda(t)] \to \frac{\mu}{1 - \frac{\alpha}{\beta}} \text{ as } t \to \infty.$$

8.4 Semi-Markov Processes

Consider a system $\{Y(t), t \geq 0\}$ with state space $S \in \{0, 1, \ldots\}$. Assume that at time t_0, the system is in the state X_0. It stays there for a non-negative random period of time (which may follow a general distribution), and then at the next transition time instant t_1, the system jumps to state X_1 (which could be the same as X_0). The process $Y(t)$ stays in that state for a non-negative random period of time before jumping to state X_2 at the next transition time instant t_2, and so on. As a result, X_n represents the state of the system at the time of the n-th transition t_n.

If $Y(t)$ denotes the state of the system at time t, then $Y(t_n) = X_n$ for $n \in \{0, 1, \ldots\}$. If the Markov property is satisfied at all the transition time instants $t_n,\ n \in \{0, 1, \ldots\}$, then the stochastic process $\{Y(t),\ t \geq 0\}$ described above is called a semi-Markov process (SMP). That is, the evolution of the system after the time instant $t = t_n$ depends only on the past history of the system till time t_n.

In other words, if $\{Y(t),\ t \geq 0\}$ is a SMP, then the process $\{Y(t + t_n),\ t > 0\}$ is independent of $\{Y(u),\ 0 \leq u \leq t_n\}$ given the complete past history $\{Y(u),\ 0 \leq u \leq t_n\}$ and $X_n = i,\ i \in S$, and is identical to the process $\{Y(t),\ t \geq 0\}$ given $X_0 = i$ because of time homogeneity.

Let T_i denote the time epoch of ith transition whose value is t_i for $i = 1, 2, \ldots$. Thus, in a SMP, the Markov property is satisfied only at each of the transition epochs $\{t_n\}$, and not at all times. The fact that Markov property holds at each of the transition epochs $t_n,\ n \in \{0, 1, \ldots\}$ of the stochastic process $\{Y(t),\ t \geq 0\}$ entails that the Markov property also holds for $\{X_n,\ n \in 0, 1, \ldots\}$. Hence, $\{X_n,\ n = 0, 1, \ldots\}$ turns out to be a time-homogeneous embedded discrete-time Markov chain with the state space S. Then, $\{(X_n, T_n), n = 0, 1, \ldots\}$ constitutes Markov renewal process. The process $\{Y(t), t \geq 0\}$ is not a Markov process, although it inherits some important properties of Markov processes. The associated process $\{X_n,\ n = 0, 1, \ldots\}$ is a Markov process. Hence, the name to $\{(X_n, T_n), n = 0, 1, \ldots\}$ as a Markov renewal process (MRP). The sample paths of SMP and embedded DTMC is shown in Fig. 8.4.

To summarize, a SMP process is a continuous-time stochastic process in which the embedded jump chain (the discrete process registering what values the process takes) is a Markov chain and the holding times (time between jumps) are RVs with any distribution, whose distribution function may depend on the two states between which the move is made. SMP has a renewal property and are non-Poissonian. This

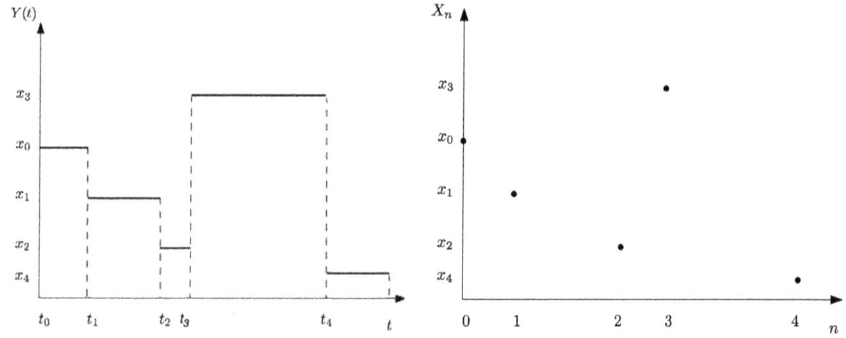

(a) SMP (b) Embedded DTMC

Fig. 8.4 Sample paths of SMP and Embedded DTMC

indicates that the likelihood of a leap from state i to state j at a given time is solely determined by the states i, j and the time t since the last jump.

Remark 8.2 A CTMC is a special case of semi-Markov process in which all of the holding times before moving into any other states are independent exponential distributed.

8.4.1 Transient Analysis

Let $0 = T_0 < T_1 < T_2 < \ldots$ be the time instants of successive state transitions in the stochastic process represented in Fig. 8.4, and the RVs $X_0, X_1, X_2, \ldots$ denote the states of the system at times T_i, $i = 0, 1, 2, \ldots$. Since at $T_0, T_1, \ldots$ the probability of the future behavior of the process, when its current state is known, is not altered by additional knowledge concerning its past, these instants are called Markov renewal moments, and the bivariate stochastic process $(\boldsymbol{X}, \boldsymbol{T}) = \{(X_n, T_n), n = 0, 1, 2, \ldots\}$ is a MRP since it satisfies

$$\begin{aligned} P\left\{X_{n+1} = j, T_{n+1} - T_n \le t \mid X_n = i, X_{n-1} = x_{n-1}, \cdots, X_0 = x_0; T_n, \ldots T_0\right\} \\ = P\left\{X_{n+1} = j, T_{n+1} - T_n \le t \mid X_n = i\right\} \end{aligned} \tag{8.2}$$

for all n = 0, 1, ... ; i, j = 1, 2, 3; and $t \ge 0$. Furthermore, since state transitions only happen at Markov renewal moments, the underlying stochastic process $\boldsymbol{Y(t)} = \{Y(t), t \ge 0\}$ is a SMP with embedded Markov renewal sequence $(\boldsymbol{X}, \boldsymbol{T})$. The states of the SMP are defined by

$$Y(t) = X_n, \quad \text{if } \ T_n \le t < T_{n+1}.$$

The time-homogeneous SMP $\boldsymbol{Y}(t)$ is determined by a vector of initial state probabilities $\boldsymbol{p}(0) = [P\{Y(0) = i\}]_{i \in S}$, and the kernel matrix $\boldsymbol{K}(t) = \left[K_{ij}(t)\right]$, where

$$K_{ij}(t) = P\{X_1 = j, T_1 \le t \mid X_0 = i\},$$

are the conditional transition probabilities of the embedded MRP.

The transient solution of the SMP proceeds by determining the matrix of conditional transition probabilities $\mathbf{V(t)} = \left[V_{ij}(t)\right]$ with

$$V_{ij}(t) = P\{Y(t) = j \mid Y(0) = i\}.$$

The conditional probabilities $V_{ij}(t)$ satisfy the equations

$$V_{ij}(t) = [1 - H_i(t)]\,\delta_{ij} + \sum_{k=1}^{|S|} \int_0^t V_{kj}(t-\tau)\mathrm{d}K_{ik}(\tau), \tag{8.3}$$

where δ_{ij} is equal to 1 if $i = j$ and is 0 otherwise, and H_i is the distribution of the sojourn time in state i

$$H_i(t) = P\{T_1 \le t | X_0 = i\} = \sum_{j=1}^{|S|} K_{ij}(t).$$

The coupled set of Volterra equations of the second kind is a Markov renewal equation and can be solved in Laplace–Stieltjes domain.

$$\boldsymbol{V}^*(s) = \left[\boldsymbol{I} - \boldsymbol{K}^*(s)\right]^{-1} \boldsymbol{G}^*(s) \tag{8.4}$$

with $\boldsymbol{G}^*(s) = \int_0^\infty e^{-st}\,\mathrm{d}\boldsymbol{G}(t)$ where $\boldsymbol{G}(t)$ is a diagonal matrix with entries $G_{ii}(t) = 1 - H_i(t)$, and $\boldsymbol{K}^*(s) = \int_0^\infty e^{-st}\,\mathrm{d}\boldsymbol{K}(t)$. Taking the inverse Laplace–Stieltjes transform of $V^*(s)$, the unconditional state probabilities $\boldsymbol{p}(t) = [p_i(t)]$ can be determined as

$$\boldsymbol{p}(t) = \boldsymbol{p}(0)\boldsymbol{V}(t).$$

8.4.2 *Steady-State Analysis*

The analysis of an SMP is performed in two stages. In the first stage, the SMP stays in a state $i,\ i \in S$ for a random amount of time. Let us consider that the time spent in state i follows a general distribution with distribution function $H_i(t)$ (sojourn time distribution). In the second stage, the SMP moves from state i to state j with probability p_{ij}

$$p_{ij} = P\{X_{n+1} = j | X_n = i\};\ i,\ j \in S.$$

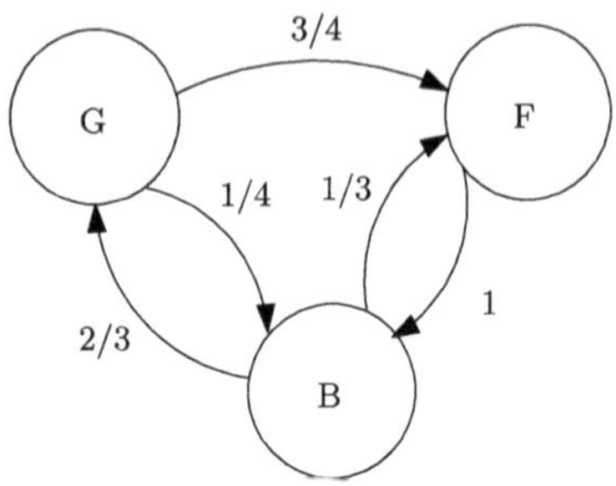

Fig. 8.5 State transition diagram for the EMC of Example 8.8

The SMP can now be completely described by the vector of sojourn time distributions $H(t)$ and the transition probability matrix $\mathbf{P} = [p_{ij}]$.

To compute the steady-state probability vector $\pi = [\pi_0, \pi_1, \ldots]$ of the SMP, the following three steps are used: the first step is to calculate the mean sojourn time $h_i = \int_0^\infty (1 - H_i(t))dt$ for each state i; the next step is to find the steady-state probability vector $\mathbf{v} = [\nu_0, \nu_1, \ldots]$ for the EMC of the SMP by solving the following system of equations:

$$\mathbf{v} = \mathbf{vP} \text{ and } \sum_{\mathbf{i}\in\mathbf{S}} \nu_\mathbf{i} = \mathbf{1}.$$

Final step is to compute the steady-state probabilities for the SMP as given below:

$$\pi_i = \frac{v_i\ h_i}{\sum_{j\in S} v_j\ h_j}\ , i \in S. \tag{8.5}$$

Example 8.8 Consider a machine that can be in one of three states: good condition (G), fair condition (F), or broken down (B). Suppose that a machine in good condition will remain this way for a mean time 5 and then will go to either the fair condition or the broken condition with respective probabilities $\frac{3}{4}$ and $\frac{1}{4}$. A machine in fair condition will remain that way for a mean time 2 and then will break down. A broken machine will be repaired, which takes a mean time 1, and when repaired will be in good condition with probability $\frac{2}{3}$ and fair condition with probability $\frac{1}{3}$. Assume that, the time spent in all the three states are non-exponential distributed (or general distribution) and are independent of other states. What proportion of time is the machine in each state in long run? (Fig. 8.5)

Solution: Since this process satisfies the conditions for steady-state distributions as given above, the steady-state probabilities can be computed. Let the states be 1, 2, 3. v_i satisfies

$$\begin{aligned} v_1 &= \frac{2}{3}v_3, \\ v_2 &= \frac{3}{4}v_1 + \frac{1}{3}v_3, \\ v_3 &= \frac{1}{4}v_1 + v_2 \end{aligned}$$

with $v_1 + v_2 + v_3 = 1$. The solution is

$$v_1 = \frac{4}{15}, \quad v_2 = \frac{1}{3}, \quad v_3 = \frac{2}{5}.$$

Hence, from (8.5), it is obtained that π_i, the proportion of time the machine is in state i, $i \in \{1, 2, 3\}$ is given by

$$\begin{aligned} \pi_1 &= \frac{4h_1}{4h_1 + 5h_2 + 6h_3}, \\ \pi_2 &= \frac{5h_2}{4h_1 + 5h_2 + 6h_3}, \\ \pi_3 &= \frac{6h_3}{4h_1 + 5h_2 + 6h_3}. \end{aligned}$$

For a given $h_1 = 5, h_2 = 2, h_3 = 1$, the machine will be in good condition $\frac{5}{9}$ of the time, in fair condition $\frac{5}{18}$ of the time, in broken condition $\frac{1}{6}$ of the time.

Example 8.9 Consider a stochastic process $\{Y(t), t \geq 0\}$ with state space $S = \{0, 1, 2\}$. $\{X_n, n = 0, 1, \ldots\}$ is a SMP process where $X_n = Y(t_n), n = 0, 1, \ldots$. Assume that the time spent in state 0 follows exponential distribution with distribution function $H_0(t)$ (sojourn time distribution) and is given by

$$H_0(t) = \begin{cases} 0, & t < 0 \\ 1 - e^{-2t}, & t \geq 0 \end{cases}$$

Assume that the time spent in states 1 and 2 follow general distributions with distribution function $H_1(t)$ and $H_2(t)$ (sojourn time distribution) respectively and are given by

$$H_1(t) = \begin{cases} 0, & t < 1 \\ t-1, & 1 \leq t < 2 \\ 1, & 2 \leq t < \infty \end{cases} ; \quad H_2(t) = \begin{cases} 0, & t < 2 \\ t-2, & 2 \leq t < 3 \\ 1, & 3 \leq t < \infty \end{cases}.$$

Therefore, the SMP moves from state i to state j with probability p_{ij}, i.e.,

$$p_{ij} = P\{X_{n+1} = j | X_n = i\}; \; i, \; j \in S.$$

The transition probability matrix $\mathbf{P} = [p_{i,j}]$ is given by

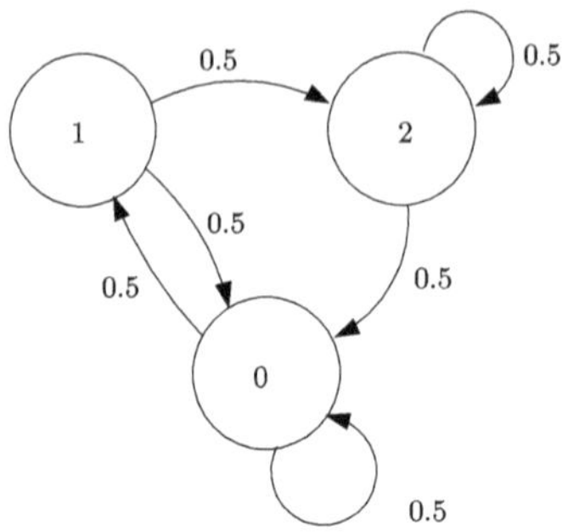

Fig. 8.6 State transition diagram of Example 8.8

$$\mathbf{P} = \begin{pmatrix} 0.5 & 0.5 & 0 \\ 0.5 & 0 & 0.5 \\ 0.5 & 0 & 0.5 \end{pmatrix}.$$

Find the steady-state probabilities for this system.
Solution: The state transition diagram for the EMC is shown in Fig. 8.6. Solving

$$\mathbf{v} = \mathbf{vP} \text{ and } \sum_{\mathbf{i} \in \mathbf{S}} \mathbf{v_i} = \mathbf{1},$$

the probabilities are obtained as

$$v_0 = 0.5, \; v_1 = v_2 = \frac{1}{4}.$$

Here, h_i can be computed through the help of distribution function $H_1(t)$, $H_2(t)$ and $H_3(t)$. Since $H_0(t)$ is exponential distributed, the mean will be $1/2$. $H_1(t)$ and $H_2(t)$ are uniformly distributed over interval $(1, 2)$ and $(2, 3)$, respectively. Therefore, the mean will be $3/2$ and $5/2$, respectively. Hence,

$$h_0 = \frac{1}{2}, \; h_1 = \frac{3}{2}, \; h_2 = \frac{5}{2}.$$

Finally, compute the steady-state probabilities for the SMP as given below

$$\pi_i = \frac{v_i \, h_i}{\sum_{j \in S} v_j \, h_j}, \quad i \in S.$$

Required steady-state probabilities are given by

$$\pi_0 = 1/5, \;\; \pi_1 = 3/10, \;\; \pi_2 = 1/2.$$

Example 8.10 Consider Example 4.16 with some modification. Assume that the transition from state 2 to state 1 follow general distribution with CDF $F(t)$ and mean 1. Find the steady-state distribution, if it exists.

Solution: Since the transition from state 2 to state 1 follows general distribution, $\{X(t), t \geq 0\}$ is a SMP with state space $S = \{1, 2, 3\}$. $\mathbf{P}$ is the one-step transition probability matrix and $\mathbf{Q}$ is the infinitesimal generator matrix. Take $q \geq max_i |q_{ii}|, i \in S$. Hence, $\mathbf{P} = \mathbf{I} + \frac{1}{q}\mathbf{Q}$. For the given generator matrix in Example 4.16, the transition probability matrix $\mathbf{P}$ is computed as

$$\mathbf{P} = \begin{pmatrix} 0 & 1/2 & 1/2 \\ 1 & 0 & 0 \\ 2/3 & 1/3 & 0 \end{pmatrix}.$$

Now, the steady-state probabilities of the system is computed by solving $\mathbf{v} = \mathbf{vP}$ with $\sum v_i = 1$, $(v_1, v_2, v_3) = (6/13, 4/13, 3/13)$. The values of (h_1, h_2, h_3) are given by using the mean sojourn time spent by the process in state i, $(h_1, h_2, h_3) = (1/2, 1, 1/3)$. Finally, compute the steady-state probabilities for the SMP as given below:

$$\pi_i = \frac{v_i\ h_i}{\sum_{j \in S} v_j\ h_j}\ , i \in S.$$

Thus,

$$\pi_1 = 3/8,\ \ \pi_2 = 4/8,\ \ \pi_3 = 1/8.$$

Note that, the steady-state distribution is the same as the steady-state distribution of Example 4.16.

Example 8.11 Consider Example 4.24 with distribution of sojourn times as given in 8.1. In this model, I_0 denotes the system is in idle period with zero packets, i.e., the system waiting for packets, W_k denotes the UE is in working period with k number of packets, i.e., $(k, 0)$th state, S_j denotes the UE is in jth short cycle, i.e., $(0, j)$th state and L_i denotes the UE is in long sleep with i number of packets to serve, i.e., $(i, N+1)$th state, where $k = 1, 2, \ldots$, $i = 0, 1, \ldots$ and $j = 1, 2, \ldots, N$. Let $F_{m,n}(t)$ denotes the CDF associated with transition from state m to n, where $m, n \in S$, $S = \{I_0, W_k, S_j, L_i, k = 1, 2, \ldots; i = 0, 1, \ldots; j = 1, 2, \ldots, N\}$. The distribution for sojourn times for states is given in Table 8.1. As the transition from one state to another state is influenced by many factors and hence exhibits random behavior. In this process, the distribution of transition from one state to another is influenced by real-time scenario of DRX mechanisms and hence, sojourn time of some states follows deterministic distribution. Hence, time epochs posses the Markov property. Therefore, this stochastic process was modeled as SMP. Find the steady-state probabilities for this model.

Solution: The steady-state probabilities for the semi-Markov model are given as:

Table 8.1 Distributions of sojourn times in a semi-Markov model

CDF	Distribution	Parameter
$F_{I_0,S_1}(t)$	Deterministic	$\frac{1}{\beta}$
$F_{S_j,S_{j+1}}(t)$	Deterministic	$\frac{1}{\alpha}$
$F_{S_N,L_0}(t)$	Deterministic	$\frac{1}{\alpha}$
$F_{W_k,W_{k+1}}(t)$	Exponential	λ
$F_{W_1,I_0}(t)$	Exponential	μ
$F_{W_k,W_{k-1}}(t)$	Exponential	μ
$F_{S_j,W_1}(t)$	Exponential	λ
$F_{L_i,L_{i+1}}(t)$	Exponential	λ
$F_{L_0,I_0}(t)$	Deterministic	$\frac{1}{\nu}$
$F_{L_i,W_i}(t)$	Deterministic	$\frac{1}{\nu}$

$$\pi_{I_0} = \frac{1-e^{-\frac{\lambda}{\beta}}}{\lambda M},$$

$$\pi_{W_1} = \frac{1-e^{-\left(\frac{\lambda}{\beta}+\frac{\lambda N}{\alpha}+\frac{\lambda}{\nu}\right)}}{\mu M},$$

$$\pi_{W_k} = \frac{1}{\mu M}\left[\left(\frac{\lambda}{\mu}\right)^{k-1}\left(1-e^{-\left(\frac{\lambda}{\beta}+\frac{\lambda N}{\alpha}+\frac{\lambda}{\nu}\right)}\right)\right.$$
$$\left.+\frac{\left(\frac{\lambda}{\mu}\right)\left(1-e^{-\frac{\lambda}{\nu}}\right)^2}{\left(\frac{\lambda}{\mu}-1+e^{-\frac{\lambda}{\nu}}\right)}e^{-\left(\frac{\lambda}{\beta}+\frac{\lambda N}{\alpha}\right)}\left[\left(\frac{\lambda}{\mu}\right)^{k-2}-\left(1-e^{-\frac{\lambda}{\nu}}\right)^{k-2}\right]\right],\ k=2,3,\ldots$$

$$\pi_{S_j} = \frac{e^{-\frac{\lambda}{\beta}}\left(e^{-\frac{\lambda}{\alpha}}\right)^{j-1}\left(1-e^{-\frac{\lambda}{\alpha}}\right)}{\lambda M},\ j=1,2,\ldots,N$$

$$\pi_{L_i} = \frac{e^{-\left(\frac{\lambda}{\beta}+\frac{\lambda N}{\alpha}\right)}\left(1-e^{-\frac{\lambda}{\nu}}\right)^{i+1}}{\lambda M},\ i=0,1,\ldots$$

The derivation of steady-state probabilities remains as an exercise for the reader.

8.5 Markov Regenerative Processes

Consider a stochastic process $Z=\{Z(t), t\geq 0\}$ with state space S which is discrete. Here, $Z(t)$ denotes the state of the system at time t. Suppose that the initial state of the system at time t_0 be $Z(0)$. It stays there for a non-negative random amount of time (which has a non-exponential distribution) and then, the next transition takes place at time t_1 to another state. It stays there for non-negative amount of time and then jumps to another state at time t_2, and continues this way forever. Thus, t_n is the time

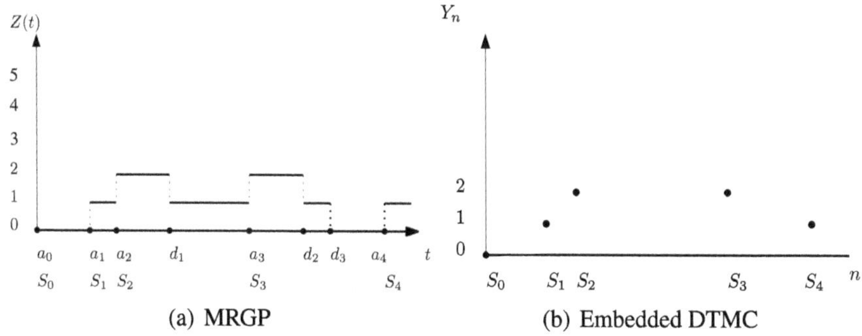

Fig. 8.7 Sample path of MRGP and embedded DTMC

instant of the nth transition. Assume that whenever a transition occurs, the process future Z after that time becomes a probabilistic replica of the process future after time zero. Such times (which are usually random) are referred to as Z regeneration time instants. The set of states where the system reaches at regeneration time instants is denoted by S' and $S' \subseteq S$. This is referred to as a regenerative process.

Markov Regenerative Theory

A stochastic process is mathematically tractable if it has the regeneration time instants where the process is memoryless. Its regeneration time instants are the names given to this time intervals. The stochastic evolution between two subsequent regeneration time instants in a Markov regenerative process (MRGP) depends only on the state at regeneration time instants, not on the history before regeneration time instants. Furthermore, because the embedded Markov renewal mechanism is time homogeneous, the evolution of the MRGP becomes a probabilistic replica after each regeneration time instant. As a result, at a regeneration time instant, all memory except the state must be reset.

For instance, consider a single-server non-Markov queueing model with interarrival follows non-exponential distribution and service follows exponential distribution. Figure 8.7 shows the sample path of this queueing model. From this sample path, it is observed that arrival time points $(S_0, S_1, \cdots)$ are regeneration time points whereas the service completion are not regeneration time points.

If S_n is the time of nth transition, then $Y_n = Z(S_n)$ for $n \epsilon \{0, 1, \cdots\}$. The concepts of an MRGP are given in the next two definitions.

A sequence of bivariate RVs $\{(Y_n, S_n), n = 0, 1, \ldots\}$ is called a *Markov renewal sequence* if:

(i) $S_0 = 0$, $S_{n+1} \geq S_n$; $Y_n \in S'$ and

(ii) for all $n = 0, 1, \ldots,$

$$\begin{aligned} P\{Y_{n+1} = j, \quad & S_{n+1} - S_n \leq t \mid Y_n = i, S_n, Y_{n-1}, S_{n-1}, \ldots, Y_0, S_0\} \\ &= P\{Y_{n+1} = j, S_{n+1} - S_n \leq t \mid Y_n = i\} \text{ (Markov Property)} \\ &= P\{Y_1 = j, S_1 \leq t \mid Y_0 = i\}. \text{ (Time Homogeneity)} \end{aligned} \tag{8.6}$$

The MRGP is defined as follows:
A stochastic process $\{Z(t), t \geq 0\}$ on S is called an MRGP if there exists a Markov renewal sequence $\{(Y_n, S_n), n = 0, 1, \ldots\}$ of RVs such that all conditional finite dimensional distributions of $\{Z(S_n + t), t > 0\}$ given $\{Z(u), 0 \leq u \leq S_n, Y_n = i\}$ are the same as those of $\{Z(t), t \geq 0\}$ given $\{Y_0 = i, i \in S' \subseteq S\}$.
Note that the above definition implies that in this case $\{Z(S_n^+), n = 0, 1, \ldots\}$ or $\{Z(S_n^-), n = 0, 1, \ldots\}$ is an *embedded discrete-time Markov chain* (DTMC) or just the *embedded Markov chain* (EMC) in $\{Z(t), t \geq 0\}$, and also that S_n is a stopping time (regeneration points) of $\{Z(t), t \geq 0\}$. Figure 8.7 shows the sample path of MRGP which arises in a single-server non-Markovian queueing model with inter-arrival follows non-exponential distribution and service follows exponential distribution. From this sample path, it is observed that arrival time points $(S_0, S_1, \ldots)$ are regeneration time points whereas the service completion are not regeneration time points.
Global and Local Kernels
As a special case, the definition implies that

$$P\{Z(S_n + t) = j \mid Z(u), 0 \leq u \leq S_n, Y_n = i\} = P\{Z(t) = j \mid Y_0 = i\}.$$

The conditional probability is obtained in (8.6) by $k_{i,j}(t),\ i, j \in S'$. The matrix $\mathbf{K(t)} = [k_{i,j}(t)]$ is called the *global kernel* of the Markov renewal sequence. Define the matrix $e_{i,j}(t),\ i \in S',\ j \in S$, as follows:

$$e_{i,j}(t) = P\{Z(t) = j, S_1 > t \mid Y_0 = i\}.$$

This matrix $\mathbf{E(t)} = [e_{i,j}(t)]$ describes the behavior of the MRGP between two transition epochs of the EMC, i.e., over the time interval $[0, S_1)$. The matrix $\mathbf{E(t)}$ is called the *local kernel*.

8.5.1 *Steady-State Analysis*

Assume that $\{Z(t), t \geq 0\}$ has ergodic property such that the limiting distribution is the same as the steady-state distribution. The limiting behavior of the MRGP is studied by taking the limit as t approaches infinity. Two new quantities are required to be defined, namely the mean time $\alpha_{i,j}$ the MRGP spends in state j between two successive regeneration instants, given that it started in state i after the last regeneration:

$$\alpha_{i,j} = E[\text{time in } j \text{ during } (0, S_1) \mid Y_0 = i] = \int_0^\infty E_{i,j}(t)dt, \tag{8.7}$$

and the steady-state probability vector $\mathbf{v} = (\nu_k)$ of the Embedded Markov Chain (EMC) satisfy

$$\mathbf{v} = \mathbf{vP}, \quad \sum_{k \in S'} \nu_k = 1, \tag{8.8}$$

where $\mathbf{P} = K(\infty)$ is the one-step transition probability matrix of the EMC.

The following theorem describes the limiting behavior of MRGP. Let $\{Z(t), t \geq 0\}$ be a MRGP on S with Markov renewal sequence $\{(Y_n, S_n), n = 0, 1, \ldots\}$ with kernel $K(\cdot)$. Let $N(t)$ denote the total number of state changes by time t, i.e., $N(t) = \sup\{n \geq 0 : S_n \leq t\}$. Suppose that

(i) The sample paths of $\{Z(t), t \geq 0\}$ are right continuous with left limits exist,
(ii) The semi-Markov process $\{Y_n, n = 0, 1, \ldots\}$ is irreducible, aperiodic, and positive recurrent,
(iii) $\nu = (\nu_k)$ is a positive solution to (8.7).

Then the steady-state probability of the MRGP is given by

$$\pi_j = \lim_{t \to \infty} P\{Z(t) = j\} = \frac{\sum_{k \in S'} \nu_k \alpha_{k,j}}{\sum_{k \in S'} \nu_k \beta_k}, \; j \in S \tag{8.9}$$

where

$$\beta_k = \sum_{l \in S} \alpha_{k,l}.$$

8.5.2 Transient Analysis

Now, let us discuss the time dependent behavior of the MRGP. Define $\mathbf{V(t)} = [V_{ij}(t)]$ as the conditional state probabilities of the MRGP given as

$$V_{ij}(t) = P\{Z(t) = j \mid Z(0) = Y_0 = i\}; \quad i, j \in S$$

where $\{Y_n, \; n = 0, 1, \ldots\}$ is the EMC of the MRGP $\{Z(t), \; t \geq 0\}$. Also,

$$\begin{aligned}\mathbf{V(t)} &= \mathbf{E(t)} + \int_0^t dK(s)V(t-s), \\ &= \mathbf{E(t)} + \mathbf{K(t)} * \mathbf{V(t)}.\end{aligned}$$

The solution method is outlined below:

1. Calculate $\mathbf{K(t)}$ and $\mathbf{E(t)}$.

2. Compute $\tilde{\mathbf{K}}(\mathbf{s})$ and $\tilde{\mathbf{E}}(\mathbf{s})$ where $\tilde{\mathbf{K}}(\mathbf{s})$ and $\tilde{\mathbf{E}}(\mathbf{s})$ are the Laplace–Stieltjes transform obtained as:

$$\tilde{\mathbf{K}}(\mathbf{s}) = \int_0^\infty e^{-st} dK(t) \text{ and } \tilde{\mathbf{E}}(\mathbf{s}) = \int_0^\infty e^{-st} dE(t).$$

3. Solve the following linear system for $\tilde{\mathbf{V}}(\mathbf{s})$

$$[\mathbf{I} - \tilde{\mathbf{K}}(\mathbf{s})]\tilde{\mathbf{V}}(\mathbf{s}) = \tilde{\mathbf{E}}(\mathbf{s}),$$

   where

$$\tilde{\mathbf{K}}(\mathbf{s}) = \int_0^\infty e^{-st} dV(t).$$

4. Invert $\tilde{\mathbf{V}}(\mathbf{s})$ to obtain $V(t)$.
5. Using $\mathbf{P(t)} = \mathbf{P(0)V(t)}$, obtain the time dependent state probabilities of the MRGP.

Example 8.12 Consider Example 2.11 with the state space $S = \{0, 1, 2\}$, where 0, 1, and 2 represent no component, one component, and two components working in the system at any time t respectively. Consider two different scenarios based on inter-arrival time and service time having non-exponential distributions with the same mean holding times. Find the steady-state probabilities for these two scenarios.
Solution:
Case 1: Consider that the interarrival time is general distributed with CDF $F_a(t)$ and service time is exponential distributed with rate μ.

Following the theory of MRGP, the global kernel $\mathbf{K(t)} = [k_{ij}(t)]$ and local kernel $\mathbf{E(t)} = [e_{ij}(t)]$ matrices for the process will be determined. The elements of the global and local kernels are defined as

$$k_{ij}(t) = P\{Y_1 = j,\ S_1 \leq t \mid X_0 = i\};\quad i, j \in S'$$

$$e_{ij}(t) = P\{Y_1 = j,\ S_1 \geq t \mid X_0 = i\}:\quad i \in S',\ j \in S$$

where $\{(Y_n,\ S_n),\ n = 0, 1, \ldots\}$ is a Markov renewal sequence, $S = \{0, 1, 2\}$ is the set of states at all time instants, and $S' = \{1, 2\}$ is the set of states only at regeneration time instants. The state transition diagram is shown in Fig. 8.8. Here, the transition with dotted arcs means that the system moves from state 0 to state 1 as well as from state 1 to state 2 with residual inter-arrival time. Thus, the state 0 is not a regeneration state which is shown within the square box.

The global kernel matrix is

$$\mathbf{K(t)} = \begin{pmatrix} k_{11}(t) & k_{12}(t) \\ k_{21}(t) & k_{22}(t) \end{pmatrix},$$

where

$$k_{ij}(t) = \begin{cases} \int_0^t P_{ij-1}(x)dF_a(x), & i, j \in S',\ i = j \neq 2 \\ \int_0^t [P_{21}(x) + P_{22}(x)]\, dF_a(x), & i, j = 2 \end{cases}$$

and $P_{ij}(t)$'s are the transition probabilities of the subordinated CTMC given as

$$\mathbf{P(t)} = \begin{pmatrix} 1 & 0 & 0 \\ 1 - e^{-\mu t} & e^{-\mu t} & 0 \\ 1 - e^{-\mu t} - \mu t e^{-\mu t} & \mu t e^{-\mu t} & e^{-\mu t} \end{pmatrix}.$$

Therefore, the elements of global kernel matrix $k_{ij}(t)$ are computed using the theory of MRGP as described above.

The local kernel matrix is

$$\mathbf{E(t)} = \begin{pmatrix} e_{10}(t) & e_{11}(t) & 0 \\ e_{20}(t) & e_{21}(t) & e_{22}(t) \end{pmatrix}$$

where $e_{ij}(t) = P_{ij}(t)(1 - F_a(t))$. The limiting behavior is obtained by taking the limit as t approaches infinity.

The steady-state probabilities v_k are computed by

$$v = vP, \quad \sum_{k=1}^{2} v_k = 1$$

where $P = K(\infty)$ is the one-step transition probability matrix of the embedded Markov chain.

$$\begin{aligned} v_1 p_{11} + v_2 p_{21} &= v_1, \\ v_1 p_{12} + v_2 p_{22} &= v_2, \\ v_1 + v_2 &= 1. \end{aligned}$$

After simplification

$$v_1 = \frac{p_{21}}{1 - p_{11} + p_{21}}, \tag{8.10}$$

$$v_2 = \frac{1 - p_{11}}{1 - p_{11} + p_{21}}, \tag{8.11}$$

where p_{ij} are obtained by using $P = K(\infty) = [p_{ij}]$. Then the steady-state probabilities of the MRGP are given by

$$\pi_j = \frac{\sum_{k=1}^{2} v_k \alpha_{kj}}{\sum_{k=1}^{2} v_k \beta_k}, \quad j = 0, 1, 2. \tag{8.12}$$

As a special case, consider the system where arrivals follow the Poisson process with rate λ. The CDF of the process is given by $F_a(t) = 1 - e^{-\lambda t}, t \geq 0$. The elements of global kernels are given as

$$\begin{aligned}
k_{11}(t) &= (1 - e^{-\lambda t})(1 - e^{-\mu t}),\\
k_{12}(t) &= (1 - e^{-\lambda t})(e^{-\mu t}),\\
k_{21}(t) &= (1 - e^{-\lambda t})(1 - e^{-\mu t} - \mu t e^{-\mu t}),\\
k_{22}(t) &= (1 - e^{-\lambda t})(\mu t e^{-\mu t}),
\end{aligned}$$

and the elements of the local kernel are given as

$$\begin{aligned}
e_{10}(t) &= (1 - e^{-\mu t})e^{-\lambda t},\\
e_{11}(t) &= e^{-\mu t}e^{-\lambda t},\\
e_{20}(t) &= (1 - e^{-\mu t} - \mu t e^{-\mu t})e^{-\lambda t},\\
e_{21}(t) &= \mu t e^{-\mu t}e^{-\lambda t},\\
e_{22}(t) &= e^{-\mu t}e^{-\lambda t}.
\end{aligned}$$

The values of steady-state probabilities are given as

$$\begin{aligned}
v_1 &= (\lambda + \mu)^2/\lambda\mu(\lambda + \mu - 1),\\
v_2 &= [(\lambda + \mu)\lambda\mu + \mu^2 + \lambda^2 + \lambda\mu](\lambda + \mu)/\lambda\mu(\lambda + \mu - 1).
\end{aligned}$$

α_{ij}'s are given as

$$\begin{aligned}
\alpha_{10} &= \int_0^\infty (1 - e^{-\mu t})e^{-\lambda t}dt = \frac{1}{\lambda + \mu} - \frac{1}{\mu},\\
\alpha_{11} &= \int_0^\infty e^{-\mu t}e^{-\lambda t}dt = -\frac{1}{\lambda + \mu},\\
\alpha_{20} &= \int_0^\infty (1 - e^{-\mu t} - \mu t e^{-\mu t})e^{-\lambda t}dt = \frac{2\mu + \lambda}{(\lambda + \mu)^2} - \frac{1}{\lambda},\\
\alpha_{21} &= \int_0^\infty \mu t e^{-\mu t}e^{-\lambda t}dt = -\frac{\mu}{(\lambda + \mu)^2},\\
\alpha_{22} &= \int_0^\infty e^{-\mu t}e^{-\lambda t}dt = -\frac{1}{\lambda + \mu}.
\end{aligned}$$

Substituting the values of v_i's and α's in the (8.12), the steady-state probabilities are obtained that

$$\pi_0 = \frac{\mu^2(\mu-\lambda)}{(\mu-\lambda)(\mu^2+\lambda^2+\mu\lambda)},$$
$$\pi_1 = \frac{\mu\lambda(\mu-\lambda)}{(\mu-\lambda)(\mu^2+\lambda^2+\mu\lambda)},$$
$$\pi_2 = \frac{\lambda^2(\mu-\lambda)}{(\mu-\lambda)(\mu^2+\lambda^2+\mu\lambda)}.$$

The above results coincide with the results of $M/M/1/2$.

Case 2: Consider that the service time is general distributed with CDF $F_s(t)$ and interarrival time is exponentially distributed with rate λ.

Following the theory of MRGP, the global kernel K(t)= $[k_{ij}(t)]$ and local kernel E(t)= $[e_{ij}(t)]$ matrices for the process will be determined. The elements of the global and local kernel are defined as

$$k_{ij}(t) = P\{X_1 = j,\ T_1 \le t \mid X_0 = i\};\ \ i, j \in S'$$
$$e_{ij}(t) = P\{X_1 = j,\ T_1 \ge t \mid X_0 = i\} :\ \ i \in S',\ j \in S$$

where $\{(X_n,\ t_n),\ n \ge 0\}$ is a Markov renewal sequence, $S = \{0, 1, 2\}$ is the set of states at all time instants, and $S' = \{0, 1\}$ is the set of states only at regeneration time instants. The state transition diagram is shown in Fig. 8.9. Here, the transition with dotted arcs means that, the system moves from state 2 to state 1 with residual service time. Thus, the state 2 is not a regeneration state which is shown within the square box.

The global kernel matrix is

$$K(t) = \begin{pmatrix} k_{00}(t) & k_{01}(t) \\ k_{10}(t) & k_{11}(t) \end{pmatrix},$$

where

$$k_{ij}(t) = \begin{cases} \int_0^t P_{ij+1}(x)dF_s(x), & i, j \in S',\ i = j \ne 2 \\ \int_0^t (P_{00}(x) + P_{01}(x))\, dF_s(x), & i, j = 0 \end{cases}$$

and $P_{ij}(t)$'s are the transition probabilities of the subordinated CTMC given as

$$\mathbf{P(t)} = \begin{pmatrix} e^{-\lambda t} & \lambda t e^{-\lambda t} & 1 - e^{-\lambda t} - \lambda t e^{-\lambda t} \\ 0 & e^{-\lambda t} & 1 - e^{-\lambda t} \\ 0 & 0 & 1 \end{pmatrix}.$$

The local kernel matrix is

$$\mathbf{E(t)} = \begin{pmatrix} e_{00}(t) & e_{01}(t) & e_{02}(t) \\ e_{10}(t) & e_{11}(t) & e_{12}(t) \end{pmatrix},$$

where $e_{ij}(t) = P_{ij}(t)(1 - F_s(t))$.

The limiting behavior is obtained by taking the limit as t approaches infinity. The steady-state probabilities v_k are computed by

$$\nu = \nu P, \quad \sum_{k=1}^{2} v_k = 1$$

where $\mathbf{P} = \mathbf{K}(\infty) = [p_{ij}]$ is the one-step transition probability matrix of the embedded Markov chain.

$$\begin{aligned} v_0 p_{00} + v_1 p_{10} &= v_0, \\ v_0 p_{01} + v_1 p_{11} &= v_1, \\ v_0 + v_1 &= 1. \end{aligned}$$

After simplification

$$v_0 = \frac{p_{10}}{1 - p_{00} + p_{10}}, \tag{8.13}$$

$$v_1 = \frac{1 - p_{00}}{1 - p_{10} + p_{10}}. \tag{8.14}$$

Then the steady-state probability of the MRGP is given by

$$\pi_j = \frac{\sum_{k=0}^{1} v_k \alpha_{kj}}{\sum_{k=0}^{1} v_k \beta_k}, \quad j = 0, 1, 2.$$

As a special case, consider the system where the service time is exponentially distributed with rate μ. The CDF of the process is given by $F_s(t) = 1 - e^{-\mu t}, t \geq 0$. The elements of global kernels are given as

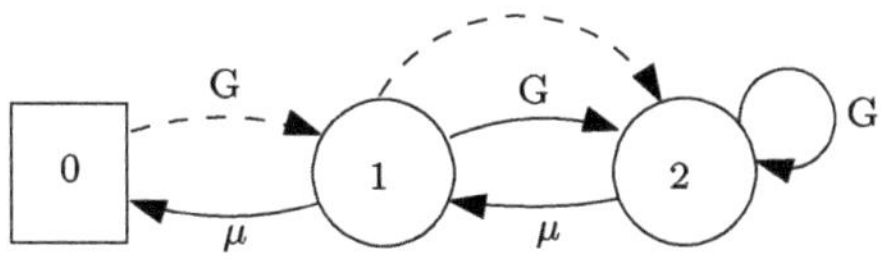

Fig. 8.8 State transition diagram for $GI/M/1/2$

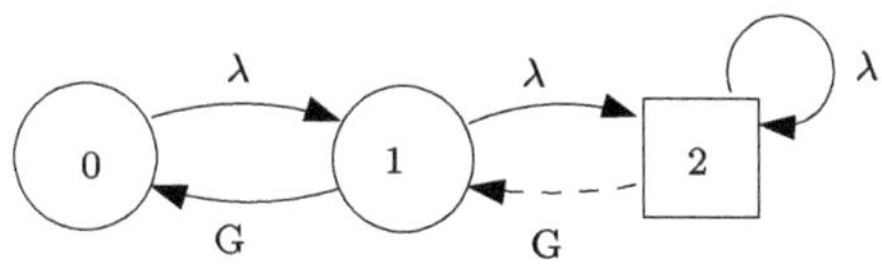

Fig. 8.9 State transition diagram for $M/G/1/2$

$$
\begin{aligned}
k_{00}(t) &= (e^{-\lambda t} + \lambda t e^{-\lambda t})\mu(e^{-\mu t}),\\
k_{01}(t) &= (1 - e^{-\lambda t} - \lambda t e^{-\lambda t})\mu(e^{-\mu t}),\\
k_{10}(t) &= e^{-\lambda t} e^{-\mu t} \mu t,\\
k_{11}(t) &= (1 - e^{-\lambda t})(\mu t e^{-\mu t}),
\end{aligned}
$$

and the elements of local kernel are given as

$$
\begin{aligned}
e_{00}(t) &= e^{-\lambda t - \mu t},\\
e_{01}(t) &= \lambda t e^{-\mu t} e^{-\lambda t},\\
e_{02}(t) &= (1 - e^{-\lambda t} - \lambda t e^{-\lambda t}) e^{-\mu t},\\
e_{11}(t) &= e^{-\mu t} e^{-\lambda t},\\
e_{12}(t) &= e^{-\mu t}(1 - e^{-\lambda t}).
\end{aligned}
$$

The values of steady-state probabilities are given as

$$
\begin{aligned}
v_0 &= \frac{1}{\lambda} + \frac{\lambda\mu}{(\lambda+\mu)^2},\\
v_1 &= \frac{\mu}{\lambda+\mu}.
\end{aligned}
$$

Substituting the values in (8.12), and after simplifying

$$
\begin{aligned}
\pi_0 &= \frac{\mu^2(\mu-\lambda)}{(\mu-\lambda)(\mu^2+\lambda^2+\mu\lambda)},\\
\pi_1 &= \frac{\mu\lambda(\mu-\lambda)}{(\mu-\lambda)(\mu^2+\lambda^2+\mu\lambda)},\\
\pi_2 &= \frac{\lambda^2(\mu-\lambda)}{(\mu-\lambda)(\mu^2+\lambda^2+\mu\lambda)}.
\end{aligned}
$$

The above results coincide with the results of $M/M/1/2$.

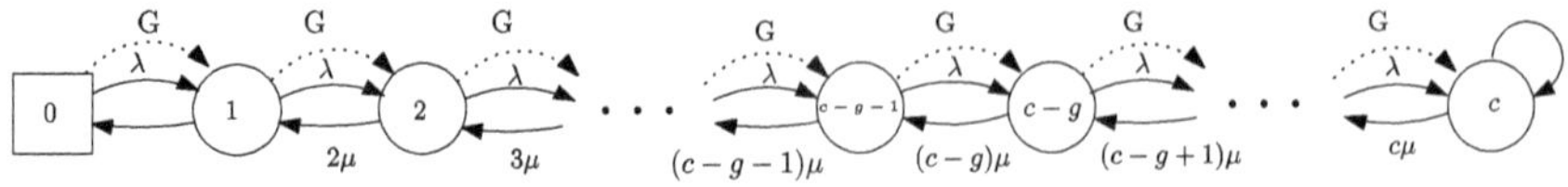

Fig. 8.10 State transition diagram for Example 8.13

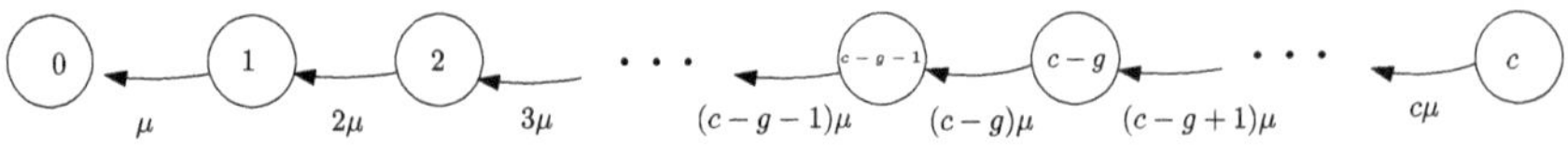

Fig. 8.11 Subordinated CTMC for Example 8.13

Example 8.13 Consider Example 5.5 with non-exponential distributed handoff interarrival times. Find the steady-state and time-dependent probabilities for the system.

Solution: Consider the cellular system state process $\{Z(t), t \geq 0\}$, where $Z(t)$ is the number of busy channels (or ongoing calls) in the system at time t. Clearly, $\{Z(t), t \geq 0\}$ is a continuous-time discrete-state stochastic process, but not a CTMC since sojourn times are not exponential distributed. Furthermore, $\{Z(t), t \geq 0\}$ is not even an SMP since between two successive handoffs call arrivals (that is, between two successive regeneration epochs) system state can change due to new call arrivals, call completions, or calls being handed out of this cell. From the stochastic behavior of handoff arrivals, it can be observed that the underlying process satisfies the Markov property at handoff arrival instants only. Figure 8.10 shows the state transition diagram of the system.

Define S_{n+1} as the time instant of the $(n+1)$th handoff call arrival, where $n \geq 0$. Let $S_0 = 0$ and let $Z(S_n) = i$, $i \in S^{'} = \{1, 2, \ldots, C\}$. Here, the elements of $S^{'}$ are called regeneration points and they occur only at handoff call arrival instants. Note that entry into state 0 of the process is not a regeneration instant. Consider the sequence of epochs $\{S_n, n \geq 0\}$ at which the process $\{Z(t), t \geq 0\}$ is observed. The state of the process $Z(t)$ can change between S_n and S_{n+1}; because of new call arrivals and ongoing call completions. Hence, the sequence $\{(Z(S_n), S_n), n = 0, 1, \ldots\}$ is a Markov renewal sequence and $\{Z(t), t \geq 0\}$ is a MRGP.

The local kernel $E(t)$ and the global kernel $K(t)$ matrices are determined. Since the cardinality of $S^{'}$ is C; the matrix $K(t)$ has entries $k_{i,j}(t)$ that denote the actual state labels according to the state transition diagram. $k_{i,j}(t)$ denotes the probability that the system will be in state $j \in \{1, 2, \ldots, C\}$ at the time of the next handoff arrival (i.e., regeneration instant) which occurs on or before time t given that the system was in state $i \in \{1, 2, \ldots, C\}$ just after the instant of previous handoff arrival. Since the new calls are not accepted when the system reaches $C - g$, the values of $k_{i,j}(t)$ for $i \in \{1, 2, \ldots, C - g\}$, $j \in \{C - g + 2, \ldots, C\}$ and for $i \in \{C - g + 1, \ldots, C - 2\}$, $j \in \{i + 2, \ldots, C\}$ are zero. Also, if let $\{Y_n, n \geq 0\}$, $Y_n \in S^{'}$ be the number of ongoing calls in the system at the time of nth handoff arrival, then $\{Y_n, n \geq 0\}$ is a DTMC with one-step transition probability matrix $K(\infty)$. The matrix $E(t)$ describes the dynamics of the process during the time between two consecutive regeneration points starting from a state at regeneration. $e_{i,j}(t)$ denotes the probability that the system will be in state $j \in \{0, 1, \ldots, C\}$ at time t and the next handoff arrival occurs after t given that the system was in state $i \in \{1, 2, \ldots, C\}$ just after the instant of previous handoff arrival. Note that, $e_{i,j}(t) = 0$ for $i \in \{1, 2, \ldots, C-g\}$ and $j \in \{C-g+1, \ldots, C\}$. Similarly, $e_{i,j}(t) = 0$ for $i \in \{C-g+1, \ldots, C\}$ and $j > i$.

Classify the elements of the matrix $K(t)$ into two sets based on their row indices, namely, $\{1, 2, \ldots, C-g\}$ and $\{C-g+1, \ldots, C\}$. Let $S(i)$ $(i = 1, 2, \ldots, C-g)$, be the set of all states reachable from state i in which the subordinated CTMC can spend a non-zero time before the next transition occurs and is given by $S(i) = \{0, 1, \ldots, C-g\}$. The state transition diagram for the subordinated CTMC is shown in Fig. 8.11. Therefore, for instance, if j new calls arrive and k ongoing calls depart during $(0, S_1)$, the new state will be $i + j - k$. The matrix Q of the corresponding subordinated CTMC with initial state i is given as

$$\mathbf{Q} = \begin{pmatrix} -\lambda & \lambda & 0 \cdots & 0 & 0 \\ \mu & -(\lambda+\mu) & \lambda \cdots & 0 & 0 \\ \vdots & \vdots & \vdots \cdots & \vdots & \vdots \\ 0 & 0 & 0 \cdots & -(\lambda+(C-g+1)\mu) & \lambda \\ 0 & 0 & 0 \cdots & (C-g)\mu & -(C-g)\mu \end{pmatrix}.$$

Let $p_{ij}(t)$ be the probability that the subordinated CTMC is in state j at time t given that it was in state i initially. Define $p(t) = [p_{i,0}(t), p_{i,1}(t), \ldots, p_{i,C-g}(t)]$. Then

$$\frac{dP(t)}{dt} = p(t)\mathbf{Q},\ p_{i,i}(0) = 1, i \in \{1, 2, \ldots, C-g\}.$$

From the definition of the kernel $\boldsymbol{K(t)}$ of an MRGP, write the first set of elements of kernel $\boldsymbol{K(t)}$

$$k_{i,j}(t) = \begin{cases} \int_0^t p_{i,j-1}(x)dG(x), & j = 1, 2, \ldots, C-g+1 \\ 0, & j = C-g+2, \ldots, C \end{cases}$$

where $p_{i,j}(t)$ are obtained by solving the above equation. The transient solution is given by

$$p_{i,j}(t) = \frac{\rho^j/j!}{\sum_k \rho^k/k!} + \frac{(C-g)!}{j!}\rho^{C-g-i} \times \sum_{r=1}^{C-g} \frac{D_i(x_r)D_j(x_r)}{rD_{C-g}(x_r)D'_{C-g}(x_r+1)} e^{x_r\mu t}$$

where $\rho = \lambda/\mu$,

$$D_n(x) = \begin{cases} 1, n = 0 \\ x + \rho, n = 1 \\ (x + \rho + n - 1)D_{n-1}(x) - (n-1)\rho D_{n-2}(x), n = 2, 3, \ldots, C - g \end{cases}$$

and $x_r, r = 1, 2, \ldots, C - g$ are the roots of $D_{C-g}(x + 1) = 0$. For the second set of rows of $K(t)$ with $i \in \{C - g + 1, \ldots, C\}$, $S(i) = \{0, 1, \ldots, i\}$. For instance, j new calls arrive and k ongoing calls depart during $(0, S_1)$, the new state will be $i + j - k$ and there is no arrival of new calls when more than $C - g$ calls are in the system.

$$\mathbf{Q}^*(i) = \begin{pmatrix} -\lambda & \lambda & 0 \cdots & 0 & 0 \\ \mu & -(\lambda+\mu) & \lambda \cdots & 0 & 0 \\ \vdots & \vdots & \vdots \cdots & \vdots & \vdots \\ 0 & 0 & 0 \cdots & -(i-1)\mu) & \lambda \\ 0 & 0 & 0 \cdots & i\mu & -i\mu \end{pmatrix}.$$

Define $q(t) = [q_{i,0}(t), q_{i,1}(t), \ldots, q_{i,i}(t)]$. Then

$$\frac{dq(t)}{dt} = q(t)\mathbf{Q}^*(i), q_{i,i}(0) = 1, i \in \{C - g + 1, \ldots, C\}.$$

$$k_{i,j}(t) = \begin{cases} \int_0^t q_{i,j-1}(x)dG(x), j = 1, 2, \ldots, i + 1 \\ 0, j = 1, 2, \ldots, C. \end{cases}$$

$$k_{C,j}(t) = \begin{cases} \int_0^t q_{C,j-1}(x)dG(x), j \neq C \\ \int_0^t [q_{C,C-1}(x) + q_{C,C}(x)]dG(x), j = C. \end{cases}$$

The transient solution $q_{i,j}(t)$ can be written as $q_{i,j}(t) = [e^{\mathbf{Q}^*(i)t}]_{i,j}$, where $[\cdot]_{i,j}$ denotes the ith row and jth column of matrix. $[e^{\mathbf{Q}^*(i)t}]_{i,j}$ can also be expressed as a weighted sum of exponential terms by using the transient solution of the BDP.

$$k_{i,j}(t) = \frac{\frac{\rho^{(j-i)}}{(j-1)!}}{\sum_k \frac{\rho^k}{k!}} + \frac{(C-g)!}{(j-1)!}\rho^{C-g-i}\sum_{r=1}^{C-g}\frac{D_i(x_r)D_{(j-1)}(x_r)}{rD_{C-g}(x_r)D'_{C-g}(x_r+1)}\widehat{g}(-x_r\mu).$$

Here $\widehat{g}(\cdot)$ denotes the Laplace transform of $g(\cdot)$. The matrix $E(t)$ describes the behavior of the process during the time between two consecutive handoff call arrivals starting from a state of the system just after the last handoff call arrival.

$$e_{i,j}(t) = \begin{cases} p_{i,j}(t)(1 - G(t)), i \in \{1, 2, \ldots, C - g\}; j \in \{0, 1, \ldots, C - g\} \\ q_{i,j}(t)(1 - G(t)), i \in \{C - g + 1, \ldots, C\}; j \in \{0, 1, \ldots, i\} \\ 0, \text{otherwise}, \end{cases}$$

where $p_{i,j}(t)$ and $q_{i,j}(t)$ are defined in the above equations. Also, the mean time $\alpha_{i,j}$ are given by

$$\alpha_{i,j} = \frac{1}{\lambda_h}\frac{\rho^j/j!}{\sum_k \rho^k/k!} + \frac{C-g}{j!}\rho^{C-g-i}\sum_{r=1}^{C-g}\frac{D_i(x_r)D_j(x_r)}{rD_{C-g}(x_r)D'_{C-g}(x_r+1)}\widehat{G_c}(-x_r\mu),$$

where $\widehat{G_c}(-x_r\mu)$ means the Laplace transform of complementary distribution function $G(\cdot)$.

Further steps to obtain the steady-state and time-dependent probabilities described in Sections 8.5.1 and 8.5.2 respectively.

For more description, readers may refer to [1–11].

8.6 Exercise Problems

1. Find the renewal function corresponding to the lifetime density $f(x) = \lambda^2 x e^{-\lambda x}$, $x \geq 0$.
2. Let $X_1, X_2 \ldots$ be the inter-occurrence times in a renewal process. Suppose $P\{X_i = 1\} = 1/3,\ P\{X_i = 2\} = 1/3,\ P\{X_i = 3\} = 1/3,\ i = 1, 2, \ldots$. Let N_n be the number of renewals up to discrete-time n. Compute $P\{N(2) = k\}$.
3. Consider a renewal process $\{N(t), t \geq 0\}$. Let $H(t)=E[N(t) - N(t-a)1_{(a\leq t)}]$. Find a renewal equation that H satisfies.
4. Let $N_1(t)$ and $N_2(t)$ be independent renewal processes with the same inter-renewal distribution, and consider the sum $N(t) = N_1(t) + N_2(t)$ (called superposition). Assuming that $N(t)$ is a renewal process. Prove that it is a Poisson process if and only if $N_1(t)$ and $N_2(t)$ are Poisson processes.
5. Prove $E[T_{N(t)}] = \mu E[N(t) + 1] - E[X_{N(t)}]$. If $N(t)$ is a Poisson process, show that $E[T_{N(t)}] < \mu E[N(t)]$.
6. Consider a renewal process $\{N(t), t \geq 0\}$ with for $n = 1, 2, \ldots$, $P\{X_n = k\} = p(1-p)^{k-1}$, $k = 1, 2, \ldots$. Prove that, $P\{N(t) = n\} = \binom{[t]}{n} p^n (1-p)^{[t]-n}$, $n = 0, 1, \ldots$.
7. Consider a renewal process $\{N(t), t \geq 0\}$ with X_i follows uniform distribution between 0 and 1. Prove that, $M(t) = e^t - 1, 0 \leq t \leq 1$.
8. Consider a sequence of independent Bernoulli trials in which each trial results in a success or failure with respective probabilities p and $q = 1 - p$. Let $N(t)$ denote the number of successes in n trials, where n is an integer. Show that N_n is a discrete time renewal process, called a Bernoulli Process. (The parameter n may denote discrete time or any integer referring to sequential information.) Justify that the inter-renewal times have the geometric distribution $P(X_1 = n) = pq^{n-1}, n \geq 1$. Find the distribution and mean of N_n, and do the same for the renewal time T_n. Show that the moment-generating function of T_n is

$$E[e^{\alpha T_n}] = \left(\frac{pe^{\alpha}}{1 - qe^{\alpha}}\right)^n, 0 < \alpha < -\log q.$$

9. A system consists of two components with independent lifetimes X_1 and X_2, where X_1 is exponential distributed with rate λ, and X_2 has a uniform distribution on [0, 1]. The components operate in parallel, and the system lifetime is $max\{X_1, X_2\}$ (the system is operational if and only if at least one component is working). When the system fails, it is replaced by another system with an identical and independent lifetime, and this is repeated indefinitely. The number of system renewals over time forms a renewal process $\{N(t), t \geq 0\}$. Find the distribution and mean of the system lifetime. Find the distribution and mean of $N(t)$. Determine the portion of time that (a) two components are working, (b) only type 1 component is working, and (c) only type 2 component is working.
10. An infinite number of jobs are to be processed one at a time by a single server. There are m types of jobs, and the probability that any job is of type i is p_i, where $p_1 + \ldots + p_m = 1$. The service time of a type i job has a distribution F_i with mean μ_i. The service times and types of the jobs are independent. Let $N(t)$ denote the number of jobs completed by time t. Show that $\{N(t), t \geq 0\}$ is a renewal process and specify its inter-renewal distribution and mean. Let $N_i(t)$ denote the number of type i jobs processed up to time t. Show that $\{N_i(t), t \geq 0\}$ is a delayed renewal process and specify $\lim_{t\to\infty} \frac{N_i(t)}{t}$.
11. The status of a system is represented by an alternating renewal process $\{X(t), t \geq 0\}$ with renewal times T_n, where the mean sojourn time in a working state 1 is λ and the mean sojourn time in a non-working state 0 is μ. The system availability is measured by the portion of time it is working, which is $\lim_{t\to\infty} \frac{\int_0^t X(s)ds}{t}$. Determine this quantity and show that it is equal to the cycle-availability measured by $\lim_{n\to\infty} \frac{\int_0^{T_n} X(s)ds}{T_n}$.
12. Consider time $T_n = \sum_{k=1}^n X_k$, where the X_k are IID with distribution F and $F(0) = P\{X_k = 0\} > 0$. The associated point process $\{N(t), t \geq 0\}$ is a renewal process with instantaneous renewals (or batch renewals). Let $\tilde{T}_n$ are the times at which batches of events arrive, and at time $\tilde{T}_n$, a batch of a_n events arrive. Hence, $N(t) = \sum_{n=1}^{\infty} a_n I_{\tilde{T}_n \leq t}$, where a_n is the number of renewals exactly at time $\tilde{T}_n$. Specify the distribution of a_n. Are the a_n IID? Are they independent of $\tilde{T}_n$? Specify the distribution of $\tilde{T}_1$ in terms of F.
13. Consider an non-homogeneous renewal process $\{N(t), t \geq 0\}$ whose interpoint times $X_n = T_n - T_{n-1}$ are independent with distributions F_n. Assuming it is finite, prove that $E[N(t)] = \sum_{n=1}^{\infty} F_1(t)F_2(t)\cdots F_n(t)$.
14. Consider a production-inventory system that produces a product at a constant rate of c units per unit time and the items are put in inventory to satisfy demands. The products may be discrete or continuous (e.g., oil, chemicals). Demands occur according to a Poisson process $\{N(t), t \geq 0\}$ with rate λ, and the demand quantities $X_1, X_2, \ldots$ are IID positive RVs with mean μ, and are independent of

the arrival times. Then, the inventory level at time t would be

$$Z_x(t) = x + ct - \sum_{n=1}^{N(t)} X_n, t \geq 0,$$

where x is the initial inventory level. Consider the probability $R(x) = P\{Z_x(t) \geq 0\}$ of never running out of inventory. Show that if $c < \lambda\mu$, then $R(x) = 0$ no matter how high the initial inventory level x is. Hint: a SLLN to show that $Z_x(t) \to \infty$ as $t \to \infty$ if $c < \lambda\mu$, where x is fixed. Find the limit of $Z_x(t)$ as $t \to \infty$ if $c > \lambda\mu$.

15. Consider a system in which customers arrive at a processing station according to a renewal process with inter-arrival distribution F and are processed by a single server. The processing or service times are independent of the common distribution G and are independent of the arrival process. Also, customer arrivals during service time are blocked from being served - they either go elsewhere or go without service. Prove that,

$$P\{\tilde{T}_1 \leq t\} = \int_0^{\infty} P\{T_{Nx(x)+1} \leq t\} dG(x).$$

Deduce the same result, when inter-arrival times are independent exponential distributions with parameter λ and processing times are independent exponential distributions with parameter μ.

References

1. Arjas, E., Haara, P., Norros, I.: Filtering the Histories of a Partially Observed Marked Point Process. Stochastic Processes and Applications **40**, 225–250 (1992)
2. Bremaud, P.: Point Processes and Queues: martingale dynamics, vol. 50. Springer-Verlag, New York (1980)
3. Bordenave, C., Luca Torrisi, G.: Large deviations of Poisson cluster processes. Stochastic Models **23**(4), 593–625 (2007). https://doi.org/10.1080/15326340701645959
4. Castaneda, L.B., Arunachalam, V., Dharmaraja, S.: Introduction to Probability and Stochastic Processes with Applications. Wiley, New Jersey (2012)
5. Çinlar, E.: Markov Renewal Theory. Advances in Applied Probability **1**, 123–187 (1969)
6. Daley, D. J., Vere-Jones, D.: An Introduction to the theory of point processes, 2nd edn. Springer Berlin/Heidelberg, Vol. I and II, Germany (2003). ISBN: 978-0-387-21564-8
7. Gerardi, A., Tardelli, P.: Filtering on a Partially Observed Ultra-High-Frequency Data Model. Acta Applicandae Mathematicae **91**, 193–205 (2006)
8. Gao, X., Zhu, L.: Functional central limit theorem for stationary Hawkes processes and its application to infinite-serve queues. Queueing Systems **90**(1), 161–206 (2018)

9. Hawkes, A.G.: Hawkes Processes and Their Applications to Finance: a Review. Quantitative Finance **18**(2), 193–198 (2018)
10. Hawkes, A.G., Oakes, D.: A cluster process representation of self-exciting process. Journal of Applied Probability **11**, 493–503 (1974). https://doi.org/10.2307/3212693
11. Janssen, J., Manca, R.: Applied Semi-Markov Processes. Springer (2006)

Chapter 9
General Queueing Models

While non-Markov processes were introduced in the previous chapter, this chapter provides a detailed analysis of non-Markov queueing models. Both transient and steady-state distributions are derived for the various non-Markov queueing models such as $M/G/1$, $GI/M/1$, $GI/G/1$, $M/G/1/N$, $GI/M/1/N$, $GI/G/1/N$, $M/G/c$, $GI/M/c$, $GI/G/c$, $M/G/c/N$, $M/G/c/c$ and $GI/G/\infty$. These queueing models and many of their variants are studied by using a variety of methodologies including embedded Markov chain, supplementary variable analysis and Markov regenerative theory.

Many examples of the above non-Markov queueing models are included in this chapter with retrial and vacation policies. At the end, $Hawkes/sdHawkes/\infty$ system is presented.

9.1 Single-Server Queueing Models

In this section, non-Markov queueing models with single server are described in detail with either non-Markov arrival flow or service process or both, such as $M/G/1$, $GI/M/1$ and $GI/G/1$.

9.1.1 $M/G/1$ Queueing Model

In this queueing model, the inter-arrival times are exponentially distributed but the service times have a non-exponential (general) distribution. Assume that the CDF of service time is $F_s(t)$ with a mean $\frac{1}{\mu}$. Let $X(t)$ be the number of customers in the system at time t. The evolution of system upon customer arrival is influenced not only by the number of customers $X(t)$, but also by the remaining service time of the client receiving service, if any.

D. Selvamuthu, *Introduction to Stochastic Processes*, University Texts in the Mathematical Sciences, https://doi.org/10.1007/978-981-97-6152-4_9

It is worth noting that the quantities of importance are the state probabilities at various time points, namely the arrival epochs. The amount of interest, from the perspective of an arriving consumer, is the number of customers a person discovers in the system, and not the number he/she leaves behind. The equilibrium distribution of the number of customers found by the arrival and the equilibrium distribution of the number of customers left behind by a departure in this queueing model are the same since the arrival follows a Poisson process. Thus, the limiting distributions of the number of customers in the system at arrival epochs, departure epochs, and arbitrary points in time are the same in the $M/G/1$ queue solely because the arrivals occur as a Poisson process. That is, the PASTA property (refer Sect. 6.2.1.9) will be satisfied by the $M/G/1$ queue.

In this section, for this $M/G/1$ queueing model, system state probabilities are obtained by three methodologies namely, the method of Embedded Markov chain, the method of supplementary variable and the method of Markov regenerative theory.

9.1.1.1 The Method of Embedded Markov Chain

Consider a single-server queue with an unlimited buffer and in which arrivals follow Poisson process with parameter λ. The customer service time, denoted by S, has an arbitrary distribution with CDF $F_s(t)$, and whose Laplace-Stieltjes transform (LST) is $\beta(s)$ Appendix A.6. The RV S has finite first and second order moments, $E(S)$ and $E(S^2)$, respectively. The stochastic process $\{X(t), t \geq 0\}$, where $X(t)$ denotes the number of customers in the system at time t, is not Markov because the behavior of the process after an arbitrary time cannot be analyzed. During the interval of service time, a random number of customers can arrive at the system and the service of one customer will be completed.

The main idea of embedded Markov chain can be described as follows. Let the number of customers $X(t_n^+)$ in the system be greater than 0 where t_n is the nth customer service completion. Since the number of customers in the system makes a jump at time t_n, Assumed that $X_n = X(t_n^+)$, i.e., the served customer exits the system and is no longer counted. For the non-Markov process $\{X(t), t \geq 0\}$, a sequence of the nth customer service completion epoch, $t_n, n = \{1, 2, \ldots\}$ forms a DTMC $\{X_n, n = 0, 1, \ldots\}$. The process $\{X_n, n = 0, 1, \ldots\}$ is a time-homogeneous DTMC which is embedded in $\{X(t), t \geq 0\}$. An effective investigation of a DTMC with a countable state space is possible only if the matrix of its one-step transition probabilities has a special structure. The matrix P of one-step transition probabilities $p_{i,j}$ of the embedded Markov chain $\{X_n, n = 0, 1, \ldots\}$ has such a specific structure. Given that the system has atleast one customer and the next customer is immediately accepted for service, also leaves the system at the next moment t_{n+1}^+. A total of $X_2 - X_1 + 1$ customers arrive at the queue during the time interval (t_n, t_{n+1}). The probability of this event is $a_{X_2 - X_1 + 1}$ where the quantities a_i are provided by

$$a_i = \int_0^{\infty} \frac{(\lambda t)^i}{i!} e^{-\lambda t} dF_s(t), i = 0, 1, 2 \ldots. \tag{9.1}$$

Therefore, the transition probability $p_{i,j}, i > 0, j \geq i - 1$ is determined from

$$p_{i,j} = a_{j-i+1}, i > 0, j \geq i - 1. \tag{9.2}$$

Stationary Distribution

Let the vector $\pi = [\pi_i]_{i \in S}$ be the stationary state distribution of the process $\{X(t), t \geq 0\}$. Assume that this distribution exists under the same conditions as the embedded distribution and is calculated through the embedded distribution as follows:

If the system becomes empty after the moment t_n, i.e., $X(t_n) = 0$, then system remains empty until the arrival of next customer. The dynamics of the system are the same as those following a service completion epoch with one customer remaining in the queue. Therefore, $p_{0j} = p_{1j}$, which implies that

$$p_{0j} = a_j, j \geq 0.$$

The one-step transition probability matrix is as follows

$$\mathbf{P} = \begin{pmatrix} a_0 & a_1 & a_2 & a_3 & \ldots \\ a_0 & a_1 & a_2 & a_3 & \ldots \\ 0 & a_0 & a_1 & a_2 & \ldots \\ 0 & 0 & a_0 & a_1 & \ldots \\ \vdots & \vdots & \vdots & \vdots & \vdots \end{pmatrix}.$$

The stationary distribution for the EMC can be computed by solving the following system of linear equations

$$\pi_j = \pi_0 a_j + \sum_{i=1}^{j+1} \pi_i a_{j-i+1}, j = 0, 1, \ldots \tag{9.3}$$

with $\sum_i \pi_i = 1$. To solve the above system of linear Equations, the probability generating function approach is used. Introduce the probability generating functions (PGFs)

$$\Pi(z) = \sum_{j=0}^{\infty} \pi_j z^j, A(z) = \sum_{j=0}^{\infty} a_j z^j, |z| < 1.$$

The explicit expression for the PGF, $A(z)$ is provided by

$$A(z) = \int_0^\infty e^{-\lambda(1-z)t} dF_s(t) = \beta(\lambda(1-z)),$$

where $\beta(.)$ is the LST of CDF $F_S(t)$ (refer Appendix A.6). Multiplying the Equations (9.3) by the corresponding powers of z and summing up them yields,

$$\Pi(z) = \pi_0 A(z) + \sum_{j=0}^{\infty} z^j \sum_{i=1}^{j+1} \pi_i a_{j-i+1},$$

Changing the order of summation,

$$\begin{aligned} \Pi(z) =& \pi_0 A(z) + \sum_{i=0}^{\infty} \pi_i z^{i-1} \sum_{j=i-1}^{\infty} a_{j-i+1} z^{j-i+1}. \\ =& \pi_0 A(z) + (\Pi(z) - \pi_0) A(z) z^{-1}. \end{aligned} \tag{9.4}$$

Due to the fact that the transition probabilities p_{ij} of the EMC for $i > 0$ depend only on $j - i$ and do not depend on i and j separately. The Eq. (9.2) can be rewritten as follows

$$\Pi(z) = \pi_0 \frac{(1-z)\beta(\lambda(1-z))}{\beta(\lambda(1-z)) - z}. \tag{9.5}$$

Equation (9.5) defines the required PGF of the embedded Markov chain in terms of π_0, the probability that the system in state 0 at an arbitrary moment of service completion. To find the probability π_0, calculate limit $z = 1$ in Eq. (9.5).

Limiting Distribution

Let π_i be the limiting distribution of the process $\{X(t), t \geq 0\}$ which is defined as:

$$\pi_i = \lim_{t \to \infty} P\{X(t) = i\}, i = 0, 1, \ldots.$$

From the theory of renewal processes, this distribution exists under the same conditions as the embedded distribution and is calculated through the embedded distribution as follows:

$$\pi_0 = \tau^{-1}\pi_0 \int_0^\infty e^{-\lambda t} \mathrm{d}t \tag{9.6}$$

$$\pi_i = \tau^{-1}\Big[\pi_0 \int_0^\infty \int_0^t e^{-\lambda v}\lambda \mathrm{d}v \frac{(\lambda(t-v))^{i-1}}{(i-1)!} e^{-\lambda(t-v)}(1-F_s(t-v))dt$$
$$+ \sum_{l=1}^{i} \pi_l \int_0^\infty \frac{(\lambda t)^{i-l}}{(i-l)!} e^{-\lambda t}(1-F_s(t))dt\Big], i = 1, 2, \ldots. \tag{9.7}$$

Here, τ is the mean time between the customer departure time points. Without loss of generality $\tau = \lambda^{-1}$. The PGF $P(z) = \sum_{i=0}^{\infty} \pi_i z^i$.

Multiplication of Eqs. (9.6) and (9.7) by the corresponding powers of z and further summation yield

$$P(z) = \tau^{-1}\Big\{\pi_0\Big[\lambda^{-1} + \int_0^\infty \int_0^t e^{-\lambda v}\lambda dv \sum_{i=1}^{\infty} \frac{(\lambda(t-v))^{i-1} z^i}{(i-1)!} e^{-\lambda(t-v)}(1-F_s(t-v))\mathrm{d}t\Big]$$
$$+ \sum_{i=1}^{\infty} z^i \sum_{l=1}^{i} \pi_l \int_0^\infty \frac{(\lambda t)^{i-l}}{(i-l)!} e^{-\lambda t}(1-F_s(t))\mathrm{d}t\Big\}.$$

Changing the order of the integration in the double integral and the order of summation in the double sum and counting the known sums,

$$P(z) = \tau^{-1}\Big\{\pi_0\Big[\lambda^{-1} + z\lambda \int_z^\infty \int_0^\infty e^{-\lambda v} dv e^{-\lambda(1-z)(t-v)}(1-F_s(t-v))\mathrm{d}t\Big]$$
$$+ \sum_{l=1}^{\infty} z^l \pi_l \int_0^\infty e^{-\lambda(1-z)t}(1-F_s(t))\mathrm{d}t\Big\}.$$

Altering the integration variable $u = t - v$ and taking into account the equality $\tau^{-1} = \lambda$ and

$$P(z) = \pi_0 \left[1 + \frac{z}{(1-z)}(1 - \beta(\lambda(1-z)))\right] + (P(z) - \pi_0)\frac{1}{1-z}(1 - \beta(\lambda(1-z))).$$

By applying the elementary transformations

$$P(z) = \pi_0 \frac{(1-z)\beta(\lambda(1-z))}{\beta(\lambda(1-z)) - z}. \tag{9.8}$$

Here, π_0 is equal to $1 - \rho$ where $\rho = \lambda E(S)$. This ensures the validity of the equation $P(z) = \Pi(z)$. Therefore, the PGF of the number of customers in the system at the end of service epochs and at arbitrary times coincide.

9.1.1.2 The Method of Supplementary Variable

This method of investigating the $M/G/1$ process consists of expanding the process state space by introducing some supplementary components so that the resulting higher dimensional process is a Markov process. In the case of an $M/G/1$ queue, the underlying stochastic process, $\{X(t), t \geq 0\}$ is not a Markov process. Let $B(t)$ be the time from the point t to the time point when the current service will complete. Analyzing the reason for the fact of Markov property, if an additional stochastic process $\{B(t), t \geq 0\}$ is included into the description of the process $\{X(t), t \geq 0\}$, the resulting two-dimensional random process $\{(X(t), B(t)), t \geq 0\}$ is a Markov process. Here, if $X(t) = 0$, a supplementary variable is not required since no customer is being served. Define the function

$$\phi_t(0) = P\{X(t) = 0\},$$

and, for $x > 0$, $i = 1, 2, \ldots$

$$\phi_t(i, x) = P\{X(t) = i, B(t) < x\}.$$

These functions $\phi_t(0)$ and $\phi_t(i, x)$ satisfies the following difference equations:

$$\begin{aligned}
\phi_{t+\Delta t}(0) =& \phi_t(0)(1 - \lambda\Delta t) + \phi_t(1, \Delta t) + o(\Delta t) \\
\phi_{t+\Delta t}(1, x) =& (\phi_t(1, x + \Delta t)) - \phi_t(1, \Delta t)(1 - \lambda\Delta t) + \phi_t(2, \Delta t)B(x) \\
&+ \phi_t(0)\lambda\Delta t B(x) + o(\Delta t) \\
\phi_{t+\Delta t}(i, x) =& (\phi_t(i, x + \Delta t)) - \phi_t(i, \Delta t)(1 - \lambda\Delta t) + \phi_t(i + 1, \Delta t)B(x) \\
&+ \phi_t(i - 1, x + \Delta t)\lambda\Delta t + o(\Delta t), \quad i = 2, 3, \ldots.
\end{aligned}$$

By dividing both sides of the above equations by Δt and allowing Δt tends to 0,

$$\frac{\partial \phi_t(0)}{\partial t} = -\lambda \phi_t(0) \frac{\partial \phi_t(1,x)}{\partial x}\mid_{x=0}$$

$$\frac{\partial \phi_t(1,x)}{\partial t} - \frac{\partial \phi_t(1,x)}{\partial x} = -\lambda \phi_t(1,x) - \frac{\partial \phi_t(1,x)}{\partial x}\mid_{x=0} + \frac{\partial \phi_t(2,x)}{\partial x}\mid_{x=0} B(x) + \lambda \phi_t(0) B(x)$$

$$\frac{\partial \phi_t(i,x)}{\partial t} - \frac{\partial \phi_t(i,x)}{\partial x} = -\lambda \phi_t(i,x) - \frac{\partial \phi_t(i,x)}{\partial x}\mid_{x=0} + \frac{\partial \phi_t(i+1,x)}{\partial x}\mid_{x=0} B(x) + \lambda \phi_t(i-1,x).$$

$$i = 2, 3, \ldots$$

The stationary distribution of the process $\{X(t), B(t)\}$ is defined as follows

$$\phi(0) = \lim_{t\to\infty} \phi_t(0), \phi(i,x) = \lim_{t\to\infty} \phi_t(i,x), i = 1, 2, \ldots, x > 0. \tag{9.9}$$

The limits (9.9) exist if the inequality $\rho = \lambda E(S) < 1$ holds. Taking the limit $t \to \infty$ in the above mentioned partial differential equations, the steady-state distribution of the process $\{(X(t), B(t)), t \geq 0\}$ is given by

$$\lambda \phi(0) = \frac{\partial \phi(1,x)}{\partial x}\mid_{x=0},$$

$$\frac{\partial \phi(1,x)}{\partial x} - \frac{\partial \phi(1,x)}{\partial x}\mid_{x=0} - \lambda \phi(1,x) + \frac{\partial \phi(2,x)}{\partial x}\mid_{x=0} + \lambda \phi(0) B(x) = 0,$$

$$\frac{\partial \phi(i,x)}{\partial x} - \frac{\partial \phi(i,x)}{\partial x}\mid_{x=0} - \lambda \phi(i,x) + \frac{\partial \phi(i+1,x)}{\partial x}\mid_{x=0} B(x) + \lambda \phi(i-1,x),$$

$i = 2, 3, \ldots.$

To solve the system of equations, PGF method is used. Defining the PGF as

$$P(z,x) = \phi(0) + \sum_{i=1}^{\infty} \phi(i,x) z^i, \quad |\, z \,| < 1.$$

After simple calculations, the following equation for the generating function $P(z,x)$ is obtained as

$$\frac{\partial P(z,x)}{\partial x} - \lambda(1-z) P(z,x) = \frac{\partial P(z,x)}{\partial x}\mid_{x=0} \left(1 - \frac{B(x)}{z}\right) - \lambda \phi(0)(1-z)(1-B(x)). \tag{9.10}$$

To solve the above partial differential equation, define the Laplace transform

$$\phi(z,s) = \int_0^{\infty} e^{-sx} P(z,x) \mathrm{d}x, \; Re(s) > 0.$$

After applying the Laplace transform on Eq. (9.10),

$$(s-\lambda(1-z))\phi(z,s) = \frac{1}{s}\left[\frac{\partial P(z,x)}{\partial x}\Big|_{x=0}\left(1-\frac{\beta(s)}{z}\right) - \lambda\phi(0)(1-z)(1-\beta(s))\right] + \phi(0),$$

which yields

$$\frac{\partial P(z,x)}{\partial x}\Big|_{x=0} = \phi(0)\lambda z \frac{\beta(1-z)(\lambda(1-z))}{\beta(\lambda(1-z))-z}.$$

$$(s-\lambda(1-z))\phi(z,s) = \frac{\lambda\phi(0)(1-z)}{s}\left[\frac{z\beta(\lambda(1-z))\lambda(1-z)}{\beta(\lambda(1-z))-z}\left(1-\frac{\beta(s)}{z}\right) - 1 + \beta(s)\right] + \phi(0).$$

The stationary distribution of the processes $\{(X(t), B(t)), t \geq 0\}$ and $\{\pi_{i_t}, t \geq 0\}$ are defined as

$$\pi_0 = \lim_{t\to\infty} P\{X(t) = 0\} = \phi(0)$$
$$\pi_i = \lim_{t\to\infty} P\{X(t) = i\} = \lim_{x\to\infty} \phi(i,x), i = 1, 2, \ldots.$$

Therefore, the PGF $P(z) = \sum_{i=0}^{\infty} \pi_i z^i$ is determined in a similar way and is as follows

$$P(z) = \lim_{x\to\infty} P(z,x) = \lim_{s\to 0} s\phi(z,s).$$

$$P(z) = \pi_0 \frac{(1-z)\beta(\lambda(1-z))}{\beta(\lambda(1-z))-z}. \tag{9.11}$$

Calculating the constant $\pi_0 = 1-\rho$ from the normalization condition $P(1) = 1$, $P(z)$ becomes

$$P(z) = (1-\rho)\frac{(1-z)\beta(\lambda(1-z))}{\beta(\lambda(1-z))-z}. \tag{9.12}$$

9.1.1.3 The Method of Markov Regenerative Theory

Let $X(t)$ denote the number of customers in the system at time t. Let $\{S_n, n = 1, 2, \cdots\}$, be the RV denoting the service time instant of the nth customer. Here, the service time instants are the only regeneration time epochs. Hence, $\{X(t), t \geq 0\}$ is

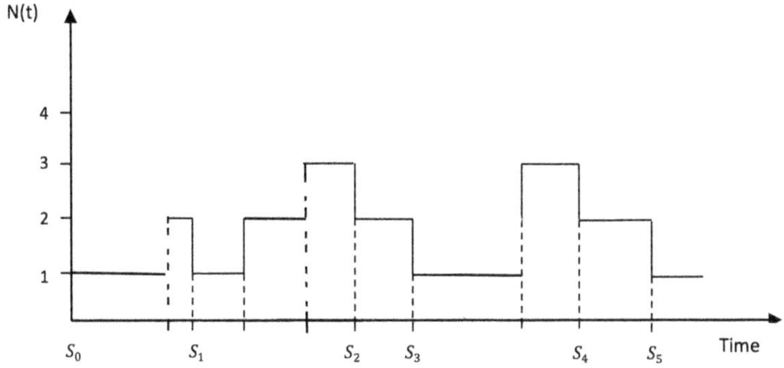

Fig. 9.1 Sample path of $M/G/1$ model for MRGP

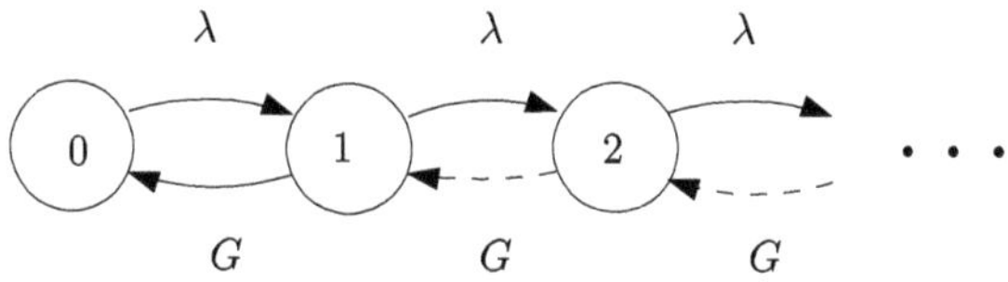

Fig. 9.2 State transition diagram of $M/G/1$ queueing system

not a semi-Markov process, but a MRGP with $\{X_n, n = 1, 2, \ldots\}$ as the embedded Markov chain. $S = \{0, 1, \cdots, \}$ is the set of states at all time instants, and $S' = \{0, 1, \ldots, \}$ is the set of states only at regeneration time instants. Figure 9.1 shows the sample path of $M/G/1$. Here, $S_0 = 0$ and S_1, S_2, S_3, S_4 and S_5 are the regeneration time epochs since, they are service completion time points.

The state transition diagram of $M/G/1$ queueing system is presented in Fig. 9.2. Note that, in this figure, the dotted arcs mean that, the system moves from state 2 to state 1 and from state n to state $n-1$ for $n > 2$ with residual service time or service time whereas the system moves from state 1 to state 0 with service complete service time.

Following the theory of MRGP, the global kernel $K(t)$ and local kernel $E(t)$ matrices for the process can be determined. The elements are defined as:

$$k_{ij}(t) = P(X_1 = j,\ T_1 \leq t \mid X_0 = i),\ \ i, j \in S'$$
$$e_{ij}(t) = P(X_1 = j,\ T_1 \geq t \mid X_0 = i),\ \ i \in S',\ j \in S\,.$$

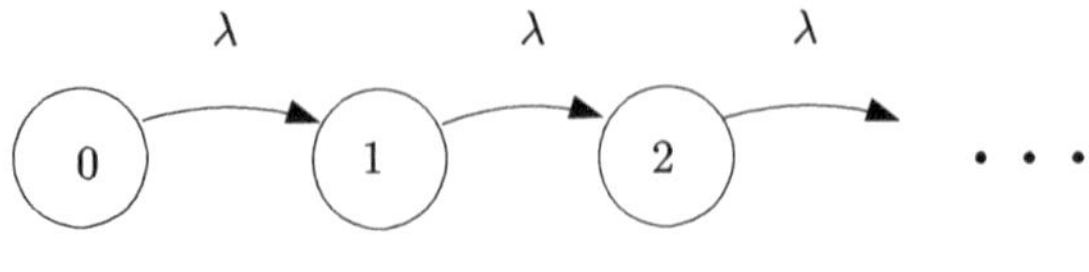

Fig. 9.3 Subordinated CTMC

The global kernel matrix is

$$\mathbf{K(t)} = \begin{pmatrix} k_{11}(t) & k_{12}(t) & k_{13}(t) & k_{14}(t) & \cdots \\ k_{21}(t) & k_{22}(t) & k_{23}(t) & k_{24}(t) & \cdots \\ 0 & k_{31}(t) & k_{32}(t) & k_{33}(t) & \cdots \\ 0 & 0 & k_{41}(t) & k_{42}(t) & \cdots \\ \cdot & \cdot & \cdot & \cdot & \cdots \\ \cdot & \cdot & \cdot & \cdot & \cdots \\ \vdots & \vdots & \vdots & \cdot & \ddots \end{pmatrix}$$

where

$$k_{ij}(t) = \int_0^t P_{ij}(x) \mathrm{d}F_s(x) , \quad i, j \in S' ,$$

and the $p_{ij}(t)$'s are the transition probabilities of the subordinated CTMC. Figure 9.3 shows the state transition diagram of subordinated CTMC.

The local kernel matrix is

$$\mathbf{E(t)} = \begin{pmatrix} e_{10}(t) & e_{11}(t) & e_{12}(t) & e_{13}(t) & e_{14}(t) & \cdots \\ e_{20}(t) & e_{21}(t) & e_{22}(t) & e_{23}(t) & e_{24}(t) & \cdots \\ 0 & e_{30}(t) & e_{31}(t) & e_{32}(t) & e_{33}(t) & \cdots \\ 0 & 0 & e_{40}(t) & e_{41}(t) & e_{42}(t) & \cdots \\ \cdot & \cdot & \cdot & \cdot & \cdot & \cdots \\ \cdot & \cdot & \cdot & \cdot & \cdot & \cdots \\ \vdots & \vdots & \vdots & \vdots & \cdot & \ddots \end{pmatrix}$$

where $e_{ij}(t) = P_{ij}(t)(1 - F_s(t))$.

9.1.1.4 Performance Measures

1. The mean number of customers in the system, denoted by $E(X)$, is calculated from Eq. (9.12)

$$E(X) = P'(1) = \left.\frac{\mathrm{d}P}{\mathrm{d}z}\right|_{z=1} = \rho + \frac{\lambda^2 E(S^2)}{2(1-\rho)}.$$

Since $E(S^2) = Var(S)^2 + (E(S))^2$, we have

$$E(X) = \rho + \frac{\lambda^2\sigma^2 + \rho^2}{2(1-\rho)} \tag{9.13}$$

where $E(S^2)$ is the second order moment of service time distribution and σ^2 is the variance of the service time distribution. This shows that for a fixed traffic intensity ρ, constant service times (i.e., $Var(S) = 0$) gives the smallest congestion in the $M/G/1/\infty$ queue.

2. The mean queue length at service completion epochs is given by

$$E(Q) = E(X) - \lambda E(S) = E(X) - \rho.$$

3. **Waiting time stationary distribution**: Assume that customers, waiting in the queue, are served in the order of their arrival. Let w_t be the waiting time and v_t, the sojourn time for a customer arriving at the queue at time t.

$$W(x) = \lim_{t\to\infty} P\{w_t < x\},\ V(x) = \lim_{t\to\infty} P\{v_t < x\}, \tag{9.14}$$

and

$$w(s) = \int_0^\infty e^{-sx} dW(x), v(s) = \int_0^\infty e^{-Sx} dV(x).$$

Since the sojourn time in the system for a customer is the sum of the waiting time and the service time, the customer's service time in the classical queueing models is assumed to be independent of the state of the system, then $v(s) = w(s)\beta(s)$. For the FIFO discipline, the number of customers still waiting for the service in the system at the time of service completion coincides with the number of customers who arrived in the system during the sojourn time that departing customers spent in the system. Hence,

$$\pi_i = \int_0^\infty \frac{(\lambda x)^i}{i!} e^{-\lambda x} dV(x),\ \ i = 0, 1, \ldots, . \tag{9.15}$$

Multiplying the Eq. (9.15) by the corresponding powers of z and summing them up,

$$\Pi(z) = \int_0^\infty e^{-\lambda(1-z)x} dV(x) = v(\lambda(1-z)).$$

Using $s = \lambda(1-z)$ in the above relation,

$$w(s) = \frac{1-\rho}{1-\lambda\frac{1-\beta(s)}{s}}. \tag{9.16}$$

This formula is called Pollaczek-Khinchine (P-K) formula in LST (refer Appendix A.6) of the waiting time distribution of $M/G/1$ queue.
The mean waiting time $E(w)$ is given by

$$E(W) = \frac{\lambda E(S^2)}{2(1-\rho)}.$$

The probability that there is no customer in the system is given by

$$P(W=0) = \lim_{s\to\infty} \frac{W(s)}{s} = 1-\rho.$$

The mean sojourn time $E(R)$ in the system is given as follows

$$E(R) = E(S) + \frac{\lambda E(S^2)}{2(1-\rho)}.$$

9.1.1.5 Special Case

Consider that the service time is exponential distributed with mean $1/\mu$ and variance $1/\mu^2$. Substitute $\beta(\lambda(1-z)) = \frac{\mu}{\mu+\lambda(1-z)}$ in Eq. 9.8;

$$P(z) = \frac{(1-\rho)(1-z)\frac{\mu}{\mu+\lambda(1-z)}}{\frac{\mu}{\mu+\lambda(1-z)}} = \frac{1-\rho}{1-z\rho} = \sum_{i=0}^{\infty}(1-\rho)(\rho z)^i,$$

Hence,

$$\pi_i = (1-\rho)\rho^i, \;\; i = 0, 1, \ldots.$$

The mean number of customers in the system simplifies to

$$\begin{aligned} E(X) = P\prime(1) &= \left.\frac{\mathrm{dP}(z)}{\mathrm{d}(z)}\right|_{z=1} \\ &= \rho + \frac{\lambda^2(1/\mu^2)^2 + \rho^2}{2(1-\rho)} = \rho + \frac{2\rho}{2(1-\rho)} = \frac{\rho}{1-\rho}, \end{aligned}$$

and the mean queue length becomes

$$E(Q) = E(X) - \rho$$
$$= \frac{\rho}{1-\rho} - \rho = \frac{\rho^2}{1-\rho}.$$

The mean waiting time $E(W)$ is given by

$$E(W) = \frac{\lambda E(S^2)}{2(1-\rho)} = \frac{\rho}{\mu(1-\rho)}$$

and the mean sojourn time is

$$E(R) = \frac{1}{\mu} + \frac{\rho}{\mu(1-\rho)}.$$

The same results have been obtained for the $M/M/1$ queue in Chap. 6.

Example 9.1 Assume that people are entering in two or four wheeler vehicles at IIT Delhi main gate to enter the campus. There is only one line to get the entry pass for the vehicles. A security person is assigned to give the entry pass who take an mean 20 s. The mean arrival rate is 2 vehicles per minute. Find the mean length of queue and mean waiting time in queue assuming $M/G/1$ queueing with service time follows uniform distribution between 10 and 30 s. What is the mean waiting time for any vehicle?
Solution: From the given information, mean service time is $1/\mu = 20$ seconds or $\mu = 3$ vehicles/minute. Arrival rate is $\lambda = 2$ vehicles/minute. Thus, $\rho = 2/3$. Further, given $S \sim U(10, 30)$, $E(S) = 1/3$ minutes. The mean waiting time of any vehicle is given by

$$E(W) = \frac{\rho}{\mu(1-\rho)} = \frac{2}{3} \text{ min} = 40 \text{ s}.$$

9.1.2 $GI/M/1$ *Queueing Model*

Suppose that the single-server queueing system follows general and independent distributed inter-arrival times with distribution $F_a(t)$ and the service time is exponential distributed. Then, this type of queueing system is called as $GI/M/1$ system. The steady-state of the system will be explored by the method of embedded Markov chain and by the method of Markov regenerative theory.

9.1.2.1 The Method of Embedded Markov Chain

When the nth customer arrives at time t_n and $Q_n = Q(t_n^-)$ denotes the number of customers in the system before nth customer arrives. Define X_n as the number of potential service completions during the inter-arrival period Z_n. Let $\{b_j, j = 0, 1, \ldots\}$ be the distribution of X_n.

Then,

$$b_j = P\{X_n = j\} = \int_0^\infty \frac{e^{-\mu t}(\mu t)^j}{j!} \mathrm{d}F_a(t), \qquad j = 0, 1, \ldots$$

The relationship between Q_n and Q_{n+1} is given by

$$Q_{n+1} = \begin{cases} Q_n + 1 - X_{n+1}, & Q_n + 1 - X_{n+1} > 0 \\ 0, & Q_n + 1 - X_{n+1} \le 0 \end{cases}.$$

Note that Q_{n+1} is independent of X_{n+1} and Q_{n+1} does not depend on $Q_0, Q_1, \ldots, Q_{n-1}$. Thus, $\{Q_n, n = 0, 1, \ldots\}$ is a time-homogeneous discrete-time Markov chain. The transition probability is given by

$$\begin{aligned} p_{ij} &= P\{Q_{n+1} = j \mid Q_n = i\} \\ &= \begin{cases} P\{X_{n+1} = i - j + 1\}, & j = 1, 2, \ldots \\ P\{X_{n+1} \ge i + 1\}, & j = 0 \end{cases}. \end{aligned}$$

In matrix form,

$$\mathbf{P} = \begin{pmatrix} \sum_{i=1}^{\infty} b_i & b_0 & 0 & 0 & \ldots \\ \sum_{i=2}^{\infty} b_i & b_1 & b_0 & 0 & \ldots \\ \sum_{i=3}^{\infty} b_i & b_2 & b_1 & b_0 & \ldots \\ \vdots & \vdots & \vdots & \vdots & \end{pmatrix}.$$

For the DTMC to be irreducible, $b_0 > 0$ and $b_0 + b_1 < 1$.

A necessary and sufficient condition for the existence of the steady-state distributions

$$\pi_i = \lim_{t \to \infty} P\{X(t) = i\}, r_i = \lim_{k \to \infty} P\{X(t_k) = i\}, i = 0, 1, \ldots$$

is the fulfillment of $\rho < 1$, where $\rho = \frac{\lambda}{\mu}$ is the system load. Equilibrium equations for the stationary state probabilities r_i of the embedded Markov chain are written as

$$r_j = \sum_{i=j-1}^{\infty} r_i \int_0^\infty \frac{(\mu t)^{i-j+1}}{(i-j+1)!} e^{-\mu t} \mathrm{d}F_a(t), \; j = 1, 2, \ldots. \tag{9.17}$$

The solution of Eq. (9.17) can be found in the form $r_j = C\sigma^j, j = 0, 1, \ldots$. By substitution

$$\sigma = \int_0^{\infty} e^{-\mu(1-\sigma)t} \mathrm{d}F_a(t) = \alpha(\mu(1-\sigma)). \tag{9.18}$$

RHS is the LST of inter-arrival time distribution. Suppose Eq. (9.18) has a unique real root and is denoted by $y(\sigma) = \sigma - \alpha(\mu(1-\sigma))$. $y(\sigma)$ has a unique real root if it satisfies

$$\begin{aligned} y(0) &= -\alpha(\mu) < 0; \quad y(1) = 0 \\ y^{'}(\sigma) &= 1 + \mu\alpha^{'}(\mu(1-\sigma)); \quad y^{'}(1) = 1 - \rho^{-1} < 0 \\ y^{''}(\sigma) &= -(\mu)^2\alpha^{''}(\mu(1-\sigma)) \le 0. \end{aligned}$$

Therefore, the function $y(\sigma)$ is concave, negative at the point $\sigma = 0$, and decreasing at the point $\sigma = 1$. This implies the uniqueness of the root.

The constant C can be obtained from the normalization condition $\sum_{i=0}^{\infty} r_i = 1$ and $C = 1 - \sigma$. Therefore, the stationary probabilities of the system state refer to arrival instants, r_i of the embedded Markov chain will be

$$r_i = (1-\sigma)\sigma^i, i = 0, 1, \ldots. \tag{9.19}$$

Note that the above result will not hold at any arbitrary time epochs since PASTA properties will not be applicable for a $GI/M/1$ queue. To find the stationary probabilities $p_i, i \ge 0$ that there are i customers in the system at an arbitrary epoch. Fix an arbitrary epoch, at which time, the system has exactly i customers. If there were $j, j \ge i-1$ customers in the system at the arrival epoch of previous customer and $i - j + 1$ customers got served and left the system during the time, say u, since the previous arrival epoch. Taking into account that the time u has the distribution function $F(t) = \lambda \int_0^t (1 - F_a(y))\mathrm{d}y$,

$$\pi_i = \sum_{j=i-1}^{\infty} r_j \lambda \int_0^{\infty} \frac{(\mu t)^{j-i+1}}{(j-i+1)!} (1 - F_a(t))\mathrm{d}t, i = 1, 2, \ldots. \tag{9.20}$$

Substituting the value of r_i in the above equation

$$\pi_i = (1-\sigma)\sigma^{i-1}\lambda \int_0^{\infty} e^{-\mu(1-\sigma)t}(1 - F_a(t))\mathrm{d}t, \tag{9.21}$$

which results in $\pi_i = \rho(1-\sigma)\sigma^i, i = 1, 2, \ldots$. π_0 is calculated by the normalization condition as $\pi_0 = 1 - \sigma$. Note that the stationary distribution on π_i is found under equilibrium conditions at the time instant immediately prior to the customers arrivals to the system. It is also valid for the departure instants (just after a customer leaves the system).

9.1.2.2 The Method of Markov Regenerative Theory

Let $X(t)$ denote the number of customers in the system at time t. Figure 9.4 shows the state transition diagram of $GI/M/1$ queueing system. Note that, in this figure, the dotted arcs mean that, the system moves from 0 to state 1 and from state n to state $n+1$ for $n > 0$ with residual inter-arrival time or inter-arrival time. Let $T_n,\ n = 1, 2, \cdots$, be the RV's denoting the arrival time instant of the nth customer. Here, the arrival time instants are the only regeneration time epochs. Let $X_n = X(t_n^-)$ be the number of customers in the system before the nth arrival time instant. Hence, $\{X(t), t \geq 0\}$ is not a semi-Markov process, but a MRGP with $\{X_n, n = 1, 2, \ldots\}$ as an embedded Markov chain for $\{X(t), t \geq 0\}$. $S = \{0, 1, 2, \cdots, \}$ is the state space, and $S^{'} = \{1, 2, \cdots, \}$ is the set of states only at regeneration time instants.

The global kernel matrix is

$$\mathbf{K(t)} = \begin{pmatrix} k_{11}(t) & k_{12}(t) & 0 & 0 & .. & 0 \\ k_{21}(t) & k_{22}(t) & k_{23}(t) & 0 & .. & 0 \\ k_{31}(t) & k_{32}(t) & k_{33}(t) & k_{34}(t) & .. & 0 \\ k_{41}(t) & k_{42}(t) & k_{43}(t) & k_{44}(t) & .. & 0 \\ . & . & . & . & .. & . \\ . & . & . & . & .. & . \\ \vdots & \vdots & \vdots & . & .. & \ddots \end{pmatrix}$$

where

$$k_{ij}(t) = \int_0^t P_{ij-1}(x) dF_a(x)\ ,\quad i, j \in S^{'},$$

and the $P_{ij}(t)$'s are the transition probabilities of the subordinated CTMC. Figure 9.5 denotes the subordinated CTMC of the proposed system and $P_{ij}(t)$ are transition probabilities.

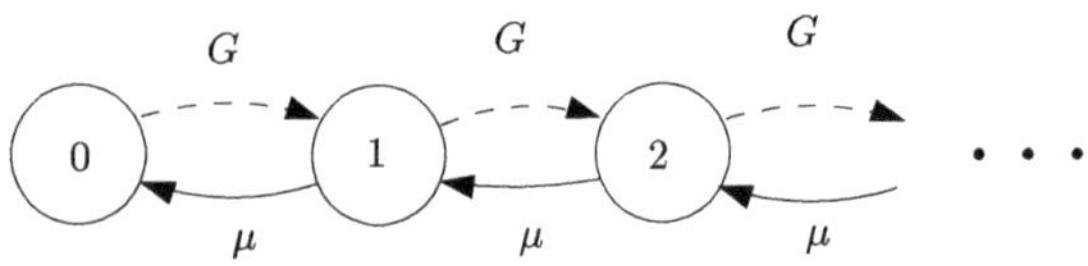

Fig. 9.4 State transition diagram for $GI/M/1$ queueing system

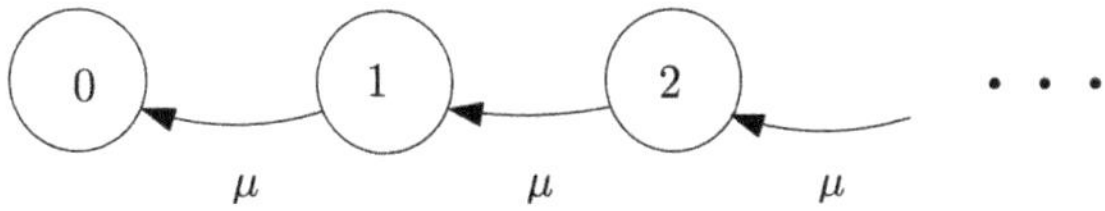

Fig. 9.5 Subordinated CTMC

The local kernel matrix is

$$\mathbf{E(t)} = \begin{pmatrix} e_{10}(t) & e_{11}(t) & 0 & 0 & 0 \ldots & 0 \\ e_{20}(t) & e_{21}(t) & e_{22}(t) & 0 & 0 \ldots & 0 \\ e_{30}(t) & e_{31}(t) & e_{32}(t) & e_{33}(t) & 0 \ldots & 0 \\ \cdot & \cdot & \cdot & \cdot & \cdots & \cdot \\ \cdot & \cdot & \cdot & \cdot & \cdots & \cdot \\ \cdot & \cdot & \cdot & \cdot & \cdots & \cdot \\ \vdots & \vdots & \vdots & \vdots & \cdots & \ddots \end{pmatrix}$$

where $e_{ij}(t) = P_{ij}(t)(1 - F_a(t))$.

9.1.2.3 Performance Measures

1. The mean number of customers in the system at an arbitrary time using Eq. (9.21) is

$$E(X) = \sum_{i=1}^{\infty} i\pi_i = \frac{\rho}{1-\sigma}.$$

2. The mean queue length

$$E(Q) = E(X) - \rho = \sigma\frac{\rho}{1-\sigma}.$$

3. Waiting time stationary distribution
A customer arriving at an empty system has a zero waiting time, i.e., $P(W = 0) = \pi_0 = 1 - \sigma$. The sum of i independent exponential distributed RV with parameter μ is an Erlang RV of order i. Thus, a customer who finds i customers in the system upon arrival have to wait for a time having Erlang distribution with parameters (i, μ). Note that the residual service time of the customer in service which is exponential distributed due to the memoryless property. Therefore,

$$F_W(t) = P(W \le t) = 1 - \sigma + (1-\sigma)\sum_{i=1}^{\infty}\sigma^i \int_0^t \frac{\mu(\mu x)^{i-1}}{(i-1)!}e^{-\mu x}\mathrm{d}x$$
$$= 1 - \sigma e^{-\mu(1-\sigma)t}.$$

The mean waiting time is given by $\frac{\sigma}{\mu(1-\sigma)}$.

Special Case: Let the arrivals follow a Poisson process with parameter λ. For a $M/M/1$ queueing system, Eq. (9.17) becomes

$$\sigma = \int_0^\infty e^{-\mu(1-\sigma)t} \mathrm{d}F_a(t)\mathrm{d}t.$$

Solving $\mu\sigma^2 - (\lambda + \mu)\sigma + \lambda = 0$, one of the values of σ is $\sigma = \rho$, and

$$\pi_j = (1-\rho)\rho^j, \ j = 0, 1, \ldots.$$

Therefore,

$$E(X) = \frac{\rho}{1-\rho}$$

and the mean queue length is

$$E(Q) = \frac{\rho^2}{1-\rho}.$$

The mean waiting time is given by

$$E(W) = \frac{\rho}{\mu(1-\rho)}.$$

Example 9.2 Suppose that, in a telephone manufacturing factory, a component arrives randomly at a testing facility once in 30 s. Assume the duration (in seconds) for testing the component has an exponential distribution with mean of 20 seconds. By considering the above system as a $GI/M/1$ queueing model, find the steady-state probability that no components are in the system for testing.
Solution: Given $\lambda = 2$ per minute. $\mu = 3$ per minute. The required probability that there are no components in the system for testing in the long run is

$$\pi_0 = 1 - \rho = 1/3.$$

9.1.3 $GI/G/1/\infty$ Queueing Model

By Little's formula, $E(X) = \lambda E(R)$, for stable system, $\rho = \lambda E(S) < 1$. $\pi_0 = 1 - \rho$. When S and A are independent RVs with FIFO or LIFO (but not for sharing processor time)

$$E(X) = \lambda E(S)E(A) + \frac{1}{2}\lambda E(S^2).$$

This results in Pallaczele-Khinchin's mean value formula.

9.2 Single-Server Finite Capacity Systems

Next a finite capacity non-Markov queueing model is discussed. The first model in this list is the $M/G/1/N$ queueing model. This is similar to $M/G/1$, but now the system capacity is restricted to N. Figure 9.6 shows the state transition diagram of $M/G/1/N$ queueing system. Here, the transition with dotted arcs means that, the system moves from state 2 to state 1 or state n to state $n-1$ for $3 \leq n \leq N$ with residual service time or service time whereas the system moves from state 1 to state 0 with service time. Further note that, when the system is in state N, a new arrival will not make any change in the state of the system.

9.2.1 $M/G/1/N$ Queueing Model

Consider a single-server queue with a limited buffer (say, N) and arrivals follow Poisson process with parameter λ. The customer's service time, denoted by S, has an arbitrary distribution with CDF $F_s(t)$. The analysis of the finite-capacity $M/G/1/N$ queue proceeds in a way very similar to that of the unlimited-waiting-room case. Let $X(t)$ denote the number of customers in the system at time t. Let $T_n, n = 1, 2, \ldots$ be the RVs denoting the departure time instant of the nth customer. The P-K formula will not hold now, since the expected number of (joined) arrivals during a service period must be conditioned on the system size. The best way to get the new result is directly from the steady-state probabilities, since there are now only a finite number of them. The Markov regenerative process $\{X(t), t \geq 0\}$ constitutes a $S = \{0, 1, \ldots, N\}$ and embedded Markov chain $\{X(t_n^+), n = 0, 1, \ldots\}$ where t_n is the nth customer departure instant and $X(t_n^+)$ be the corresponding number of customers left behind in the system by departing customer. $\{X_n, n = 0, 1, 2, \ldots\}$ is an embedded Markov chain with state space $S' = \{0, 1, 2, \ldots, N-1\}$.

The single-step transition matrix must be truncated at $N-1$ such that

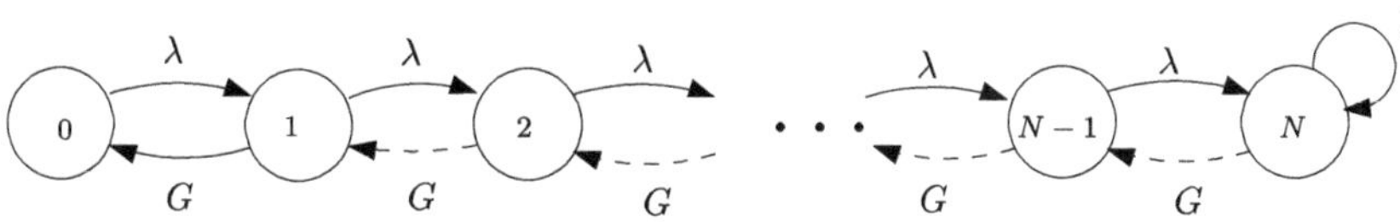

Fig. 9.6 State transition diagram of $M/G/1/N$ queueing system

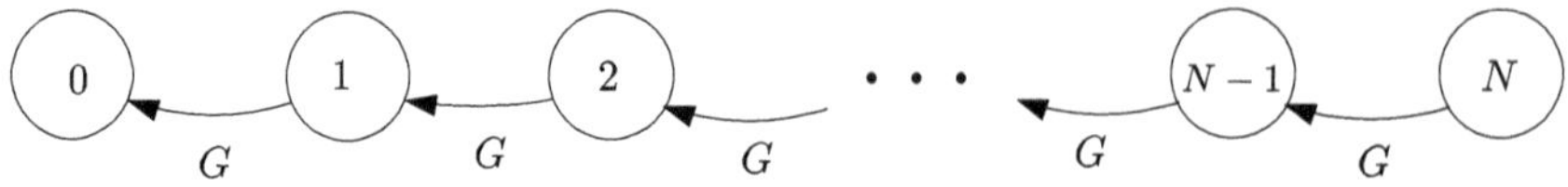

Fig. 9.7 Subordinated CTMC

$$\mathbf{K(t)} = \begin{pmatrix} k_{11}(t) & k_{12}(t) & k_{13}(t) & k_{14}(t) & \dots & k_{1N}(t) \\ k_{21}(t) & k_{22}(t) & k_{23}(t) & k_{24}(t) & \dots & k_{2N}(t) \\ 0 & k_{31}(t) & k_{32}(t) & k_{33}(t) & \dots & k_{3N}(t) \\ 0 & 0 & k_{41}(t) & k_{42}(t) & \dots & k_{4N}(t) \\ \vdots & \vdots & \vdots & \vdots & \ddots & \vdots \\ . & . & . & . & \dots & k_{NN}(t) \end{pmatrix}$$

where

$$k_{ij}(t) = \int_0^t P_{ij-1}(x) dF_s(x) \,, \quad i, j \in S^{'},$$

and the $P_{ij}(t)$'s are the transition probabilities of the subordinated CTMC. Figure 9.7 depicts the subordinated CTMC of the queueing system.

The local kernel matrix is

$$\mathbf{E(t)} = \begin{pmatrix} e_{10}(t) & e_{11}(t) & e_{12}(t) & e_{13}(t) & e_{14}(t) & \dots & e_{1N}(t) \\ e_{20}(t) & e_{21}(t) & e_{22}(t) & e_{23}(t) & e_{24}(t) & \dots & e_{2N}(t) \\ 0 & e_{30}(t) & e_{31}(t) & e_{32}(t) & e_{33}(t) & \dots & e_{3N}(t) \\ \vdots & \vdots & \vdots & \vdots & \vdots & \ddots & \vdots \\ . & . & . & . & . & \dots & e_{NN}(t) \end{pmatrix}$$

where $e_{ij}(t) = P_{ij}(t)(1 - F_s(t))$.

Example 9.3 Consider a manufacturing factory which can keep at most 2 items at a time for service of components. Components arrive as a Poisson process with rate $\lambda = 2$. The time required for serving a component follows a general distribution. Find the steady-state probability of the system when the time required for serving follows a $Weibull(2, 1/2)$ distribution.

Solution: The system can be modeled as a $M/G/1/2$ queueing system where each state represents a number of items for serving. In this model, service completion time instants are the only regeneration time points. Hence, the time points of entering into states 0 and 1 are the only regeneration time instants. Therefore, $\Omega^{'} = \{0, 1\}$ and $\Omega = \{0, 1, 2\}$. When service time follows a $Weibull(2, 1/2)$

$$F_s(t) = 8te^{-4t^2}, t \geq 0.$$

The global kernel can be defined as

$$\mathbf{K(t)} = [k_{ij}(t)], i, j \in \Omega',$$

where

$$k_{ij}(t) = \begin{cases} \int_0^t (P_{00} + P_{01}(x))dF_s(x), i, j = 0 \\ \int_0^t P_{ij}(x)dF_s(x), i, j \in \Omega'. \end{cases}$$

The local kernel is $E(t) = [e_{ij}(t)], i, j \in \Omega'$

$$e_{ij}(t) = P(X_1 = j, T_1 \geq t \mid X_0 = i) = p_{ij}(t)(1 - F_s(t)), i, j \in \Omega',$$

where $P_{ij}(t)$ are the time-dependent probabilities of subordinated CTMC which are computed as follows

$$\mathbf{P(t)} = \begin{pmatrix} e^{-\lambda t} & \lambda t e^{-\lambda t} & 1 - e^{-\lambda t} - \lambda t e^{-\lambda t} \\ 0 & e^{-\lambda t} & 1 - e^{-\lambda t} \\ 0 & 0 & 1 \end{pmatrix}.$$

The elements of $\mathbf{K(t)}$ are evaluated as

$$k_{00}(t) = \int_0^\infty 8x(\lambda x + 1)e^{-\lambda x - 4x^2}\mathrm{d}x,$$

$$k_{01}(t) = \int_0^\infty 8x(1 - \lambda x e^{-\lambda x} - e^{-\lambda x})e^{-4x^2}\mathrm{d}x,$$

$$k_{10}(t) = \int_0^\infty 8xe^{-\lambda x - 4x^2}\mathrm{d}x,$$

$$k_{11}(t) = \int_0^\infty 8x(1 - e^{-\lambda x})e^{-4x^2}\mathrm{d}x.$$

Substituting $\lambda = 2$, the values are $k_{00}(t) = 0.7728, k_{01}(t) = 0.2272, k_{10}(t) = 0.4544$, $k_{11}(t) = 0.5456$. The steady-state probability is computed by $\mathbf{v}=\mathbf{vP}$, $\sum v_k = 1$, to obtain $v_0 = v_1 = 1/2$. The components of the matrix $\mathbf{E}$ are

$$\begin{aligned} e_{00}(t) &= e^{-\lambda t - 4t^2}, \\ e_{01}(t) &= \lambda t e^{-\lambda t - 4t^2}, \\ e_{02}(t) &= (1 - e^{-\lambda t} - \lambda t e^{-\lambda t})e^{-4t^2}, \\ e_{10}(t) &= 0, \\ e_{11}(t) &= e^{-\lambda t - 4t^2}, \\ e_{12}(t) &= (1 - e^{-\lambda t})e^{-4t^2}. \end{aligned}$$

Substituting $\lambda = 2$, and using $\alpha_{ij} = \int_0^\infty e_{ij}(t)dt$, $\alpha_{00} = 0.2728, \alpha_{01} = 0.1136, \alpha_{02} = 0.0567, \alpha_{10} = 0, \alpha_{11} = 0.2728, \alpha_{12} = 0.1703$. Finally π_j's are computed using

$$\pi_j = \frac{\sum\limits_{k\in\Omega'} v_k\alpha_{kj}}{\sum\limits_{k\in\Omega'} v_k\beta_k}, \quad j = 0, 1, 2.$$

Here, $\beta_0 = \beta_1 = \sqrt{2\pi}/8$. Hence $\pi_0 = 0.3078, \pi_1 = 0.436, \pi_2 = 0.2562$.

9.2.2 $GI/M/1/N$ Queueing Model

This queueing model is similar to $GI/M/1$, but now the system capacity is restricted to N.

9.2.2.1 The Method of Markov Regenerative Process

Let X_t denote the number of customers in the system at time t. Figure 9.8 demonstrates the state transition diagram of $GI/M/1/N$ queueing system. Note that, in this figure, the dotted arcs mean that, the system moves from 0 to state 1 and from state n to state $n+1$ for $1 \le n \le N$ with residual inter-arrival time or inter-arrival time. Further note that, when the system is in state N, the system will not change the state if a new arrival occurs. Let T_n, $n = 1, 2, \ldots$, be the RVs denoting the arrival time instant of the nth customer. Here, the arrival time instants are the only regeneration time epochs. Note that the system changes to different states either by service or by inter-arrival instances. Hence, $\{X_t, t \ge 0\}$ is not a semi-Markov process, but a MRGP with $\{X_n, n = 1, 2, \ldots\}$ as an embedded Markov chain. $S = \{0, 1, \ldots, N\}$ is the state space, and $S' = \{1, 2, \ldots, N\}$ is the set of states only at regeneration time instants.

The global kernel matrix is

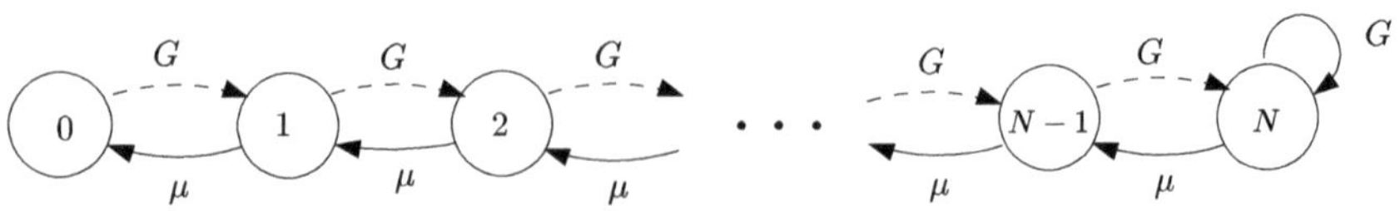

Fig. 9.8 State transition diagram of $GI/M/1/N$ queueing system

$$\mathbf{K(t)} = \begin{pmatrix} k_{11}(t) & k_{12}(t) & 0 & 0 & \dots & 0 \\ k_{21}(t) & k_{22}(t) & k_{23}(t) & 0 & \dots & 0 \\ k_{31}(t) & k_{32}(t) & k_{33}(t) & k_{34}(t) & \dots & 0 \\ k_{41}(t) & k_{42}(t) & k_{43}(t) & k_{44}(t) & \dots & 0 \\ \vdots & \vdots & \vdots & \vdots & \ddots & \vdots \\ k_{N1}(t) & k_{N2}(t) & k_{N3}(t) & . & \dots & k_{NN}(t) \end{pmatrix}$$

where

$$k_{ij}(t) = \begin{cases} \int_0^t P_{ij-1}(x) dF_a(x)\,, & i, j \in S^{'} \\ \int_0^t (P_{NN-1}(x) + P_{NN}(x))\, dF_a(x)\,, & i, j = N \end{cases}$$

and the $P_{ij}(t)$'s are the transition probabilities of the subordinated CTMC. The state transition diagram of subordinate CTMC is given in Fig. 9.9.

The local kernel matrix is

$$\mathbf{E(t)} = \begin{pmatrix} e_{10}(t) & e_{11}(t) & 0 & 0 & 0 & \dots & 0 \\ e_{20}(t) & e_{21}(t) & e_{22}(t) & 0 & 0 & \dots & 0 \\ e_{30}(t) & e_{31}(t) & e_{32}(t) & e_{33}(t) & 0 & \dots & 0 \\ \vdots & \vdots & \vdots & \vdots & \vdots & \ddots & \vdots \\ e_{N0}(t) & e_{N1}(t) & e_{N2}(t) & e_{N3}(t) & e_{N4}(t) & \dots & e_{NN}(t) \end{pmatrix},$$

where $e_{ij}(t) = P_{ij}(t)(1 - F_a(t))$.

Example 9.4 Consider a single-chair unisex hair-styling shop in which at most two customers can accommodate. The arrivals of customers follow the Poisson process with parameter λ. The hair-styling time for a customer follows a general distribution with a mean $1/\mu$. Compute the steady-state probability when the service time distribution follows a $Weibull(2, 1/2)$ distribution.

Solution: This system can be analyzed as a $GI/M/1/N$ queue with $N = 2$. Here, $\Omega^{'} = \{1, 2\}$ and $\Omega = \{0, 1, 2\}$. When service time follows a $Weibull(2, 1/2)$

$$F_s(t) = 8te^{-4t^2}, t \geq 0.$$

The global kernel can be defined as

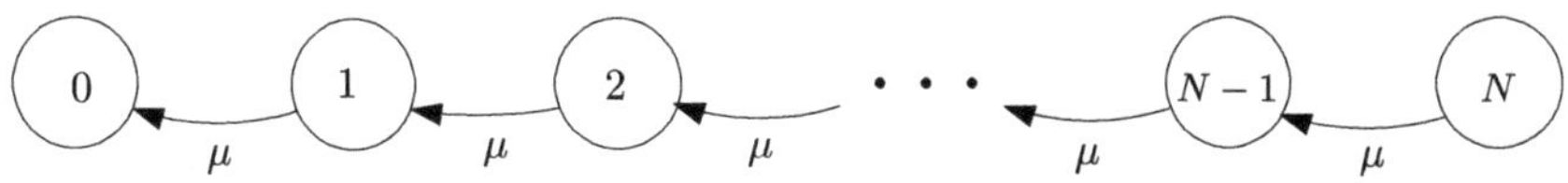

Fig. 9.9 Subordinated CTMC

$$\mathbf{K(t)} = [k_{ij}(t)], i, j \in \Omega',$$

where

$$k_{ij}(t) = \begin{cases} \int_0^t (P_{ij-1}) dT(x), \ (i, j) = (1, 1)(1, 2)(2, 1) \\ \int_0^t (P_{21}(x) + P_{22}(x)) dT(x), \ (i, j) = (2, 2). \end{cases}$$

The local kernel is $E(t) = [e_{ij}(t)], i, j \in \Omega'$

$$e_{ij}(t) = P\{X_1 = j, T_1 \geq t \mid X_0 = i\} = P_{ij}(t)(1 - F_s(t)), \ \ i, j \in \Omega'$$

where $P_{ij}(t)$ are the time-dependent probabilities of subordinated CTMC which are computed as follows

$$\mathbf{P(t)} = \begin{pmatrix} 1 & 0 & 0 \\ e^{-\mu t} & 1 - e^{-\mu t} & \\ e^{-\mu t} & \mu t e^{-\mu t} & 1 - e^{-\mu t} - \mu t e^{-\mu t} \end{pmatrix}.$$

The elements of $\mathbf{K(t)}$ are evaluated as

$$\begin{aligned} k_{11}(t) &= \int_0^t (1 - e^{\mu x}) 8x e^{4x^2} dx, \\ k_{12}(t) &= \int_0^t 8x e^{-\mu x - 4x^2} dx, \\ k_{21}(t) &= \int_0^t 8x(1 - e^{-\mu x} - \mu x e^{-\mu x}) e^{-4x^2} dx, \\ k_{22}(t) &= \int_0^t 8\mu x^2 e^{-\mu x - 4x^2} dx + \int_0^t 8x e^{-\mu x - 4x^2} dx. \end{aligned}$$

Substituting $\mu = 2$, $k_{11}(t) = 0.5456$, $k_{12}(t) = 0.4544$, $k_{21}(t) = 0.2172$, $k_{22}(t) = 0.728$. The steady-state probability is computed by $v = vP, \sum v_k = 1$. Solving, it is obtained that $v_1 = v_2 = 1/2$. The matrix $\mathbf{E}$ is

$$\begin{aligned} e_{10}(t) &= (1 - e^{-\mu t}) e^{-4t^2}, \\ e_{11}(t) &= e^{-\mu t - 4t^2}, \\ e_{20}(t) &= (1 - e^{-\mu} t - \mu t e^{-\mu t}) e^{-4t^2}, \\ e_{12}(t) &= 0, \\ e_{21}(t) &= \mu t e^{-\mu t - 4t^2}, \\ e_{22}(t) &= e^{-\mu t - 4t^2}. \end{aligned}$$

Substituting $\mu = 2$, and using $\alpha_{ij} = \int_0^\infty e_{ij}(t) \mathrm{d}t$ one gets $\alpha_{10} = 0.1703$, $\alpha_{11} = 0.2728$, $\alpha_{12} = 0$, $\alpha_{20} = 0.0567$, $\alpha_{21} = 0.1136$, $\alpha_{22} = 0.2728$. Finally, π_j are computed using

$$\pi_j = \frac{\sum\limits_{k\in\Omega'} v_k \alpha_{kj}}{\sum\limits_{k\in\Omega'} v_k \beta_k}, \quad j = 0, 1, 2.$$

Here, $\beta_1 = \beta_2 = \sqrt{2\pi}/8$. Hence $\pi_0 = 0.2561$, $\pi_1 = 0.436$, $\pi_2 = 0.3078$.

9.3 Multi-server Queueing Models

Next, the multi-server queueing models will be described where either the arrival or service process is generally distributed. Figure 9.10 shows the state transition diagram of $M/G/c$ queueing system. Note that, in this figure, the dotted arcs mean that, the system moves from state 2 to state 1 and from state n to state $n-1$ for $n \geq 3$ with residual service time or service time whereas the system moves from state 1 to state 0 with service time.

9.3.1 *M/G/c Queueing Model*

For a $M/G/c$ queue, the result of interest can be obtained through the relationship between the kth factorial moment of system size and regular kth moment of system delay and is given by $E(N)_k = \lambda^k E(R)_k$. It is known that

$$\pi_n = P\{n \text{ in system just after a departure}\} = \frac{1}{n!}\int_0^\infty (\lambda t)^n e^{-\lambda t} dE(R)(t), \; n = 0, 1, \ldots.$$

In the modified form for $M/G/c$

$$\pi_n^q = P\{n \text{ in queue just after a departure}\} = \frac{1}{n!}\int_0^\infty (\lambda t)^n e^{-\lambda t} dE(R)_q(t), \; n = 0, 1, \ldots.$$

with the mean queue length at departure points is given by

Fig. 9.10 State transition diagram of $M/G/c$ queueing system

$$E(Q) = \sum_{n=1}^{\infty} n\pi_n^q = \int_0^{\infty} \lambda t dE(R)_q(t) = \lambda E(R)_q.$$

$L_{q(k)}$ denote the kth factorial moment of the departure point queue size. Then

$$\begin{aligned} L_{q(k)} &= \sum_{n=1}^{\infty} n(n-1)\ldots(n-k+1)\pi_n^q \\ &= \int_0^{\infty} \sum_{n=1}^{\infty} \frac{n(n-1)\ldots(n-k+1)(\lambda t)^n e^{-\lambda t}}{n!} dE(R)_q(t). \end{aligned}$$

But the summand is nothing more than the kth factorial moment of the Poisson distribution which can be shown to be $(\lambda t)^k$. Hence $L_{q(k)} = \lambda^k E(R)_{q,k}$ where $E(R)_{q,k}$ is the regular kth moment of the waiting time.

Example 9.5 Consider a $M/D/c$ model, with service follows deterministic distribution. The service will be of constant time and $\mu = 1$. Let π_n denote the CDF of the system size in the steady-state. Therefore,

$$\begin{aligned} \pi_0 =& P\{c \text{ or less system at arbitrary instant in steady state}\} \\ & \times\ P\{0 \text{ arrivals in subsequent unit of time}\}, \\ \pi_1 =& P\{c \text{ or less in system }\} \times\ P\{1 \text{ arrival}\} + P\{c+1 \text{ in system}\} \times\ P\{0 \text{ arrival}\}, \\ \pi_2 =& P\{c \text{ or less in system}\}\ \times\ P\{2 \text{ arrival}\} + P\{c+1 \text{ in system}\} \\ & \times\ P\{1 \text{ arrival}\} + P\{c+2 \text{ in system}\} \times\ P\{0 \text{ arrival}\}, \\ & \vdots \\ \pi_n =& P\{c \text{ or less in system}\} \times r\{n \text{ arrival}\} + P\{c+1 \text{ in system}\} \times P\{n-1 \text{ arrival}\} \\ & + \ldots + P\{c+n \text{ in system}\} \times\ P\{0 \text{ arrival}\}, \end{aligned}$$

or

$$\begin{aligned} \pi_0 &= \pi_c e^{-\lambda}, \\ \pi_1 &= \pi_c \lambda e^{-\lambda} + \pi_{c+1} e^{-\lambda}, \\ &\vdots \\ \pi_n &= \frac{\pi_c \lambda^n e^{-\lambda}}{n!} + \frac{\pi_{c+1}\lambda^{n-1} e^{-\lambda}}{(n-1)!} + \ldots + \pi_{c+n} e^{-\lambda},\ \ n = 1, 2, \ldots \end{aligned}$$

Define the PGF $P(z) = \sum_{n=0}^{\infty} \pi_n z^n$. Therefore

$$P(z) = \frac{\sum_{n=0}^{c} \pi_n z^n - \pi_c z^c}{1 - z^c e^{\lambda(1-z)}} = \frac{\sum_{n=0}^{c-1} \pi_n (z^n - z^c)}{1 - z^c e^{\lambda(1-z)}}. \tag{9.22}$$

The numerator is a polynomial of degree c, and may be written as

$$N(z) = K(z-1)(z-z_1)\dots(z-z_{c-1}),$$

where $1, z_1, z_2, \dots, z_{c-1}$ are the coincident roots of the denominator and numerator. To obtain K, use the fact that $P(1) = 1$, By L'Hospital rule,

$$1 = \lim_{z \to 1} P(z) = \frac{K(1-z_1)\dots(1-z_{c-1})}{\lambda - c},$$

$$K = \frac{\lambda - c}{(1-z_1)\dots(1-z_{c-1})}.$$

Hence,

$$P(z) = \frac{(z-1)(z-z_1)\dots(z-z_{c-1})}{1 - z^c e^{\lambda(1-z)}}. \tag{9.23}$$

The roots $\{z_i, i = 1, 2, \dots, c-1\}$ can be obtained as $z^c = e^{-\lambda(1-z)}$. To obtain π_n, π_0 is evaluated from the generating function at $z = 0$,

$$\pi_0 = \frac{(c-\lambda)(-1)^{c-1} \Pi_{i=1}^{c-1} z_i}{\Pi_{i=1}^{c-1}(1-z_i)}, \quad c = 2, 3, \dots. \tag{9.24}$$

Next, $\{\pi_1, \pi_2, \dots, \pi_{c-1}\}$ are obtained from a $(c-1)$ ordered complex valued system of equations which are obtained from the numerator of Equation (9.23) set equal to zero at each of the $c-1$ roots as z_i. The resultant matrix of coefficient has an (i, j) entry given by

$$a_{ij} = \pi_0 = z_i^j - z_i^c, i, j = 1, 2, \dots, c-1,$$

and right hand side elements equal

$$b_i = \pi_0(z_i^c - 1), i = 1, 2, \dots, c-1.$$

The remaining probabilities are obtained as $\pi_0 = \pi_c e^{-\lambda}$ and

$$\pi_c = (e^{\lambda} - 1)\pi_0 - \sum_{i=1}^{c-1} \pi_i.$$

9.3.2 $GI/M/c$ Queueing Model

In the $GI/M/c$ queueing system the arrivals are considered to be generally distributed and service time is assumed to be exponentially distributed with multiple servers.

9.3.2.1 The Embedded Markov Chain Method

The embedded Markov chain analysis of the queue $GI/M/1$ can easily be extended to the multi-server queue $GI/M/c$. Let $X(t)$ be the number of customers in the system at time t. Let $X_n = X(t_n^-)$ be the number of customers in the system just before the arrival time epoch t_n of the nth customer. Therefore, $\{X_n, n \geq 0\}$ is an embedded Markov chain of $\{X(t), t \geq 0\}$. Here, $X_{n+1} = X_n + 1 - B_{n+1}$, where B_{n+1} is the number of departures during $T_n = (t_{n+1} - t_n)$, i.e., the interval between the nth and $(n+1)$th arrivals. The transition probabilities are defined as follows

$$p_{ij} = P\{X_n = j | X_{n-1} = i\}.$$

Consider three cases for the initial value i and the final value j:

$$i + 1 \geq j \geq c; \quad i + 1 \leq c \text{ and } j \leq c \quad \text{and} \quad i + 1 > c \text{ but } j < c.$$

Case 1: When $i + 1 \geq j \geq c$ The system serves at the mean rate $c\mu$, since all servers are busy. Therefore,

$$p_{ij} = b_{i+1-j}, c \leq j \leq i + 1,$$

here

$$b_n = \int_0^\infty e^{-c\mu t} \frac{(c\mu t)^n}{n!} \mathrm{d}A(t).$$

Thus, the transition probability matrix for the multi-server problem has the same kind of layout from its cth column to the right as the $G/M/1$ does from its first column to the right.

Case 2: When $i + 1 \leq c$ and $j \leq c$, everyone is being served, and the probability that anyone has completed service by a time t is the CDF of service time by an individual server, $1 - e^{-\mu t}$. Therefore,

$$p_{ij} = \binom{i+1}{i+1-j} \int_0^\infty e^{-j\mu t} (1 - e^{-\mu t})^{i+1-j} \mathrm{d}A(t).$$

This represents $i + 1 - j$ out of $i + 1$ customers complete service.

Case 3: When $i + 1 > c$ but $j < c$. Here the system starts out with all servers busy, since $i \geq c$, and sometime during the inter-arrival time, servers become idle, until finally only j servers are busy. Let's assume that at a time, denoted by V, after the arrival comes $(0 < V < T)$, it goes into service with $H(v)$, the CDF of V. Therefore, to go from state i to j in time t, $c - j$ service should be completed from V to T or during a time interval of length $T - V$.

$$p_{ij} = \int_{t=0}^{\infty} \int_{0}^{t} e^{-c\mu t} \frac{(c\mu t)^{i-c} c\mu}{(i-c)!} \binom{c}{c-j} e^{-j\mu(t-v)} (1 - e^{-\mu(t-v)})^{c-j} dH(v) dA(t).$$

This represents $i + 1 - c$ customers complete service with rate $c\mu$, and then $c - j$ out of c customers complete service. The RV V represents the time until $i - c + 1$ people have been served with all c servers working, which is the $i - c + 1$ fold convolution of the exponential distributions with parameter $c\mu$. Hence, the PDF of V, $h(v) = dH(v)/dv$ is same as the PDF of Erlang distributed with parameters $i - c + 1$ and $c\mu$ and is given by

$$h(v) = \frac{c\mu(c\mu v)^{i-c} e^{-c\mu v}}{(i-c)!}, v > 0.$$

Substituting for $dH(v)$ for $j < c < i + 1$

$$p_{ij} = \binom{c}{c-j} \frac{(c\mu)^{i-c+1}}{(i-c)!} \int_{0}^{\infty} \int_{0}^{t} (1 - e - \mu(t - v))^{c-j} e^{-\mu(t-v)j} v^{i-c} e^{-c\mu v} \mathrm{d}v \mathrm{d}A(t).$$

The stationary equation is given by

$$\pi_j = \sum_{i=0}^{\infty} p_{ij} \pi_i, \; j \geq 0,$$

where the p_{ij}'s are given as throughput. For $j \geq c$

$$\pi_j = \sum_{k=0}^{\infty} b_k \pi_{j+k-1}, \; j \geq c.$$

The above equation is identical to

$$\pi_j = K r_0^j, \; j \geq c$$

where r_0 is the root of $\beta(z) = A^*[c\mu(1 - z)] = z$ and A^* the LST of $A(t)$. The constant K and π_j can be obtained from the boundary condition and the first $c - 1$ stationary equations

$$1 = \sum_{j=0}^{\infty} \pi_j = \sum_{j=0}^{c-1} \pi_j + \sum_{j=c}^{\infty} K r_0^j$$

$$K = \frac{1 - \sum_{j=0}^{c-1} \pi_j}{\sum_{j=c}^{\infty} r_0^j} = \frac{1 - \sum_{j=0}^{c-1} \pi_j}{r_0^c (1 - r_0)^{-1}}.$$

Through a recursive relation:

$$\begin{aligned} \pi_j &= \sum_{i=0}^{\infty} p_{ij} \pi_i = \sum_{i=0}^{c-1} p_{ij} \pi_i + \sum_{i=c}^{\infty} p_{ij} K r_0^i \\ &= \sum_{i=j-1}^{c-1} p_{ij} \pi_i + K \sum_{i=c}^{\infty} p_{ij} r_0^i. \end{aligned}$$

Since $p_{ij} = 0$ for $j > i + 1$. After rewriting

$$\pi_{j-1} = \frac{\pi_j - \sum_{i=j}^{c-1} p_{ij} \pi_i - K \sum_{i=c}^{\infty} p_{ij} r_0^i}{p_{j-1,j}}.$$

From the stationary equations,

$$\pi_c = \sum_{i=0}^{\infty} p_{ic} \pi_i = \sum_{i=c-1}^{c} p_{ic} \pi_i + \sum_{i=c+1}^{\infty} b_{i+1-c} K r_0^i.$$

Hence,

$$\pi_{c-1} = \frac{(1 - p_{cc}) \pi_c - K \sum_{i=c+1}^{\infty} b_{i+1-c} r_0^i}{p_{c-1,c}}$$

and

$$\pi_{c-1}^{'} = \frac{(1 - b_1) \pi_c^{'} - \sum_{i=c+1}^{\infty} b_{i+1-c} r_0^i}{b_0}.$$

Here $\pi_c^{'} = r_0^c$ and

$$K = \left(\sum_{j=0}^{\infty} \pi_j^{'} + \frac{r_0^c}{1 - r_0} \right)^{-1} .$$

9.3.2.2 The Method of Markov Regenerative Process

Let X_t denote the number of customers in the system at time t. Figure 9.11 depicts the state transition diagram of $GI/M/c$ queueing system. Note that, in this figure, the dotted arcs mean that, the system moves from state 0 to state 1 and from state n to state $n+1$ for $n > 0$ RVs with residual inter-arrival time or inter-arrival time. Let $T_n,\ n = 1, 2, \ldots$, be the RVs denoting the arrival time instant of the nth customer. Let $X_n = X(t_n^-)$ denote the number of customers in the system before nth customer arrives. Here, the arrival time instants are the only regeneration time epochs. Hence, $\{X_t, t \geq 0\}$ is not a semi-Markov process, but a MRGP with $\{X_n, n = 1, 2, \ldots\}$ as an embedded Markov chain. $S = \{0, 1, \ldots\}$ is the state space, and $S^{'} = \{1, 2, \ldots\}$ is the set of states only at regeneration time instants.

The global kernel matrix is

$$\mathbf{K(t)} = \begin{pmatrix} k_{11}(t) & k_{12}(t) & 0 & 0 & \ldots 0 \\ k_{21}(t) & k_{22}(t) & k_{23}(t) & 0 & \ldots 0 \\ k_{31}(t) & k_{32}(t) & k_{33}(t) & k_{34}(t) & \ldots 0 \\ k_{41}(t) & k_{42}(t) & k_{43}(t) & k_{44}(t) & \ldots 0 \\ \vdots & \vdots & \vdots & \vdots & \ddots \vdots \end{pmatrix},$$

where

$$k_{ij}(t) = \int_0^t P_{ij-1}(x) dF_a(x)\ , \quad i, j \in S^{'},$$

and the $P_{ij}(t)$'s are the transition probabilities of the subordinated CTMC.

Figure 9.4 denotes the subordinated CTMC of the proposed system and $P_{ij}(t)$ are transition probabilities (Fig. 9.12).

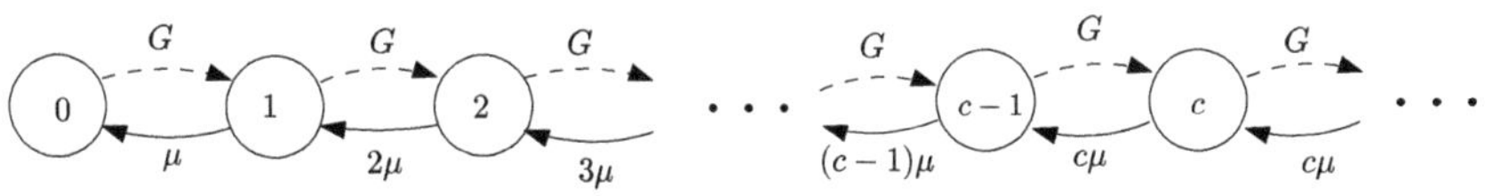

Fig. 9.11 State transition diagram of $GI/M/c$ queueing system

The local kernel matrix is

$$\mathbf{E(t)} = \begin{pmatrix} e_{10}(t) & e_{11}(t) & e_{12}(t) & e_{13}(t) & e_{14}(t) & \ldots & . \\ e_{20}(t) & e_{21}(t) & e_{22}(t) & e_{23}(t) & e_{24}(t) & \ldots & . \\ 0 & e_{30}(t) & e_{31}(t) & e_{32}(t) & e_{33}(t) & \ldots & . \\ 0 & 0 & e_{40}(t) & e_{41}(t) & e_{42}(t) & . & . \\ . & . & . & . & . & \ldots & . \\ . & . & . & . & . & \ldots & . \\ \vdots & \vdots & \vdots & \vdots & . & & \ddots \end{pmatrix}$$

where $e_{ij}(t) = P_{ij}(t)(1 - F_a(t))$.

9.3.2.3 Performance Measures

1. The mean waiting time in the system is given by

$$E(R) = \frac{\pi_c}{c\mu(1 - r_0)^2} = \frac{K r_0^c}{c\mu(1 - r_0)^2}.$$

Example 9.6 Consider an $E_k/M/1$ queueing model where $A^* = (\lambda k/(s + \lambda k))^k$. Find the steady-state probability with $k = 1$.
Solution: The characteristic equation becomes

$$\beta(z) = z - \left(\frac{\lambda k}{\mu - \mu z + \lambda k}\right)^k = 0.$$

For $k = 1$, the equation reduces to $z^2 - (1 + \rho)z + \rho = 0$ with $r_0 = \rho$ as the unique root inside $|z| = 1$. Therefore, $\pi_n = (1 - \rho)\rho^n, n \geq 0$.

9.4 Multi-server Loss Systems

In this section, the multi-server queueing models will be described where the service process is generally distributed with finite capacity c.

Fig. 9.12 Subordinated CTMC

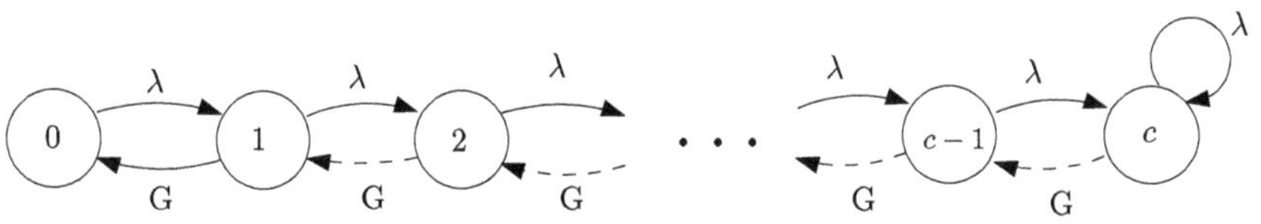

Fig. 9.13 State transition diagram of $M/G/c/c$ loss system

9.4.1 *M/G/c/c System*

In the $M/G/c/c$ system, the arrivals follow the Poisson process and the service time is exponentially distributed. This model is similar to the $M/G/c$ queueing system but here the capacity is finite. Figure 9.13 shows the state transition diagram of the $M/G/c/c$ loss system. Note that, in this figure, the dotted arcs means that, the system moves from state 2 to state 1 and from state n to state $n-1$ for $n \geq 3$ with residual service time or service time. Whereas the system moves from state 1 to state 0 with service time. Further note that, when the system is in state c, a new arrival will not change the state.

9.4.1.1 The Embedded Markov Chain Method

Let us consider a $M/G/c/c$ loss system. Let us assume that at some epoch t, $X(t)$, the number of busy server equals $k, 0 \leq k \leq c$ and $X(s) \neq k$ when $s < t$. Let the busy servers at any epoch t be given numbers $1, 2, \ldots, k$ in random order and let $f_i(t)$ denote the time elapsed from epoch t until the server with number i completes the service ($f_i(t)$ is the remaining service time from epoch t of server number i). The vector stochastic process

$$V(t) = \{X(t), f_1(t), \ldots, f_k(t)\}$$

is a Markov process. Denote

$$F_k(t; x_1, \ldots, x_k) = P\{X(t) = k; f_1(t) \leq x_1, \ldots, f_k(t) \leq x_k\}.$$

Then

$$F_k(t; \infty, \ldots, \infty) = P\{X(t) = k\} = p_k(t)$$
$$p_k = \lim_{t \to \infty} F_k(t; \infty, \ldots, \infty).$$

Let

$$F_k(x_1, \ldots, x_k) = \lim_{t \to \infty} F_k(t; x_1, \ldots, x_k).$$

The Chapman–Kolmogorov equations could be obtained. Further,

$$F_k(x_1, \ldots, x_k) = \frac{\lambda^k}{k!} F_0 \prod_{i=1}^{k} \int_0^x [1 - F_s(u)]du$$

satisfies the equations and $F_s(.)$ is the CDF of service time. Take $\rho = \frac{\lambda}{\mu}$. Therefore,

$$\pi_k = \lim_{x \to \infty} F(x_1, \ldots, x_k) = \frac{\rho^k}{k!} F_0, \quad 0 \leq k \leq c.$$

Using the normalization condition, Erlang's formula can be deduced.

Erlang B Formula

Because of PASTA property, the long-run probability of the arriving customer who sees c servers busy, denoted by Erlang B formula $B(c, \rho)$, is given by

$$B(c, \rho) = \frac{\rho^c / c!}{\sum_{i=0}^{c} \rho^i / i!}.$$

when ρ and c are large, the computation becomes difficult due to the presence of factorials and potentially very large powers. One can use the following recursive formula

$$B(k, \rho) = \frac{\rho B(k-1, \rho)}{k + \rho B(k-1, \rho)}, \quad k = 1, 2, \ldots, c$$

where $B(0, \rho) = 1$.

The Erlang B formula is a fundamental result in telephone traffic engineering problems and can be used to select the appropriate number of trunks (servers) needed to ensure a small proportion of lost calls (customers).

For fixed ρ, the blocking probability $B(c, \rho)$ monotonically decreases to zero as c increases, and for c fixed, $B(c, \rho)$ monotonically increases to unity as ρ increases. When the service times are IID. exponential RVs with mean $1/\mu$, the system is a $M/M/c/c$ loss system. In this case,

$$B(c, \rho) = \frac{\rho^c / c!}{\sum_{i=0}^{c} \rho^i / i!}. \tag{9.25}$$

Erlang C Formula

Related to the Erlang B formula is the Erlang C formula (or Erlang delay formula) for the $M/M/c$ system (or Erlang delay system), which includes an infinite-capacity queue to accommodate arriving customers who find all c servers busy. For this model, π_c is interpreted as the long-run proportion of customers who experience a delay before their service begins. The model assumes that the customers are willing to wait as long as needed to receive service. The Erlang C formula is given by

$$C(c,\rho) = \frac{\frac{\rho^c}{c!(1-\rho)}}{\sum_{i=0}^{c-1} \frac{\rho^i}{i!} + \frac{\rho^c}{c!(1-\rho)}}.$$

The above result does not hold for arbitrary service time distributions, and it requires that the traffic intensity ρ does not exceed unity.

Example 9.7 With $\lambda = 6$, $\mu = 3$, and $c = 4$, compute the fraction of customers blocked for $M/G/c/c$ queueing system.

Solution: The fraction of customers blocked for an $M/G/c/c$ queue is $B(4, 2)$. This can be calculated from the Erlang B formula as given in Equation (9.25).

$$B(4,2) = \frac{2^4/4!}{1+2+2^2/2!+2^3/3!+2^4/4!} = \frac{16/24}{24+48+48+32+16/24} = \frac{2}{21}.$$

9.4.2 $GI/M/c/c$ System

In the $GI/M/c/c$ system, the arrival follows general distribution and the service time is exponentially distributed. This model is similar to $GI/M/c/c$ queueing system but here the capacity is finite.

9.4.2.1 The Method of Markov Regenerative Process

Let X_t denote the number of customers in the system at time t. Figure 9.14 shows the state transition diagram of the $GI/M/c/c$ queueing system. Note that, in this figure, the dotted arcs means that, the system moves from state 0 to state 1 and from state n to state $n+1$ for $n > 0$ with residual inter-arrival time or inter-arrival time. Further, when the system is in state c, if a new arrival occurs, the system will remain

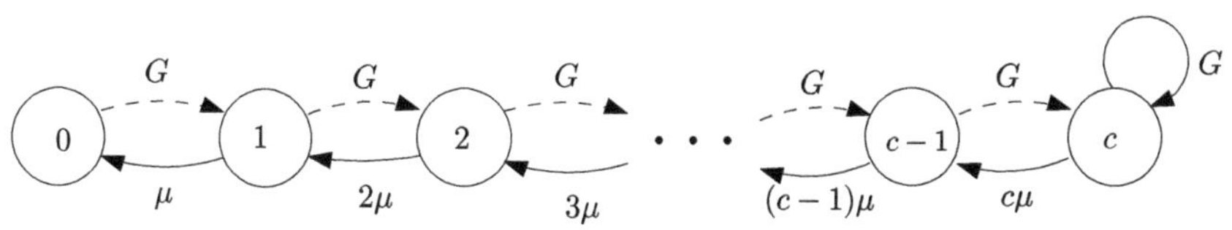

Fig. 9.14 State transition diagram of $G/M/c/c$ queueing system

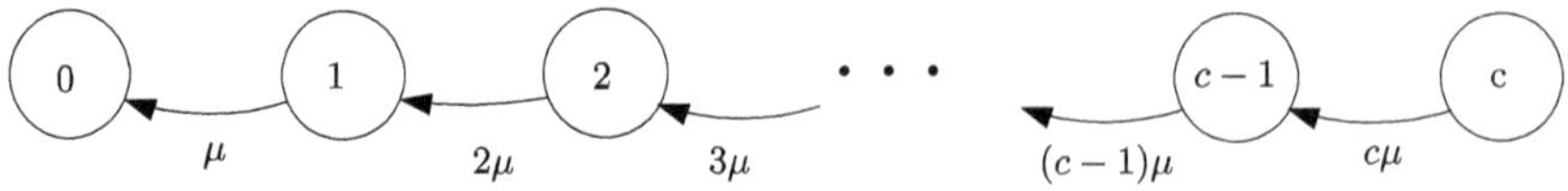

Fig. 9.15 Subordinated CTMC

in state c. Let $T_n,\ n = 1, 2, \ldots$, be the RVs denoting the arrival time instant of the nth customer. Let $X_n = X(t_n^-)$ denote the number of customers in the system before the nth customer arrival time instant. Here, the arrival time instants are the only regeneration time epochs. Hence, $\{X_t, t \geq 0\}$ is not a semi-Markov process, but a MRGP with $\{X_n, n = 1, 2, \ldots\}$ as an embedded Markov chain. $S = \{0, 1, \ldots\}$ is the state space, and $S^{'} = \{1, 2, \ldots\}$ is the set of states only at regeneration time instants.

The global kernel matrix is

$$\mathbf{K(t)} = \begin{pmatrix} k_{11}(t) & k_{12}(t) & 0 & 0 & \ldots & 0 \\ k_{21}(t) & k_{22}(t) & k_{23}(t) & 0 & \ldots & 0 \\ k_{31}(t) & k_{32}(t) & k_{33}(t) & k_{34}(t) & \ldots & 0 \\ k_{41}(t) & k_{42}(t) & k_{43}(t) & k_{44}(t) & \ldots & 0 \\ \vdots & \vdots & \vdots & \vdots & \ddots & \vdots \\ k_{c1}(t) & k_{c2}(t) & k_{c3}(t) & k_{c4}(t) & \ldots & k_{cc}(t) \end{pmatrix}$$

where

$$k_{ij}(t) = \begin{cases} \displaystyle\int_0^t P_{ij-1}(x)\mathrm{d}F_a(x)\,, & i, j \in S^{'} \\ \displaystyle\int_0^t (P_{cc-1}(x) + P_{cc}(x))\,\mathrm{d}F_a(x)\,, & i, j = c \end{cases}$$

and the $P_{ij}(t)$'s are the transition probabilities of the subordinated CTMC. Figure 9.15 denotes the subordinated CTMC of the proposed system and $P_{ij}(t)$ are transition probabilities.

The local kernel matrix is

$$\mathbf{E(t)} = \begin{pmatrix} e_{10}(t) & e_{11}(t) & 0 & 0 & 0 & \ldots & 0 \\ e_{20}(t) & e_{21}(t) & e_{22}(t) & 0 & 0 & \ldots & 0 \\ e_{30}(t) & e_{31}(t) & e_{32}(t) & e_{33}(t) & 0 & \ldots & 0 \\ \vdots & \vdots & \vdots & \vdots & \vdots & \ddots & \vdots \\ e_{c0}(t) & e_{c1}(t) & e_{c2}(t) & e_{c3}(t) & e_{c4}(t) & \ldots & e_{cc}(t) \end{pmatrix}$$

where $e_{ij}(t) = P_{ij}(t)(1 - F_a(t))$.

Fig. 9.16 State transition diagram of $M/G/\infty$ queueing system

9.5 Infinite Server Systems

The infinite-server system (also know as the self-service system) is a multi-server queueing model where every arrival experiences immediate service and does not wait. ISS often serves as a surprisingly good and useful approximations for multi-server queueing systems. Moreover, the asymptotic results for ISS can be useful tools for proving the corresponding asymptotic results for queueing systems with only finitely many servers.

Infinite-server systems are common in real life. Indeed, in practice, many situations are modeled by an ISS. For instance: the production systems, computer systems, computer communications and telecommunications, aerodrome management of an airport, etc.

The underlying stochastic process $\{X(t), t \geq 0\}$ is a continuous parameter discrete-state stochastic process whose state space is the set $S = \{0, 1, \ldots\}$ where $X(t)$ denotes the number of customers currently being served. Since, the number of servers in parallel is infinite, there is no queue and the number of customers in the systems coincides with the number of customers being served at any moment.

Further, one of the important self-service systems, $M/M/\infty$ system was discussed in detail in Chap. 6. In this section, $M/G/\infty$ and $GI/G/\infty$ systems will be discussed.

9.5.1 $M/G/\infty$ System

In the infinite-server system, the arrivals follow the Poisson process with parameter λ and service times follow a non-exponential distribution with CDF F_S and mean $1/\mu$.

Let $X(t)$ be the number of customers in the system at any time epoch t. Figure 9.16 shows the state transition diagram for the $M/G/\infty$ system. In this figure, the dotted arcs means that, the system moves from 2 to state 1 and from state n to state $n-1$ for $n > 2$ with residual service time or service time whereas the system moves from state 1 to state 0 with service time. Let $A(t)$ be the number of arrivals by time t, i.e., in the interval $(0, t)$, and $D(t)$ be the number of departures by time t. Here, $p_n(t) = P\{X(t) = n\}, t \geq 0, n = 0, 1, \ldots$ such that $A(t) = X(t) + D(t)$. Assume that, $\{A(t), t \geq 0\}$ is a Poisson process with parameter λ. Assume that $p_0(0) = 1, p_n(0) = 0, n \neq 0$. By conditioning on $A(t)$,

$$\begin{aligned}\pi_n(t) &= P\{X(t)=n\}\\ &= \sum_{k=n}^{\infty} P\{X(t)=n|A(t)=k\}P\{A(t)=k\}\\ &= \sum_{k=n}^{\infty} P\{X(t)=n|A(t)=k\}\frac{e^{-\lambda t}(\lambda t)^k}{k!}, \quad n=0,1,\ldots\end{aligned}$$

since the arrival process is a Poisson process with parameter λ. Now, for $n = 0, 1, \ldots$

$$\begin{aligned}P\{X(t)=n|A(t)=k\} &= P\{\text{out of } k \text{ arrivals in } (0,t) \text{ the no. in service at } t \text{ is } n\}\\ &= \binom{k}{n}(r(t))^n[1-r(t)]^{k-n}, k \geq n,\end{aligned}$$

where

$$\begin{aligned}r(t) = &\quad P\{\text{a unit arriving in } (0,t), \text{ is still in service at epoch } t\}\\ = &\int_0^{\infty} P\{\text{service time of a unit exceeds } (t-x) \text{ given that unit arrives at}\\ &\quad \text{epoch}\} \times P\{\text{the unit's arrival epoch is } x\}dx.\end{aligned}$$

Given that an arrival occurs in $(0, t)$, the arrival x in $(0, t)$ is uniformly distributed over $(0, t)$. Hence, $r(t)$ equals the probability that a unit's service time does not terminate by time t, given that the arrival is uniformly distributed over the interval $(0, t)$. Therefore,

$$r(t) = \frac{1}{t}\int_0^t P\{v(t-x)\}dx = \frac{1}{t}\int_0^t \{1-B(t-x)\}dx = 1-\frac{1}{t}\int_0^t F_S(u)du,$$

where $F_s(t)$ is the distribution of service time. Therefore, for $n = 0, 1, \ldots$

$$\begin{aligned}\pi_n(t) &= \sum_{k=n}^{\infty}\binom{k}{n}(r(t))^n[1-r(t)]^{k-n}\frac{e^{-\lambda t}(\lambda t)^{k-n+n}}{k!}\\ &= \sum_{k=n}^{\infty}\frac{e^{-\lambda t}}{n!(k-n)!}\{\lambda t r(t)\}^n\{\lambda t(1-r(t))\}^{k-n}\\ &= \frac{\{\lambda t r(t)\}^n}{n!}e^{-\lambda t}e^{\lambda t(1-r(t))}.\end{aligned}$$

Finally,

$$\pi_n(t) = \frac{\{\lambda t r(t)\}^n}{n!}e^{-\lambda t r(t)}, \quad n = 0, 1, \ldots. \tag{9.26}$$

The distribution of $X(t)$ for fixed t is a non-homogeneous Poisson with mean

$$\{\lambda tr(t)\} = \lambda \int_0^t \{1 - F_S(u)\}\mathrm{d}u. \tag{9.27}$$

As $t \to \infty$, $\lambda tr(t)$ becomes $\frac{\lambda}{\mu}(= \rho)$. Hence, the stationary distribution of $X(t)$ is given by

$$\pi_n = \frac{\rho^n}{n!}e^{-\rho}, \quad n = 0, 1, 2, \ldots. \tag{9.28}$$

As a special case, when the service time follows an independent exponential distribution with mean $1/\mu$, i.e., in $M/M/\infty$ system,

$$\lambda tr(t) = \rho.$$

As $t \to \infty$ in equation (9.27),

$$\lambda tr(t) = \lambda \int_0^t \{1 - F_S(u)\}du \to \rho.$$

Thus, the steady-state distribution of the system size is Poisson d with mean ρ. Therefore, when the number of channels is large, the steady-state distribution of the system size is Poisson, irrespective of the form of the service time distribution and of magnitude ρ.

9.5.1.1 Performance Measures

1. Expected busy period with at least one server busy

$$E(R) = \frac{1 - \pi_0}{\pi_0} = e^{\rho} - 1.$$

2. The distribution of the number of busy servers is Poisson distributed with mean ρ. The number of busy servers N_b in steady-state is approximately normal distributed with mean a

$$P\{N_b \geq x\} = 1 - \Phi\left(\frac{x - \rho}{\sqrt{\rho}}\right),$$

where Φ is the CDF of standard normal RV.

Example 9.8 Consider customers arriving and spending time in a store as an $M/G/\infty$ system. Since the $M/G/\infty$ system has an infinite number of servers, there is no queue. Each arrival immediately starts service and the service time S_i of customer i is IID with some given distribution function $F(t)$. The service time is independent of arrival epochs. Let $N(t)$ be the number of customers who have exited the store by time t. Assume that $N(0) = 0$. Find the distribution function of the

number of customers being served at a given epoch τ. Show that, $\{N(t), 0 \le t \le \tau\}$ is a non-homogeneous Poisson process with rate $\lambda(1 - F(\tau - t)), t \le \tau$. Further, discuss the distribution of the number of customers in service when $\tau \to \infty$.
Solution: Given that $\{N(t), t \ge 0\}$ is a Poisson counting process with rate λ. Consider the arrival times of those customers who are still in service at some fixed time τ. In an arbitrary small interval $(t, t + \delta]$, the probability of arrival is $\delta\lambda + o(\delta)$. The probability that a customer arrives in $(t, t + \delta]$ and is still being served at time $\tau > t$ is $\delta\lambda[1 - F(t)] + o(\delta)$. Consider a counting process $\{N_1(t), 0 \le t \le \tau\}$ where $N_1(t)$ is the number of arrivals between 0 and t that are still in service at τ. This counting process has an independent increment property and the arrivals in $\{N(t), t \ge 0\}$ have independent service times, and thus are independently in or not in $\{N_1(t), 0 \le t \le \tau\}$. It follows that $\{N_1(t), 0 < t \le \tau\}$ is a non-homogeneous Poisson process with rate $\lambda(1 - F(\tau - t)), t \le \tau$.

The expected number of arrivals still in service at time τ is then

$$\mu(\tau) = \lambda \int_0^{\tau} (1 - F(t)\mathrm{d}t.$$

Note that as $\tau \to \infty$, the above integral approaches the mean of the service time distribution. That is, in steady state (as $\tau \to \infty$), the distribution of the number of customers in service at τ depends on the service time distribution only through its mean.

9.5.2 $GI/G/\infty$ *Self Service System*

Consider a $GI/G/\infty$ self service system in which the inter-arrival times and service times in the system are independent and generally distributed. The results of this system can be deduced from related stochastic models such as branching processes with immigration. If mean inter-arrival times and mean service times are finite, then $GI/G/\infty$ self service system is stable. In this section, the infinite server systems with both arrival and service follow *Hawkes* process will be analyzed.

9.5.2.1 Hawkes/sdHawkes/∞ model

In this infinite server system, the arrival process $M = \{M_t, t \ge 0\}$ follows *Hawkes* process having arrival intensity given by

$$\lambda_t = \lambda^* + e^{-rt}(\lambda_0 - \lambda^*) + \int_0^t B_u e^{-r(t-u)} dM_u \tag{9.29}$$

where $\lambda_0 > 0$ is the initial intensity and λ^* is the baseline intensity. For $i > 0$, if t_i denotes the arrival times of ith customer, then B_{t_i} is positive RV representing the size of the jump.

For the service requirement, it is considered $S = \{S_t, t \geq 0\}$, another Hawkes process having serving intensity μ_t, with initial intensity $\mu_0 > 0$, baseline intensity μ^* and, exponential excitation function. Taking N_t as the number of customers in the system at t, for a constant $s > 0$, and for $t > 0$, given by

$$\mu_t = N_t \left(\mu^* + e^{-st}(\mu_0 - \mu^*) + \int_0^t C_u e^{-s(t-u)} \mathrm{d}S_u \right). \tag{9.30}$$

For $i > 0$, τ_i denotes the time epoch of i-th customer departure after the service completion, and C_{τ_i} is a positive RV representing the size of that jump. The number of customers in the system whose service is completed on, or before, time t is given by S_t and $N_t = M_t - S_t$.

This form of the serving intensity μ_t allows us to take into account the crowding of the system, which is the number of customers in the system at time t, and about the experience gained by the server process, which is given by the Hawkes structure.

This is the first time in literature that Hawkes processes are introduced to model the experience in arrivals and services of an infinite-server system. Furthermore, it is noted that at time t, state dependent Hawkes process ($sdHawkes$ process), for service will start as new, whenever the number of customers in the system become one, i.e., $N_t = 1$. Therefore, this queueing model is called as $Hawkes/sdHawkes/\infty$ in Kendall's notation.

9.5.2.2 Markov Property of the $Hawkes/sdHawkes/\infty$ System

Recall that, if the excitation function of a Hawkes process is exponential, then the process jointly with its intensity is a Markov process. In order to characterize $\mathcal{L}(N_t, \lambda_t, \mu_t)$, the law of N_t, the Hawkes arrival intensity λ_t and the state dependent Hawkes server intensity μ_t, the Markov property of the $Hawkes/sdHawkes/\infty$ system is proved.

Let the processes M, S and N be defined on the same filtered probability space $(\Omega, \mathfrak{F}, \{\mathfrak{F}_t\}_{t\geq 0}, P)$. The definition of filtered probability space will be discussed in detail in Definition 10.3. Given the sub σ-algebras of $\mathfrak{F}_t$

$$\mathfrak{F}_t^M := \sigma\{M_u, u \leq t\}, \qquad \mathfrak{F}_t^S := \sigma\{S_u, u \leq t\}, \qquad \text{and} \qquad \mathfrak{F}_t^{M,S} := \mathfrak{F}_t^M \vee \mathfrak{F}_t^S.$$

Observe that $\mathfrak{F}_t^{M,S}$ contains, all the informations related to the process $N = \{N_t, t \geq 0\}$ until time t. This concept along with martingale will be discussed in the next chapter in details.

Note that, $\mathcal{F}_t^M \subseteq \mathcal{F}_t^{M,S}$ and that $\mathcal{F}_t^S \subseteq \mathcal{F}_t^{M,S}$. Consequently, every $\mathcal{F}_t^{M,S}$-martingale is a $\mathcal{F}_t^M$-martingale and, at the same time, every $\mathcal{F}_t^{M,S}$-martingale is a $\mathcal{F}_t^S$-martingale. But, in general, it is not true that a $\mathcal{F}_t^M$-martingale is a $\mathcal{F}_t^{M,S}$-martingale and that a $\mathcal{F}_t^S$-martingale is a $\mathcal{F}_t^{M,S}$-martingale, unless the immersion property is assumed. This is a classical topic in this context, the so-called problem of the enlargement of filtrations.

The joint transient distribution of (N, λ, μ) is uniquely defined by the transformation

$$\zeta(t, z, u, v) = E\left[z^{N_t} e^{-u\lambda_t} e^{-v\mu_t}\right], \tag{9.31}$$

where $t \geq 0$, $0 \leq z \leq 1$, $u \geq 0$ and $v \geq 0$. In this section, ζ is categorized in terms of the solution of a system of ordinary differential equations (ODEs).

Theorem 9.1 *Let the arrival process be a Hawkes process and let the service process be a sdHawkes process. Let $\beta(u) := E[e^{-uB}]$ and let $\gamma(v) := E[e^{-vC}]$.*
If $(N_t, \lambda_t, \mu_t)|_{t=0} = (0, \lambda_0, 0)$, then the couple $(u(\cdot), v(\cdot))$ solves the system of ODEs

$$\begin{cases} u'(t) + ru(t) - 1 + z\beta\left(u(t)\right) e^{-s\mu^* \int_0^t v(x)\, dx} = 0, \\ v'(t) + sv(t) - 1 + \frac{\gamma(v(t))}{z} e^{s\mu^* \int_0^t v(x)\, dx} = 0, \end{cases} \tag{9.32}$$

with boundary conditions $u(0) = u$ and $v(0) = v$. Furthermore,

$$\zeta(t, z, u, v) = e^{-u(0)\lambda_0} \exp\left\{-\lambda^* r \int_0^t u(x)\, dx\right\}, \tag{9.33}$$

where ζ depends on v through the coupling with u given by system of ODEs (9.32).

In order to prove the above theorem, we make use of two Propositions 9.1 and 9.2 given below without proofs.

Proposition 9.1 *The joint distribution of (N, λ, μ), for $t > 0$,*

$$F(t, k, \lambda, \mu) = P\{N_t = k, \lambda_t \leq \lambda, \mu_t \leq \mu\}$$

is such that

$$\begin{aligned} &\frac{\partial^3 F(t,k,\lambda,\mu)}{\partial\mu\partial\lambda\partial t} + (\lambda+\mu)\frac{\partial^2 F(t,k,\lambda,\mu)}{\partial\mu\partial\lambda} \\ &= \int_0^\lambda x \frac{\partial^2 F(t,k-1,x,\mu)}{\partial\mu\partial\lambda}\, dP[B \leq \lambda - x] - \frac{\partial}{\partial\lambda}\left[r\left(\lambda^* - \lambda\right)\frac{\partial^2 F(t,k,\lambda,\mu)}{\partial\mu\partial\lambda}\right] \\ &+ \int_0^\mu y \frac{\partial^2 F(t,k+1,\lambda,y)}{\partial\lambda\partial\mu}\, dP[C \leq \mu - y] - \frac{\partial}{\partial\mu}\left[s\left(k\mu^* - \mu\right)\frac{\partial^2 F(t,k,\lambda,\mu)}{\partial\mu\partial\lambda}\right]. \end{aligned} \tag{9.34}$$

Proposition 9.2 *Taking*

$$\xi(t,k,u,v)=\int_0^\infty\int_0^\infty e^{-u\lambda}e^{-v\mu}\,\frac{\partial^2 F(t,k,\lambda,\mu)}{\partial\lambda\partial\mu}\,d\lambda\,d\mu,$$

implies

$$\begin{aligned}&\frac{\partial\xi(t,k,u,v)}{\partial t}+(ur-1)\frac{\partial\xi(t,k,u,v)}{\partial u}+\beta(u)\frac{\partial\xi(t,k-1,u,v)}{\partial u}\\&+(vs-1)\frac{\partial\xi(t,k,u,v)}{\partial v}+\gamma(v)\frac{\partial\xi(t,k+1,u,v)}{\partial v}=-(ur\lambda^*+vsk\mu^*)\xi(t,k,u,v).\end{aligned}\tag{9.35}$$

Proof of Theorem 9.1 Recalling the definition of ζ given in Eq. (9.31), which in turn implies that

$$\zeta(t,z,u,v)=\sum_{k=0}^{\infty}z^k\xi(t,k,u,v),$$

and taking into account that $F(t,-1,\lambda,\mu)=0$, note that

$$\begin{aligned}\sum_{k=0}^{\infty}z^k\frac{\partial\xi(t,k-1,u,v)}{\partial u}&=\frac{\partial\xi(t,-1,u,v)}{\partial u}+z\sum_{k=1}^{\infty}z^{k-1}\frac{\partial\xi(t,k-1,u,v)}{\partial u}\\&=z\frac{\partial}{\partial u}\sum_{k=0}^{\infty}z^k\xi(t,k,u,v)=z\frac{\partial\zeta(t,z,u,v)}{\partial u}.\end{aligned}$$

Moreover,

$$\begin{aligned}\sum_{k=0}^{\infty}z^k\frac{\partial\xi(t,k+1,u,v)}{\partial v}&=\frac{1}{z}\frac{\partial}{\partial v}\sum_{k=0}^{\infty}z^{k+1}\xi(t,k+1,u,v)\\&=\frac{1}{z}\frac{\partial}{\partial v}\int\left[\sum_{k=0}^{\infty}kz^{k-1}\xi(t,k,u,v)\right]\,\mathrm{d}z=\frac{1}{z}\frac{\partial\zeta(t,z,u,v)}{\partial v},\end{aligned}$$

and

$$\sum_{k=0}^{\infty}kz^k\xi(t,k,u,v)=z\frac{\partial\zeta(t,z,u,v)}{\partial z}.$$

Substituting all these in Eq. (9.35), the partial differential equation (PDE) satisfied by ζ is given by

$$
\begin{aligned}
&\frac{\partial\zeta(t,z,u,v)}{\partial t} + vs\mu^* z\frac{\partial\zeta(t,z,u,v)}{\partial z} + (ur + z\beta(u) - 1)\frac{\partial\zeta(t,z,u,v)}{\partial u}\\
&+\left(vs - 1 + \frac{1}{z}\gamma(v)\right)\frac{\partial\zeta(t,z,u,v)}{\partial v} = -ur\lambda^*\zeta(t,z,u,v).
\end{aligned} \tag{9.36}
$$

Applying the method of the characteristics, the PDE is reduced to a system of ODEs, along which the solutions are integrated with some initial data given on a suitable hypersurface.

To this end, let z, u and v be parameterized by $w, 0 < w < t$, and with the boundary conditions $z(t) = z$, $u(t) = u$, $v(t) = v$. A comparison with Eq. (9.36), taking into account the chain rule, gives us

$$
\begin{aligned}
\frac{\mathrm{d}\zeta\,(t, z(w), u(w), v(w))}{\mathrm{d}w} &= \frac{\partial\zeta}{\partial t}\frac{\mathrm{d}t}{\mathrm{d}w} + \frac{\partial\zeta}{\partial z}\frac{\mathrm{d}z}{\mathrm{d}w} + \frac{\partial\zeta}{\partial u}\frac{\mathrm{d}u}{\mathrm{d}w} + \frac{\partial\zeta}{\partial v}\frac{\mathrm{d}v}{\mathrm{d}w}\\
&= -u(w)r\lambda^*\zeta\,(t, z(w), u(w), v(w))\,.
\end{aligned} \tag{9.37}
$$

This in turn implies that, for a real constant c, it can be deduced that

$$
\zeta\,(t, z(w), u(w), v(w)) = c\exp\left\{-r\lambda^*\int_0^w u(x)\,\mathrm{d}x\right\},
$$

where $c = \zeta\,(0, z(0), u(0), v(0)) = \zeta\,(0, z, u, v) = e^{-u(0)\lambda_0}$, and

$$
\zeta\,(t, z(t), u(t), v(t)) = c\exp\left\{-r\lambda^*\int_0^t u(x)\,\mathrm{d}x\right\},
$$

which implies Equation (9.33).

Furthermore, by Eq. (9.36) and the chain rule written in Eq. (9.37), it is deduced that

$$
\frac{\mathrm{d}z(w)}{\mathrm{d}w} = s\mu^* v(w)z(w).
$$

Recalling that $0 < w < t$, and that $z(t) = z$,

$$
z(w) = C\exp\left\{s\mu^*\int_0^w v(x)\,\mathrm{d}x\right\},
$$

for a real constant $C = z\exp\left\{-s\mu^*\int_0^t v(x)\,\mathrm{d}x\right\}$, which allows us to write that

$$z(w) = z \exp\left\{-s\mu^* \int_w^t v(x)\,\mathrm{d}x\right\}.$$

Similarly, by Equation (9.36) and the chain rule written in Eq. (9.37), and substituting $z(w)$, implies

$$\begin{aligned}\frac{du(w)}{\mathrm{d}w} &= ru(w) - 1 + z(w)\beta(u(w)) \\ &= ru(w) - 1 + z\beta(u(w)) \exp\left\{-s\mu^* \int_w^t v(x)\,\mathrm{d}x\right\},\end{aligned}$$

and

$$\begin{aligned}\frac{\mathrm{d}v(w)}{\mathrm{d}w} &= sv(w) - 1 + \frac{1}{z(w)}\gamma(v(w)) \\ &= sv(w) - 1 + \frac{1}{z}\gamma(v(w)) \exp\left\{s\mu^* \int_w^t v(x)\,\mathrm{d}x\right\}.\end{aligned}$$

By substituting t for $t - w$, and by taking into account that a change of sign in the LHS of both $\frac{du(w)}{dw}$ and $\frac{\mathrm{d}v(w)}{\mathrm{d}w}$, Eq. (9.32) is obtained, with boundary conditions $u(0) = u$ and $v(0) = v$.

Special Cases:

1. $M/sdHawkes/\infty$ **system:**

Theorem 9.2 *For an $M/sdHawkes/\infty$ system, let*

$$\zeta(t, z, v) = E\left[z^{N_t} e^{-v\mu_t}\right] \tag{9.38}$$

where $t \geq 0$, $0 \leq z \leq 1$, $v \geq 0$. Given the constant intensity $\lambda_t = \lambda$, and the initial values, $(N_t, \mu_t)|_{t=0} = (0, 0)$, implies

$$\zeta\left(t, z(t), v(t)\right) = \exp\left\{-\lambda \int_0^t \left(z(x) + 1\right)\, dx\right\}, \tag{9.39}$$

where

$$z(w) = z \exp\left\{-s\mu^* \int_w^t v(x)\, dx\right\},$$

and, given the boundary condition $v(0) = v$, $v(\cdot)$ solves the ODE

$$v'(t) + sv(t) - 1 + \frac{\gamma\,(v(t))}{z} e^{s\mu^* \int\limits_0^t v(x)\,dx} = 0. \tag{9.40}$$

The proof of Theorem 9.2 is left as an exercise problem but can be obtained in three steps using Propositions 9.3 and 9.4 given below.

Proposition 9.3 *The joint distribution of* (N, μ)*, for* $t > 0$*, and* $\lambda_t = \lambda$*, positive constant,*

$$F(t, k, \lambda, \mu) = F(t, k, \mu) 1_{\lambda_t = \lambda} = P\{N_t = k, \mu_t \leq \mu\} 1_{\lambda_t = \lambda}$$

is such that

$$\frac{\partial^2 F(t, k, \mu)}{\partial \mu \partial t} + \frac{\partial}{\partial \mu}\left[s\left(k\mu^* - \mu\right)\frac{\partial F(t, k, \mu)}{\partial \mu}\right] = \tag{9.41}$$
$$\lambda \frac{\partial F(t, k-1, \mu)}{\partial \mu} - (\lambda + \mu)\frac{\partial F(t, k, \mu)}{\partial \mu} + \int_0^\mu y \frac{\partial F(t, k+1, y)}{\partial \mu} dP[C \leq \mu - y].$$

Proposition 9.4 *Taking*

$$\xi(t, k, v) = \int\limits_0^\infty e^{-v\mu} \frac{\partial F(t, k, \mu)}{\partial \mu} d\mu,$$

implies

$$\frac{\partial \xi(t, k, v)}{\partial t} + (vs - 1)\frac{\partial \xi(t, k, v)}{\partial v} + \gamma(v)\frac{\partial \xi(t, k+1, v)}{\partial v}$$
$$= -\lambda \xi(t, k-1, v) - (\lambda + vsk\mu^*)\xi(t, k, v). \tag{9.42}$$

2. $Hawkes/M/\infty$ system: For a $Hawkes/M/\infty$ system,

$$\zeta(t, z, u) = E\left[z^{N_t} e^{-u\lambda_t}\right] = e^{-u(t)\lambda_0} e^{-\lambda_0 r \int\limits_0^t u(w)\,dw},$$

   where $u(\cdot)$ solves the ODE

$$u'(w) = -ru(w) - (1 + (z-1)e^{-\mu^* w})\beta(u(w)) + 1.$$

   In Theorem 9.1, these known results have been extended taking into account the state-dependent Hawkes service times.
3. $M/M/\infty$ system: In Theorem 9.1, if $r = s = 0$ and $E[B_t] = E[C_t] = 0$, Eqs. (6.64) and (9.30) are simplified and $\lambda_t = \lambda_0$ and $\mu_t = N_t \mu_0$. This means that the arrival process M_t turns out to be a Poisson process with parameter λ_0 and, if there are N_t customers in the system at time t, the service time has an exponential

distribution with parameter $N_t\mu_0$, i.e., this is an $M/M/\infty$ system. Therefore, as usual, the joint distribution is

$$F(t,k,\lambda,\mu) = P\{N_t = k, \lambda_t = \lambda_0, \mu_t = N_t\mu_0\} = P\{N_t = k\}\mathbb{I}_{\lambda_t=\lambda_0}\mathbb{I}_{\mu_t=N_t\mu_0},$$

and taking $F(t,k,\lambda,\mu) = F(t,k) = P\{N_t = k\}$

$$\frac{\mathrm{d}F(t,0)}{\mathrm{d}t} = \lim_{\Delta t\to 0}\frac{F(t+\Delta t,0) - F(t,0)}{\Delta t} = -\lambda_0 F(t,0) + \mu_0 F(t,1),$$

while, for $k = 1, 2, \ldots$

$$\frac{\mathrm{d}F(t,k)}{\mathrm{d}t} = \lambda_0 F(t,k-1) - (\lambda_0 + k\mu_0)F(t,k) + (k+1)\mu_0 F(t,k+1).$$

Solving the above system of difference-differential equations, we obtain the distribution of N_t, as

$$F(t,k) = \frac{1}{k!}\left(\frac{\lambda_0}{\mu_0}(1 - e^{-\mu_0 t})\right)^k \exp\left\{-\frac{\lambda_0}{\mu_0}(1 - e^{-\mu_0 t})\right\},$$

and by noting that

$$\zeta(t,z,u,v) = e^{-u\lambda_0}E\left[(ze^{-v\mu_0})^{N_t}\right] = e^{-u\lambda_0}\sum_{k=0}^{\infty}(ze^{-v\mu_0})^k P\{N_t = k\},$$

we have that

$$\zeta(t,z,u,v) = e^{-u\lambda_0}\exp\left\{-\frac{\lambda_0}{\mu_0}(1 - e^{-\mu_0 t})(1 - ze^{-v\mu_0})\right\}.$$

Furthermore, note that $F(t,k)$, for $t \to \infty$, turns out to be

$$\pi_k = P\{N_\infty = k\} = \frac{1}{k!}\left(\frac{\lambda_0}{\mu_0}\right)^k e^{-\lambda_0/\mu_0}, \quad k = 0, 1, \ldots.$$

These results of distribution of N_t and of limiting distribution coincide with the results presented in [6] and Gross et al. [11].

For more description, readers may refer to [1–10].

9.6 Exercise Problems

1. Consider $M/G/1/\infty$ system with multiple independent vacation. Show that the mean waiting time is

$$E(W) = \frac{\lambda(E(S))^2}{2(1-\rho)} + \frac{E(V^2)}{2E(V)}.$$

2. Consider $M/G/1/\infty$ system with LIFO preemptive policy. Assume that $\rho = \lambda E(S) < 1$. Prove that the stationary distribution π is

$$\pi_n = (1-\rho)\rho^n, \; n = 0, 1, \ldots.$$

3. Consider a single-server queue with arrival following the Poisson process with a mean rate of 10 per hour. Currently, the server has an average service time of 5 minutes, and service times follow an exponential distribution. A training program is expected to reduce the variability in service times, although the average service time will increase slightly. After the training, the average service time is expected to be 5.5 minutes, and the standard deviation will decrease from 5 minutes (the exponential case) to 4 minutes. Management wants to know if it is worth giving the server additional training.
4. Let $X(t)$ be the number of customers in the system at time t in an $M/G/1$ queueing model. Assume that, the arrival times $0 < \tau_1 < \tau_2 < \cdots$, with the initial condition $X(0) = 0$. Further, assume that $\rho < 1$. Let $T = \inf\{t > \tau_1 : X(t) = 0\}$, denoting the time when the system first reaches an empty. Now, $T = \tau_1 + Y$, where Y represents the duration for which the server is busy. Find $E[T]$. Also, show that $E[Y] = \frac{\rho}{\lambda(1-\rho)}$.
5. Compute the transition probabilities of an embedded Markov chain for an $M/G/1$ queue where service is uniformly distributed on $(a, b), 0 < a < b < \infty$.
6. Compute the mean number of customers in the system and mean waiting time for $M/G/1$ queue where service time is beta distributed with parameters α and β.
7. A certain assembly-line operation is assumed to be a $M/G/1$ queueing system, with input rate $5/h$ and service times with mean 9 minutes and variance 90 minutes. Find $E(X)$ and $E(R)$. Is the operation improved or degraded if the service times are forced to be exponential distribution with the same mean?
8. Customers arrive at an ATM machine according to a Poisson process with rate $\lambda = 60/h$. The following transaction times are observed (in seconds):

$$28, 71, 0, 70, 51, 62, 36, 25, 35, 87, 69, 27, 56, 25, 36.$$

   (a) Would an $M/M/1$ queueing system be an appropriate model for this system?
   (b) Estimate the mean number of people waiting in line at ATM.

(c) The bank wishes to keep the number in queue less than or equal to one. What is the mean transaction time needed to achieve this goal?

9. Consider a $GI/M/1$ queueing system, where the inter-arrival times are uniformly distributed from 0 to 6 minutes and the expected service time is 2 minutes. Find, the following measures in steady-state.

   (a) Mean queue length.
   (b) Mean time spent in queue for each customer.

10. Consider a $GI/G/1$ system. Let $W_n^{'}$ denote the length of time the nth customer waits in the queue prior to obtaining service. Using Little law, find the mean wait time

$$W^{'} = \lim_{n\to\infty} \frac{\sum_{k=1}^{n} W_k^{'}}{n-1}.$$

11. Find the steady-state probabilities of $M/G/1/2$. Assume that G follows Erlang distribution with parameters 2 and $1/3$.
12. Derive the stationary system-size probabilities for the $M/M/2/2$ loss system using only Little's formula and fundamental steady-state identities.
13. Consider a system where trucks provide coils to a steelworks warehouse. They arrive with random inter-arrival time, distributed according an exponential distribution with a mean of 20 minutes. The time needed by a team of workers to unload a truck is a RV and it has an deterministic distribution with a mean of half an hour. Evaluate how many teams must work in parallel to reduce the mean time of each truck in the system to less than one hour and the mean number of trucks at the steelworks to less than three.
14. Consider an $GI/M/c$ queueing system with service in order of arrival. Prove that, the equilibrium conditional probability that a blocked customer will still be waiting in the queue when the next customer arrives is equal to the generalized occupancy i.e., the state of having all servers busy and a queue of length n.
15. Suppose that, in a production plant a certain number of identical machines will be installed. These machines require maintenance after 5 days, i.e., 120 hours, in average. This working time is a RV with an exponential distribution. Some operators will be in charge for the maintenance. The time needed by each of them for each machine is a RV with deterministic distribution and a mean of 2 hours. Owing to a suitable rotation of work-shifts a team of operators will be available 24h a day. According to the design the number of machines should be 20 and the number of operators in the maintenance team should be 20. Consider this scenario as $M/G/c/c$ loss system and find out what is the mean fraction of time during which each operator would be busy? How long will a machine stop last, in average?
16. Determine the lowest value of the parameter k of Erlang distribution which will allow us to approximate the mean line delay of a $D/M/c/c$ loss system to no worse than 2.0% by an $E_k/M/c/c$ loss system when $\lambda = 4$, $\mu = 1.5$, and $c = 3$.

17. Consider an M/G/∞ self-service system in which customers arrive at rate λ and in which the customer service times have distribution $F_S(t)$. Suppose that the system is empty at time 0 . Let $N_1(t)$ be the number of customers who have completed their service by time t, and let $N_2(t)$ be the number of customers still in the system at time t. What is the steady-state probabilities for the number of customers in this system?
18. Consider a $GI/G/\infty$ self-service system with stable condition. Prove that

    (a) Mean number of customers in the system is

$$\frac{E(S)}{E(A)} = E(N) = \lambda \int_0^{\infty} (1 - F_s(t))dt$$

    where $\lambda = \frac{1}{E(A)}$.

    (b) Mean number of busy servers greater than or equal to r is given by

$$\lambda \int_r^{\infty} (1 - F_s(t))dt.$$

19. Consider Proposition 9.1. Prove that the joint distribution of (N, λ, μ), for $t > 0$,

$$F(t, k, \lambda, \mu) = P\left[N_t = k, \lambda_t \leq \lambda, \mu_t \leq \mu\right]$$

satisfies

$$\begin{aligned}
&\frac{\partial^3 F(t, k, \lambda, \mu)}{\partial \mu \partial \lambda \partial t} + (\lambda + \mu)\frac{\partial^2 F(t, k, \lambda, \mu)}{\partial \mu \partial \lambda} \\
&= \int_0^{\lambda} x \frac{\partial^2 F(t, k-1, x, \mu)}{\partial \mu \partial \lambda} dP[B \leq \lambda - x] - \frac{\partial}{\partial \lambda}\left[r\left(\lambda^* - \lambda\right)\frac{\partial^2 F(t, k, \lambda, \mu)}{\partial \mu \partial \lambda}\right] \\
&\quad + \int_0^{\mu} y \frac{\partial^2 F(t, k+1, \lambda, y)}{\partial \lambda \partial \mu} dP[C \leq \mu - y] \\
&\quad - \frac{\partial}{\partial \mu}\left[s\left(k\mu^* - \mu\right)\frac{\partial^2 F(t, k, \lambda, \mu)}{\partial \mu \partial \lambda}\right].
\end{aligned}$$

20. Consider Proposition 9.2. Let

$$\xi(t, k, u, v) = \int_0^{\infty}\int_0^{\infty} e^{-u\lambda} e^{-v\mu} \frac{\partial^2 F(t, k, \lambda, \mu)}{\partial \lambda \partial \mu} d\lambda \, d\mu.$$

Prove that $\xi(t, k, u, v)$ satisfies

$$
\begin{aligned}
&\frac{\partial \xi(t,k,u,v)}{\partial t} + (ur-1)\frac{\partial \xi(t,k,u,v)}{\partial u} + \beta(u)\frac{\partial \xi(t,k-1,u,v)}{\partial u} \\
&+(vs-1)\frac{\partial \xi(t,k,u,v)}{\partial v} + \gamma(v)\frac{\partial \xi(t,k+1,u,v)}{\partial v} \\
&= -(ur\lambda^* + vsk\mu^*)\xi(t,k,u,v).
\end{aligned}
$$

References

1. L. Zhu, Process-level large deviations for nonlinear Hawkes point processes, Annales de l'Institut Henri Poincaré Probability and Statistics, 50 (3) (2014) 845–871
2. L. Zhu, Large deviations for Markovian nonlinear Hawkes processes, Annales of Applied Probability 25 (2015) 548–581
3. T. Tanaka, 0. Hashida, Y. Takahashi, Transient analysis of fluid models for ATM statistical multiplexer, Performance Evaluation 23 (1995) 145–162
4. S. Dharmaraja, A. Aggarwal and R. Sudhesh, Analysis of Energy Saving in User Equipment in LTE-A using Stochastic Modelling, Telecommunication Systems, 80, 123–140 (2022)
5. S. Dharmaraja and P. Tardelli: Infinite-server System with Hawkes Arrivals and Hawkes Services, Queueing Systems, 101 (2022), pp. 329 - 351. doi.org:10.1007/s11134-022-09813-3
6. Dharmaraja, S., Tardelli, P.: Infinite-server system with Hawkes arrivals and Hawkes services. Queueing Syst. **101**, 329–351 (2022)
7. S. Dharmaraja, Kishor Trivedi and Dimitris Logothetis, Performance modelling of wireless networks with generally distributed handoff inter-arrival times. Computer Communications, 26, 1747–1755 (2003)
8. L. B. Castaneda, V. Arunachalam, S. Dharmaraja, Introduction to Probability and Stochastic Processes with Applications. Wiley, New Jersey; 2012
9. N. Barbot, and B. Sericola, (2001). Transient analysis of a fluid queue driven by an $M/M/1$ queue. In Proc. 9th International Conference on Telecommunication Systems: Modeling and Analysis (ICTS'9), Dallas, TX, USA
10. E. Bacry, S. Delattre, M. Hoffmann, J. F. Muzy, Scaling limits for Hawkes processes and application to financial statistics, Stochastic Processes and Applications 123 (7) (2012) 2475–2499. doi: https://doi.org/10.1016/j.spa.2013.04.007

Chapter 10
Martingales

A martingale is also known as a fair game, for the following reason. In a gambling situation, if X_s is the amount of money the gambler has at time s, then the gambler's expected wealth at time $t > s$ is also X_s. This concept of martingale applies equally to discrete and continuous parameter.

Martingale is an important concept used for studying stochastic processes. In this chapter, after giving the definition and some examples of martingale, the Doob's martingale process, the optional stopping theorem, Jensen's inequality, maximum inequality theorem, and martingale convergence theorem are presented. The martingale theory of stochastic integrals and stochastic differential equations, are discussed in later chapters.

10.1 Filtration

This section discusses the definition of filtration which is introduced to represent the information available at each time, and then, we discuss the adapted process. Both the definitions play an essential role in the concept of martingales.

Definition 10.1 **(σ-field generated by a RV)** Let X be a RV. Then the σ-field generated by X denoted as $\sigma(X)$ consists of all sets of the form $\{\omega \in \Omega : X(\omega) \in B\}$, where B is a Borel set in $\mathbb{R}$.

Definition 10.2 **(σ-field generated by a stochastic process)** Let $\{X_t, t \in T\}$ be a stochastic process. The σ-field generated by $\{X_t, t \in T\}$ denoted as $\sigma(\{X_t, t \in T\})$ is the smallest σ-field containing all sets of the form

$$\{\omega : \text{the sample path } (X_t(\omega), t \in T) \in B\}$$

where B is a Borel set in $\mathbb{R}$ and $t \in T$.

D. Selvamuthu, *Introduction to Stochastic Processes*, University Texts in the Mathematical Sciences, https://doi.org/10.1007/978-981-97-6152-4_10

Definition 10.3 (Filtration) Let $(\Omega, \mathcal{F}, P)$ be a probability space. A filtration $\{\mathcal{F}_t\}_{t \in T}$ is a collection of sub σ-algebras of $\mathcal{F}$ such that $\mathcal{F}_s \subseteq \mathcal{F}_t$ for all $s < t \in T$.

The probability space taken together with the filtration, i.e., $(\Omega, \mathcal{F}, \{\mathcal{F}_t\}_{t \in T}, P)$ is called a filtered probability space.

Here, $\{\mathcal{F}_t\}_{t \in T}$ represents the set of events observable by time t. It consists of all events $\mathcal{E}$ such that at time t it is possible to determine whether $\mathcal{E}$ has occurred or not. As t increases, there will be more of these events $\mathcal{E}$. The filtration $\{\mathcal{F}_t\}_{t \in T}$ representing our knowledge will grow in size. If $T = [0, \infty)$, we will denote the filtration by $\{\mathcal{F}_t\}_{t \geq 0}$ and if $T = \{0, 1, ...\}$, we will denote the filtration by $\{\mathcal{F}_n\}_{n \geq 0}$.

Example 10.1 Consider $\Omega = \{a, b, c, d\}$. Construct σ-fields $\mathcal{F}_i, i = 0, 1, 2,$ such that $\mathcal{F}_0 \subset \mathcal{F}_1 \subset \mathcal{F}_2$.

Solution: Let $\mathcal{F}_0 = \{\emptyset, \Omega\}$, let $\mathcal{F}_1 = \{\emptyset, \{a, b\}, \{c, d\}, \Omega\}$ and $\mathcal{F}_2 = \{\emptyset, \{a\}, \{b\}, \{c\}, \{d\}, \{a, b\}, \{c, d\}, \{a, c\}, \{b, c\}, \{b, d\}, \{a, d\}, \{a, b, c\}, \{a, c, d\}, \{b, c, d\}, \{a, b, d\}, \Omega\}$. Then $\mathcal{F}_i, i = 1, 2, 3$ satisfies the given condition.

Definition 10.4 (Natural Filtration) Let $\{X_n, n = 0, 1, \ldots\}$ be a discrete parameter stochastic process and $\mathcal{F}_n = \sigma(X_m, m \leq n)$. Then, $\{\mathcal{F}_n\}_{n \geq 0}$ is known as the natural filtration of the stochastic process $\{X_n, n = 0, 1, \ldots\}$. Similarly, let $\{X_t, t \geq 0\}$ be a continuous parameter stochastic process and $\mathcal{F}_t = \sigma(X_s, s \leq t)$. Then, $\{\mathcal{F}_t\}_{t \geq 0}$ is known as the natural filtration of the stochastic process $\{X_t, t \geq 0\}$.

Example 10.2 Let $\{S_t, t \geq 0\}$ be a stochastic process where S_t is the price of a stock at time t. Consider $\{\mathcal{F}_t\}_{t \geq 0}$ to be the filtration consisting of stock price information till time t. Then $\{\mathcal{F}_t\}_{t \geq 0}$ is the natural filtration of the stochastic process $\{S_t, t \geq 0\}$.

Definition 10.5 (Adapted Process) The continuous parameter stochastic process $X = \{X_t, t \geq 0\}$ is said to be adapted to a given filtration $\{\mathcal{F}_t\}_{t \geq 0}$ if the σ-field generated by X_t is a subset of $\{\mathcal{F}_s\}_{0 \leq s \leq t}$, for all $t \geq 0$. For discrete parameter, the adapted process can be defined analogously.

Remark 10.1
1. We say that the RV X_t is $\mathcal{F}_t$-measurable, if for every Borel set $B \in \mathcal{B}(\mathbb{R})$ and for every $t \geq 0$, $\{X_t \in B\} \in \mathcal{F}_t$, i.e., $\{X_t, t \geq 0\}$ is adapted to $\{\mathcal{F}_t\}_{t \geq 0}$.
2. The intuitive idea behind an adapted process is that $\mathcal{F}_t$ should contain all the information needed to predict the behaviour of X up to and including time t.

Example 10.3 The stock price process in Example 10.2 is adapted to the filtration $\{\mathcal{F}_t\}_{t \geq 0}$ as $\mathcal{F}_t$ consists of all information about S_t till time t.

Example 10.4 Consider a stochastic process $\{X_t, t \geq 0\}$ which has independent increment property. Now consider the natural filtration of X_t as $\mathcal{F}_t = \sigma\ (X_s, 0 \leq s \leq t)$. Define a new stochastic process $\{Y_t, t \geq 0\}$ such that $Y_t = X_{t+1}$, then Y_t is not a adapted process.

10.2 Review of Conditional Expectation

The study of martingales depends on the notion of conditional expectation and their related properties which are discussed in this section. In this chapter, the inequalities and equalities for the conditional expectations are assumed to hold almost surely.

Definition 10.6 (Conditional Expectation given Discrete Random Variable) Let X be an integrable RV and let Y be a discrete RV. Then the conditional expectation of X given Y is a RV defined as

$$E[X|Y](\omega) = E[X|Y(\omega) = y_n] \text{ if } Y(\omega) = y_n$$

If X is a discrete RV, then, $E[X|Y(\omega) = y_n] = \sum_i x_i P(X = x_i|Y(\omega) = y_n)$.

Definition 10.7 (Conditional Expectation given Arbitrary Random Variable) Let X be an integrable RV and let Y be an arbitrary RV. Then the conditional expectation of X given Y is defined to be a RV $E[X|Y]$ such that

1. $E[X|Y]$ is $\sigma(Y)$-measurable.
2. For any $A \in \sigma(Y)$
$$\int_A E[X|Y]\mathrm{d}P = \int_A X\mathrm{d}P.$$

Definition 10.8 (Conditional Expectation given σ-Field) Let X be an integrable RV on a probability space $(\Omega, \mathcal{F}, P)$, and let $\mathcal{G}$ be a σ-field contained in $\mathcal{F}$. Then the conditional expectation of X given $\mathcal{G}$ is defined to be a RV $E[X|\mathcal{G}]$ such that

1. $E[X|\mathcal{G}]$ is $\mathcal{G}$-measurable.
2. For any $A \in \mathcal{G}$
$$\int_A E[X|\mathcal{G}]\mathrm{d}P = \int_A X\mathrm{d}P.$$

10.2.1 Properties of Conditional Expectation

Let X_1 and X_2 be integrable RVs on $(\Omega, \mathcal{F}, P)$ and $\mathcal{G}, \mathcal{H}$ be σ-fields on Ω contained in $\mathcal{F}$. Then the following equalities and inequalities hold a.s..

1. (Linearity property) Given the integrable RVs X_1 and X_2 and the real constants a_1 and a_2,
$$E[(a_1X_1 + a_2X_2)|\mathcal{G}] = a_1E[X_1|\mathcal{G}] + a_2E[X_2|\mathcal{G}].$$

2. (Taking out what is known) Given the integrable RVs X_1 and X_2 such that $X_1 X_2$ is integrable and X_1 is $\mathcal{G}$-measurable, then

$$E[X_1 X_2|\mathcal{G}] = X_1 E[X_2|\mathcal{G}].$$

3. (Tower property) If X is an integrable RV and $\mathcal{H}$ is a sub σ-field of $\mathcal{G}$, then

$$E[E[X|\mathcal{G}]|\mathcal{H}] = E[X|\mathcal{H}].$$

4. (Independence) If X is an integrable and independent RV with respect to σ-field $\mathcal{G}$, then

$$E[X|\mathcal{G}] = E[X].$$

5. (Conditional Jensen's inequality) If $\varphi(x)$ is a convex and integrable function and X is an integrable RV, then

$$E[\varphi(X)|\mathcal{G}] \geq \varphi(E[X|\mathcal{G}]).$$

10.3 Martingales

This section defines the concept of martingales. Let us first start with the definition of martingales in discrete parameter stochastic process.

Definition 10.9 (Martingale in Discrete Parameter) Let $(\Omega, \mathcal{F}, \{\mathcal{F}_n\}_{n\geq 0}, P)$ be a filtered probability space. A stochastic process $\{X_n, n = 0, 1, \ldots\}$ is called a martingale with respect to the filtration $\{\mathcal{F}_n\}_{n\geq 0}$, or $\mathcal{F}_n$-martingale if for every $n \in \{0, 1, \ldots\}$,

1. X_n is $\mathcal{F}_n$-measurable, i.e., $\{X_n, n = 0, 1, \ldots\}$ is adapted to $\{\mathcal{F}_n\}_{n\geq 0}$.
2. $E[X_n]$ exists, i.e., $E[|X_n|] < \infty$.
3. $$E[X_{n+1} \mid \mathcal{F}_n] = X_n, \text{ a.s..} \tag{10.1}$$

Now, the martingale property for the continuous parameter stochastic process is defined as follows.

Definition 10.10 (Martingale in Continuous Parameter) Let $(\Omega, \mathcal{F}, \{\mathcal{F}_t\}_{t\geq 0}, P)$ be a filtered probability space. For every $t \geq 0$, if

1. X_t is $\mathcal{F}_t$-measurable, i.e., $\{X_t, t \geq 0\}$ is adapted to the filtration $\{\mathcal{F}_t\}_{t\geq 0}$.
2. $E[X_t]$ exists, i.e., $E[|X_t|] < \infty$.
3. For $0 \leq s < t$,
$$E[X_t \mid \mathcal{F}_s] = X_s, \text{ a.s.,} \tag{10.2}$$

then $\{X_t, t \geq 0\}$ is called a martingale w.r.t. $\{\mathcal{F}_t\}_{t\geq 0}$ or $\mathcal{F}_t$-martingale.

Remark 10.2 1. The martingale property is identical to

$$E[X_n - X_m | \mathcal{F}_m] = 0 \text{ for every } m < n,$$

indicating that the increment $X_n - X_m$ given the "past" has expected value 0. A martingale is nothing but a stochastic process which on the average, given the past, does not increase or decrease.
2. A martingale has a constant expected value, i.e., $E[X_n] = E[E[X_n | \mathcal{F}_0]] = E[X_0]$ for all n.
3. The Markov process concept deals with the distributions, while the martingale concept concerns with the expectations.
4. A Markov process need not necessarily be a martingale.

Example 10.5 Let $\{Z_i, i = 1, 2, \ldots\}$ be a sequence of IID RVs each with mean 0 and finite variance. Define a sequence of RVs $\{X_n, n = 0, 1, \ldots\}$ as

$$X_n = \sum_{i=1}^{n} Z_i, \text{ for } n = 1, 2, \ldots \quad \text{and } X_0 = 0.$$

Let $\mathcal{F}_n = \sigma(Z_1, Z_2, \ldots, Z_n),\ n = 1, 2, \ldots$ with $\mathcal{F}_0 = \{\emptyset, \Omega\}$. Prove that, $\{X_n, n = 0, 1, \ldots\}$ is a martingale.
Solution: The following assumptions need to be checked for a martingale.

1. By the assumption on $Z_1, Z_2, \ldots, Z_n$ and linearity property of the expectation,

$$E[|X_n|] \leq \sum_{i=1}^{n} E[|Z_i|] < \infty.$$

2. Since $Z_1, Z_2, \ldots, Z_n$ are $\mathcal{F}_n$-measurable and X_n is a function of $Z_1, Z_2, \ldots, Z_n$, then X_n is $\mathcal{F}_n$-measurable.
3. Since

$$\begin{aligned} E[X_{n+1} | \mathcal{F}_n] &= E[Z_1 + Z_2 + \cdots + Z_{n+1} | \mathcal{F}_n] \\ &= E[X_n | \mathcal{F}_n] + E[Z_{n+1} | \mathcal{F}_n] \\ &= X_n + 0 \\ &= X_n. \end{aligned}$$

Since Z_{n+1} is independent of $\mathcal{F}_n$ and X_n is $\mathcal{F}_n$-measurable, it can be concluded that $\{X_n, n = 0, 1, \ldots\}$ is a martingale w.r.t. $\{\mathcal{F}_n\}_{n\geq 0}$.

Example 10.6 Let $(\Omega, \mathcal{F}, \{\mathcal{F}_n\}_{n\geq 0}, P)$ be a filtered probability space. Let X be an integrable RV on this filtered probability space. Define a sequence for $n = 0, 1, \ldots$

$$X_n = E[X|\mathcal{F}_n].$$

Prove that, $\{X_n, n = 0, 1, \ldots\}$ is a martingale.
Solution: X_n is $\mathcal{F}_n$-measurable,

$$|X_n| = |E[X|\mathcal{F}_n]| \leq E[|X||\mathcal{F}_n],$$

which implies that

$$E[|X_n|] \leq E[E[|X||\mathcal{F}_n]] = E[|X|] < \infty,$$

and

$$E[X_{n+1}|\mathcal{F}_n] = E[E[X|\mathcal{F}_{n+1}]|\mathcal{F}_n] = E[X|\mathcal{F}_n] = X_n, \text{ a.s..}$$

Therefore, $\{X_n, n = 0, 1, \ldots\}$ is a martingale with respect to $\{\mathcal{F}_n\}_{n\geq 0}$.

Example 10.7 Let $\{H_n, n = 0, 1, \ldots\}$ be a bounded and adapted process w.r.t. the natural filtration $\{\mathcal{F}_n\}_{n\geq 0}$. Prove that for any $\mathcal{F}_n$-martingale $\{Y_n, n = 0, 1, \ldots\}$, the stochastic process $\{X_n, n = 0, 1, \ldots\}$ which is defined as $X_n = \sum_{i=0}^{n-1} H_i\,(Y_{i+1} - Y_i)$ is also martingale w.r.t. $\{\mathcal{F}_n\}_{n\geq 0}$.
Solution: Since, Y_n is martingale w.r.t. $\{\mathcal{F}_n\}_{n\geq 0}$, Y_n is $\mathcal{F}_n$-measurable, which implies X_n is $\mathcal{F}_n$-measurable. Now consider,

$$\begin{aligned} E[|X_n|] &= E\left[\left|\sum_{i=0}^{n-1} H_i\,(Y_{i+1} - Y_i)\right|\right] \\ &\leq \sum_{i=0}^{n-1} E\left[|H_i||Y_{i+1} - Y_i|\right] < \infty, \end{aligned}$$

as H_n is a bounded. Now for any $n = 0, 1, \ldots$ we have

$$E[(X_{n+1} - X_n)|\mathcal{F}_n] = E[H_n\,(Y_{n+1} - Y_n)\,|\mathcal{F}_n].$$

Since, H_n is $\mathcal{F}_n$-measurable,

$$\begin{aligned} E[(X_{n+1} - X_n)|\mathcal{F}_n] &= H_n E[(Y_{n+1} - Y_n)\,|\mathcal{F}_n] \\ &= H_n\,(E[Y_{n+1}|\mathcal{F}_n] - Y_n) = 0, \end{aligned}$$

which implies $E[X_{n+1}|\mathcal{F}_n] = X_n$ a.s. for any $n = 0, 1, \ldots$. Hence, $\{X_n, n = 0, 1, \ldots\}$ is a martingale w.r.t. $\{\mathcal{F}_n\}_{n\geq 0}$.

Example 10.8 Let $\{N(t), t \geq 0\}$ be a Poisson process with parameter $\lambda > 0$. Show that $\{N(t), t \geq 0\}$ is not a martingale.

Solution: We know that, if $\{X(t), t \geq 0\}$ is a martingale, then $E[X(0)] = E[X(t)]$ for all $t \geq 0$. But for the Poisson process, $E[N(t)] = \lambda t, t \geq 0$, which implies each expectation is different. Hence, the Poisson process is not a martingale.

Example 10.9 Define $X(t) = e^{-\lambda\mu t}(\mu + 1)^{N(t)},\ t > 0$, where $\{N(t), t \geq 0\}$ is a Poisson process with parameter $\lambda > 0$ and $\mu > -1$ is a constant. Prove that $\{X(t), t \geq 0\}$ is a martingale.

Solution: Let, $\{\mathcal{F}_t\}_{t\geq 0}$ is a natural filtration of $\{N(t), t \geq 0\}$. For $0 \leq s < t$, using the properties of conditional expectations we have

$$\begin{aligned}
E[X(t)|\mathcal{F}_s] &= E[e^{-\lambda\mu t}(\mu + 1)^{N(t)}|\mathcal{F}_s] \\
&= e^{-\lambda\mu t}E[(\mu + 1)^{N(t)}|\mathcal{F}_s] \\
&= e^{-\lambda\mu t}E[(\mu + 1)^{N(t)-N(s)}(\mu + 1)^{N(s)}|\mathcal{F}_s] \\
&= e^{-\lambda\mu t}(\mu + 1)^{N(s)}E[(\mu + 1)^{N(t)-N(s)}] \\
&= e^{-\lambda\mu t}(\mu + 1)^{N(s)}\left(\sum_{n=0}^{\infty} e^{-\lambda(t-s)}(\mu + 1)^n\frac{(\lambda(t - s))^n}{n!}\right) \\
&= e^{-\lambda\mu t}(\mu + 1)^{N(s)}e^{-\lambda(t-s)}e^{\lambda(t-s)(\mu+1)} \\
&= e^{-\lambda s\mu}(\mu + 1)^{N(s)} = X(s).
\end{aligned}$$

Hence, $\{X(t), t \geq 0\}$ is a martingale with respect to $\{\mathcal{F}_t\}_{t\geq 0}$.

10.3.1 Sub-martingales and Super-martingales

Definition 10.9 defined a martingale via an equality for certain conditional expectation. If this equality is replaced with inequality, we obtain sub-martingales and super-martingales.

Definition 10.11 (Sub-Martingales and Super-Martingales) Consider a sequence of RVs $\{X_n, n = 0, 1, \ldots\}$ defined on a probability space $(\Omega, \mathcal{F}, P)$ and $\{\mathcal{F}_n\}_{n\geq 0}$ be a filtration in $\mathcal{F}$. Suppose that $\{X_n, n = 0, 1, \ldots\}$ is adapted to the filtration $\{\mathcal{F}_n\}_{n\geq 0}$ (i.e., X_n is $\mathcal{F}_n$-measurable) and $E[|X_n|]$ exists for all n. Then,

1. $\{X_n, n \geq 0\}$ is a sub-martingale if and only if $E[X_n|\mathcal{F}_m] \geq X_m$ a.s. for all $m \leq n$.
2. $\{X_n, n \geq 0\}$ is a super-martingale if and only if $E[X_n|\mathcal{F}_m] \leq X_m$ a.s. for all $m \leq n$.

Similarly, in continuous parameter, sub-martingales and super-martingales are defined by replacing the equality in Eq. 10.2 with $\geq$ and $\leq$, respectively.

Example 10.10 Modify Example 10.5. If $E[Z_n] > 0$, then $\{X_n, n = 0, 1, \ldots\}$ is a sub-martingale. On the other hand, $E[Z_n] < 0$, then $\{X_n, n = 0, 1, \ldots\}$ is a super-martingale.

Example 10.11 Consider a discrete parameter stochastic process $\{X_n, n = 0, 1, \ldots\}$ which is defined as

$$X_{n+1} = aX_n + b\epsilon_{n+1},$$

where each n taking positive integer values, and $\{\epsilon_n, n = 0, 1, \ldots\}$ is a sequence of IID RVs with mean 0 and finite variance and a and b are constants such that $a \neq 1$. Prove that this stochastic process is not a martingale.

Solution: Let $\mathfrak{F}_n = \sigma\{X_i, i \leq n\}$, the so-called natural filtration of X_n, i.e., the σ-algebra containing all the information of X_i, for $i \leq n$. Recall that X_n is a martingale, if

$$E[X_{n+1}|\mathfrak{F}_n] = X_n.$$

Fix a value of n, the conditional expectation of X_{n+1} given the information at time n is given by

$$\begin{aligned} E[X_{n+1}|\mathfrak{F}_n] &= E[aX_n + b\epsilon_{n+1}|\mathfrak{F}_n] \\ &= aX_n + bE[\epsilon_{n+1}] \\ &= aX_n. \end{aligned}$$

Since,

$$E[X_{n+1}|\mathfrak{F}_n] = aX_n \neq X_n.$$

Therefore, this implies that $\{X_n, n = 1, 2, \ldots\}$ is not a martingale.

Remark 10.3
1. A stochastic process $\{X_n, n = 1, 2, \ldots\}$ is a sub-martingale with respect to the filtration $\{\mathfrak{F}_n\}_{n\geq 1}$ if and only if $\{-X_n, n = 1, 2, \ldots\}$ is a super-martingale with respect to the filtration $\{\mathfrak{F}_n\}_{n\geq 1}$.
2. A stochastic process $\{X_n, n = 1, 2, \ldots\}$ is a martingale with respect to the filtration $\{\mathfrak{F}_n\}_{n\geq 1}$ if and only if $\{X_n, n = 1, 2, \ldots\}$ is both sub-martingale and super-martingale with respect to the filtration $\{\mathfrak{F}_n\}_{n\geq 1}$.
3. If $\{X_n, n = 1, 2, \ldots\}$ is a martingale, then $E[X_n] = E[X_1]\ \forall\ n$. If $m < n$ and $\{X_n, n = 1, 2, \ldots\}$ is a sub-martingale, then $E[X_m] \leq E[X_n]$. On the other hand, if $\{X_n, n = 1, 2, \ldots\}$ is a super-martingale, then $E[X_m] \geq E[X_n]$.

10.3.2 Doob's Martingale Process

Doob's martingale is a specific type of martingale process introduced by American mathematician Joseph L. Doob (February 27, 1910–June 7, 2004).

The martingale considered in Example 10.6 is called Doob's Martingale Process.

In essence, as n increases X_n approximates X and the approximation becomes more accurate because more information has been assembled and included in the consideration.

The next theorem which will be used many times, has been stated as a result and is called Doob-Lévy martingale.

Theorem 10.1 (Doob-Lévy Martingale) *Let* $\{\mathcal{F}_t\}_{t\geq 0}$ *be a filtration and let* X *be a RV with* $E[|X|] < \infty$*, and let* $M_t = E[X|\mathcal{F}_t]$*. Then* $\{M_t, t \geq 0\}$ *is a martingale.*

Proof Since $E[|M_t|] \leq E[E[|X||\mathcal{F}_t]] = E[|X|] < \infty$, the integrability is clear, and the adaptability is clear from the definition of M_t itself. For $s \leq t$,

$$E[M_t|\mathcal{F}_s] = E[E[X|\mathcal{F}_t]|\mathcal{F}_s] = E[X|\mathcal{F}_s] = M_s, \text{ a.s..}$$

Therefore, $\{M_t, t \geq 0\}$ is a martingale.

Super-martingales typically decrease while sub-martingales tends to increase. The nomenclature may appear to go in the wrong direction, but Doob defined these terms by comparing them to the concepts of subharmonic and superharmonic functions in analysis. Keep in mind that the set of σ-fields affects how martingale is defined. One can say that $\{(M_t, \mathcal{F}_t)\}$ is a martingale or that M_t is a $\mathcal{F}_t$-martingale when it is needed. A martingale also depends on the probability because probability is required to define conditional expectation. The phrase $\{M_t\}$ is a martingale with respect to the probability P will be used as needed.

10.4 Stopping Times, Inequalities and Convergence Theorems

Stopping times idea work well with martingales. A stopping time is a specific type of "random time", a RV whose value is interpreted as the time at which a given stochastic process exhibits a certain behavior of interest. It is also known as Markov time or optional time. A stopping time is often defined by a stopping rule, a mechanism for deciding whether to continue or stop a process on the basis of the present position and past events, and which will almost always lead to a decision to stop at some finite time.

Definition 10.12 (Stopping Time or Markov Time or Optional Time) Let $(\Omega, \mathcal{F}, P)$ be a probability space, and $\{\mathcal{F}_t\}_{t\in T}$ be a filtration on this probability space for an index set $T \subseteq \mathbb{R}$. A mapping τ from Ω into T is a stopping time if the event $\{\tau \leq t\} \in \mathcal{F}_t$ for all $t \in T$.

For the special case, when $T = 1, 2, \ldots$ is a countable index set, a stopping time with respect to a filtration $\{\mathcal{F}_n\}_{n\geq 1}$ is a non-negative RV τ with values in the set $\mathbb{N} \bigcup \{\infty\}$ s.t. for each $n \in \mathbb{N}$, the event $\{\tau = n\} = \{\tau \leq n\} \cap \{\tau \leq n-1\}^c \in \mathcal{F}_n$.

Example 10.12 Let $\{X_i, i \in \mathbb{N}\}$ be a sequence of IID RVs with $P(X_i = 1) = p = 1 - P(X_i = 0)$ where $0 < p < 1$. Define $\tau_r = \min\{n \in \mathbb{N} : X_1 + X_2 + \ldots + X_n = r\}$, i.e., τ_r is the number of trials needed until there have been a total of r successes. Since, $\{\tau_r = n\} \in \mathcal{F}_n$ for each $n \in \mathbb{N}$, τ_r is the stopping time for this sequence of RVs.

Definition 10.13 (Local-Martingale) A stochastic process $\{X_t, t \geq 0\}$ is said to be a local martingale if there exists an increasing sequence of stopping times $\{\tau_n, n = 1, 2, \ldots\}$ such that

1. $\lim_{n\to\infty} \tau_n = \infty$ a.s,
2. for each $n \geq 1$, $X_{t\wedge\tau_n}$ is an $\mathcal{F}_t$-martingale.

The family $\{\tau_n, n = 1, 2, \ldots\}$ is a collection of stopping times of $\{X_t, t \geq 0\}$.

Remark 10.4 Here, $a \wedge b = \min\{a, b\}$ and $a \vee b = \max\{a, b\}$.

Remark 10.5 The basic result for the stochastic integrals (will be discussed in Chap. 12), is that the integral with respect to a local martingale is a local martingale.

Example 10.13 Prove that, the first time the process $\{Y_0, Y_1, \ldots\}$ reaches a subset A of the state space is a Markov time, i.e., $T(A) = \min\{n : Y_n \in A\}$ is a Markov time.
Solution: Let

$$\{T = n\} = \{Y_0 \notin A\} \cap \{Y_1 \notin A\} \cap \cdots \cap \{Y_{n-1} \notin A\} \cap \{Y_n \in A\}$$

which implies that $\{T = n\}$ is determined by $(Y_0, Y_1, \ldots, Y_n)$ and hence $\{T = n\}$ is a Markov time.

Remark 10.6 Let S and T be two Markov times with respect to the process $\{Y_n, n = 0, 1, \ldots\}$. Then the following holds

1. $S + T$ is a Markov time.
2. $S \wedge T = \min\{S, T\}$, is a Markov time.
   Also, for any fixed $n = 0, 1, \ldots$ $T \wedge n = \min\{n, T\}$ is a Markov time.
3. $S \vee T = \max\{S, T\}$ is a Markov time.

Example 10.14 Let $\{X_n, n = 1, 2, \ldots\}$ be a sequence of RVs adapted to a filtration $\{\mathcal{F}_n\}_{n\geq 1}$ and let $B \subset \mathbb{R}$ be a Borel set. Then the time of first entry of X_n into B, i.e.,

$$\tau = \min\{n : X_n \in B\}$$

is a stopping time.

Example 10.15 Let us consider the scenario when two coins are tossed, i.e., one is followed by the other and consider the natural filtration of the coin tosses. $\Omega = \{HH, HT, TH, TT\}$. Let $\mathcal{F}_0 = \{\emptyset, \Omega\}$, $\mathcal{F}_1 = \{\emptyset, \{HH, HT\}, \{TH, TT\}, \Omega\}$ and $\mathcal{F}_2 = \mathcal{P}(\Omega)$. Consider a RV τ with $\tau(\omega) = 1$, if $\omega \in \{HH, HT\}$, and $\tau(\omega) = 2$, if $\omega \in \{HH, TT\}$. To show that τ is a stopping time, as τ is discrete, it is required to show that $\{\tau = k\} \in \mathcal{F}_k$ for $k = 0, 1, 2$. Now, $\{\tau = 0\} = \emptyset \in \mathcal{F}_0$, $\{\tau = 1\} = \{HH, HT\} \in \mathcal{F}_1$, $\{\tau = 2\} = \{HH, TT\} \in \mathcal{F}_2$. This shows that τ is a stopping time (with respect to this filtration), i.e., the information available at time n is sufficient to determine the events connected with τ. Let us consider another RV, $\rho(\omega) = 1$, if $\omega \in \{HH, TT\}$, and $\rho(\omega) = 2$, if $\omega \in \{HT, TH\}$, then it is not a stopping time, since $\{\rho = 1\} = \{HH, TT\} \notin \mathcal{F}_1$. According to our observation, the information provided at time $t = 1$ is insufficient to determine the event $\{\rho = 1\}$, since it also requires knowledge of the second throw.

Definition 10.14 (Hitting Time) Let $\{X_t, t \geq 0\}$ be a stochastic process. Let B be a Borel set in $\mathbb{R}$. Define

$$\tau = \inf\{t > 0 : X_t \in B\}.$$

Then, τ is called hitting time of B for the stochastic process $\{X_t, t \geq 0\}$.

From this point onwards, we will concentrate on discrete parameter martingales.

10.4.1 Optional Stopping Theorem

The optional stopping theorem says that, under some specific conditions, the expected value of a martingale at a stopping time is equivalent to its initial expected value. The concept of martingales is used to represent the amount of money a gambler has which is engaged in a fair game. Thus, optional stopping theorem states that, on average, there is no advantage to be gained by discontinuing play based on the already available information. Let us discuss some of the necessary conditions for this result to hold true.

Theorem 10.2 (Optional Stopping Theorem) *Let $\{X_n, n = 1, 2, \ldots\}$ be a martingale and τ be a stopping time. Let*

1. $P\{\tau < \infty\} = 1$,
2. $E[|X_\tau|] < \infty$,
3. $\lim_{n\to\infty} E[X_n I_{(\tau>n)}] = 0$,

then $E[X_\tau] = E[X_0]$.

Proof Since,

$$X_\tau = X_{\tau\wedge n} + (X_\tau - X_n)1_{\{\tau>n\}}$$

it follows that

$$E[X_\tau] = E[X_{\tau\wedge n}] + E[X_\tau 1_{\{\tau>n\}}] - E[X_n 1_{\{\tau>n\}}].$$

Since $X_{\tau\wedge n}$ is a martingale (X_n is a martingale, hence is $X_{\tau\wedge n}$), thus

$$E[X_{\tau\wedge n}] = E[X_0].$$

By Condition 3, $E[X_n 1_{\{\tau>n\}}] \to 0$. The term

$$E[X_\tau 1_{\{\tau>n\}}] = \sum_{k=n+1}^{\infty} E[X_k 1_{\{\tau=k\}}]$$

tends to zero as $n \to \infty$ because the series

$$E[X_\tau] = \sum_{k=1}^{\infty} E[X_k 1_{\{\tau=k\}}]$$

is convergent by Condition 2. It follows that $E[X_\tau] = E[X_0]$, as required.

Here,

$$E[X_0] = \lim_{n\to\infty} E[X_{\tau\wedge n}] = E[\lim_{n\to\infty} X_{\tau\wedge n}] = E[X_\tau].$$

As an application, the following example is discussed.

Example 10.16 (Unrestricted Random Walk) Consider the stochastic process $\{S_n, n = 0, 1, \ldots\}$ where $S_0 = 0$ and $S_n = \sum_{i=1}^{n} Y_i, n = 1, 2, \ldots$. Here, $Y_0 = 0$ and for $i = 1, 2, \ldots$, $Y_i's$ are IID random variables with

$$P(Y_i = 1) = p \text{ and } P(Y_i = -1) = q \text{ where } \quad p + q = 1.$$

The stochastic process $\{S_n, n = 0, 1, \ldots\}$ is referred to as an unrestricted random walk (URW) (also discussed in Chap. 2). Discuss the martingale property for this random walk.
Solution:The sample path of the random walk $\{S_n, n = 0, 1, \ldots\}$ is shown in Fig. 10.1.

Based on the possible values of p, the random walk $\{S_n, n = 0, 1, \ldots\}$ is discussed as follows.

1. When $p = q(= 1 - p) = 1/2$, $\{S_n, n = 0, 1, \ldots\}$ is a martingale.
   If $T = \min\{n : S_n = 1\}$, then $P(T < \infty) = 1$. Since $E[T] = \infty$, optional stopping theorem does not hold, and hence $E[S_T] \neq E[S_0] = 0$. Also, $\{S_n, n = 0, 1, \ldots\}$ is called a symmetric random walk.
2. When $p > q$, $\{S_n - n(p - q), n = 0, 1, \ldots\}$ is a martingale.
   If $T = \min\{n : S_n = -a \ \ or \ \ S_n = b\}$ $(a, b$ positive integers), then $P(T < \infty) = 1$ and $E[T] = ab$.
   Since $E[T] = ab$, optional stopping theorem holds, and $E[S_T] = (p - q)E[T]$ for any Markov time T satisfying $E[T] < \infty$.

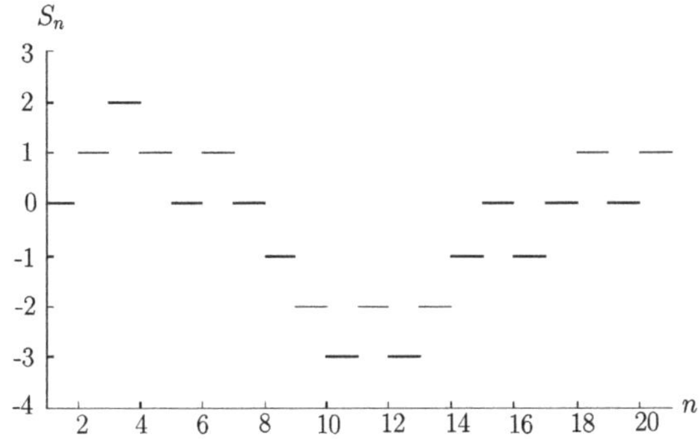

Fig. 10.1 Sample path of a random walk with $p = 0.5$.

3. When $p < q$, $\{S_n - n(p-q), n = 0, 1, \ldots\}$ is a martingale.
   If $T = \min\{n : S_n = -a \ \ or \ \ S_n = b\}$ $(a, b$ positive integers), by similar argument optional stopping theorem holds, and $E[S_T] = (p-q)E[T]$ for any Markov time T such that $E[T] < \infty$.

Definition 10.15 (Predictable Process) A discrete parameter process $\{M_n\}_{n\geq 0}$ is said to be predictable with respect to the filtration $\{\mathfrak{F}_n\}_{n\geq 0}$ if M_n is $\mathfrak{F}_{n-1}$-measurable for every $n = 0, 1, \ldots$.

Now, the concept of a discrete integral is introduced which is a useful tool in martingale theory and it is explained in the subsequent theorem.

Definition 10.16 (Martingale Transform) Let $M = \{M_n, n = 0, 1, \ldots\}$ and $X = \{X_n, n = 0, 1, \ldots\}$ be two discrete parameter processes. Now, a process $(X \cdot M)$ is defined with $(X \cdot M)_0 = 0$ and

$$(X \cdot M)_n = \sum_{i=1}^{n} X_i(M_i - M_{i-1}).$$

It is said that $X \cdot M$ is the discrete integral of X with respect to M and is also called the martingale transform of M by X, if M is a (sub-, super-) martingale.

Lemma 10.1 *If M is a sub-martingale (resp. super-martingale) and X is a bounded, positive predictable process, then the discrete integral of X with respect to M, i.e., $X \cdot M$ is also a sub-martingale (resp. super-martingale). On the other hand, if M is a martingale and X is bounded, predictable process, then $X \cdot M$ is a martingale.*

Proof Let $Y = X \cdot M$, then the process Y is adapted. Since X is bounded, i.e., $|X_n| \leq K$ for all n, $E|Y_n| \leq 2K \sum_{k\leq n} E|M_k| < \infty$. Now suppose that M is a sub-martingale and X is positive. Then,

$$\begin{aligned} E[Y_n|\mathfrak{F}_{n-1}] &= E[Y_{n-1} + X_n(M_n - M_{n-1})|\mathfrak{F}_{n-1}] \\ &= Y_{n-1} + X_n E[M_n - M_{n-1}|\mathfrak{F}_{n-1}] \geq Y_{n-1}, \text{ a.s.,} \end{aligned}$$

which implies Y is a sub-martingale.
If M is a martingale, then the last inequality is converted to equality, irrespective of the sign of X_n, which implies that Y is a martingale.

With the use of this lemma, it is easy to verify that a stopped (sub-, super-) martingale is again a (sub-, super-) martingale.

Theorem 10.3 *Let M be a (sub-, super-) martingale and τ be the stopping time. Then, the stopped process, i.e., $M^\tau = \{M_{n\wedge\tau}\}$ is also a (sub-, super-) martingale.*

Proof Define the process X by $X_n = I_{\{\tau\geq n\}}$ and it can be seen that $M^\tau = M_0 + X \cdot M$. Since τ is the stopping time, $\{\tau \geq n\} = \{\tau \leq n-1\}^c \in \mathfrak{F}_{n-1}$, this shows that the process X is predictable. Since, it is also a bounded process, hence by using the above lemma the statement of the theorem holds.

Definition 10.17 (Convex Function) A real-valued function $\varphi : [a, b] \to \mathbb{R}$ is said to be a convex function if, for all $x, y \in [a, b]$,

$$\varphi(\alpha x + (1-\alpha)y) \leq \alpha\varphi(x) + (1-\alpha)\varphi(y) \;\text{ for }\; 0 \leq \alpha \leq 1.$$

For examples, $\varphi(x) = e^{kx}$ and $\varphi(x) = (x-k)^+$ where k is a constant.

Example 10.17 Let X be a RV defined on $(\Omega, \mathfrak{F}, P)$ with $E[\mid X \mid] < \infty$. Consider a concave function f on $\mathbb{R}$. Then

$$E[f(X)] \leq f(E[X]).$$

Solution: If f is concave, then $-f$ is convex. Using Jensen's inequality,

$$E[-f(X)] \geq -f(E[X])$$

which implies

$$E[f(X)] \leq f(E[X]).$$

Now, Jensen's inequality is presented in the following theorem without proof.

Theorem 10.4 (Jensen's Inequalities) *Let X and Y be the RVs on $(\Omega, \mathfrak{F}, P)$ with $E[\mid X \mid] < \infty$. Consider a convex function φ on $\mathbb{R}$. Then*

$$E[\varphi(X)] \geq \varphi(E[X]), \text{ and}$$

$$E[\varphi(X)|Y] \geq \varphi(E[X|Y])$$

provided all the expectations exist.

Corollary 10.1 *Consider X is a RV on $(\Omega, \mathcal{F}, P)$ and let φ be a convex function on $\mathbb{R}$. Assume that both X and φ are integrable and $\mathcal{G}$ is a sub σ-field of $\mathcal{G}$. Then,*

$$E[\varphi(X)|\mathcal{G}] \geq \varphi(E[X|\mathcal{G}])$$

provided the expectations exist.

Example 10.18 Suppose that a strictly positive RV X has mean 0. What can you say about the expected value of e^X?
Solution: Using Jensen's inequality

$$\begin{aligned} E[e^X] &\geq e^{E[X]} \\ &= e^0 = 1, \end{aligned}$$

which implies expected value of e^X is greater than or equal to 1.

Example 10.19 Let X be a positive RV. Then compare $E[X^p]$ with $(E[X])^p$ for all values of $p \in \mathbb{R}$.
Solution:

(i) If $p = 0$, then

$$E[X^p] = (E[X])^p = 1,$$

(ii) If $p = 1$, then

$$E[X^p] = (E[X])^p = E[X],$$

(iii) Consider $f(x) = x^p, x > 0$. When $p \neq 0, 1$, then it can be easily verified that for $p < 0$ or $p > 1$, $f(x)$ is convex, and for $0 < p < 1$, $f(x)$ is concave. Using Jensen's inequality

$$E[X^p] \geq (E[X])^p \text{ for } p < 0 \text{ or } p > 1, \text{ and}$$
$$E[X^p] \leq (E[X])^p \text{ for } 0 < p < 1.$$

Theorem 10.5 *Suppose that the process $\{X_n, n = 1, 2, \ldots\}$ is a martingale with respect to the filtration $\{\mathcal{F}_n\}_{n\geq 1}$ and φ is a convex function on $\mathbb{R}$. If $\varphi(X_n)$ is integrable for all n, then $\{\varphi(X_n), n = 1, 2, \ldots\}$ is a sub-martingale with respect to the filtration $\{\mathcal{F}_n\}_{n\geq 1}$.*

Proof For $m \leq n$, by using Jensen's inequality for conditional expectation, we have

$$\varphi(X_m) = \varphi(E[X_n|\mathcal{F}_m]) \leq E[\varphi(X_n)|\mathcal{F}_m].$$

Other properties like integrability and adaptability are trivial to verify.

Example 10.20 Let $\{X_n, n = 1, 2, \ldots\}$ be a sequence of square integrable RVs, and $\{X_n, n = 1, 2, \ldots\}$ is a martingale with respect to filtration $\{\mathfrak{F}_n\}_{n\geq 1}$, then show that X_n^2 is a sub-martingale with respect to the same filtration.
Solution:Using Jensen's inequality

$$\begin{aligned} E[X_{n+1}^2|\mathfrak{F}_n] &\geq (E[X_{n+1}|\mathfrak{F}_n])^2 \\ &= (X_n)^2 \text{ a.s. using martingale property.} \end{aligned}$$

Theorem 10.6 *Suppose $\{X_n, n = 1, 2, \ldots\}$ is a sub-martingale with respect to the filtration $\{\mathfrak{F}_n\}_{n\geq 1}$ and φ is a convex and an increasing function on $\mathbb{R}$. Then, if $\varphi(X_n)$ is integrable for all n, $\{\varphi(X_n), n = 1, 2, \ldots\}$ is a sub-martingale with respect to the filtration $\{\mathfrak{F}_n\}_{n\geq 1}$.*
Proof For $m \leq n$,

$$X_m \leq E[X_n|\mathfrak{F}_m],$$

and since φ is an increasing function,

$$\varphi(X_m) \leq \varphi(E[X_n|\mathfrak{F}_m]) \leq E[\varphi(X_n)|\mathfrak{F}_m].$$

Now, Doob's decomposition is presented in the following theorem.

Theorem 10.7 *Let $X = \{X_n, n = 0, 1, \ldots\}$ be a sub-martingale with respect to the filtration $\{\mathfrak{F}_n\}_{n\geq 0}$. Then, there exists a unique martingale $M = \{M_n, n = 0, 1, \ldots\}$ and a stochastic process $A = \{A_n, n = 0, 1, \ldots\}$, such that*

1. *$X_n = M_n + A_n, n = 0, 1, \ldots$.*
2. *A_n is an increasing stochastic process, i.e., $A_n \leq A_{n+1}, n = 0, 1, \ldots$ a.e. with $A_0 = 0$.*
3. *A_n is $\mathfrak{F}_{n-1}$-measurable, for all n (or predictable).*
4. *$\{M_n, n = 0, 1, \ldots\}$ is a martingale with respect to the filtration $\{\mathfrak{F}_n\}_{n\geq 0}$.*

Proof Inductively, define A_n as $A_0 = 0$, and

$$A_n = A_{n-1} + \left[E[X_n|\mathfrak{F}_{n-1}] - X_{n-1}\right] \quad \text{for } n \geq 1,$$

Clearly, A_n satisfies all the required properties. Define $M_n = X_n - A_n$. Then, M_n is integrable, and $E[M_n|\mathfrak{F}_{n-1}] = M_{n-1}$. Thus, $\{M_n, n = 0, 1, \ldots\}$ is a martingale. For uniqueness, suppose X has another decomposition as

$$X_n = \bar{M}_n + \bar{A}_n.$$

From the definition of A_n,

$$A_n - A_{n-1} = E[X_n|\mathfrak{F}_{n-1}] - X_{n-1} = \bar{A}_n - \bar{A}_{n-1}.$$

Moreover, $A_0 = \bar{A}_0 = 0$, which implies $A_n = \bar{A}_n$ and $M_n = \bar{M}_n$.

When $X = \{X_n, n = 0, 1, \ldots\}$ is a super-martingale, the following result holds.

Corollary 10.2 *Let* $X = \{X_n, n = 0, 1, \ldots\}$ *be a super-martingale with respect to the filtration* $\{\mathfrak{F}_n\}_{n\geq 0}$. *Then, there exists a unique martingale* $M = \{M_n, n = 0, 1, \ldots\}$ *and a stochastic process* $A = \{A_n, n = 0, 1, \ldots\}$, *such that*

1. $X_n = M_n - A_n, n = 0, 1, \ldots.$
2. A_n *is an increasing stochastic process, i.e.,* $A_n \leq A_{n+1}, n = 0, 1, \ldots$ *a.e..*
3. A_n *is* $\mathfrak{F}_{n-1}$*-measurable, for all n (or predictable).*
4. $\{M_n, n = 0, 1, \ldots\}$ *is a martingale with respect to the filtration* $\{\mathfrak{F}_n\}_{n\geq 0}$.

Example 10.21 Let $\{M_n, n = 0, 1, \ldots\}$ be a martingale such that $E[M_n^2] < \infty \, \forall \, n$. Show that there exists a stochastic process $\{A_n, n = 0, 1, \ldots\}$ such that $\{M_n^2 - A_n, n = 0, 1, \ldots\}$ is a martingale and $A_n - A_{n-1} = E[(M_n - M_{n-1})^2 | \mathfrak{F}_{n-1}]$ a.s.. Furthermore,

$$E[M_n^2] = E[M_0^2] + \sum_{i=1}^{n} E[(M_i - M_{i-1})^2].$$

Solution: Since $\{M_n, n = 0, 1, \ldots\}$ is a martingale, by Jensen's inequality $\{M_n^2, n = 1, 2, \ldots\}$ is a sub-martingale. Using Doob's decomposition, there exists a unique martingale $\{Y_n, n = 0, 1, \ldots\}$ and a stochastic process $\{A_n, n = 0, 1, \ldots\}$ such that

$$M_n^2 = Y_n + A_n, n = 0, 1, \ldots$$

which implies $\{M_n^2 - A_n, n = 0, 1, \ldots\}$ is a martingale.

Consider,

$$\begin{aligned} E[(M_n - M_{n-1})^2 | \mathfrak{F}_{n-1}] &= E[M_n^2 + M_{n-1}^2 - 2M_n M_{n-1} | \mathfrak{F}_{n-1}] \\ &= E[M_n^2 | \mathfrak{F}_{n-1}] + M_{n-1}^2 - 2M_{n-1} E[M_n | \mathfrak{F}_{n-1}] \\ &= E[M_n^2 | \mathfrak{F}_{n-1}] - M_{n-1}^2 \\ &= E[Y_n + A_n | \mathfrak{F}_{n-1}] - M_{n-1}^2 \\ &= Y_{n-1} + A_n - M_{n-1}^2 = A_n - A_{n-1}, \end{aligned}$$

which implies $E[A_n - A_{n-1}] = E[(M_n - M_{n-1})^2] \, \forall \, n.$
Now consider,

$$\begin{aligned} E[M_i^2 - M_{i-1}^2] &= E[(M_i^2 - A_i) + (A_i - A_{i-1}) - (M_{i-1}^2 - A_{i-1})] \\ &= E[Y_i] + E[A_i - A_{i-1}] - E[Y_{i-1}] \\ &= E[A_i - A_{i-1}] = E[(M_i - M_{i-1})^2] \end{aligned}$$

which implies,

$$\sum_{i=1}^{n} E[M_i^2 - M_{i-1}^2] = \sum_{i=1}^{n} E[(M_i - M_{i-1})^2]$$

$$\implies E[M_n^2] = E[M_0^2] + \sum_{i=1}^{n} E[(M_i - M_{i-1})^2].$$

Example 10.22 Let, $\{X_n, n = 0, 1, \ldots\}$ be a sequence of IID square integrable RVs, i.e., $E[X_n^2] < \infty$ and $E[X_n] = 0 \,\forall\, n$. Let, $M_n = \sum_{i=0}^{n} X_i$ and $\mathfrak{F}_n = \sigma\{X_0, X_1, \ldots\}$. Calculate the predictable process $\{A_n, n = 0, 1, \ldots\}$ such that $\{M_n^2 - A_n, n = 0, 1, \ldots\}$ is a martingale.

Solution: It can be easily shown that $\{M_n, n = 0, 1, \ldots\}$ is a martingale and by Jensen's inequality $\{M_n^2, n = 0, 1, \ldots\}$ is a sub-martingale. Using Doob's decomposition there exists a stochastic process $\{A_n, n = 0, 1, \ldots\}$ such that $\{M_n^2 - A_n, n = 0, 1, \ldots\}$ is also a martingale.

$$E[M_n^2 - M_{n-1}^2 | \mathfrak{F}_{n-1}] = E[(M_n - M_{n-1})^2 | \mathfrak{F}_{n-1}] = A_n - A_{n-1} \quad \forall\, n$$

which implies,

$$\begin{aligned} A_n &= \sum_{i=1}^{n} E[M_n^2 - M_{n-1}^2 | \mathfrak{F}_{n-1}] \\ &= \sum_{i=1}^{n} E[M_n^2 | \mathfrak{F}_{n-1}] - M_{n-1}^2. \end{aligned}$$

Thus, $A_n = \sum_{i=1}^{n} E[M_n^2 | \mathfrak{F}_{n-1}] - M_{n-1}^2$.

Now, Maximal Inequality Theorem is presented below.

Theorem 10.8 (Maximal Inequality Theorem) *Let $\{X_n, n = 0, 1, \ldots\}$ be a sub-martingale for which $X_n \geq 0$ for all n. Then, for any positive λ,*

$$\lambda P(\max_{0 \leq k \leq n} X_k > \lambda) \leq E[X_n].$$

Proof Using

$$T = \begin{cases} \min\{k \geq 0; X_k > \lambda\}, & \text{if } X_k > \lambda, \text{ for } k = 0, 1, \ldots, n \\ n, & \text{if } X_k \leq \lambda, \text{ for } k = 0, 1, \ldots, n \end{cases}$$

and Optional Stopping Theorem,

$$E[X_n] \geq E[T] = E\left[X_T I\left\{\max_{0 \leq k \leq n} X_k > \lambda\right\}\right] \geq \lambda P(\max_{0 \leq k \leq n} X_k > \lambda).$$

Theorem 10.9 (Martingale Convergence Theorem) *Let $\{X_n, n = 1, 2, \ldots\}$ be a sub-martingale satisfying $\sup_{n \geq 1} E[| X_n |] < \infty$. Then there exists a RV X_∞ to which $\{X_n, n = 1, 2, \ldots\}$ converges with probability one, that is*

$$P(\lim_{n\to\infty} X_n = X_\infty) = 1. \tag{10.3}$$

If $\{X_n, n = 1, 2, \ldots\}$ *is a martingale and is uniformly integrable, such that,*

$$\lim_{c\to\infty} \sup_{n\geq 1} E[| X_n | I\{| X_n |> c\}] = 0,$$

then, in addition to Equation (10.3), $\{X_n, n = 1, 2, \ldots\}$*, converges in the mean, that is,*

$$\lim_{n\to\infty} E[| X_n - X_\infty |] = 0,$$

and

$$E[X_\infty] = E[X_n], \quad \textit{for all } n.$$

Example 10.23 Consider Doob's martingale process. Let $\{Y_n, n = 0, 1, \ldots\}$ be an arbitrary sequence of RVs and suppose X is a RV with $E[| X |] < \infty$. Define $X_n = E[X|Y_0, Y_1, \ldots, Y_n], n = 0, 1, \ldots$. Show that $E[| X_n - X_\infty |] \to 0$ as $n \to \infty$.
Solution:Since $\{X_n, n = 0, 1, \ldots\}$ is a martingale and is uniformly integrable, it can be concluded that $\lim_{n\to\infty} X_n = X_\infty$ for some RV X_∞. Moreover, $E[X_\infty] = E[X_0]$ and $E[| X_n - X_\infty |] \to 0$ as $n \to \infty$.

Example 10.24 Let $\{X_n, n = 1, 2, \ldots\}$ be a non-negative super-martingale, then show that it converges almost surely, to an integrable RV.
Solution: Since $\{X_n, n = 1, 2, \ldots\}$ is a non-negative super-martingale, then,

$$\begin{aligned}\sup_{n\geq 1} E[|X_n|] &= \sup_{n\geq 1} E[X_n] \\ &\leq E[X_1] = E[|X_1|] < \infty,\end{aligned}$$

thus Martingale Convergence Theorem implies that X_n converges almost surely with probability one to an integrable RV.

For more description, readers may refer to [1–4].

10.5 Exercise Problems

1. Let $\{X_t, t \geq 0\}$ be a martingale with respect to a given filtration. Show that

$$E[X_t] = E[X_0] \quad \forall\ t \geq 0.$$

   Is the converse true?
2. If X and Y are two martingales with respect to a given filtration $\mathfrak{F}$, then can you say that the sum $X + Y$ is also a martingale with respect to $\mathfrak{F}$.

3. Let S_n be a symmetric random walk such that $S_n = Y_1 + Y_2 + \ldots$ where $\{Y_i, i = 1, 2, \ldots\}$ are IID RVs with

$$P(Y_i = 1) = 1 - P(Y_i = -1) = \frac{1}{2}.$$

Let $\mathfrak{F}_n = \sigma(Y_1, Y_2, \ldots, Y_n)$. Show that

$$X_n = (-1)^n \cos(\pi S_n), \quad n = 1, 2, \ldots$$

is a martingale with respect to $\{\mathfrak{F}_n\}_{n\geq 1}$.

4. Let $\{N(t), t \geq 0\}$ be a Poisson process with rate 1. Which of the following are martingales.
   (a) $\{N(t) - t, t \geq 0\}$.
   (b) $\{(N(t))^2 - t, t \geq 0\}$.
5. Let $\{M(t), t \geq 0\}$, where $M(t) = N(t) - \lambda t$ be the compensated Poisson process. Show that $\{M(t), t \geq 0\}$ is a martingale.
6. A gambler wins or loses one rupee in each round of betting, with equal chances and independently of the past events. She starts betting with the determination that she will stop when either she wins m rupees or she loses n rupees.
   (a) What is the probability that she will be winning when she stops playing further?
   (b) What is the expected number of rounds of betting before she stops playing?
7. Let $\{X_n, n = 1, 2, \ldots\}$ be IID RVs with $E[X_n] = 0$ and $E[X_n^2] = \sigma^2$ for $n \geq 1$. Define

$$Y_n = \left(\sum_{k=1}^{n} X_k\right)^2 - n\sigma^2.$$

Then, show that $\{Y_n, n = 1, 2, \ldots\}$ is a martingale.

8. Let $X_1, X_2, \ldots$ be independent RVs with

$$X_n = \begin{cases} 1 & \text{with probability } \frac{1}{2n} \\ 0 & \text{with probability } 1 - \frac{1}{n} \\ -1 & \text{with probability } \frac{1}{2n}. \end{cases}$$

Let $Y_1 = X_1$, and for $n \geq 2$,

$$Y_n = \begin{cases} X_n & Y_{n-1} = 0 \\ nY_{n-1}\,|X_n| & Y_{n-1} \neq 0 \end{cases}.$$

(a) Show that $\{Y_n, n = 1, 2, \ldots\}$ is a martingale with respect to $\mathcal{F}_n = \sigma(Y_1, Y_2, \ldots, Y_n)$.
(b) Show that $\lim_{n\to\infty} P(Y_n = 0) = 1$.

9. Let $\left(\Omega, \mathcal{F}, \{\mathcal{F}_n\}_{n\geq 0}, P\right)$ be a filtered probability space and $\{Y_n, n = 0, 1, \ldots\}$ be a sequence of absolutely integrable RVs adapted to the filtration $\{\mathcal{F}_n\}_{n\geq 0}$. Assume that there exist real numbers $u_n, v_n, n = 0, 1, \ldots$, such that

$$E\left[Y_{n+1} \mid \mathcal{F}_n\right] = u_n Y_n + v_n.$$

Find two real sequences a_n and $b_n, n = 0, 1, \ldots$, such that the sequence $M_n = a_n Y_n + b_n, n = 0, 1, \ldots$, is martingale w.r.t. the same filtration.
10. Let $\{X_n, n = 0, 1, \ldots\}$ be a stochastic process defined on a filtered probability space $(\Omega, \mathcal{F}, \{\mathcal{F}_n\}_{n\geq 0}, P)$ such that $E[|X_n|] < \infty$ for all $n = 0, 1, \ldots$. Also, X_n is $\mathcal{F}_n$-measurable and for all bounded stopping times τ, $E[X_\tau] = E[X_0]$. Prove that $\{X_n, n = 0, 1, \ldots\}$ is a martingale.
11. (**Jensen's Inequality**) Let $\varphi : \mathbb{R} \to \mathbb{R}$ be a convex function and let X be an integrable RV defined on a probability space $(\Omega, \mathcal{F}, P)$ such that $\varphi(x)$ is also integrable. Prove that $\varphi(E[X|\mathcal{G}]) \leq E[\varphi(X)|\mathcal{G}]$ a.s. for any $\sigma-$ field $\mathcal{G}$ contained in $\mathcal{F}$.
12. Suppose $X, X_1, X_2, \ldots$ are square integrable RVs such that $X_n \to X$ in L^2 as $n \to \infty$. Prove that $E[X_n|\mathcal{G}] \to E[X|\mathcal{G}]$ in L^2 as $n \to \infty$ for any sub $\sigma-$ field $\mathcal{G}$ on Ω contained in $\mathcal{F}$. (Hint: Use Jensen inequalities, Problem 11, with $\varphi(x) = x^2$)
13. Use Jensen's inequality to show that for any convex, measurable function φ

$$E\left(\varphi(\sigma t + W(t)e^{-\sigma W(t)}\right) \geq \varphi(t)e^{\frac{\sigma^2}{2}t},$$

where $\{W(t), t \geq 0\}$ is a Brownian motion.

References

1. Castaneda, L.B., Arunachalam, V., Dharmaraja, S.: Introduction to Probability and Stochastic Processes with Applications. Wiley, New Jersey (2012)
2. Karatzas, I., Shreve, S.E.: Brownian Motion and Stochastic Calculus. Springer, New York (1991)
3. Shreve, S.E.: Stochastic calculus for finance II: Continuous-time models, vol. 11. Springer, New York (2004)
4. Mikosch Thomas, Elementary stochastic calculus with finance in view. World Scientific (1998)

Chapter 11
Diffusion and Jump-Diffusion Processes

In this chapter, diffusion and jump-diffusion processes are discussed. Diffusion processes are continuous parameter continuous state processes with sample paths that are continuous everywhere but differentiable nowhere. The most common diffusion process is Brownian motion. Brownian motion is a continuous parameter continuous state stochastic process having stationary and independent increments. Diffusion processes can be a function of Brownian motion or integral with respect to Brownian motion. In this chapter, the construction of the Brownian motion and its various properties are presented. The main discussion will be about some of the important processes derived from Brownian motion and their properties.

Further, this chapter introduces jump-diffusion processes. Jump-diffusion processes are the stochastic processes that involve jumps and diffusion components. The fundamental pure jump process is the Poisson process which serves as starting point for jump processes. The jumps of Poisson process are of fixed size one. Another important jump process which is widely used is compound Poisson process. Compound Poisson process can be seen as a generalized version of Poisson process with jump size as value of a RV. It is similar to Poisson process, but its jumps are of random size.

This chapter also introduces Lévy processes. Lévy processes are processes with independent and stationary increments. Brownian motion, Poisson processes, and compound Poisson process are examples of Lévy processes and they also form building blocks for Lévy processes. The general properties of Lévy processes, Lévy-Khintchin formula and Lévy Itô decomposition will also be discussed. Lévy-Khintchin formula helps to study distributional properties of Lévy processes and Lévy Itô decomposition supports in visualization of sample paths of Lévy processes.

At the end, a special type of jump process is introduced named Hawkes jump process. Hawkes process is a self-exciting point process which takes into consideration the clustering of jumps phenomena. In this process when a jump occurs in the stochastic process, it increases the probability of the occurrence of future jumps, thus resulting in jump clustering. In Poisson jump processes, due to independent increments, this self-exciting behavior is not observed.

D. Selvamuthu, *Introduction to Stochastic Processes*, University Texts in the Mathematical Sciences, https://doi.org/10.1007/978-981-97-6152-4_11

11.1 Brownian Motion and Multi-dimensional Brownian Motion

The Brownian motion (also called Wiener process) has applications in many areas and remains a basic element in financial modeling. This stochastic process is also a Markov process, a martingale, and a process with independent increments.

11.1.1 Random Walk

Consider a Bernoulli trial with probabilities of p and $1-p$ for success and failure, respectively. Repeat the trial a large number of times (for example, throw a fair coin a large number of times). The successive outcomes are labeled as $\omega = (\omega_1, \omega_2, \omega_3, \ldots)$ e.g., $\omega = (\omega_1, \omega_2, \omega_3, \ldots) = (H, T, T, \ldots)$ or $(T, H, T, \ldots)$ and the set of all possible outcomes is Ω. Here 'H' represents success whereas 'T' is for failure. Define

$$X_j = \begin{cases} 1 & \text{if } \omega_j = H \\ -1 & \text{if } \omega_j = T \end{cases}.$$

The PMF of X_j is given by

$$P(X_j = 1) = p; \quad P(X_j = -1) = 1 - p = q; \quad .$$

Set $S_0 = 0$ and define

$$S_n = \sum_{j=1}^{n} X_j, n = 1, 2, \ldots,$$

where X_j's are IID RVs. Then, $\{S_n, n = 0, 1, \ldots\}$ is known as a random walk.

Sample path of the random walk is shown in Fig. 11.1 for an outcome $\omega \in \Omega$, where $\omega_1 = H, \omega_2 = H, \omega_3 = T, \ldots$ with $p = 0.50$.

Properties of Random Walk

1. Choose non-negative integers $0 = k_0 < k_1 < \ldots < k_n$. Then

$$S_{k_{i+1}} - S_{k_i} = \sum_{j=k_i+1}^{k_{i+1}} X_j,$$

Since X_j's are IID RVs, $S_{k_1} - S_{k_0}, S_{k_2} - S_{k_1}, \ldots, S_{k_n} - S_{k_{n-1}}$ are mutually independent random variables. Hence, $\{S_n, n = 0, 1, \ldots\}$ has the independent increment property.

2. Similarly, for $0 \le i < j \le k < l$ such that $j - i = l - k$

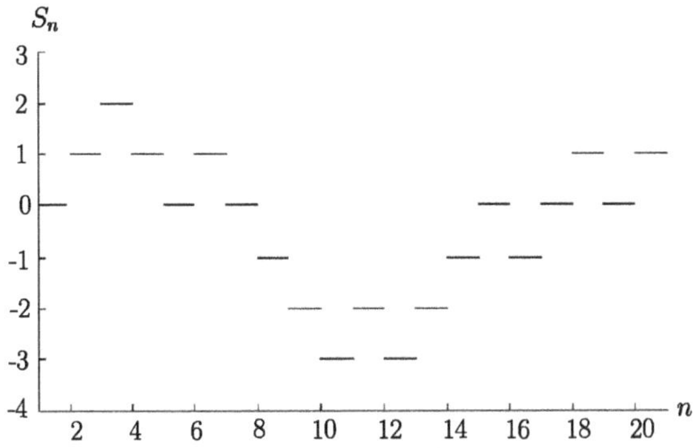

Fig. 11.1 Sample path of a random walk with $p = 0.5$

$$S_j - S_i \stackrel{d}{=} S_l - S_k.$$

Hence, $\{S_n, n = 0, 1, \ldots\}$ has stationary increment property.

3. Also, suppose $S_{k_{i+1}} - S_{k_i} = X_1 + X_2 + \ldots + X_n$, one can observe that,

$$E[S_{k_{i+1}} - S_{k_i}] = n(p - q) \text{ and } \mathrm{var}(S_{k_{i+1}} - S_{k_i}) = 4npq.$$

11.1.2 Brownian Motion

This section discusses the Brownian motion (also discussed in Chapter 1) and its related properties.

Definition 11.1 (Brownian Motion/Wiener Process) Let $(\Omega, \mathcal{F}, P)$ be a probability space. A stochastic process $\{W(t), t \geq 0\}$ defined on the probability space is said to be a Wiener Process (or Brownian motion) with respect to filtration $\{\mathcal{F}_t\}_{t \geq 0}$ if

1. $W(0) = 0$ a.s..
2. For $0 \leq t_0 < t_1 < \ldots < t_n$ and for all n, increments $W(t_i) - W(t_{i-1}), i = 1, 2, \ldots, n$ are independent and stationary RVs.
3. For $0 \leq s < t < \infty$, every increment $W(t) - W(s)$ has normal distribution with mean $\mu(t - s)$ and variance $\sigma^2(t - s)$.
4. For $t > 0$, the sample paths of $W(t)$ are almost everywhere continuous functions.

Remark 11.1 1. For each $t \geq 0$, $W(t)$ is $\mathcal{F}_t$-measurable.

2. A Brownian motion $\{W(t), t \geq 0\}$ with $\mu = 0, \sigma = 1$ is called a standard Brownian motion.
3. The stationary increment property means that $W(t) - W(s)$ has the same distribution as $W(t - s)$ for $s \leq t$.

4. All the states of Brownian motion are recurrent. It means that the Brownian motion visits an infinite number of times each of its states, i.e., all real numbers.
5. When filtration is not specified, then the filtration is assumed to be the natural filtration generated by $\{W(t), t \geq 0\}$, i.e., $\mathcal{F}_t = \sigma(W(s), s \leq t)$.

Theorem 11.1 *Prove that the limiting distribution of a random walk is Brownian motion using Central Limit Theorem (CLT).*

Proof Consider, a random walk $\{S_n, n = 1, 2, \ldots\}$, which is defined as

$$S_n = X_1 + X_2 + \ldots + X_n,$$

where X_i's are independently and identically distributed RVs whose PMF is

$$P(X_i = 1) = P(X_i = -1) = \frac{1}{2}.$$

Define the normalized random walk as $Z_n = \frac{S_n}{\sqrt{n}}$. Thus by CLT, Z_n converges in distribution to $N(0, 1)$. Also, S_n has independent and stationary increment.

Now define scaled random walk as

$$W^{(n)}(t) = \frac{S_{\lfloor nt \rfloor}}{\sqrt{n}} \quad \text{for} \quad t \geq 0,$$

where $\lfloor nt \rfloor$ is the integer part of nt. Now using CLT, $W^{(n)}(t)$ converges in distribution to $N(0, t)$, and has independent and stationary increments. Therefore, the limiting distribution of a random walk is a Brownian motion using CLT.

Example 11.1 Prove that $\{X(t), 0 \leq t \leq T\}$ is a Brownian motion, where $X(t) = tW\left(\frac{1}{t}\right), t > 0$ and $X(0) = 0$, and $\{W(t), 0 \leq t \leq T\}$ is a standard Brownian motion on $[0, T]$.

Solution: Clearly $X(t)$ is continuous almost everywhere for $t > 0$,

$$\lim_{t \to 0} X(t) = \lim_{t \to 0} tW\left(\frac{1}{t}\right) = \lim_{n \to \infty} \frac{1}{n} W(n) = \lim_{n \to \infty} \frac{1}{n} \sum_{j=1}^{n} \{W(t_j) - W(t_{j-1})\}.$$

Note that, $W(t_j) - W(t_{j-1})$ are IID RVs with mean zero where $0 = t_0 \leq t_1 \leq \ldots t_n = T$. Hence by Strong Law of Large Numbers (SLLN),

$$\lim_{n \to \infty} \frac{1}{n} \sum_{j=1}^{n} \{W(t_j) - W(t_{j-1})\} = 0 \ \ a.s..$$

Hence the sample path is continuous almost surely. For $t > s$, consider,

$$X(t) - X(s) = tW\left(\frac{1}{t}\right) - sW\left(\frac{1}{s}\right) = (t-s)W\left(\frac{1}{t}\right) + s\left(W\left(\frac{1}{t}\right) - W\left(\frac{1}{s}\right)\right)$$

Now,

$$W\left(\frac{1}{t}\right) - W\left(\frac{1}{s}\right) \sim N\left(0, \left(\frac{1}{s} - \frac{1}{t}\right)\right), \quad W\left(\frac{1}{t}\right) \sim N\left(0, \frac{1}{t}\right),$$

and $s\left(W\left(\frac{1}{t}\right) - W\left(\frac{1}{s}\right)\right)$ and $(t-s)W\left(\frac{1}{t}\right)$ are independent. Hence $X(t) - X(s)$ is a normally distributed RV with mean zero and variance $t - s$, i.e., for $0 \le s < t < \infty$, $X(t) - X(s) \sim N(0, t-s)$.
For any $s > 0, t > 0$,

$$\begin{aligned} \text{cov}(X(t), X(s)) &= E[X(t)X(s)] = stE\left[W\left(\frac{1}{t}\right)W\left(\frac{1}{s}\right)\right] \\ &= st\min\left(\frac{1}{t}, \frac{1}{s}\right) = \min(s, t). \end{aligned}$$

Let $s < t$. Then

$$\text{cov}(X(s), X(t) - X(s)) = \text{cov}(X(s), X(t)) - \text{cov}(X(s), X(s)) = 0.$$

Since $X(s)$ and $X(t) - X(s)$ are normally distributed RVs, and $\text{cov}(X(s), X(t) - X(s)) = 0$, this implies that they are independent. Hence $\{X(t), t \ge 0\}$ is a Brownian motion.

Example 11.2 For $a > 0$, define $Y(t) = aW\left(\frac{t}{a^2}\right)$. Show that $\{Y(t), t \ge 0\}$ is a Brownian motion.
Solution: We know that $Y(t)$ has continuous paths and $Y(0) = 0$ a.s.. Take, $\mathcal{G}_t = \mathcal{F}_{\frac{t}{a^2}}$. Thus, $Y(t)$ is $\mathcal{G}_t$-measurable. Let $s < t$. Then, $Y(t) - Y(s) = a\left(W\left(\frac{t}{a^2}\right) - W\left(\frac{s}{a^2}\right)\right)$ is independent of $\mathcal{F}_{\frac{s}{a^2}}$, hence independent of $\mathcal{G}_s$. $Y(t) - Y(s)$ is normally distributed with mean 0 and $\text{var}(Y(t) - Y(s)) = a^2\text{var}(W(\frac{t}{a^2}) - W(\frac{s}{a^2})) = t - s$.
Hence $\{Y(t), t \ge 0\}$ is a Brownian motion with respect to the filtration $\left\{\mathcal{F}_{\frac{t}{a^2}}\right\}_{t \ge 0}$.

11.1.3 Joint Distribution of Wiener Process

Let $\{W(t), t \ge 0\}$ be a Brownian motion. For any $t_1, t_2 \ge 0$, consider the joint distribution of $(W(t_1), W(t_2))$. Here, $W(t_1)$ and $(W(t_2) - W(t_1))$ are independent. Also, $W(t_1) \sim N(0, t_1)$ and $W(t_2) - W(t_1) \sim N(0, t_2 - t_1)$. Then the joint PDF of $(W(t_1), W(t_2))$ is

$$f(x_1, x_2) = p(x_1, t_1)p(x_2 - x_1, t_2 - t_1),$$

where the marginal PDF $p(x, t)$ is given by

$$p(x, t) = \frac{1}{\sqrt{2\pi t}}\exp\left(- \frac{x^2}{2t}\right), t > 0; -\infty < x < \infty.$$

Note that, this is the solution of the diffusion equation (also known as heat equation) given below which also satisfies

$$\frac{\partial p}{\partial t} = \frac{1}{2}\frac{\partial^2 p}{\partial x^2}$$

Thus,

$$f(x_1, x_2) = \frac{1}{2\pi\sqrt{t_1(t_2 - t_1)}}\exp\left(- \frac{1}{2}\left(\frac{x_1^2}{t_1} + \frac{(x_2 - x_1)^2}{(t_2 - t_1)}\right)\right). \tag{11.1}$$

The RV $X = (W(t_1), W(t_2), \ldots, W(t_n))$ is jointly normally distributed whose CDF for $0 < t_1 < t_2 < \cdots < t_n$ and $-\infty < a_i < \infty,\ i = 1, 2, \ldots, n$, is given by

$$P(W(t_1) \le a_1, W(t_2) \le a_2, \ldots, W(t_n) \le a_n)$$

$$= \frac{1}{\sqrt{(2\pi)^n t_1(t_2 - t_1)\cdots(t_n - t_{n-1})}}\int_{-\infty}^{a_n}\cdots\int_{-\infty}^{a_1}$$
$$\exp\left(- \frac{1}{2}\left(\frac{x_1^2}{t_1} + \frac{(x_2 - x_1)^2}{t_2 - t_1} + \cdots + \frac{(x_n - x_{n-1})^2}{t_n - t_{n-1}}\right)\right)dx_1dx_2\cdots dx_n.$$

The joint PDF of $(W(t_1), W(t_2), \ldots, W(t_n))$ is given by

$$f_X(x) = \frac{1}{(2\pi)^{n/2}(det(\Sigma))^{1/2}}\exp\left(- \frac{1}{2}(X - \mu)\Sigma^{-1}(X - \mu)^{'}\right)$$

where

$$\mu = E[X] = (E[W(t_1)], E[W(t_2)], \ldots, E[W(t_n)]) = (0, 0, \ldots, 0),$$

$$\Sigma = \text{cov}(W(t_i), W(t_j)); i, j = 1, 2, \ldots, n) = \begin{pmatrix} t_1 & t_1 & \ldots & t_1 \\ t_1 & t_2 & \ldots & t_2 \\ \vdots & \vdots & \vdots & \vdots \\ t_1 & t_2 & \ldots & t_n \end{pmatrix}.$$

Remark 11.2 1. Every Standard Brownian motion is a Gaussian process (discussed in Chap. 1) with mean 0 and covariance function, $\text{cov}(W(t), W(s)) = s,\ s \leq t$. But the converse is not true in general.
2. If $\{X(t), t \geq 0\}$ is a Gaussian process and $E[X(t)X(s)] = \min(t, s)$, then $\{X(t), t \geq 0\}$ is also a Brownian motion.

Example 11.3 Let $\{W(t), t \geq 0\}$ be a Brownian motion. Find the conditional distribution of $W\left(\frac{t_1}{2}\right)$ given that $W(t_1) = x_1$.
Solution: From (11.1), the joint probability density function of $(W(s), W(t))$ for $s < t$ is given by

$$f(x, y) = \frac{1}{2\pi\sqrt{s(t-s)}} \exp\left(-\frac{1}{2}\left(\frac{x^2}{s} + \frac{(y-x)^2}{t-s}\right)\right).$$

Take $s = t_1/2$ and $t = t_1$. We get

$$f(x, y) = \frac{2}{2\pi t_1} \exp\left(-\frac{1}{t_1}\left(x^2 + (y-x)^2\right)\right).$$

The conditional distribution of $W\left(\frac{t_1}{2}\right)$ given that $W(t_1) = x_1$ is, for $-\infty < x < \infty$

$$\begin{aligned} f\left(W(t_1/2)|\{W(t_1) = x_1\}\right)(x) &= \frac{\frac{2}{2\pi t_1} \exp\left(-\left(\frac{x^2}{t_1} + \frac{(x_1-x)^2}{t_1}\right)\right)}{\frac{1}{\sqrt{2\pi t_1}} \exp\left(-\frac{x_1^2}{2t_1}\right)} \\ &= \frac{1}{\sqrt{2\pi t_1/4}} \exp\left(-\frac{1}{2}\left(\frac{(x-x_1/2)^2}{t_1/4}\right)\right). \end{aligned}$$

This implies that the conditional distribution of $W\left(\frac{t_1}{2}\right)$ given $W(t_1) = x_1$ is normal with mean $\frac{x_1}{2}$ and variance $\frac{t_1}{4}$.

Example 11.4 Suppose $\{W(t), t \geq 0\}$ is a stochastic process such that all the finite dimensional distributions are jointly normal. $E[W(t)] = 0$, for all t, $\text{cov}(W(t), W(s)) = s$ if $s \leq t$ and the paths of $W(t)$ are continuous. Show that $\{W(t), t \geq 0\}$ is a Brownian motion.
Solution: Take the filtration $\{\mathcal{F}_t\}_{t\geq 0}$ as the filtration generated by $\{W(t), t \geq 0\}$. Suppose $s = t$, $\text{var}(W(t)) = \text{cov}(W(t), W(t)) = t$. In particular, $\text{var}(W(0)) = 0$ and since $E[W(0)] = 0$, then $W(0) = 0$ a.s.. We have,

$$\begin{aligned} \text{var}(W(t) - W(s)) &= \text{var}(W(t)) - 2\text{cov}(W(t), W(s)) + \text{var}(W(s)) \\ &= t - 2s + s = t - s. \end{aligned}$$

Now, if $r \leq s < t$,

$$\text{cov}(W(t) - W(s), W(r)) = \text{cov}(W(t), W(r)) - \text{cov}(W(s), W(r)) = 0.$$

Hence $W(t) - W(s)$ is independent of $W(s)$. Hence, $\{W(t), t \geq 0\}$ is a Brownian motion.

Example 11.5 Find the characteristic function $E\left[e^{iuW(t)}\right]$, for any $u \in \mathbb{R}$. Also, find $E[(W(t))^4]$.
Solution:

$$E[e^{iuW(t)}] = \int_{-\infty}^{\infty} e^{iux} p(x,t)dx = \frac{1}{\sqrt{2\pi t}} \int_{-\infty}^{\infty} e^{iux} e^{-\frac{x^2}{2t}} dx = e^{-\frac{u^2 t}{2}}.$$

$$E\left[W(t)^4\right] = \frac{d^4}{du^4} E\left[e^{iuW(t)}\right]\bigg|_{u=0} = 3t^2.$$

11.1.4 Multi-dimensional Brownian Motion

Multi-dimensional Brownian motion, also known as a multivariate or n-dimensional Brownian motion, is a mathematical model used to describe the random motion of particles or variables in multiple dimensions over time. It is a fundamental concept in stochastic calculus and is extensively used in various fields, including finance, physics, biology, and engineering.

Definition 11.2 (n-**Dimensional Brownian Motion**) An n-dimensional Brownian motion $\{W(t), t \geq 0\}$ is a stochastic process where

$$W(t) = (W_1(t), W_2(t), \ldots, W_n(t))$$

with the following properties.

1. Each $\{W_i(t), t \geq 0\}$, $1 \leq i \leq n$, is a one-dimensional Brownian motion.
2. For $i \neq j$, $W_i(t)$ and $W_j(t)$ are independent for any t.

Proposition 11.1 *If $\{W(t), t \geq 0\}$ is an n-dimensional Brownian Motion, then the following holds.*

1. *$(W_1(0), W_2(0), \ldots, W_n(0)) = (0, 0, \ldots 0)$ a.s..*
2. *For $0 < s < t$, the increments $W(t) - W(s)$ are independent RVs.*
3. *For any $0 \leq s < t$, $(W_1(t) - W_1(s), W_2(t) - W_2(s), \ldots, W_n(t) - W_n(s))$ has multivariate normal distribution.*

Example 11.6 Let $(W_1(t), W_2(t))$ be a 2-dimensional Brownian motion. Define

$$X(t) = a(W_1(t) + W_2(t)), \text{ for any } t \geq 0, \ a > 0.$$

Find $a > 0$ such that $\{X(t), t \geq 0\}$ is a Brownian motion.

Solution: Since, $\{X(t), t \geq 0\}$ is a Brownian motion, i.e., $X(t) \sim N(0, t)$.
Also $\{W_1(t), W_2(t), t \geq 0\}$ is a 2-dimensional Brownian motion which implies

$$\text{cov}(W_1(t), W_2(t)) = 0.$$

Consider

$$\text{var}(X(t)) = a^2\big(\text{var}(W_1(t)) + \text{var}(W_2(t))\big)$$

$$\implies a^2\big(2t\big) = t \implies a^2 = \frac{1}{2} \implies a = \frac{1}{\sqrt{2}}.$$

Consequently, it can be verified that for $a = \frac{1}{\sqrt{2}}$, $X(t)$ satisfies all the conditions of Brownian motion.

Example 11.7 Let $\mathbf{x} = (x_1, \ldots, x_n)$ be a non-random vector in $\mathbb{R}^n$ with length 1, i.e., $\sqrt{x_1^2 + \ldots x_n^2} = 1$, and let $\{W(t), t \geq 0\}$ be an n-dimensional Brownian motion. Let $X(t) = x_1 W_1(t) + x_2 W_2(t) + \ldots + x_n W_n(t)$. Prove that $\{X(t), t \geq 0\}$ is a one-dimensional Brownian motion.
Solution: We have,

$$X(t) = x_1 W_1(t) + x_2 W_2(t) + \ldots + x_n W_n(t).$$

As each $W_i(0) = 0$, $X(0) = 0$.
Also, $x_i(W_i(t) - W_i(s)) \sim N(0, x_i^2(t - s))$ and for $i \neq j$, and for any t, $W_i(t)$ and $W_j(t)$ are independent RV. Hence

$$x_1(W_1(t) - W_1(s)) + x_2(W_2(t) - W_2(s)) + \ldots + x_n(W_n(t) - W_n(s)) \sim N(0, (t - s)),$$

i.e.,

$$X(t) - X(s) \sim N(0, (t - s)).$$

That is, it has been proved that for $0 \leq s < t < \infty$, $X(t) - X(s)$ is normally distributed with mean 0 and variance $(t - s)$. Again using the fact that, for $i \neq j$, and for any t, $W_i(t)$ and $W_j(t)$ are independent RVs, we get

$$E[X(t)X(s)] = \sum_{i,j=1}^{n} E[x_i x_j W_i(t) W_j(s)] = \sum_{i=1}^{n} E[x_i^2 W_i(t) W_i(s)] = \min(s, t).$$

Now,

$$\text{cov}(X(t), X(s)) = E[X(t)X(s)] - E[X(t)]E[X(s)] = \min(s, t).$$

Let $s < t$, then

$$\text{cov}(X(s), X(t) - X(s)) = \text{cov}(X(s), X(t)) - \text{cov}(X(s), X(s)) = 0.$$

Since $X(s)$ and $X(t) - X(s)$ are normally distributed RVs, and $\text{cov}(X(s), X(t) - X(s)) = 0$, this implies that they are independent RVs. Hence $\{X(t), t \geq 0\}$ is a Brownian motion.

11.2 Properties of Brownian Motion

A Brownian motion is continuous everywhere, but differentiable nowhere. Due to its randomness, Brownian motion does not behave well enough to be integrated using conventional techniques. We have already discussed some properties of Brownian motion like strict-sense stationarity, wide-sense stationarity and Markov property in Chap. 1.

11.2.1 Nowhere Differentiable Property

Figure 11.2 depicts a sample path of a Brownian motion.

It is not possible to define a tangent line at any point in the sample path. Using second order moment convergence of RVs, let's calculate for any $t \geq 0$,

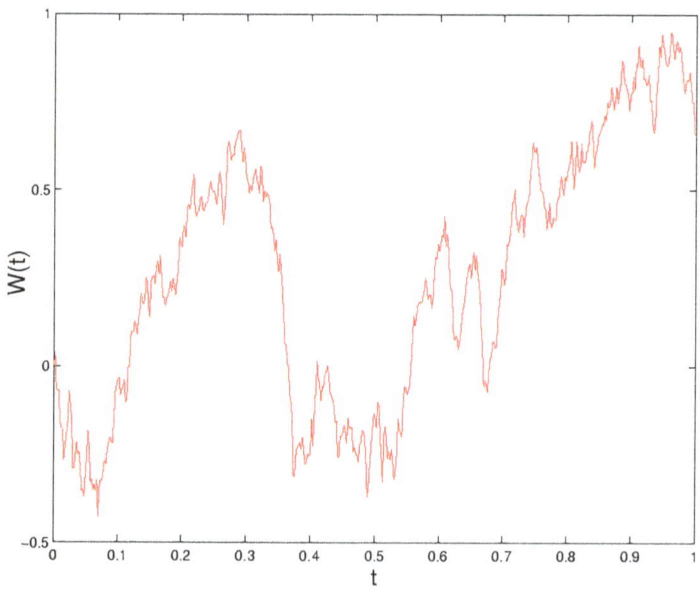

Fig. 11.2 Realization of Wiener process

$$\lim_{\Delta t \to 0} \text{var}\left(\frac{W(t_0 + \Delta t) - W(t_0)}{\Delta t}\right) \text{ or } \lim_{\Delta t \to 0} E\left(\frac{W(t_0 + \Delta t) - W(t_0)}{\Delta t}\right)^2.$$

Since, $W(t_0 + \Delta t) - W(t_0)$ has normal distribution with mean 0 and variance Δt, the limit is

$$\lim_{\Delta t \to 0} \frac{1}{(\Delta t)^2} \Delta t = \infty.$$

Hence, sample path is not differentiable.

The non-differentiability of the Wiener process may seem surprising since it arose as a model for the Brownian motion of a particle, but the trajectory of a particle undergoing Brownian motion, observed under a microscope, actually does appear to be differentiable nowhere.

11.2.2 Independent Increment Property

Before discussing independent increment property of Brownian motion, let us recall that it is a strict-sense stationary process but not a wide-sense stationary process (discussed in Chap. 1).

Theorem 11.2 *For any* $0 = t_0 < t_1 < \ldots < t_n$ *the increments*

$$W(t_1) - W(t_0), \ldots, W(t_n) - W(t_{n-1})$$

are independent.

Proof Since, $W(t_{i+1}) - W(t_i)$ for $i = 0, 1, \ldots, n-1$ are normally distributed with mean 0 and variance $t_{i+1} - t_i$. Let $X_i = W(t_{i+1}) - W(t_i)$. Further, we have to show that $\text{cov}\left(X_j, X_i\right) = 0 \, \forall \, 0 \le i < j < n-1$, to get that $W(t_1) - W(t_0), \ldots, W(t_n) - W(t_{n-1})$ are independent. To this end, recalling

$$\text{cov}\left(X_j, X_i\right) = E[X_j X_i] - E[X_j]E[X_i],$$

note that

$$\begin{aligned} E[X_j X_i] = & E[W(t_{j+1})W(t_{i+1})] - E[W(t_{j+1})W(t_i)] \\ & - E[W(t_j)W(t_{i+1})] + E[W(t_j)W(t_i)]. \end{aligned}$$

Since, $E[W(t)W(s)] = \min(t, s)$ for $i < j$, this implies for $i + 1 \le j$

$$E[X_j X_i] = (i + 1) - i - (i + 1) + i = 0$$

and $E[X_j] = 0 \; \forall \; 0 \le j < n - 1$. Hence $\text{cov}\left(X_j, X_i\right) = 0$.

11.2.3 Self-Similar Property

Definition 11.3 (**H-Self-Similar**) A stochastic process is said to be H-self-similar for some $H > 0$, if each finite dimensional random vector satisfies the condition, for every $a > 0$, any choice of $t_i \geq 0, i = 1, 2, \ldots, n$ and $n \geq 1$

$$(a^H X(t_1), a^H X(t_2), \ldots, a^H X(t_n)) \stackrel{d}{=} (X(at_1), X(at_2), \ldots, X(at_n)).$$

Here, H is the Hurst parameter which is a dimensionless factor used to estimate the magnitude of the self-similarity property.
For example, Brownian motion and random walk are self-similar with $H = 0.5$.

11.2.4 Martingale Property

Proposition 11.2 *Brownian motion is a martingale w.r.t. the natural filtration.*

Proof Let $\{\mathcal{F}_t\}_{t\geq 0}$ be the natural filtration. For every $t > 0$

1. $E[W(t)]$ exists (by definition).
2. $W(t)$ is $\mathcal{F}_t$-measurable (by definition).

Also, for $0 \leq s \leq t$

$$\begin{aligned} E[W(t)|\mathcal{F}_s] &= E[W(t) - W(s) + W(s)|\mathcal{F}_s] \\ &= E[W(t) - W(s)|\mathcal{F}_s] + E[W(s)|\mathcal{F}_s] \\ &= 0 + W(s). \end{aligned}$$

Therefore, $\{W(t), t \geq 0\}$ is a martingale.

Example 11.8 Show that $\{(W(t))^2, t \geq 0\}$ is a sub-martingale.
Solution:

$$E\left[(W(t))^2|\mathcal{F}_s\right] = E\left[(W(t) - W(s) + W(s))^2|\mathcal{F}_s\right].$$

Using

$$(W(t) - W(s) + W(s))^2 = (W(t) - W(s))^2 + 2W(s)(W(t) - W(s)) + (W(s))^2,$$

$$E\left[(W(t))^2|\mathcal{F}_s\right] = (t - s) + 0 + (W(s))^2.$$

Hence, $E\left[(W(t))^2|\mathcal{F}_s\right] \geq (W(s))^2$, a.s.. Therefore, $\{(W(t))^2, t \geq 0\}$ is a sub-martingale.

Definition 11.4 (Exponential Martingale) Let $\{W(t), t > 0\}$, be a Brownian motion w.r.t. the filtration $\{\mathcal{F}_t\}_{t\geq 0}$, then the exponential martingale is defined by the process $\{Z(t), t \geq 0\}$,

$$Z(t) = \exp\left\{\sigma W(t) - \frac{1}{2}\sigma^2 t\right\},$$

where σ is a constant.

Example 11.9 Prove that the exponential martingale $\{Z(t), t \geq 0\}$ is a martingale.
Solution: For $0 \leq s \leq t$,

$$\begin{aligned}
E[Z(t)|\mathcal{F}_s] &= E[\exp\{\sigma W(t) - \frac{1}{2}\sigma^2 t\}|\mathcal{F}_s] \\
&= E[\exp\{\sigma(W(t) - W(s))\} \cdot \exp\{\sigma W(s) - \frac{1}{2}\sigma^2 t\}|\mathcal{F}_s] \\
&= \exp\{\sigma W(s) - \frac{1}{2}\sigma^2 t\} \cdot E[\exp\{\sigma(W(t) - W(s))\}|\mathcal{F}_s] \\
&= \exp\{\sigma W(s) - \frac{1}{2}\sigma^2 t\} \cdot E[\exp\{\sigma(W(t) - W(s))\}].
\end{aligned}$$

Since, $W(t) - W(s)$ is normally distributed RV with mean zero and variance $t - s$, then

$$E[\exp\{\sigma(W(t) - W(s))\}] = \exp\left(\frac{1}{2}\sigma^2(t - s)\right).$$

Substituting this, implies the martingale property.

$$E[Z(t)|\mathcal{F}_s] = \exp\{\sigma W(s) - \frac{1}{2}\sigma^2 s\} = Z(s).$$

Thus every exponential martingale is a martingale.

Example 11.10 Assume that $\{S(t) = e^{\sigma W(t)}, t \geq 0\}$ gives the price of a stock at time t (in days). Assume that an investor owns 500 shares of the stock at time $t = 0$. He plans to sell the shares as soon as their price hits \$3. What is the probability that he will have to wait more than 30 days to sell the shares, given that $\sigma = 2$?
Solution: Let

$$T = \min\{t \geq 0 : S(t) = 3\} = \min\left\{t \geq 0 : W(t) = \frac{\ln(3)}{2}\right\}.$$

Thus $T = \min\{t \geq 0 : W(t) = 0.5493\}$

$$P(T > 30) = P(T(0.5493) > 30) = 2\Phi(0.5493/\sqrt{30}) - 1 = 0.4172,$$

where, $\Phi(\cdot)$ is the CDF of standard normal distribution. Hence, the investor must probably wait longer than 30 days with a probability 0.4172.

11.2.5 Reflection Principle

This principle states that there are the same number of paths that reach level m before a final time but are below m at the final time as there are paths that reach level m before a final time but are above m at the final time.
This enables one to count the pathways that eventually reach level m by the final time. There is a "reflected path" that is at level $2m - w$ at time t for any Brownian motion path that reaches level m before time t but is at a level w below m at time t. By alternating the up and down motions of the Brownian motion starting at time τ_m, this reflected path is created. Of course, the likelihood that a Brownian motion route finishes at precisely w or at exactly $2m - w$ is nil. The paths are analyzed that reach level m before time t and are at or below level w at time t, as well as their reflections, which are at or above $2m - w$ at time t, in order to have non-zero probability. This leads to the essential reflection equality,

$$P(\tau_m < t, W(t) \leq w) = P(W(t) \geq 2m - w), w \leq m, m > 0.$$

Another application of reflection principle is that it enables us to regulate the maximum of a Brownian motion. Define the maximum to date for Brownian motion to be

$$M(t) = \max_{0 \leq s \leq t} W(s).$$

For positive m, $M(t) \geq m$ if and only if $\tau_m \leq t$. This observation permits us to rewrite the reflection equality as

$$P(M(t) \geq m, W(t) \leq w) = P(W(t) \geq 2m - w), \quad w \leq m, m > 0.$$

Fig. 11.3 shows a sample path of a Wiener process and its reflection upon reaching a crossing point.

11.2.6 First Passage Time and Reflection Principle

First passage time is the first time Brownian motion hits a level m. Define the first passage time for any real number m as

$$\tau_m = \min\{t \geq 0 : W(t) = m\}.$$

Theorem 11.3 *Let τ_m be the first passage time of Brownian motion. Then*

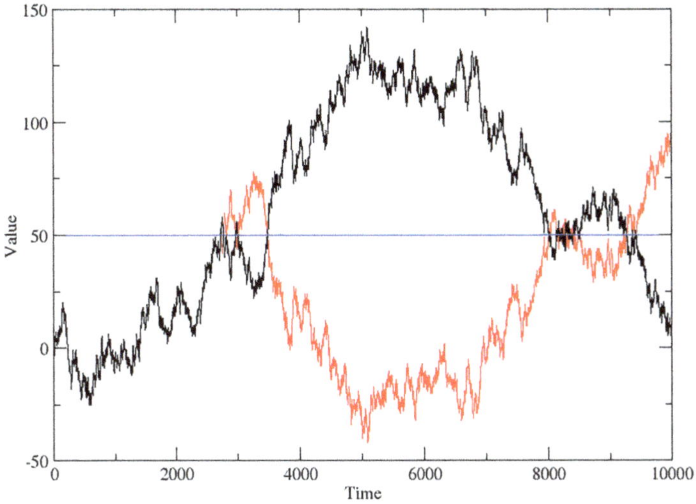

Fig. 11.3 A sample path of a Wiener process and its reflection upon reaching a crossing point

$$P(\tau_m \leq t) = 2P(W(t) \geq |m|) = 2(1 - \Phi(|m|/\sqrt{t})),$$

where $\Phi(\cdot)$ *is the CDF of standard normal distribution.*

Proof Assume $m > 0$. Then, using the Law of Total Probability, consider,

$$P(\tau_m \leq t) = P(\tau_m \leq t, W(t) \geq m) + P(\tau_m \leq t, W(t) < m).$$

However, $\{W(t) \geq m\}$ implies $\{\tau_m \leq t\}$. Hence

$$P(\tau_m \leq t, W(t) \geq m) = P(W(t) \geq m).$$

Next, suppose $\tau_m = s \leq t$. Since, $W(s) = m$, hence, by using independence of increments, it can be observed that $W(t) = W(t) - W(s) + m \sim N(m, (t-s))$ (This implicitly uses the fact that $\tau_m = s$ depends only on $\{W(u), 0 \leq u \leq s\}$). Thus, using the symmetry of $N(m, (t-s))$ around m, it implies,

$$P(W(t) < m|\tau_m = s) = P(W(t) > m|\tau_m = s), \quad 0 \leq s \leq t.$$

Hence,

$$P(W(t) < m, \tau_m \leq t) = P(W(t) > m, \tau_m \leq t) = P(W(t) \geq m, \tau_m \leq t).$$

The use of symmetry above is called the reflection principle.

$$\begin{aligned}P(\tau_m \le t) &= P(\tau_m \le t, W(t) \ge m) + P(\tau_m \le t, W(t) < m)\\ &= P(W(t) \ge m) + P(\tau_m \le t, W(t) \ge m)\\ &= P(W(t) \ge m) + P(W(t) \ge m).\end{aligned}$$

Thus proving the theorem. The result for $m < 0$ follows by a similar argument.

Theorem 11.4 *For $m > 0$, τ_m is finite almost surely.*

Proof For any $\sigma > 0$, consider the exponential martingale $\{Z(t), t \ge 0\}$. Since τ_m is a stopping time as defined in Definition (10.12), which implies,

$$1 = Z(0) = E[Z(t \wedge \tau_m)] = E[\exp\{\sigma W(t \wedge \tau_m) - \frac{1}{2}\sigma^2(t \wedge \tau_m)\}]. \quad (11.2)$$

It is easy to see that if $\sigma > 0$ and $m > 0$, Brownian motion is at or below level m for $t \le \tau_m$, since

$$0 \le \exp\{\sigma W(t \wedge \tau_m)\} \le e^{\sigma m}$$

$$\lim_{t\to\infty} \exp\{-\frac{1}{2}\sigma^2(t \wedge \tau_m)\} = 1_{\{\tau_m < \infty\}} \exp\{-\frac{1}{2}\sigma^2 \tau_m\}.$$

Notice that, if $\tau_m < \infty$, then for large t, $\exp\{\sigma W(t \wedge \tau_m)\} = e^{\sigma m}$, but for $\tau_m = \infty$, as $t \to \infty$ $\exp\{\sigma W(t \wedge \tau_m) - \frac{1}{2}\sigma^2(t \wedge \tau_m)\}$ converges to zero. Combining these analysis, it implies that

$$\lim_{t\to\infty} \exp\{\sigma W(t \wedge \tau_m) - \frac{1}{2}\sigma^2(t \wedge \tau_m)\} = 1_{\{\tau_m < \infty\}} \exp\{\sigma m - \frac{1}{2}\sigma^2 \tau_m\}.$$

Taking the limit in (11.2) as $t \to \infty$ implies

$$\begin{aligned}1 &= \lim_{t\to\infty} E[\exp\{\sigma W(t \wedge \tau_m) - \frac{1}{2}\sigma^2(t \wedge \tau_m)\}]\\ &= E[\lim_{t\to\infty} \exp\{\sigma W(t \wedge \tau_m) - \frac{1}{2}\sigma^2(t \wedge \tau_m)\}]\\ &= E[1_{\{\tau_m < \infty\}} \exp\{\sigma m - \frac{1}{2}\sigma^2 \tau_m\}].\end{aligned}$$

Since this holds for all $\sigma > 0$, let $\sigma \to 0$, therefore,

$$E[1_{\{\tau_m < \infty\}}] = 1 \implies P(\tau_m < \infty) = 1,$$

i.e., τ_m is finite almost surely.

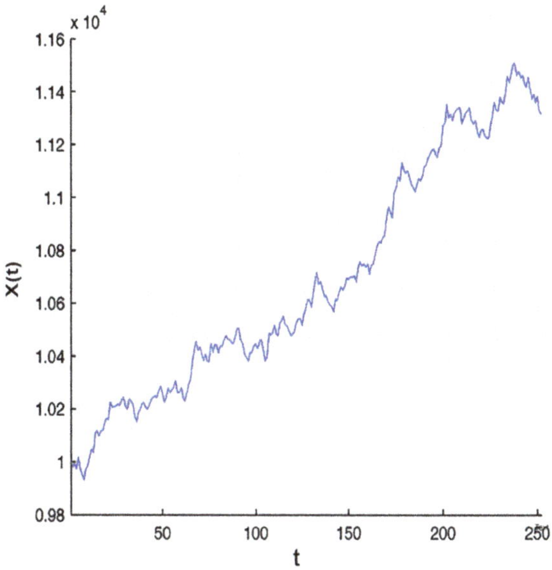

Fig. 11.4 Sample path of a Geometric Brownian motion

11.3 Processes Derived from Brownian Motion

In this section, two important stochastic processes which are derived from Brownian motion - geometric Brownian motion and Brownian bridge, and their properties are discussed.

11.3.1 Geometric Brownian Motion

Geometric brownian motion is a continuous-time stochastic process whose logarithm follows a Brownian motion with drift. It is a key concept in finance and is widely used to model the dynamics of stock prices, exchange rates, and other financial instruments.

Definition 11.5 (Geometric Brownian Motion) A stochastic process $\{X(t), t \geq 0\}$ is said to be a geometric Brownian motion if $X(t) = X(0)e^{W(t)}$ where $\{W(t), t \geq 0\}$ is a Brownian motion.

Sample path of geometric Brownian motion is shown in Fig. 11.4.

11.3.1.1 Markov Property

The geometric Brownian motion has the Markov property which is discussed below. Consider,

$$\begin{aligned} X(t+h) &= X(0)e^{W(t+h)} \\ &= X(0)e^{W(t)+W(t+h)-W(t)} \\ &= X(t)e^{W(t+h)-W(t)}. \end{aligned}$$

Since Brownian motion has independent increments, given $X(t)$, the future $X(t+h)$ only depends on the future increment of the Brownian motion. This future is independent of the past. Hence, the Markov property is satisfied. Thus, $\{X(t), t \geq 0\}$ is a Markov process.

11.3.1.2 Mean and Variance of Geometric Brownian Motion

The moment generating function of a normal RV X with mean μ and variance σ^2 is given by

$$M_X(s) = E[e^{sX}] = \exp(\mu s + \frac{1}{2}\sigma^2 s^2), \quad -\infty < s < \infty.$$

For Brownian motion with drift, since $W(t) \sim N(\mu t, \sigma^2 t)$,

$$M_{W(t)}(s) = E[e^{sW(t)}] = \exp(\mu t s + \frac{1}{2}\sigma^2 t s^2), \quad -\infty < s < \infty.$$

By using the values $s = 1, 2$,

$$E[X(t)] = X(0)M_{W(t)}(1)$$

and

$$E[X^2(t)] = X^2(0)M_{W(t)}(2).$$

After simplifications,

$$E[X(t)] = X(0)e^{(\mu+\frac{\sigma^2}{2})t}$$

$$\text{var}(X(t)) = X^2(0)e^{2\mu+\sigma^2 t}(e^{\sigma^2 t} - 1).$$

Letting $\overline{r} = \mu + \frac{\sigma^2}{2}$, implies that

$$E[X(t)] = X(0)e^{\overline{r}t}$$

and more generally

$$E[X(t)|X(s)] = e^{\overline{r}(t-s)}.$$

Because the geometric Brownian motion is non-negative, it provides for a more realistic model of stock prices. Suppose that the stock price $S(t)$ at time t is given by

$$S(t) = S(0)e^{H(t)}, t \geq 0$$

where $S(0)$ is the initial price and $H(t) = \sigma W(t) + \mu t$ is a Brownian motion with drift. In this case, $H(t)$ represents a continuously compounded rate of return of the stock price over the period of time $[0, t]$. Here, $H(t)$ refers to the logarithmic growth of the stock price, satisfying $H(t) = \ln\left(\frac{S(t)}{S(0)}\right)$. Therefore, $H(t)$ has a normal distribution with mean μt and variance $\sigma^2 t$. Letting $\overline{r} = \mu + \frac{1}{2}\sigma^2$, it implies,

$$E[S(t)] = S(0)e^{\overline{r}t}$$

$$\text{var}(S(t)) = (S(0)e^{(\mu+\frac{\sigma^2}{2})t})^2(e^{\sigma^2 t} - 1).$$

The predicted stock price is influenced not only by the drift μ, but also by the volatility σ, also the predicted price grows in the same way that a fixed-income security with a continually compounded interest rate $\overline{r}$. In the real world, r is significantly lower than the real fixed-income interest rate $\overline{r}$, which is why one buys in stocks. However, the stock is volatile due to the randomness of the underlying Brownian motion, therefore, there is some risk involved.

Example 11.11 Suppose that stock price $\{S(t), t \geq 0\}$ follows geometric Brownian motion with drift $\mu = 0.08$ per year and variance $\sigma^2 = 8.5\%$ per annum. Assume that, the current price of the stock is $S(0) = 60$. Find (i) $E[S(3)]$ (ii) $P(S(3) > 100)$.
Solution:

$$E[S(3)] = S(0)e^{\left(\mu+\frac{\sigma^2}{2}\right)t} = 60e^{0.36}.$$

Hence, we get $E[S(3)] = 60 \times 1.43 = 85.8$.

$$P(S(3) > 100) = P\left(W(3) > \frac{\ln(\frac{100}{S(0)}) - \left(\mu - \frac{\sigma^2}{2}\right)t}{\sigma}\right) = P(W(3) > 1.366) = 0.21.$$

Example 11.12 A stock's price moves according to a geometric Brownian motion with the parameters μ and σ^2 set at 0.12 and 0.1, respectively. What is the likelihood that a call option with a four-month expiration date and a strike price of K = Rs. 42 would be executed if the stock's current price is Rs. 40?
Solution: A call option on a stock with an exercise time of T and a strike price of K gives the holder the right but not the obligation to purchase a certain number of shares at T for K. If this number of shares has a market price of $S(t)$, then the option's value at time T is $S(T) - K$ if $S(T) > K$. This is because the shares can be purchased for K and then immediately sold in the market for $S(T)$. However, if $S(T) \leq K$, the option is worthless and will not be exercised. Thus

$$P(S(1/3) > 42) = P\left(W(1/3) > \frac{\ln(\frac{K}{S(0)}) - \left(\mu - \frac{\sigma^2}{2}\right)t}{\sigma}\right) = P(W(1/3) > 0.09).$$

Hence, $P(S(1/3) > 42) = 0.43$.

Example 11.13 Let $S(t)$ be the stock price at any time t. $S(t)$ follows geometric Brownian motion with parameters r and σ. Let K be the strike price which is a positive constant. Show that for maturity date $T > 0$,

$$E[e^{-rT}(S(T)-K)^+] = S(0)\Phi(d_+(T, S(0))) - Ke^{-rT}\Phi(d_-(T, S(0)),$$

where $X^+ = \max(X, 0)$,

$$d_\pm(T, S(0)) = \frac{1}{\sigma\sqrt{T}}\left[\ln\frac{S(0)}{K} + \left(r \pm \frac{\sigma^2}{2}\right)T\right]$$

and $\Phi(\cdot)$ is the CDF of standard normal distribution.
Solution: Consider,

$$\begin{aligned}
S(T) - K > 0 &\iff S(T) > K \\
&\iff S(0)\exp\big((r-\sigma^2/2)T + \sigma W(T)\big) - K > 0 \\
&\iff (r-\sigma^2/2)T + \sigma W(T) > \ln\left(\frac{K}{S(0)}\right) \\
&\iff \sigma W(T) > \ln\left(\frac{K}{S(0)}\right) - (r-\sigma^2/2)T \\
&\iff W(T) > -\frac{\ln\left(\frac{S(0)}{K}\right) + (r-\sigma^2/2)T}{\sigma}.
\end{aligned}$$

Since, $W(T) \sim N(0, T)$, which implies $W(T) = \sqrt{T}Y$ where $Y \sim N(0, 1)$. Thus we obtain

$$S(T) - K > 0 \iff Y > -d_-(T, S(0)).$$

Now consider,

$$E[e^{-rT}(S(T)-K)^+] = e^{-rT}E[\max\{0, S(0)\exp\big((r-\sigma^2/2)T + \sigma\sqrt{T}Y\big) - K\}].$$

Thus we get

$$
\begin{aligned}
E[(S(T)-K)^{+}] &= \int_{-d_-(T,S(0))}^{\infty} \left(S(0)\exp\left((r-\sigma^2/2)T+\sigma\sqrt{T}y\right)-K\right)\phi(y)dy \\
&= S(0)e^{rT}\int_{-d_-(T,S(0))}^{\infty} e^{-\frac{\sigma^2}{2}T+\sigma\sqrt{T}y}\phi(y)dy - K\int_{-d_-(T,S(0))}^{\infty}\phi(y)dy \\
&= S(0)e^{rT}\int_{-d_-(T,S(0))}^{\infty} e^{-\frac{\sigma^2}{2}T+\sigma\sqrt{T}y}\phi(y)dy - K\int_{-\infty}^{d_-(T,S(0))}\phi(y)dy \\
&= S(0)e^{rT}\frac{1}{\sqrt{2\pi}}\int_{-d_-(T,S(0))}^{\infty} e^{-\frac{\sigma^2}{2}T+\sigma\sqrt{T}y}e^{-y^2/2}dy - K\Phi(d_-(T,S(0)) \\
&= S(0)e^{rT}\frac{1}{\sqrt{2\pi}}\int_{-d_-(T,S(0))}^{\infty} e^{-\frac{1}{2}(y-\sigma\sqrt{T})^2}dy - K\Phi(d_-(T,S(0)) \\
&= S(0)e^{rT}\Phi(d_-(T,S(0)-\sigma\sqrt{T}) - K\Phi(d_-(T,S(0)).
\end{aligned}
$$

Therefore, we obtain

$$
\begin{aligned}
E[e^{-rT}(S(T)-K)^{+}] &= e^{-rT}\left(S(0)e^{rT}\Phi(d_+(T,S(0)) - K\Phi(d_-(T,S(0))\right) \\
&= S(0)\Phi(d_+(T,S(0)) - Ke^{-rT}\Phi(d_-(T,S(0)).
\end{aligned}
$$

Now, some remarks are presented as follows.

Remark 11.3 1. For fixed t, $W(t)$ has normal distribution while $X(t)$ has lognormal distribution.
2. Product of independent lognormal distributions is also lognormal distribution.
3. It removes the negativity problem. Also, it justifies from basic economic principles as a reasonable model for stock prices in an ideal non-arbitrage world.

11.3.2 Brownian Bridge

Brownian bridge refers to a process that connects two points in the Brownian motion path with certain properties. Specifically, it is a conditional Brownian motion between two given points at different times, with the condition that the process passes through a specified intermediate point.

Definition 11.6 (Brownian Bridge) A stochastic process $\{X(t), 0 \le t \le 1\}$ is said to be a Brownian bridge if

$$X(t) = W(t) - tW(1), t \in [0, 1]$$

where $\{W(t), t \ge 0\}$ is a standard Brownian motion.

Remark 11.4 Let, $\{W(t), t \ge 0\}$ be a standard Brownian motion. Then the Brownian bridge $\{X(t), t \in [0, T]\}$ is defined as

$$X(t) = W(t) - \frac{t}{T}W(T).$$

Mean and Variance of Brownian Bridge

$$E[X(t)] = E[W(t)] - \frac{t}{T}E[W(T)] = 0,$$

$$\text{var}(X(t)) = E[X^2(t)] = E[W^2(t)] + \frac{t^2}{T^2}E[W^2(T)] - 2\frac{t}{T}E[W(t)W(T)] = \frac{t(T-t)}{T}.$$

Covariance of Brownian Bridge
The covariance function $C(s,t)$, is given by

$$\begin{aligned}
C(s,t) &= E[(X(t) - E[X(t)])(X(s) - E[X(s)])] \\
&= E[X(t)X(s)] \\
&= E[W(t)W(s)] - \frac{s}{T}E[W(t)W(T)] - \frac{t}{T}E[W(s)W(T)] + \frac{st}{T^2}E[W^2(T)] \\
&= \min(s,t) - \frac{st}{T}.
\end{aligned}$$

Hence, $C(s,t) = \min(s,t) - \frac{st}{T}$.

Example 11.14 Let $\{X(t), t \geq 0\}$ be a Brownian bridge in the interval $[0, T]$, then show that $U^{-1/2}X(tU)$ is a Brownian bridge in the interval $[0, T/U]$.
Solution: As, $\{X(t), t \geq 0\}$ is a Brownian bridge, therefore

$$X(t) = W(t) - \frac{t}{T}W(T), t \in [0, T]$$

where $\{W(t), t \geq 0\}$ is a standard Brownian motion. Let $Y(t) = U^{-1/2}X(tU), 0 \leq t \leq T/U$ and $t' = tU$ then $X(t') = \sqrt{U}Y(t'/U)$ where $0 \leq t' \leq T$.

$$\sqrt{U}Y(t/U) = W(t) - \frac{t}{T}W(T), t \in [0, T].$$

Define $W'(t) = \sqrt{U}W(\frac{t}{U})$. One can easily prove that $\{W'(t), t \in [0, \frac{T}{U}]\}$ is standard Brownian motion and

$$Y(t) = W'(t) - \frac{Ut}{T}W'(T/U), t \in [0, T/U].$$

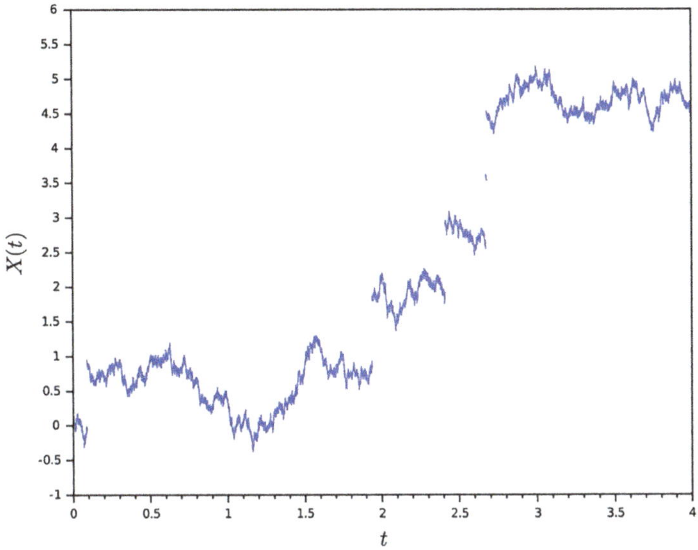

Fig. 11.5 Sample path of Lévy process

11.4 Lévy Processes

The Brownian motion or the Wiener process was discussed in the previous section, which are both continuous in space and time. However, there is a family of processes that are both continuous in time and discontinuous in space, admitting jumps. The Poisson process is the most basic of these processes (for more information, see Chap. 5). The Lévy Process is the most generic process that allows both continuous and discontinuous movements. Paul Pierre Lévy (15 September 1886–15 December 1971) was a French mathematician and was active in probability theory.

Definition 11.7 (Lévy Process) A Lévy process $X = \{X(t), t \geq 0\}$ is a stochastic process taking values in $\mathbb{R}$ such that

1. $X(t) - X(s)$ is independent of $\{X(u), u \leq s\}$ for any $s < t$. (independent increments)
2. $X(s+t) - X(s)$ has the same distribution as $X(t) - X(0)$ for any $s, t > 0$. (stationary increments)
3. $X(s) \to X(t)$ in probability as $s \to t$ for any $s < t$. (continuity in probability)

It is often required that $X(0)$ is zero and that X has cádlág sample paths (a sample path which is right continuous and has left limit).

A sample path of Lévy process is shown in Fig. 11.5.

The most common example of a Lévy process is Brownian motion. Other examples of Lévy processes are Poisson process, compound Poisson process, the Cauchy process, gamma process and the variance gamma process.

Example 11.15 Show that Poisson process and Brownian motion are Lévy processes.

Solution:

1. Poisson Process $\{N(t), t \geq 0\}$:
   From the definition of Poisson process, $N(0) = 0$ and $N(t)$ has independent and stationary increments. Also, for any $s < t$,

$$P(|N(t) - N(s)| > a) = P(N(|t-s|) > a) = \sum_{k=[a]+1}^{\infty} \frac{e^{-\lambda|t-s|}(\lambda|t-s|)^k}{k!}.$$

$$\text{As } s \to t,\ \lambda|t-s| \to 0, \text{ i.e., } \sum_{k=[a]+1}^{\infty} \frac{e^{-\lambda|t-s|}(\lambda|t-s|)^k}{k!} \to 0.$$

   Thus for $a > 0$, $P(|N(t) - N(s)| > a) \to 0$ as $s \to t$.
2. Brownian Motion $\{W(t), t \geq 0\}$:
   From the definition of Wiener process, $W(0) = 0$ and $W(t)$ has independent and stationary increments. Also the sample path of $\{W(t), t \geq 0\}$ is continuous almost surely and $W(t) - W(s) \sim N(0, t-s)$. Finally, for any $s < t$,

$$P(|W(t) - W(s)| > a) = \int_a^{\infty} \frac{e^{\frac{-x^2}{2(t-s)}}}{\sqrt{2\pi(t-s)}} dx = \frac{e^{\frac{-a^2}{2(t-s)}}}{\sqrt{2\pi(t-s)}}.$$

$$\text{As } s \to t,\ \frac{e^{\frac{-a^2}{2(t-s)}}}{\sqrt{2\pi(t-s)}} \to 0.$$

   Thus for $a > 0$, $P(|W(t) - W(s)| > a) \to 0$ as $s \to t$.

11.5 Jump-Diffusion Processes

Jump-diffusion processes are a class of stochastic processes that incorporate both continuous random movements, i.e., diffusion and sudden discontinuous movements, i.e., jumps. These processes are widely used in finance, physics, and other fields and is defined as follows.

Definition 11.8 (Jump-Diffusion Process) A jump-diffusion process $\{X(t), t \geq 0\}$ is a combination of a Brownian motion $\{W(t), t \geq 0\}$ with drift μ and of an independent compound Poisson process $\sum_{i=1}^{N(t)} Y_i$, where $Y_i's$ are IID RVs, i.e., a process which sometimes jumps and has a continuous but random evolution between the jump times, which is given by

$$X(t) = \mu t + \sigma W(t) + \sum_{i=1}^{N(t)} Y_i, \ t \geq 0.$$

Remark 11.5 The jump-diffusion process is also a Lévy process.

11.5.1 Compound Poisson Jump-Diffusion Processes

Compound Poisson jump-diffusion process combines elements of a Poisson process, representing the occurrence of jumps, with a diffusion process to account for continuous random motion. The compound Poisson jump-diffusion process is defined as follows.

Definition 11.9 (Jump Processes) Jump processes are stochastic processes whose trajectories have discontinuities called jumps, that can occur at random times.

Definition 11.10 (Compound Poisson Process) Consider a Poisson process $\{N(t), t \geq 0\}$. Let X_i's be IID RVs which are also independent of $N(t)$. Also, define $X(t) = X_1 + X_2 + \cdots + X_{N(t)}$. Then, $\{X(t), t \geq 0\}$ is called a compound Poisson process. In particular, if $P(X_i = 1) = 1, \forall i$, then $\{X(t), t \geq 0\}$ is a Poisson process.

Example 11.16 Insurance claims occur according to a Poisson process of rate $\lambda = 10$ per month and the amounts of claims are IID with mean 500 and standard deviation 200. Find the expected total claims in a three month period. Based on the normal approximation, estimate the probability that the total claim during 3 months will not exceed 20000.
Solution: Let $\{N(t), t \geq 0\}$ be the Poisson process that counts the number of claims by time t and let $\{Y_i, i = 1, 2, \ldots\}$ be the IID claim amounts with $E[Y_i] = 500$ and $\text{var}(Y_i) = (200)^2$. Then the total amount $X(t)$ of claims by time t is a compound Poisson process which is given by

$$X(t) = \sum_{i=1}^{N(t)} Y_i.$$

The expected total claim during 3 months $= E[X(3)] = 10 \times 3 \times 500 = 15000$. $\text{var}(X(3)) = 10 \times 3 \times (200^2 + 500^2) = 8700000$. Assuming $X(3)$ has approximately the normal distribution $N(15000, 8700000)$ and Z being the corresponding standard normal RV, then using the CDF of standard normal distribution table in Appendix A.4 it implies,

$$P(X(3) \leq 20000) = P\left(Z \leq \frac{20000 - 15000}{\sqrt{8700000}}\right) = 0.9554.$$

Example 11.17 In an Airline, the number of flights that are delayed follows a Poisson process with rate of $\lambda = 5$ per month. The passengers on each delayed flights get compensation of Rs. 500, where the number of passengers on each flights is independently and identically distributed with mean 30 and variance 1600. Calculate the expected total amount and standard deviation of the total annual compensation for delayed flights.

Solution: Let $\{N(t), t \geq 0\}$ be the Poisson process that counts the number of flights delayed by time t and let $\{Y_i, i = 1, 2, \ldots\}$ be the independently and identically distributed number of passengers in flight with $E[Y_i] = 30$ and $\text{var}(Y_i) = 1600$. Then the total number of customers $X(t)$ that are eligible for compensation by time t is a compound Poisson process.

The expected total amount of compensation in 12 months is

$$500 \times E[X(12)] = 500 \times 5 \times 12 \times 30 = 900000.$$

$\text{var}(X(12)) = 5 \times 12 \times (1600 + 900) = 1500000$. The standard deviation of the total amount of compensation in 12 months is $500 \times 500\sqrt{6} = 250000\sqrt{6}$.

Example 11.18 The number of customers arriving in a store follows a Poisson process with rate 5 per hour and let $\{Y_i, i = 1, 2, \ldots\}$ be the total time spent by the i-th customer has the distribution (independent)

$$P(Y_i = k) = \frac{1}{2^k}, \quad k = 1, 2, \ldots.$$

Find the mean and the variance of the time spend by the customers per day.

Solution: Note that, for $i = 1, 2, \ldots$

$$E[Y_i] = \sum_{k=1}^{\infty} \frac{k}{2^k} = 2$$

and

$$E[{Y_i}^2] = \sum_{k=1}^{\infty} \frac{k^2}{2^k} = 6.$$

Then $\text{var}(Y_i) = 2$. Let $X(t)$ be the time spend by the customers per day in the store. Then $X(t)$ is a compound Poisson process. $E[X(24)] = \lambda t E[Y_i] = 240$ and $\text{var}(Y_i) = \lambda t E[Y_i^2] = 720$.

Example 11.19 The faculty members of IIT Delhi Mathematics Department send emails through an email server according to a Poisson process with rate $\lambda = 10$ per hour. All the emails sent through the server are stored on a local hard drive, which

has the capacity 100 GB. If the emails have an average size of 1 MB then how long on average it will take for storage space to become filled with emails.
Solution: Let $\{N(t), t \geq 0\}$ be the Poisson process counting the number of faculty members who send email through email server by time t. Let $\{Y_i, i = 1, 2, \ldots\}$ be the size of an email sent by the ith faculty. Then, $E[Y_i] = 1$ MB. The total storage used in local hard drive $X(t)$ that are eligible for compensation by time t is a compound Poisson process.
The expected total amount of data used in time t is given by

$$E[X(t)] = 10 \times 1 \times t = 10t \text{ MB } = 102400.$$

Hence, on an average it will take 10240 hours, i.e., 426 days, approximately 1 year and 61 days to fill the storage.

Example 11.20 Let u be the initial capital of an insurance company and c be the premium rate. Let $N(t)$ be total number of claims which occur on or before at time t that arrive according to Poisson process. Let the ith claim cause a payment X_i. The sequence $\{X_i, i = 1, 2, \ldots\}$ is assumed to be an IID sequence of non-negative RVs. The claim size process $\{X_i, i = 1, 2, \ldots\}$ and $\{N(t), t \geq 0\}$ are assumed to be independent. Let $\{C(t), t \geq 0\}$ denotes the surplus process. Then

$$C(t) = u + ct - \sum_{i=1}^{N(t)} X_i.$$

Find $P(C(5) \leq 50)$ for fixed value of other parameters such as $u = 1000$ and $c = 50$. Assume that, $\lambda = 3$. Further, assume that, X_i follows Bernoulli distribution with $p = 1/2$.
Solution: Let, $N(t)$ be total number of claims which occur before or at time t follows a Poisson process with rate $\lambda = 3$, and X_i be an IID sequence of non-negative RVs having Bernoulli distribution with $p = 1/2$. Consider,

$$C(5) = 1000 + 50 \cdot 5 - \sum_{i=1}^{N(5)} X_i = 1250 - \sum_{i=1}^{N(5)} X_i.$$

Consider,

$$
\begin{aligned}
P\left(C(5) \leq 50\right) &= P\left(1250 - \sum_{i=1}^{N(5)} X_i \leq 50\right) \\
&= P\left(\sum_{i=1}^{N(5)} X_i \geq 1200\right) \\
&= \sum_{k=0}^{\infty} P\left(\sum_{i=1}^{N(5)} X_i \geq 1200 \mid N(5) = k\right) P\left(N(5) = k\right) \\
&= \sum_{k=0}^{\infty} P\left(\sum_{i=1}^{k} X_i \geq 1200\right) P\left(N(5) = k\right) \\
&= \sum_{k=0}^{\infty} P\left(\sum_{i=1}^{k} X_i \geq 1200\right) e^{-15} \frac{(15)^k}{k!}.
\end{aligned}
$$

Since X_i is an IID sequence of RVs which follows Bernoulli distribution with $p = 1/2$, implies $\sum_{i=1}^{k} X_i$ follows Binomial distribution with parameter k and $p = 1/2$. Thus

$$
P\left(\sum_{i=1}^{k} X_i \geq 1200\right) = \sum_{j=1200}^{k} \binom{k}{j} \left(\frac{1}{2}\right)^k.
$$

Hence,

$$
P\left(C(5) \leq 50\right) = \sum_{k=0}^{\infty} \sum_{j=1200}^{k} \binom{k}{j} \left(\frac{1}{2}\right)^k e^{-15} \frac{(15)^k}{k!},
$$

which implies,

$$
P\left(C(5) \leq 50\right) = \sum_{k=1200}^{\infty} \sum_{j=1200}^{k} \frac{1}{k!} \binom{k}{j} e^{-15} \left(\frac{15}{2}\right)^k.
$$

The complementary CDF plot with changes in surplus amount for various values of t is shown in Fig. 11.6.

Theorem 11.5 *For compound Poisson process* $\{X(t), t \geq 0\}$, *prove that*

$$
E[X(t)] = \lambda t E[X_1], \quad \text{var}(X(t)) = \lambda t E[X_1^2].
$$

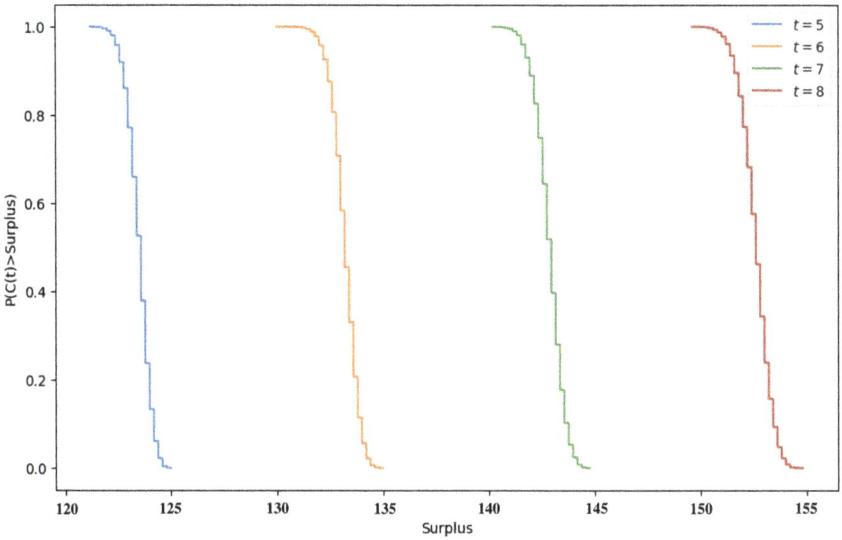

Fig. 11.6 Complementary CDF plot with change in surplus amount

Proof

$$
\begin{aligned}
E[X(t)] &= E[E[X_1 + X_2 + \cdots + X_N \mid N(t)]] \\
&= E[N(t)E[X_1]] = E[X_1]E[N(t)] \\
&= \lambda t E[X_1]. \\
E[X(t)^2] &= E[E[(X_1 + X_2 + \cdots + X_N)^2 \mid N(t)]] \\
&= E[N(t)E[X_1]^2 + N(t)(N(t) - 1)(E[X_1])^2] \\
&= \lambda t E[(X_1)^2] + (\lambda t)^2 (E[X_1])^2.
\end{aligned}
$$

Hence,

$$
\text{var}(X(t)) = E[(X(t))^2] - (E[X(t)])^2 = \lambda t E[(X_1)^2].
$$

11.6 Hawkes Jump Processes

Hawkes processes are in general non-Markov extension of homogeneous Poisson processes whose intensity is dependent on both time and history, implying that an arrival has an impact on the probability of next arrival.

11.6.1 Hawkes Process

Alan Geoffrey Hawkes (19 September 1938–9 November 2023), a British statistician, first proposed these processes in 1971 with the goal of modeling earthquakes. His plan was to simulate the replica effects after an earthquake. However, these processes have a wide range of application in various fields including finance, criminology, neurology, genome analysis, and sociology.

Definition 11.11 (Hawkes Process) A counting process (discussed in Chapter 8) $\{N(t), t \geq 0\}$ with natural filtration $\{\mathcal{F}_t\}_{t\geq 0}$ is a Hawkes process if $N(0) = 0$ and

$$\begin{aligned} P(N(t+\delta t) - N(t) = 0|\mathcal{F}_t) &= 1 - \lambda(t)\delta t + o(\delta t), \\ P(N(t+\delta t) - N(t) = 1|\mathcal{F}_t) &= \lambda(t)\delta t + o(\delta t), \\ P(N(t+\delta t) - N(t) = k|\mathcal{F}_t) &= o(\delta t), \ k \geq 2, \end{aligned}$$

where the conditional intensity function λ is given by

$$\lambda(t) = \psi\left(\int_0^t h(t-s)dN(s)\right),$$

where $\psi(\cdot) : \mathbb{R}^+ \to \mathbb{R}^+$ is a locally integrable, left continuous function and $h(\cdot) : \mathbb{R}^+ \to \mathbb{R}^+$ is assumed to be integrable, $||h||_{L^1} = \int_0^\infty h(t)dt < \infty$.
Note that, $h(\cdot)$ and $\psi(\cdot)$ are commonly referred in literature as the exciting function and the intensity function, respectively. Moreover, if $\psi(\cdot)$ is linear, the Hawkes process is linear, otherwise, it is non-linear.

In the linear case

$$\lambda(t) = \mu + \int_0^t h(t-s)dN(s),$$

where $h : [0, \infty) \to [0, \infty)$ is called exciting function or memory kernel and μ is the baseline intensity, and reflects the influence of past events on the conditional intensity. This definition is in more general form than Definition 8.2 in Section 8.

Remark 11.6 Note that,

$$\int_0^t h(t-s)dN(s) = \sum_{i:t_i<t} h(t-t_i)$$

where $0 = t_0 < t_1 < \ldots$ are the arrival times of the Hawkes process.

Hawkes process is a self-exciting process as the arrival of an event causes the conditional intensity function to increase.

Some common choices for memory kernel functions are as follows.

1. Exponential kernel: $h(t) = \alpha e^{-\beta t}$ for $\alpha, \beta \in \mathbb{R}^+$.

2. Gamma decay kernel: $h(t) = te^{-t}$.
3. Power law kernel: $h(t) = \frac{\alpha}{(\beta+t)^p}$ for $\alpha, \beta,$ and $p \in \mathbb{R}^+$.
4. Rayleigh kernel: $h(t) = \eta t e^{-\gamma t^2}$ for $\gamma, \eta \in \mathbb{R}^+$.

Remark 11.7 The exponential memory kernel is mainly used in Hawkes processes because of its decay, and the Markov property.

Empirical observations in several fields like earthquakes, finance distress, crime and riots, and social media trends, motivate us to develop models to capture feedback effects and contagion effects. A natural choice in this context is to model these phenomena using Hawkes processes. These are a type of point processes suitable for modeling phenomena of self-excitation and mutual excitation, i.e., situations in which the occurrence of an event makes future events more likely to happen. To capture self-exciting properties as well as clustering effects, Hawkes processes have wide applications in cosmology, ecology, epidemiology, seismology, neuroscience, and DNA modeling.

On the contrary, the simplest case of point processes, the so-called time-homogeneous Poisson processes, assumes that the arrival of an event follows a constant rate, which is the intensity function, and that the occurrence of a new event is independent of the occurrence of previous events. This is the reason why a Hawkes process is a non-Markov, in general, extension of a Poisson process.

Generally, the Hawkes process is non-Markov in nature since the timing of the past events determine the future growth. But, under some particular assumptions on the exciting function, such as, for instance, if the exciting function is exponential or sum of exponential functions, the couple (N, λ), given by the Hawkes process and its intensity, is proven to be a Markov process (see for instance Dharmaraja, Pasricha and Tardelli [8], Errais, Giesecke and Goldberg [9], Zhu [10], and references therein). However, the difficulty to prove the Markovianity arises when h is not an exponential function, and an open problem in literature is the identification of the necessary conditions for the Markovianity.

When $\lambda(\cdot)$ is linear, say $\lambda(\xi) = \upsilon + \xi$, $\upsilon > 0$, and $h(\cdot)$ is integrable, the Hawkes process has a nice immigration-birth representation (Hawkes and Oakes [12])

For the linear Hawkes process, LLN (Daley and Vere-Jones [6]), allow us to write that,

$$\frac{N(t)}{t} \to \frac{\upsilon}{1 - ||h||_{L^1}}, \quad \text{a.s. as } t \to \infty.$$

Under some regularity assumptions, Bacry, Delattre, Hoffmann and Muzy [14], proved the CLT, namely,

$$\frac{N(t) - \mu t}{\sqrt{t}} \to \sigma N(0, 1), \text{ as } t \to \infty,$$

where $\mu = \frac{\upsilon}{1-||h||_{L^1}}$ and $\sigma^2 = \frac{\upsilon}{(1-||h||_{L^1})^3}$. Note that in this case there is a weak convergence on $D[0, 1]$, the space of cádlág functions on $[0, 1]$, equipped with Skorokhod topology. Furthermore, Bordenave and Torrisi [15], proved a large deviation principle (LDP), for $P(\frac{N(t)}{t} \in \cdot)$.

When $\lambda(\cdot)$ is non-linear, however, the usual immigration-birth representation no longer works, therefore, to establish limit theorems, a different approach must be considered. For non-linear Hawkes processes, Zhu [16], proved the functional central limit theorem. Assuming that $h(\cdot)$ is exponential or sum of exponential functions, and consequently the Hawkes process is a Markov process, Zhu [10], developed a LDP for $P(\frac{N(t)}{t} \in \cdot)$.

Moreover, Zhu [17] showed a process-level, i.e., level-3 LDP for the general $h(\cdot)$, and, hence, obtained the level-1 LDP for $P(\frac{N(t)}{t} \in \cdot)$ by Contraction Principle.

11.6.2 Exponential Exciting Function: Asymptotic Behavior

If $h(t) = ae^{-bt}$, $a > 0$ and $b > 0$, the couple (N, λ) satisfies the Markov property. Now define $\{\xi(t), t \geq 0\}$, a Markov process which fulfills the dynamics,

$$d\xi(t) = -b\xi(t)dt + adN(t), \qquad t \geq 0.$$

The intensity of the process $\{N(t), t \geq 0\}$, for each t, is given by $\lambda(t) = v + \xi_{t-}$, where the so-called baseline intensity v takes positive real values. Furthermore, looking to the existing literature, some important results related to the asymptotic behavior of an Hawkes process are mentioned.

Gao and Zhu [11], as a consequence of the law of large numbers, deduced that

$$\frac{1}{t}\int_0^t \xi(s)ds \to \frac{v}{b-a} \quad \text{a.s. as } t \to \infty.$$

As a consequence of the CLT,

$$\frac{1}{\sqrt{t}}\left[\int_0^t \xi(s)ds - \frac{v}{b-a}t\right] \to N\left(0, \frac{a^2vb}{(b-a)^3}\right) \quad \text{in distribution as } t \to \infty.$$

For more description, readers may refer to [1–7].

11.7 Exercise Problems

1. Let $\{W(t), t \geq 0\}$ be a Brownian motion. Calculate $P(W(6) - W(2) < 0, W(1) > 0)$.
2. Show that $f(W(t))$ is continuous in mean square where f is a Lipschitz continuous function and $\{W(t), t \geq 0\}$ is a Brownian motion.

3. Find $E[W(1) + 2W(2) + 3W(3)]^2$ where $\{W(t), t \geq 0\}$ is a Wiener process.
4. Let $\{W(t), t \geq 0\}$ be a Brownian motion and $\{\mathcal{F}_t\}_{t\geq 0}$ be its natural filtration. Show that the process $\{X(t), t \geq 0\}$ defined by

$$X(t) = 2e^{3W(t)+4t}, \ t \geq 0,$$

is a Markov process.
5. For $r \in \mathbb{R}^+$, consider the stochastic process $\{X(t), t \geq 0\}$

$$X(t) = 3\exp\{(r-2)t + 2W(t)\}, \quad t \geq 0,$$

where $\{W(t), t \geq 0\}$ is the Brownian motion. Show that for any $t > 0$,

$$E\left[e^{-rt}(X(t)-3)^+\right] = 3\Phi(m_+) - 3e^{-rt}\Phi(m_-)$$

where $\Phi(\cdot)$ is the CDF of standard normal distribution and $m_\pm = \frac{\sqrt{t}}{2}(r \pm 2)$.
6. Define $X(t) = e^{-t}W\left(e^{2t}\right)$, where $\{W(t), t \geq 0\}$ is a Brownian motion. Show that

$$E[X(t)X(s)] = e^{-|t-s|}, \quad \forall -\infty < s, t < \infty.$$

7. Show that for any $T > 0$, $V(t) = W(t+T) - W(T)$ is a Wiener process if $\{W(t), t \geq 0\}$ is a Wiener process.
8. Show that $-W(t)$ is a Wiener process if $\{W(t), t \geq 0\}$ is a Wiener process.
9. Prove that, for a Brownian motion $\{W(t), t \geq 0\}$, $E\left(|W(t) - W(s)|^2\right) = |t-s|$.
10. Let $\{W(t), t \geq 0\}$ be a Brownian motion and $\{\mathcal{F}_t\}_{t\geq 0}$ be its natural filtration. Consider, the Brownian motion $\{X(t), t \geq 0\}$ with drift μ then there holds that

$$X(t) = \mu t + W(t)$$

is a Markov process.
11. Let $\{\mathcal{F}_t\}_{t\geq 0}$ be the natural filtration of $\{W(t), t \geq 0\}$. Let b be a positive real number. Show that the process Y defined by $Y(t) = W^2(bt) - bt$ is a martingale.
12. Let $\{W(t), t \geq 0\}$ be a Brownian motion. Let $T = \inf\{t > 0 : W(t) \notin [-a, b]\}$. Let $a, b > 0$. Prove that (i) $P(W(T) = -a) = \frac{b}{a+b}$ (ii) $P(W(T) = b) = \frac{a}{a+b}$ (iii) $E[T] = ab$.
    Hint: $W(t \wedge T)$ is a martingale and $P(T < \infty) = 1$.
13. Let $\{W(t), t \geq 0\}$ be a Brownian motion. For $b > 0$, let $T = \inf\{t > 0 : W(t) = b\}$. Show that $T < \infty$ a.s.. Further, show that $E[T] = \infty$.
    Hint: Take a limit in Problem 12.
14. Let $\{W(t), t \geq 0\}$ be a Brownian motion. Prove that, for $\lambda > 0$
    (a) $P(\sup_{s\leq t} W(s) \geq \lambda) \leq e^{-\lambda^2/2t}$;
    (b) $P(\sup_{s\leq t} W(s) \geq \lambda) \leq 2e^{-\lambda^2/2t}$.
    Hint: $E[e^{cW(t)}|\mathcal{F}_s] \geq e^{cW(t)}$. Use Doob's inequality.

15. Consider a two-dimensional Wiener process $\{W(t), t \geq 0\}$, where $W(t) = (W_1(t), W_2(t))$. Find the probability that $|W(t)| < R$ where $R > 0$. Here, $|(x_1, x_2)|^2 = (x_1)^2 + (x_2)^2$.
16. Suppose that $\left(W^1(t), W^2(t), \ldots, W^d(t)\right)$ is a d-dimensional Wiener process. Show that if $i \neq j$, then $W^i(t)W^j(t)$ is a martingale.
17. Let $\{W(t), t \geq 0\}$ be the Brownian motion. Prove that the following processes are martingales.

    (a) $\{(W(t))^3 - 3tW(t), t \geq 0\}$.
    (b) $\{(W(t))^4 - 6t(W(t))^2 + 3t^2, t \geq 0\}$.

18. Show that Wiener process is 0.5-self-similar.
19. Consider a stochastic process $\{X(t), t \geq 0\}$,

$$X(t) = 2t + W(t), \quad t \geq 0,$$

where $\{W(t), t \geq 0\}$ is a Brownian motion. For $m > 0$, define the first passage time of $X(t)$ in level m as

$$\tau_m = \min\{t \geq 0 : X(t) = m\}.$$

(a) Show that the process $\{Z(t), t \geq 0\}$,

$$Z(t) = \exp\left\{\sigma X(t) - \left(2\sigma + \frac{1}{2}\sigma^2\right)t\right\}, \quad t \geq 0,$$

is a martingale with respect to the filtration $\mathcal{F}_t = \sigma(W(s), 0 \leq s \leq t)$.
(b) Show that

$$E\left[\exp\left\{\sigma m - \left(2\sigma + \frac{1}{2}\sigma^2\right)\tau_m\right\} 1_{\{\tau_m < +\infty\}}\right] = 1.$$

(c) Show that

$$E\left[e^{-\alpha\tau_m}\right] = \exp\{2m - m\sqrt{2\alpha + 4}\}, \quad \forall \alpha > 0.$$

Moreover, prove that $E[\tau_m] < +\infty$.

20. Use the reflection principle to prove that if $\{W(t), t \geq 0\}$ is a Brownian motion and $T = \inf\{t > 0 : W_t \in (0, \infty)\}$, then $P(T = 0) = 1$. In other words, Brownian motion enters the interval $(0, \infty)$ immediately. By symmetry it enters the interval $(-\infty, 0)$ immediately. Conclude that Brownian motion hits 0 infinitely often in every time interval $[0, t]$.

21. Let $\{X(t), t \geq 0\}$ be a compound Poisson process. Let $0 = t_0 < t_1 < \ldots < t_n$. Prove that the increments $X(t_1) - X(t_0), X(t_2) - X(t_1), \ldots, X(t_n) - X(t_{n-1})$ are stationary and independent. Also, show that the distribution of $X(t_j) - X(t_{j-1})$ is same as the distribution of $X(t_j - t_{j-1})$.
22. Prove that, any Lévy process is a Markov process.
23. Consider, a Hawkes process $\{N(t), t \geq 0\}$ with exponential exciting function where the intensity function is given by

$$\lambda(t) = \mu + \int_0^t \alpha e^{-\beta(t-s)} dN(s), \quad \alpha, \beta > 0.$$

Under, the assumptions that $\frac{\alpha}{\beta} < 1$, prove

(a) Law of large numbers for $\{N(t), t \geq 0\}$.
(b) Central limit theorem for $\{N(t), t \geq 0\}$.

References

1. Calin, O.: An informal introduction to stochastic calculus with applications. World Scientific, Second Edition (2022)
2. Castaneda, L.B., Arunachalam, V.S., Dharmaraja: Introduction to Probability and Stochastic Processes with Applications. Wiley, New Jersey (2012)
3. Dudley, R.M., and Norvaiša, R.: Differentiability of Six Operators on Non smooth Functions and p-Variation, Springer, First edition (2006). Berlin/Heidelberg, Vol.I. ISBN: 978-3-540-48814-9
4. Gianin, E.R., and Sgarra, C.: Mathematical Finance Theory Review and Exercises, Springer Cham, Second Edition (2023)
5. Karatzas, I. and Shreve, S.E.: Brownian Motion and Stochastic Calculus. New York, Springer (1991)
6. Shreve, S.E.: Stochastic calculus for finance II: Continuous-time models. Vol. 11. New York: springer (2004)
7. Mikosch, T.: Elementary stochastic calculus with finance in view. World scientific (1998)
8. Dharmaraja, S., Pasricha, P., Tardelli, P.: Hedging and utility valuation of a defaultable claim driven by Hawkes processes. Appl. Stochast. Models Bus. Ind. **38**(2), 334–352 (2022)
9. Errais, E., Giesecke, K., Goldberg, L.R.: Affine point processes and portfolio credit risk. SIAM J. Financ. Math. **1**(1), 642–665 (2010)
10. Zhu, L.: Large deviations for Markovian nonlinear Hawkes processes. Ann. Appl. Probab. **25**, 548–581 (2015)
11. Gao, X., Zhu, L.: Functional central limit theorem for stationary Hawkes processes and its application to infinite-serve queues. Queueing Syst. **90**(1), 161–206 (2018)
12. Hawkes, A.G., Oakes, D.: A cluster process representation of self-exciting process. J. Appl. Probab. **11**, 493–503 (1974). https://doi.org/10.2307/3212693
13. Daley, D.J., Vere-Jones, D.: *An Introduction to the Theory of Point Processes*, 2nd ed., vol. I and II. (Springer, Germany, Berlin/Heidelberg, 2003). ISBN: 978-0-387-21564-8
14. Bacry, E., Delattre, S., Hoffmann, M., Muzy, J.F.: Scaling limits for Hawkes processes and application to financial statistics. Stochast. Process. Appl. **123**(7), 2475–2499 (2012)

15. Bordenave, C., Luca Torrisi, G.: Large deviations of poisson cluster processes. Stochast Models **23**(4), 593–625 (2007)
16. Zhu, L.: Central limit theorem for nonlinear Hawkes processes. J. Appl. Probab. **50**(3), 760–771 (2013)
17. Zhu, L.: Process-level large deviations for nonlinear Hawkes point processes. Ann. l'Institut Henri Poincaré Probab. Statist. **50**(3), 845–871 (2014)

Chapter 12
Stochastic Calculus

This chapter describes the stochastic calculus for diffusion processes and jump processes. In this chapter, integrals and differentials of stochastic processes are introduced.

The foundation of many stochastic processes is a continuous function that is nowhere differentiable. Due to their inability to be defined on non-smooth functions, derivative-based differential equations are ruled out. An integration theory is necessary in place of integral equations that do not require the direct formulation of derivative terms. This theory is referred to as Itô calculus.

Itô calculus provides a theory of calculus for stochastic processes. It is further divided into stochastic integrals and stochastic differentials (SDEs), where stochastic processes are used in place of functions. The results of the integration or differentiation are also stochastic processes.

The discussion will start with the quadratic variation of stochastic processes, which is similar to taking the sum of squares of increments of stochastic processes over a partition of the time domain and then decreasing the norm of the partition to zero. This quadratic variation will form the building blocks to solve a stochastic integral. Later on properties of Itô integrals and different methods to solve them will be discussed.

A stochastic process can be written as the sum of an Itô integral, a Riemann integral, and a pure jump process. Itô integral is an integral with respect to Brownian motion or its function, whereas Riemann integral is an integral with respect to dt. One or more of these components can be zero resulting in a pure diffusion or a pure jump process. Similar to the fundamental theorem of calculus for Riemann integration, here the fundamental theorem of stochastic calculus for stochastic integration will be discussed.

D. Selvamuthu, *Introduction to Stochastic Processes*, University Texts in the Mathematical Sciences, https://doi.org/10.1007/978-981-97-6152-4_12

12.1 Quadratic Variation

The quadratic variation of stochastic processes is an extension of the notion of the total variation of a function. Here, rather than summing the absolute values of the changes of the function sampled at a sequence of times, the squares are summed. Quadratic variation has important applications in stochastic calculus, appearing in the integration by parts and change of variables formulas for stochastic integration. The quadratic variation also has applications in the study of martingales, occurring in Itô isometry. In this section, the quadratic variation of Brownian motion is presented in detail. First, the variations for real-valued functions will be given, and then it will be extended to stochastic processes.

12.1.1 Variations of a Real-Valued Function

Definition 12.1 (Variations of a Real-Valued Function) Let g be a real-valued right-continuous function on the time interval $[a, b]$ where $0 \leq a < b < \infty$. Let Π be a set of all finite partitions π of the interval $[a, b]$ with $a = t_0 < t_1 < \ldots < t_n = b$. Define $\| \Pi \| = \max_i |t_{i+1} - t_i|$, the length of the largest sub-interval in the partition. The variation (or 1st variation) of $g(t)$ over the interval $[a, b]$ is defined as

$$V_g([a, b]) = \sup_{\pi \in \Pi} \sum_{i=0}^{n-1} | g(t_{i+1}) - g(t_i) | .$$

When $g(t)$ is a continuous function for all t in $[a, b]$, define the Riemann integral $\int_a^b g(t)dt$ as follows: For each sub-interval $[t_{i-1}, t_i]$, we get

$$M_i = \max_{t_{i-1} \leq t \leq t_i} g(t) \text{ and } m_i = \min_{t_{i-1} \leq t \leq t_i} g(t).$$

The upper sum is

$$RS_{\Pi}^{+}(g) = \sum_{i=1}^{n} M_i (t_i - t_{i-1})$$

and the lower sum is

$$RS_{\Pi}^{-}(g) = \sum_{i=1}^{n} m_i (t_i - t_{i-1}).$$

As $\| \Pi \|$ converges to zero (i.e., more and more partition) the upper Riemann sum $RS_{\Pi}^{+}(g)$ and the lower sum $RS_{\Pi}^{-}(g)$ converge to the same limit. This is called as the Riemann integral $\int_a^b g(t)dt$.

Definition 12.2 (Bounded Variation) The real-valued function $g(t)$ is of finite variation if, for every t,

$$V_g(t) = V_g([0, t]) < \infty.$$

In other words, the $y-$axis distance (e.g., up-down distance) traveled by g in any finite interval of time is finite. Moreover, for all t, if $V_g(t) < K$, where K is a constant independent of t, then $g(t)$ is of bounded variation.

Example 12.1 If g is a differentiable function, then from the Mean Value Theorem,

$$V_g([a, b]) = \sup_{\pi \in \Pi} \sum_{i=0}^{n-1} | g(t_{i+1}) - g(t_i) | = \sup_{\pi \in \Pi} \sum_{i=0}^{n-1} | g^{'}(t_i^*) | (t_{i+1} - t_i)$$

where $g(t_{i+1}) - g(t_i) = g^{'}(t_i^*)(t_{i+1} - t_i)$ for some value $t_i^* \in [t_i, t_{i+1}]$. But this is a Riemann sum for the integral $\int_a^b | g^{'}(x) | \, dx$. Thus, when g is differentiable,

$$V_g([a, b]) = \int_a^b | g^{'}(t) | \, dt < \infty.$$

Therefore, differentiable functions are of bounded variation.

Remark 12.1 A function g is of bounded variation over $[a, b]$ if and only if $g(t)$ can be rewritten as $g(t) = g_1(t) - g_2(t)$ where g_1 and g_2 are bounded monotone increasing functions on $[a, b]$.

It should be somewhat intuitive that a typical Brownian motion path can't possibly be expressed as the difference of monotone functions as in Remark 12.1; the paths seem to exhibit rapid infinitesimal movement up and down in any interval of time. This is most apparent when recalling how Brownian motion can be obtained by a limiting procedure by scaling a simple symmetric random walk (refer Section 11.1.1). In any time interval, no matter how small, the number of moves of the random walk tends to infinity, while the size of each move to 0.

Definition 12.3 (p-Variation of Real-Valued Function) For $p \geq 1$, p-variation of a real-valued function $g(t)$ in the interval $[0, t]$ is defined as

$$V_g^p(t) = \sup_{\pi \in \Pi} \sum_{i=0}^{n-1} | g(t_{i+1}) - g(t_i) |^p .$$

Definition 12.4 (Quadratic Variation) The quadratic variation of a function g on the interval $[a, b]$ is defined as

$$Q(g) = \lim_{\|\Pi\| \to 0} Q_\Pi(g),$$

where $Q_\Pi(g) = \sum_{i=1}^{n} |g(x_i) - g(x_{i-1})|^2$. Following the definition, it can be observed that if g is differentiable, then $Q(g) = 0$. We have

$$\begin{aligned}\sum_{i=1}^{n} |g(x_i) - g(x_{i-1})|^2 &= \sum_{i=1}^{n} \left|g^{'}(x_i^*)\right|^2 |x_i - x_{i-1}|^2 \\ &\leq \| \Pi \| \sum_{i=1}^{n} \left|g^{'}(x_i^*)\right|^2 |x_i - x_{i-1}| \\ &\to 0 \text{ as } \| \Pi \| \to 0.\end{aligned}$$

Now, the so-called squared (quadratic) variation of diffusion processes is introduced, and later on it will be proved that Brownian motion has finite non-zero quadratic variation.

12.1.2 Quadratic Variation of Diffusion Processes

It has been already established that the sample paths of $W(t)$ are nowhere differentiable. Because of the above-mentioned argument, the first variation does not make sense. As a result, there is no first variation for Brownian motion.

Following is an important result on p-variation without proof (readers can refer to Dudley and Norvaiša [2]).

Theorem 12.1 *A function that has bounded variation has a quadratic variation that is equal to zero. Furthermore, it generalizes to higher dimensions. Variation can be defined in higher dimensions d, i.e., $V^{(d)}(g)$:*

$$V^{(d)}(g) = \lim_{\|\Pi\| \to 0} V_\Pi^{(d)}(g)$$

where

$$V_\Pi^{(d)}(g) = \sum_{i=1}^{n} |g(x_i) - g(x_{i-1})|^d .$$

If the dth dimensional variation of a function g is bounded, the variation in any higher dimensions is equal to zero.

Example 12.2 Consider $g(t) = t^4$. Then $g'(t) = 4t^3$. and $\int_0^1 g'(t)dt = 1$.

$$\int_0^1 | g'(t) |^2 \, dt = 16 \int_0^1 t^6 dt = 16 \left. \frac{t^7}{7} \right|_0^1 = \frac{16}{7}.$$

But

$$\lim_{\|\Pi\|\to 0} \| \Pi \| \int_0^1 |g'(t)|^2 dt = 0,$$

$V_g(1) = 1$ and $V_g^2(1) = 0$. By applying Theorem 12.1,

$$V_g^p(1) = \begin{cases} 1, & p = 1 \\ 0, & p > 1 \end{cases}.$$

The quadratic variation for Brownian motion over the interval $[0, T]$,

$$[W, W](T) = V_{W(t)}^2(T) = \lim_{\|\Pi\|\to 0} Q_\Pi,$$

where

$$Q_\Pi = \sum_{i=0}^{n-1} (W(t_{i+1}) - W(t_i))^2.$$

Q_Π is obviously a function of the sample points $\omega \in \Omega$. As a result, the quadratic variation estimated for a Brownian motion for each division Π, Q_Π is a RV. It's worth noting that the limit applies to all partitions of $[0, T]$, with $\| \Pi \| \to 0$ as $n \to \infty$. How one can obtain the limiting distribution for large n when each partition Π, Q_Π is a RV? What is the right mode of convergence for these RVs? To answer these, convergence in mean square (convergence in L^2) will be used, which is well defined as follows:

Definition 12.5 (**L^2 Convergence**) Let $\{X_n, n \geq 1\}$ and X be RVs defined on a common probability space $(\Omega, \mathcal{F}, P)$. X_n converges to X in L^2 (see Appendix A.1.3) if

$$\lim_{n\to\infty} E\left(|X_n - X|^2\right) = 0.$$

In the case of Brownian motion, it will be shown that

$$\lim_{\|\Pi\|\to 0} E\left(|Q_\Pi - T|^2\right) = 0. \tag{12.1}$$

When the preceding result holds, it can be almost certainly asserted that the quadratic variation of the Brownian motion throughout the interval $[0, T]$ converges to T in L_2 and indicated as $[W, W](T) = T$. Let us prove the above result in the following theorem.

Theorem 12.2 *The variation of the path of Brownian motion does not converge with probability one; in other words, the following limit is a.s. infinite*

$$\lim_{n\to\infty} \sum_{k=1}^{2^n} \left| W\left(\frac{kt}{2^n}\right) - W\left(\frac{(k-1)t}{2^n}\right) \right|.$$

Proof Note that the following inequality is true:

$$\sum_{k=1}^{2^n}\left|W\left(\frac{kt}{2^n}\right)-W\left(\frac{(k-1)t}{2^n}\right)\right| \geq \frac{\sum_{k=1}^{2^n}\left|W\left(\frac{kt}{2^n}\right)-W\left(\frac{(k-1)t}{2^n}\right)\right|^2}{\max_{k=1,2,\ldots,2^n}\left|W\left(\frac{kt}{2^n}\right)-W\left(\frac{(k-1)t}{2^n}\right)\right|}.$$

Consider the numerator and denominator separately. In the numerator, this is the quadratic variation of Brownian motion: After this, it will be shown that the quadratic variation of Brownian motion is t. Meanwhile, the denominator will converge to zero because as the time increments indexing Brownian motion become arbitrarily close, the distance between them also converges to zero. Thus, Brownian motion does not have bounded variation.

Example 12.3 The quadratic variation of standard Brownian motion on the interval $[a, b]$ is $b - a$.
Solution. To prove this, for the sake of simplicity, we will only consider the case on the interval $[0, t]$, and prove that the quadratic variation L^2 converges to t. In other words, it can be proved that

$$\lim_{\|\Pi\|\to 0} E\left[|Q_\Pi - t|^2\right] = 0.$$

Note that for any partition $\{t_1, t_2, \ldots, t_n\}$, we have

$$t = \sum_{i=1}^{n} t_i - t_{i-1}.$$

Let us use the notation:

$$W(t_i) - W(t_{i-1}) = \Delta W(t_i) \text{ for } i = 1, 2, \ldots, n-1.$$

It is known that $\Delta W(t_i) \sim N(0, t_i - t_{i-1})$. The following equality is obtained

$$Q_\Pi - t = \sum_{i=1}^{n}\left((\Delta W(t_i))^2 - (t_i - t_{i-1})\right).$$

The expectation of the square of $Q_\Pi - t$ is given by

$$E\left[|Q_\Pi - t|^2\right] = \sum_{i=1}^{n} E\left[\left((\Delta W(t_i))^2 - (t_i - t_{i-1})\right)^2\right].$$

First, the terms inside of the expectation are squared and simplified. To simplify, separate the sum using linearity of expectation,

$$= \sum_{i=1}^{n} E\left[\left((\Delta W(t_i))^4 - 2\,(\Delta W(t_i))^2\,(t_i - t_{i-1}) + (t_i - t_{i-1})^2\right)\right],$$

$$= \sum_{i=1}^{n} E\left[(\Delta W(t_i))^4\right] - 2\sum_{i=1}^{n} E\left[(\Delta W(t_i))^2\right](t_i - t_{i-1}) + \sum_{i=1}^{n} (t_i - t_{i-1})^2 .$$

Note that $(t_i - t_{i-1})$ is removed from inside of the expectation, since $(t_i - t_{i-1})$ is not random. To simplify this further, remember that $(\Delta W(t_i))$ is normal distributed. In probability, higher powers of expectation can be found with moment generating functions. In particular, for the normal distribution, where $X \sim N\left(0, \sigma^2\right)$, $E\left[X^2\right] = \sigma^2$ and $E\left[X^4\right] = 3\sigma^4$. Using this fact,

$$\sum_{i=1}^{n} 3\,(t_i - t_{i-1})^2 \qquad -2\sum_{i=1}^{n} (t_i - t_{i-1})\,(t_i - t_{i-1})$$
$$+\sum_{i=1}^{n} (t_i - t_{i-1})^2 = 2\sum_{i=1}^{n} (t_i - t_{i-1})^2 .$$

Consider the supremum Π^* of distances between points $(t_i - t_{i-1})$ of the partition Π. The following inequality is obtained

$$= 2\sum_{i=1}^{n} (t_i - t_{i-1})^2 \le 2\Pi^* \sum_{i=1}^{n} (t_i - t_{i-1}) = 2\Pi^* t.$$

To find the quadratic variation, we take the limit as the length of the distances between the points in the partitions (for any partition Π) approaches zero. In particular, Π^* will approach zero, meaning that this product also converges to zero:

$$\lim_{||\Pi||\to 0} 2\Pi^* t = 0.$$

And clearly, this bound means that:

$$E\left[(Q_\Pi - t)^2\right] = \lim_{\|\Pi\|\to 0} E\left[|Q_\Pi - t|^2\right] = 0.$$

This shows L^2 convergence. L^2 convergence also implies that the quadratic variation of standard Brownian motion converges to t with probability 1.

Thus (via Chebychev's inequality) convergence holds in probability as well; for each $t > 0$, it holds that for all $\epsilon > 0$, as $|| \Pi || \to 0$, $P(| Q_\Pi(W)[0, t] - t | > \epsilon) \to 0$. It always holds for any collection of partitions with $|| \Pi || \to 0$ but a.s., convergence depends on having $|| \Pi ||$ tend to 0 sufficiently fast.

It has unit quadratic variation per unit time in this way. To put it another way, the situation described above can be expressed in differential form as $dW(t)dW(t) = dt$. By applying Theorem 12.1,

$$V_W^p(T) = \begin{cases} \infty, & p = 1 \\ T, & p = 2 \\ 0, & p > 2 \end{cases}.$$

This concludes that the Brownian motion $\{W(t), t \geq 0\}$ is of unbounded variation and finite quadratic variation for every t.

Many stochastic processes are disguised as Brownian motions. How can we recognise them? Without proof, a classical theorem by Paul Lévy can be stated, which is a useful tool to show that when does a one-dimensional process is a Brownian motion.

Theorem 12.3 (Lévy's Theorem) *Let $\{X(t), t \geq 0\}$ be a continuous martingale with respect to the filtration $\mathcal{F}_t$ with $X(0) = 0$ and $[X, X](t) = t$ for all $t \geq 0$. Then, $\{X(t), t \geq 0\}$ is a Brownian motion.*

Example 12.4 Let $X(t) = \int_0^t sign(W(s))dW(s)$. Here, $sign(x) = 1$, if $x > 0$ and $= -1$ if $x \leq 0$. Prove that $\{X(t), t \geq 0\}$ is a Brownian motion.
Solution: Since

$$dX(t) = sign(W(t))dW(t).$$

$dX(t)dX(t) = dt$. Also, $X(t)$ is a continuous martingale and its quadratic variation is given by

$$Q(X(t), X(t)) = \int_0^t (dX(s))^2 = \int_0^t ds = t$$

then Lévy's theorem implies that $\{X(t), t \geq 0\}$ is a Brownian motion.

When discussing processes, there is a filtration and the process is adapted (refer 10.5 in Chap. 10); the minimal filtration can always be used for the given process. For martingale theory, it is assumed that $\mathcal{F}_0$ is a trivial σ-field, i.e., for all $A \in \mathcal{F}_0$, either $P(A) = 0$ or $P(A) = 1$. As the process will be adapted, this implies X_0 is constant, a.s..

12.2 The Itô Integral

The Itô calculus, often referred to as the stochastic calculus, was developed in 1940 by the Japanese mathematician Kiyosi Itô (1915–2008). Itô calculus is comparable to Newton's classical calculus, which deals with the differentials and integrals of deterministic functions. This section will explain the stochastic integral of the process $\{X(t), t \geq 0\}$ with respect to a Brownian motion, focusing on the definition of the following expression

$$I(t) = \int_0^t X(s)dW(s). \tag{12.2}$$

The above Itô integral cannot be defined as a Lebesgue-Stieltjes integral because the path of Brownian motion is nowhere differentiable.

Definition 12.6 (Itô integral) Let $\{X(t), 0 \leq t \leq T\}$ be a stochastic process. Let $\{X(t), 0 \leq t \leq T\}$ be adapted to the natural filtration $\{\mathfrak{F}_t\}_{t\geq 0}$ of Wiener process $\{W(t), 0 \leq t \leq T\}$, i.e., $X(t)$ be $\mathfrak{F}_t$-measurable. Define

$$I(t) = \int_0^t X(u)dW(u), \quad 0 \leq t \leq T \tag{12.3}$$

a stochastic integral with respect to a Wiener process. The above integral is called an Itô integral.

12.2.1 *Itô Process*

Itô process is a continuous-time stochastic process that generalizes the notion of Brownian motion and is the essential aspect in the formulation of stochastic differential equations.

Definition 12.7 (Itô process) Let $\{W(t), t \geq 0\}$ denote a Brownian motion and $\{\mathfrak{F}_t\}_{t\geq 0}$ denote the corresponding natural filtration. A stochastic process of the type $\{X(t), t \geq 0\}$

$$X(t) = X_0 + \int_0^t \Delta(u)dW(u) + \int_0^t \Theta(u)du \tag{12.4}$$

is called an Itô process. Here X_0 is a non-RV and $\Delta(u)$ and $\Theta(u)$ are adapted processes such that $\Delta(u)$ is mean square integrable.

It is now easy to write a stochastic differential equation form (which will be discussed in Chap. 13) of an Itô process which is

$$dX(t) = \Delta(t)dW(t) + \Theta(t)dt.$$

All stochastic processes with no jumps are actually Itô processes. Itô integral and Brownian motion are examples of Itô processes.

Consider the case if $\{X(t)\}_{t\geq 0}$ and $\{W(t)\}_{t\geq 0}$ are not stochastic processes, but rather deterministic functions. Assuming $f(s)$ is a differential function and $g(s)$ is a smooth function,

$$\int_0^t g(s)df(s) = \int_0^t g(s)f'(s)ds$$

which is a Riemann integral. But what happens if $f(s)$ is not differentiable.

The integral can still be defined. When $f(s)$ is of bounded variation, it can be proven that the integral is well-defined. Further, if the function $g(s)$ is extremely fluctuating at different points in time, the limit may still diverge.

Since $f(s)$ has bounded variation, one can prove that the limit exists as long as $g(s)$ does not vary too much and is given by

$$\int_0^t g(s)df(s) = \lim_{n\to\infty} \sum_{i=0}^{n-1} g(s_i)\left(f(s_{i+1}) - f(s_i)\right).$$

Remember that, in the integral (12.3), $\{X(t), t \geq 0\}$ and $\{W(t), t \geq 0\}$ are stochastic processes, as well as $W(t)$ is of unbounded variation and is nowhere differentiable. Hence, it is different from Riemann integral. Now, rewrite the Itô integral (12.3) in the above form as:

$$I(t) = \lim_{n\to\infty} \sum_{i=0}^{n-1} X_{s_i}\left(W(s_{i+1}) - W(s_i)\right).$$

12.2.1.1 Itô Integrable

A stochastic process $\{X(t), t \geq 0\}$ is called Itô integrable on $[0, t]$ if it satisfies

1. $X(t)$ is adapted, i.e., $X(s)$ is $\mathcal{F}_s$ - measurable, $0 \leq s \leq t$.
2. $\int_0^t E\left[(X(s))^2\right])ds < \infty$.

For instance, the following stochastic integrals exist:

$$\int_0^T (W(t))^2 dW(t), \quad \int_0^T \sin(W(t))dW(t), \quad \int_a^b \frac{\cos(W(t))}{t} dW(t).$$

When a function f has paths of bounded variation, then one can define integrals $\int_0^t g(s)df(s)$ by using partitions Π of $[0, t]$, forming Riemann-Stieltjes sums,

$$\sum_{k=1}^{n} g(t_k)(f(t_k) - f(t_{k-1})),$$

and letting $\| \Pi \| \to 0$ as is calculated in elementary calculus. Also, if f is differentiable, then the integral would be given by $\int_a^b g(s)f'(s)ds$ since $df(s) = f'(s)ds$. Such integrals are also independent of what partitions are used when letting $\| \Pi \| \to 0$; any collection is to be allowed. Since Brownian motion does not have paths of bounded variation, we need to be defined $\int_0^t X(s)dW(s)$ differently, it can not be merely done path by path in the Riemann-Stieltjes sense. Differential equations are recognised in classical calculus as equations consisting of terms of the type dx. These equations are referred as stochastic differential equations if the term dx is substituted by an expression of the form $dX(t)$. A stochastic differential equation has the following structure.

$$dX(t) = \mu(t, X(t))\,dt + \sigma(t, X(t))dW(t), \qquad (12.5)$$

where $\mu(t, x)$ and $\sigma(t, x)$ are given functions. Equation (12.5) can be written in integral form:

$$X(t, \omega) = X_0(\omega) + \int_0^t \mu(s, X(s, \omega))ds + \int_0^t \sigma(s, X(s, \omega))dW(s, \omega). \qquad (12.6)$$

A Riemann integral makes up the first integral. What meaning does the second integral have? At first, this integral is defined as the limit of partial sums, as

$$\sum_{i=1}^{n} \sigma(t_i^*, X(t_i^*))(W(t_i) - W(t_{i-1})) , \qquad t_i^* \in [t_{i-1}, t_i].$$

provided the sum exists. In contrast to Riemann sums, here the value of the summation is determined by the points t_i's picked. The crucial concept in the context of stochastic integrals is to take into account Riemann sums in which the integrand is determined at the left ends of the subintervals, which is:

$$\sum_{i=1}^{n} \sigma(t_{i-1}, X(t_{i-1}))(W(t_i) - W(t_{i-1})).$$

The challenge is to demonstrate that the limit of the aforementioned sum exists in an acceptable sense after noting that the sum of RVs will always be another RV. The stochastic integral is defined in terms of the mean square convergence. The class of stochastic processes is identified that allow for the definition of the Itô integral.

Let L^2 be the collection of all stochastic processes $\{X(t), t \geq 0\}$ that satisfy the following conditions:

(a) With regard to the filtration $\{\mathcal{F}_t\}_{t\geq 0}$, the process $\{X(t), t \geq 0\}$ is continuously measurable. This indicates that, the mapping $(s, \omega) \to X(s, \omega)$ on any set $[0, t] \times \Omega$ is measurable for every t.

(b) $E\left(\int_0^T (X(t))^2 dt\right) < \infty$ for all $T > 0$.

Let define $\mathcal{Y} = \mathcal{Y}[0, T]$ be the class of functions $f(t, \omega) : [0, T] \times \Omega \to \mathbb{R}$ such that $f(t, \omega)$ is jointly measurable, $f(t)$ is $\mathcal{F}_t$-adapted and $E\left[\int_0^T f^2(t)dt\right] < \infty$.

Definition 12.8 (Stochastic integral) Assume $\{X(t), t \geq 0\}$ is a stochastic process in L^2 with a fixed $T > 0$. The stochastic integral of $X(t)$ with respect to Brownian motion $W(t)$ across the interval $[0, T]$ is defined as follows:

$$\int_0^T X(t, \omega)dW(t, \omega) = \lim_{n\to\infty} \sum_{j=1}^{n} X(t_{j-1}, \omega)\left(W(t_j, \omega) - W(t_{j-1}, \omega)\right). \qquad (12.7)$$

where Π_n is a partition of the interval $[0, T]$ such that

$$\Pi_n : 0 = t_0 < t_1 < \cdots < t_n = T$$

with

$$\|\Pi_n\| = \max_{1\le i\le n} (t_i - t_{i-1}) \to 0 \text{ as } n \to \infty.$$

Notation: Let

$$I_T(X)(\omega) = \int_0^T X(t)(\omega)dW(t,\omega) .$$

12.2.2 *Properties of the Itô Integral*

The stochastic integral (12.7) for all $T > 0$ satisfies the following properties:

1. *Zero mean:*

$$E\left[\left(\int_0^T X(t)dW(t)\right)\right] = 0.$$

2. *Itô isometry*:

$$E\left[\left(\int_0^T X(t)dW(t)\right)^2\right] = E\left(\int_0^T (X(t))^2 dt\right).$$

3. *Martingale*: For $t \le T$,

$$E\left[\int_0^T X(s)dW(s) \mid \mathcal{F}_t\right] = \int_0^t X(s)dW(s), \quad a.s.$$

4. *Linearity:* For $\{X(t), t \ge 0\}, \{Y(t), t \ge 0\} \in L^2$,

$$\int_0^T (\alpha X(t) + \beta Y(t))\, dW(t) = \alpha \int_0^T X(t)dW(t) + \beta \int_0^T Y(t)dW(t).$$

5. *Additivity:* For $0 < t < T$,

$$\int_0^T X(t)dW(t) = \int_0^t X(s)dW(s) + \int_t^T X(s)dW(s).$$

6. $var(I(t)) = \int_0^t E([X(s))^2]ds.$
7. Quadratic variation is given by

$$[I, I](t) = \int_0^t (X(s))^2 ds.$$

The Itô integral is a RV $I(t)$, for all $\omega \in \Omega$

$$I(t, \omega) = \int_0^t X(s, \omega)dW(s, \omega) = \lim_{n\to\infty} \sum_{i=0}^{n-1} X(s_i, \omega)(W(s_{i+1}, \omega) - W(s_i, \omega)).$$

Actually the convergence takes place only for a sub-sequence. Itô integral depends on the sample path. Note that

$$dW(t)dW(t) = dt; \quad dtdt = 0; \quad dW(t)dt = 0.$$

8. For a set $A \subset [0, T]$, then

$$\int_0^T X(t)1_A dW(t) = \int_A X(t)dW(t).$$

Proof Here we have merely verified the Itô integral's martingale property. The reader can find proofs of the other properties in Karatzas and Shreve [4]. Consider for any $0 \le t \le T$

$$\begin{aligned} E\left(\int_0^T X(s)dW(s) \mid \mathcal{F}_t\right) &= E\left(\int_0^t X(s)dW(s) + \int_t^T X(s)dW(s) \mid \mathcal{F}_t\right) \\ &= E\left(\int_0^t X(s)dW(s) \mid \mathcal{F}_t\right) + E\left(\int_t^T X(s)dW(s) \mid \mathcal{F}_t\right) \\ &= \int_0^t X(s)dW(s) + E\left(\int_t^T X(s)dW(s) \mid \mathcal{F}_t\right) \\ &= \int_0^t X(s)dW(s) \end{aligned}$$

where the aforementioned equality proceeds from the zero mean property.

Example 12.5 Let $X(t) = \int_0^t e^{W(s)}dW(s)$ denotes an Itô integral. Find the variance of $X(t)$.

Solution: Using the zero mean property of (12.2.2), $E\,(X(t)) = 0$. The variance is determined using Itô isometry

$$E\left(\left[\int_0^t e^{W(s)}dW(s)\right]^2\right) = \int_0^t E\left(e^{2W(s)}\right)ds = \int_0^t e^{2s}ds = \frac{1}{2}\left(e^{2t}-1\right).$$

Example 12.6 Let $I(t) = \int_0^t X(u)dW(u)$, where $\{X(t), t \geq 0\}$ is a Poisson process with $\lambda = 1$. Find (i) $E[I(t)]$ (ii) $var[I(t)]$.
Solution: $I(t)$ is an Itô integral. Hence, $E[I(t)] = 0$.

$$E[(I(t))^2] = E\left[\left(\int_0^t X(u)dW(u)\right)^2\right].$$

By Itô isometry

$$E\left[\left(\int_0^t X(u)dW(u)\right)^2\right] = E\left(\int_0^t (X(u))^2du\right).$$

Therefore,

$$E[(I(t))^2] = E\left(\int_0^t (X(u))^2du\right) = \int_0^t E[(X(u))^2]du = \int_0^t (u+u^2)du = \frac{t^2}{2}+\frac{t^3}{3}.$$

Now, $var[I(t)] = E[(I(t))^2] - (E[I(t)])^2 = \frac{t^2}{2} + \frac{t^3}{3}$.

Remark 12.2 The following Novikov's condition

$$E\left[\exp\left(\frac{1}{2}\int_0^T u^2(s)ds\right)\right] < \infty$$

is a sufficient to guarantee that the process

$$M(t) := \exp\left\{\int_0^t u(s)dW(s) - \frac{1}{2}\int_0^t u^2(s)ds\right\}$$

is a martingale on $[0, T]$.

Example 12.7 Let $M(t) = \int_0^t W(u)dW(u)$. Find $dM(t)dM(t)$.
Solution: $dM(t) = W(t)dW(t)$. Hence, $dM(t)dM(t) = (W(t))^2dt$.

Example 12.8 Let $\{W(t), t \geq 0\}$ be a Brownian motion and let $g(s)$ be a non-random function of time. Define $I(t) := \int_0^t g(s)dW(s)$. Prove that, for each $t \geq 0$, the RV $I(t)$ is normal distributed with mean 0 and variance $\int_0^t g^2(s)ds$.
Solution: It can be seen that $I(t)$ is a martingale and hence $E[I(t)] = I(0) = 0$. Moreover, thanks to Itô isometry,

$$\text{var}[I(t)] = E\left[(I(t))^2\right] = \int_0^t g^2(s)ds.$$

It remains to show that $I(t)$ is normal distributed. To do so, it is shown that $I(t)$ has moment-generating function of a normal RV with mean 0 and variance $\int_0^t g^2(s)ds$, i.e., it can be shown that

$$E\left[e^{uI(t)}\right] = \exp\left\{\frac{1}{2}u^2 \int_0^t g^2(s)ds\right\} \quad \forall\ u \in \mathbb{R}$$

$$\Longleftrightarrow E\left[\exp\left\{uI(t) - \frac{1}{2}u^2 \int_0^t g^2(s)ds\right\}\right] = 1.$$

This can be written as

$$E\left[\exp\left\{\int_0^t ug(s)dW(s) - \frac{1}{2}\int_0^t (ug(s))^2 ds\right\}\right] = 1. \tag{12.8}$$

In view of above remark, the process

$$Z(t) := \exp\left\{\int_0^t ug(s)dW(s) - \frac{1}{2}\int_0^t (ug(s))^2 ds\right\}$$

is a martingale and hence $E[Z(t)] = Z(0) = 1$ which gives us (12.8). This completes the solution.

Example 12.9 Consider the Itô Integral for simple integrand function where price may be negative. Let $W(t)$ be the price per share of an asset at time (position) t, $X(t)$ be the number of shares taken in the asset at a time t and t_i be the trading dates of an asset. Assume $X(t)$ be the simple process. It means $X(t)$ is a constant in each $[t_i, t_{i+1}]$. The gain/loss from trading at each time t can be viewed as $I(t)$.
Since $\{X(t), 0 \leq t \leq T\}$ is a simple process, it can be written as

$$X(t, \omega) = \phi_0(\omega)1_{\{0\}}(t) + \lim_{n\to\infty} \sum_{i=1}^{n} \phi_i(\omega)1_{(t_{i-1}, t_i]}(t)\ \forall\ \omega \in \Omega$$

where ϕ_i are bounded RVs such that ϕ_0 is $\mathcal{F}_0$-measurable, ϕ_i is $\mathcal{F}_{t_{i-1}}$-measurable and $0 = t_0 < t_1 < \ldots < t_n = T, n \in \mathbb{N}$. For a simple process $\{X(t), t \in [0, T]\}$, the stochastic integral $I(t)$ is defined by

$$I(t) = \int_0^t X_s dW(s) = \lim_{n\to\infty} \sum_{i=1}^{n} \phi_i(W(t_i) - W(t_{i-1})).$$

In particular $X(t) = 1,\ t \in [0, T]$, hence $\phi_i = 1,\ 1 \leq i \leq n,\ n \in \mathbb{N}$, for $t \in [0, T]$

$$I(t) = \int_0^t dW(s) = \lim_{n\to\infty} \sum_{i=1}^{n} (W(t_i) - W(t_{i-1})) = W(t).$$

Example 12.10 Evaluate the stochastic integral

$$I(T) = \int_0^T W(t)dW(t).$$

Solution: The solution will be provided by two different methods.

Let $0 = t_0 < t_1 < t_2 < \cdots < t_n = T$ be a partition of the interval $[0, T]$. From the definition of the stochastic integral:

$$\int_0^T W(t)dW(t) = \lim_{n\to\infty} \sum_{i=1}^{n} W(t_{i-1}) \left(W(t_i) - W(t_{i-1})\right) .$$

The ambiguity resulting from different choices of the intermediate points s_i in each sub-interval $[t_{i-1}, t_i]$ can be removed by insisting that the approximation of the integrand should consist only of process adapted to the underlying filtration $\mathfrak{F}_t$. This amount to taking $s_i = t_{i-1}$ for each i.

Method I: Use the identity

$$a(b - a) = \frac{1}{2}(b^2 - a^2) - \frac{1}{2}(b - a)^2.$$

$$\begin{aligned}
\int_0^T W(t)dW(t) &= \lim_{n\to\infty} \sum_{i=1}^{n} \left[\frac{1}{2}\left(W(t_i)^2 - W(t_{i-1})^2\right) - \frac{1}{2}\left(W(t_i) - W(t_{i-1})\right)^2\right] \\
&= \frac{1}{2} \lim_{n\to\infty} \sum_{i=1}^{n} \left(W(t_i)^2 - W(t_{i-1})^2\right) - \frac{1}{2} \lim_{n\to\infty} \sum_{i=1}^{n} \left(W(t_i) - W(t_{i-1})\right)^2 \\
&= \frac{1}{2} W(T)^2 - \frac{1}{2} T.
\end{aligned}$$

Method II. Note that, for each i, $W(t_i)$ and $W(t_{i+1}) - W(t_i)$ are independent variables and are having normal distributions. Let Π be the set of all finite subdivisions of π of the interval $[0, T]$ with $0 = t_0 < t_1 < \ldots < t_n = T$.

$$
\begin{aligned}
Q_{\Pi} &= \sum_{i=0}^{n-1} (W(t_{i+1}) - W(t_i))^2 \\
&= \sum_{i=0}^{n-1} \left((W(t_{i+1})^2 - W(t_i)^2\right) - 2\, W(t_i)(W(t_{i+1}) - W(t_i)) \\
&= (W(T))^2 - (W(0))^2 - 2 \sum_{i=0}^{n-1} W(t_i)(W(t_{i+1}) - W(t_i)).
\end{aligned}
$$

Since, $\lim_{n\to\infty} Q_{\Pi} = T$ which implies

$$
\int_0^T W(t) dW(t) = \frac{(W(T))^2 - T}{2}.
$$

It can be seen that there is an extra term $\frac{T}{2}$, unlike the Riemann integral. This is due to the finite quadratic variation of the Brownian motion. The integral above is defined not only for the upper limit of integration T, but for all upper limits of integration between 0 and t. The Itô integral is constructed similarly to the Stieltjes integral. It is integral with respect to a random function, more accurately the path of a Brownian motion, rather than integrity with respect to a deterministic function (as in Stieltjes integral).

Example 12.11 Evaluate

$$
\int_0^t W(1) dW(t), \quad 0 \le t \le 1.
$$

Solution: Note that, $W(1)$ is not adapted to the filtration $\sigma\{W(s), 0 < s \le t\}, 0 \le t < 1$. Hence, Itô integral does not exist. This example shows that the assumption that the integrand is adapted to the filtration $\{\mathcal{F}_t\}_{t\ge 0}$ is needed to have the existence of the Itô integral.

12.3 Itô's Formula for Diffusion Processes

The most important result in the theory of stochastic integration is Itô's formula. This is also known as the change of variables formula.

12.3.1 Itô's Formula Version 1

Theorem 12.4 (Itô's Formula) *Let $f : \mathbb{R} \to \mathbb{R}$ represents a function which is twice-differentiable and let $W = \{W(t), t \geq 0\}$ denotes a Brownian motion that begins at x_0, i.e., $W(0) = x_0$. Then*

$$f(W(t)) = f(W(0)) + \int_0^t f'(W(s))\,dW(s) + \frac{1}{2}\int_0^t f''(W(s))\,ds.$$

In the differential form:

$$df(W(t)) = f'(W(t))dW(t) + \frac{1}{2}f''(W(t))dt.$$

Proof Let $0 = t_0 < t_1 < \cdots < t_n = t$ be a partition of $[0, t]$ for fixed $t > 0$. Taylor's theorem provides us with

$$\begin{aligned} f(W(t)) - f(W(0)) &= \sum_{j=1}^{n}\left[f\left(W(t_j)\right) - f\left(W(t_{j-1})\right)\right] \\ &= \sum_{j=1}^{n}\Bigg[f'(W(t_{j-1}))\left(W(t_j) - W(t_{j-1})\right) \\ &\quad + \frac{1}{2}f''(W(t_{j-1}))\left(W(t_j) - W(t_{j-1})\right)^2 + \cdots\Bigg]. \end{aligned}$$

Using the limit $n \to \infty$, it is discovered that the right-hand side's first sum converges to the Itô integral and because of mean square convergence, the second sum converges to $\frac{1}{2}\int_0^t f''(W(s))ds$. Therefore, we obtain

$$f(W(t)) - f(W(0)) = \int_0^t f'(W(s))dW(s) + \frac{1}{2}\int_0^t f''(W(s))ds.$$

Thus

$$f(W(t)) = f(x_0) + \int_0^t f'(W(s))dW(s) + \frac{1}{2}\int_0^t f''(W(s))ds.$$

Example 12.12 Show that $\{(W(t))^3, t \geq 0\}$ is an Itô process and find the stochastic differential of $(W(t))^3$.

Solution: Let $f(x) = x^3$. Apply Itô's formula,

$$(W(T))^3 = \frac{1}{2}\int_0^T 6\, W(t)dt + \int_0^T 3(W(t))^2\, dW(t).$$

One can prove that $3W(t)$ and $3(W(t))^2$ are adapted processes. Further, $3(W(t))^2$ is mean square integrable. Hence, $(W(t))^3$ is an Itô process. In differential form

$$d(W(t))^3 = 3\, W(t)dt + 3(W(t))^2 dW(t).$$

Example 12.13 Let us apply Itô's formula to $(W(t))^4$. Obtain

$$d((W(t))^4) = 4(W(t))^3 dW(t) + \frac{1}{2}12(W(t))^2 dt = 4(W(t))^3 dW(t) + 6(W(t))^2 dt,$$

namely, written in an integral form

$$(W(t))^4 = 4\int_0^t (W(s))^3 dW(s) + 6\int_0^t (W(s))^2 ds.$$

Taking expectations of both sides and recalling that Itô integral is a martingale, obtain

$$E\left[\int_0^t (W(s))^2 ds\right] = \frac{1}{6}E\left[(W(t))^4\right] = \frac{1}{2}t^2.$$

12.3.2 Itô's Formula Version 2 or Itô-Doeblin Formula

Let $f(t, x)$ have continuous partial derivatives of at least second order. Start with $X(t) = f(t, W(t))$. Therefore, expression of $f(t, x)$ using Taylor's equation up to the terms with the partial derivatives of order two is given by

$$\begin{aligned} f(t+\Delta t, x+\Delta x) = f(t,x) &+ \frac{\partial f(t,x)}{\partial t}dt + \frac{\partial f(t,x)}{\partial x}dx \\ &+ \frac{1}{2}\left[\frac{\partial^2 f(t,x)}{\partial t^2}(dt)^2 + 2\frac{\partial^2 f(t,x)}{\partial t \partial x}dtdx + \frac{\partial^2 f(t,x)}{\partial x^2}(dx)^2\right] \\ &+ \ldots \end{aligned}$$

$$\begin{aligned} f(t+\Delta t, W(t+\Delta t)) = f(t, W(t)) &+ \frac{\partial f(t, W(t))}{\partial W(t)}dW(t) \\ &+ \frac{\partial f(t, W(t))}{\partial t}dt + \frac{1}{2}[\ldots]. \end{aligned}$$

Using the so-called Itô multiplication table $dtdt = 0, dtdW(t) = 0$ and $dW(t)dW(t) = dt$.

$$f(t + \Delta t, W(t + \Delta t)) = f(t, W(t)) + \frac{\partial f(t, W(t))}{\partial W(t)} dW(t) + \left[\frac{\partial f(t, W(t))}{\partial t} + \frac{1}{2}\frac{\partial^2 f(t, W(t))}{\partial (W(t))^2}\right] dt.$$

An easy way to remember the Itô's formula is to write down the Taylor's equation of $f(t, x)$ up to the terms with the partial derivatives of order two, and using the so-called Itô multiplication table $dtdt = 0, dtdW(t) = 0$ and $dW(t)dW(t) = dt$. In general, Itô process $\{X(t), t \geq 0\}$, where $X(t)$ is given by

$$X(t) = X(0) + \int_0^t A_{(1)}(s)ds + \int_0^t A_{(2)}(s)dW(s),$$

where $\{A_{(1)}(t), t \geq 0\}$ and $\{A_{(2)}(t), t \geq 0\}$ are stochastic processes which are adapted to filtration $\{\mathcal{F}(t), t \geq 0\}$ and $A_{(2)}(t)$ is mean square integrable.

$$df(t, X(t)) = \frac{\partial f(t, X(t))}{\partial X(t)} dX(t) + \frac{\partial f(t, X(t))}{\partial t} dt + \frac{1}{2}\frac{\partial^2 f(t, X(t))}{\partial (X(t))^2} dX(t)dX(t).$$

Example 12.14 A stochastic process $\{S(t), t \geq 0\}$ is governed by,

$$dS(t) = \mu S(t)dt + \sigma S(t)dW(t).$$

Find the SDE of $\sqrt{S(t)}$.
Solution: Let $f(t, x) = \sqrt{x}$. Then $f_t(t, x) = 0$, $f_x(t, x) = \frac{1}{2\sqrt{x}}$, $f_{xx}(t, x) = \frac{-1}{4x^{3/2}}$
By Itô-Doeblin formula

$$\begin{aligned} df(t, S(t)) &= \frac{\partial f(t, S(t))}{\partial S(t)} dS(t) + \frac{\partial f(t, S(t))}{\partial t} dt + \frac{1}{2}\frac{\partial^2 f(t, S(t))}{\partial (S(t))^2} dS(t)dS(t) \\ &= \frac{1}{2\sqrt{S(t)}} [\mu S(t)dt + \sigma S(t)dW(t)] - \frac{1}{8S(t)^{3/2}} \sigma^2 S^2(t)dt \\ &= \frac{1}{2}\left(\mu - \frac{1}{4}\sigma^2\right)\sqrt{S(t)}dt + \frac{\sigma}{2}\sqrt{S(t)}dW(t). \end{aligned}$$

Example 12.15 Let $h(s)$ be a real valued function which is differentiable such that $\int_0^t h^2(s)ds < \infty$. Use Itô's formula to prove that $\int_0^t h(s)dW(s) = h(t)W(t) - \int_0^t h'(s)W(s)ds$. Also, find the distribution of $\int_0^t h(s)dW(s)$.
Solution: Take $f(t, x) = h(t)x$. Then $f_t = h'(t)W(t)$, $f_x = h(t)$ and $f_{xx} = 0$.
By Itô's formula,

$$d(h(t)W(t)) = h'(t)W(t)dt + h(t)dW(t).$$

Integrating from 0 to t, we get the required result. Now, $h(s)$ is a deterministic function. Hencc solving similarly as Example 12.8, $\int_0^t h(s)dW(s)$ follows a normal distribution with mean 0 and variance $\int_0^t h^2(s)ds$.

Example 12.16 Solve the SDE

$$dX(t) = X(t)dW(t), \text{ with } X(0) = 1.$$

Solution: Let

$$f(t,x) = e^x e^{\frac{-t}{2}};\ f_t'(t,x) = \frac{-1}{2}e^x e^{\frac{-t}{2}};\ f_x'(t,x) = e^x e^{\frac{-t}{2}};\ f_{xx}''(t,x) = e^x e^{\frac{-t}{2}}.$$

By using Itô-Doeblin formula,

$$\begin{aligned} df(t, W(t)) &= \left(f_t'(t, W(t)) + \frac{1}{2} f_{xx}'(t, W(t)) \right) dt + f_x'(t, X(t))dW(t) \\ &= \left(\frac{-1}{2} f(t, W(t)) + \frac{1}{2} f(t, W(t)) \right) dt + f(t, W(t))dW(t) \\ &= f(t, W(t))dW(t). \end{aligned}$$

Hence, the solution is $X(t) = f(t, W(t)) = e^{W(t)}e^{-\frac{t}{2}}$. It is an Itô process. Hence, a martingale.

Example 12.17 Consider the process given by,

$$S(t) = S(0)\exp\left\{ \int_0^t \sigma(s)dW(s) + \int_0^t \left(\alpha(s) - \frac{1}{2}\sigma^2(s) \right) ds \right\}$$

where $S(0)$ is non-random and positive and $\alpha(\cdot)$ and $\sigma(\cdot)$ are adapted processes such that integrals are well-defined. Show that $S(t)$ is an Itô process that satisfies the differential form

$$dS(t) = \alpha(t)S(t)dt + \sigma(t)S(t)dW(t).$$

Solution: First define an Itô process

$$X(t) = \int_0^t \sigma(s)dW(s) + \int_0^t \left(\alpha(s) - \frac{1}{2}\sigma^2(s) \right) ds.$$

Then by the given condition $S(t) = S(0)e^{X(t)}$. $S(t)$ can be written in the form $S(t) = f(X(t))$ where $f(x) = S(0)e^x$. Note that $X(t)$ in the differential form is given by

$$dX(t) = \left(\alpha(t) - \frac{1}{2}\sigma^2(t)\right) dt + \sigma(t)dW(t).$$

By Itô's formula,

$$\begin{aligned} dS(t) &= df(X(t)) \\ &= S(0)\left(e^{X(t)}\left(\alpha(t) - \frac{1}{2}\sigma^2(t)\right) + \frac{1}{2}e^{X(t)}\sigma^2(t)\right) dt + S(0)e^{X(t)}\sigma(t)dW(t) \\ &= S(0)e^{X(t)}\alpha(t)dt + S(0)e^{X(t)}\sigma(t)dW(t) \\ &= \alpha(t)S(t)dt + \sigma(t)S(t)dW(t). \end{aligned}$$

Observe that if $\alpha(t) = 0$, then $S(t)$ is given by $S(t) = S(0) + \int_0^t \sigma(s)S(s)dW(s)$. Since the second term in the right hand side is a martingale and $S(0)$ is non-random, it can be concluded that $S(t)$ is a martingale provided $E\left[\int_0^T \sigma^2(s)(S(s))^2 ds\right] < \infty$ for each $T > 0$.

Example 12.18 Let $X(t)$ and $Y(t)$ be real-valued Itô processes. Then show that $X(t)Y(t)$ is again an Itô process and its differential form is given by

$$d(X(t)Y(t)) = X(t)dY(t) + Y(t)dX(t) + dX(t)dY(t).$$

The above formula is known as Itô product rule. To check this, let $X(t)$ and $Y(t)$ be Itô processes of the form

$$dX(t) = a(t)dt + \sigma(t)dW(t), \quad dY(t) = b(t)dt + \gamma(t)dW(t).$$

Then by applying Itô's formula to the function $f(x) = x^2$,

$$\begin{aligned} d\left[(X(t) + Y(t))^2\right] &= \left\{2(X(t) + Y(t))(a(t) + b(t)) + (\sigma(t) + \gamma(t))^2\right\} dt \\ &\quad + 2(X(t) + Y(t))(\sigma(t) + \gamma(t))dW(t) \\ dX(t)^2 &= \left\{2X(t)a(t) + \sigma^2(t)\right\} dt + 2X(t)\sigma(t)dW(t) \\ dY(t)^2 &= \left\{2Y(t)b(t) + \gamma^2(t)\right\} dt + 2Y(t)\gamma(t)dW(t). \end{aligned}$$

The above equations are used along with the fact that $xy = \frac{1}{2}\left((x + y)^2 - x^2 - y^2\right)$,

$$\begin{aligned} d[X(t)Y(t)] &= \{X(t)b(t) + Y(t)a(t) + \sigma(t)\gamma(t)\}dt \\ &\quad +\{X(t)\gamma(t) + Y(t)\sigma(t)\}dW(t) \\ &= X(t)\{b(t)dt + \gamma(t)dW(t)\} \\ &\quad +Y(t)\{a(t)dt + \sigma(t)dW(t)\} + \sigma(t)\gamma(t)dt \\ &= X(t)dY(t) + Y(t)dX(t) + dX(t)dY(t). \end{aligned}$$

12.4 Itô's Formula for Jump-Diffusion Processes

The Itô's formula, also known as the Itô's lemma, is a fundamental result in stochastic calculus that provides a formula for the differential of a function of a stochastic process. The Itô's formula has extensions to jump processes, such as compound Poisson processes or more general jump-diffusion processes. For jump processes, the Itô's formula takes into account both the continuous and jump components of the process.

Definition 12.9 (Jump process) Let $(\Omega, \mathcal{F}, \{\mathcal{F}_t\}_{t\geq 0}, P)$ be a filtered probability space and $\{W(t), t \geq 0\}$ be a Brownian motion. Let $X = \{X(t), t \geq 0\}$ be a process such that

$$X(t) = X_0 + \int_0^t \Gamma(s)dW(s) + \int_0^t \theta(s)ds + J(t)$$

with

1. X_0 as deterministic initial condition,
2. $I(t) = \int_0^t \Gamma(s)dW(s)$ is an Itô integral called the Itô integral part,
3. $R(t) = \int_0^t \theta(s)ds$ is the Riemann-integral of the process θ, called the Riemann-integral part,
4. $J(t)$ is an adapted right-continuous pure jump process with $J(0) = 0$.

This process X is called a jump process with continuous part $X^c(t) = X_0 + I(t) + R(t)$. The quadratic variation of $X^c(t)$ is

$$\left[X^c, X^c\right](t) = \int_0^t \Gamma^2(s)ds.$$

Definition 12.10 (Stochastic integral of jump process) Let X be a jump process and Φ be an adapted process. The stochastic integral of Φ with respect to X is defined as

$$\int_0^t \Phi(s)dX(s) = \int_0^t \Phi(s)\Gamma(s)dW(s) + \int_0^t \Phi(s)\theta(s)ds + \sum_{0\leq s\leq t} \Phi(s)\triangle J(s).$$

In differential notation, write

$$\begin{aligned}\Phi(t)dX(t) &= \Phi(t)dI(t) + \Phi(t)dR(t) + \Phi(t)dJ(t)\\ &= \Phi(t)dX^c(t) + \Phi(t)dJ(t).\end{aligned}$$

Example 12.19 Let$\{X(t), t \geq 0\}$ be a stochastic process such that $X(t) = N(t) - \lambda t$, where $\{N(t), t \geq 0\}$ is a Poisson process with intensity λ. Therefore, $I(t) = 0$, $R(t) = -\lambda t = X_t^c$, $J(t) = N(t)$. Let

$$\Phi(s) = \Delta N(s) = 1_{\{\Delta N(s) \neq 0\}} = (N(s) - N(s-)),$$

$$\int_0^t \Phi(s) dX_s^c = -\lambda \int_0^t \Phi(s) ds = 0.$$

Since $\Phi(s) = 0$ except for finitely many points.

$$\int_0^t \Phi(s) dN(s) = \sum_{0 \le s \le t} \Phi(s) \Delta N(s) = \sum_{0 \le s \le t} (\Delta N(s))^2 = N(t).$$

Corollary 12.1 *Let $\{W(t), t \geq 0\}$ be a Brownian motion and take a compensated Poisson process $\{M(t), t \geq 0\}$ where $M(t) = N(t) - \lambda t$. Then $[W, M](t) = 0$ for $t = 0$.*

Theorem 12.5 (Itô's formula for Jump Process) *Let $\{X(t), t \geq 0\}$ denote a jump process and assume that $g(x)$ is a function for which the derivatives $g'(x)$ and $g''(x)$ are defined and are continuous. Then*

$$\begin{aligned} g(X(t)) =& g(X(0)) + \int_0^t g'(X(s)) dX^c(s) \\ &+ \frac{1}{2} \int_0^t g''(X(s)) dX^c(s) dX^c(s) + \sum_{0 < s \le t} [g(X(s)) - g(X(s-))]. \end{aligned} \tag{12.9}$$

Proof The path of the process $\{X(t), t \geq 0\}$ can be fixed by fixing $\omega \in \Omega$, and the jump times in $[0, t)$ of this path can be represented $0 < \tau_1 < \tau_2 < \cdots < \tau_{n-1} < t$. Set $\tau_0 = 0$ ($\tau_0 = 0$ is not a jump time) and $\tau_n = t$ ($\tau_n = t$ is possibly or possibly not a jump time) respectively. When u and v ($u < v$) are in the same interval (τ_j, τ_{j+1}), then there is no jump between the times u and v, and applying Itô-Doeblin formula for continuous processes, we get

$$g(X(v)) - g(X(u)) = \int_u^v g'(X(s)) dX^c(s) + \frac{1}{2} \int_u^v g''(X(s)) dX^c(s) dX^c(s).$$

Letting $u \downarrow \tau_j$ and $v \uparrow \tau_{j+1}$ and using the right-continuity of X, it is concluded that

$$\begin{aligned} g\left(X(\tau_{j+1}-)\right) - g\left(X(\tau_j)\right) =& \int_{\tau_j}^{\tau_{j+1}} g'(X(s)) dX^c(s) \\ &+ \frac{1}{2} \int_{\tau_j}^{\tau_{j+1}} g''(X(s)) dX^c(s) dX^c(s). \end{aligned} \tag{12.10}$$

Adding the jump in $g(X)$ at time τ_{j+1} into (12.10), obtaining thereby

$$g\left(X(\tau_{j+1})\right) - g\left(X(\tau_j)\right) = \int_{\tau_j}^{\tau_{j+1}} g'(X(s))dX^c(s) + g\left(X(\tau_{j+1})\right) - g\left(X(\tau_{j+1}-)\right) + \frac{1}{2}\int_{\tau_j}^{\tau_{j+1}} f''(X(s))dX^c(s)dX^c(s).$$

Summing over $j = 0, 1, \ldots, n-1$, and take $n \to \infty$,

$$g(X(t)) - g(X(0)) = \int_0^t g'(X(s))dX^c(s) + \frac{1}{2}\int_0^t g''(X(s))dX^c(s)dX^c(s) + \sum_{0<s\leq t} \left[g\left(X(s)\right) - g\left(X(s-)\right)\right]$$

which is (12.9).

Definition 12.11 (Jump-diffusion process) A jump-diffusion process X is the sum of a Brownian motion and of an independent compound Poisson process. That is, for $t \geq 0$

$$X(t) = \alpha(t) + \beta W(t) + \sum_{i=0}^{N(t)} Y_i,$$

where $\{W(t), t \geq 0\}$ is a Brownian motion, Y_i is a normal distributed RV with mean μ and variance σ^2 and $\{N(t), t \geq 0\}$ is a Poisson process.

12.4.1 *Itô's formula for The Poisson Process*

The next theorem gives the simplest instance of the Itô's formula for jump process, in the case of a Poisson process $\{N(t), t \geq 0\}$.

Theorem 12.6

$$f\left(N(t)\right) = f(0) + \int_0^t \left(f\left(N(s)\right) - f\left(N(s-)\right)\right) dN(s), \quad t \geqslant 0,$$

where $N(s-)$ denotes the left limit $N(s-) = \lim_{h\to 0} N(s-h)$.

Proof Note that

$$N(s) = N(s-) + 1 \text{ if } dN(s) = 1 \text{ and } k = N(T_k) = 1 + N(T_k-), \quad k \geqslant 1.$$

Hence, there is the telescopic sum

$$
\begin{aligned}
f(N(t)) &= f(0) + \sum_{k=1}^{N(t)} (f(k) - f(k-1)) \\
&= f(0) + \sum_{k=1}^{N(t)} (f(N(T_k)) - f(N(T_k-))) \\
&= f(0) + \int_0^t (f(N(s)) - f(N(s-)))\, dN(s).
\end{aligned}
$$

where $N(s-)$ denotes the left limit $N(s-) = \lim_{h \to 0} N(s-h)$.

Itô Multiplication with Jumps

For a stochastic process $\{X(t), t \geq 0\}$ given by

$$X(t) = \int_0^t u_s dW(s) + \int_0^t v_s ds + \int_0^t \eta_s dN(s), \quad t \geqslant 0,$$

the Itô's formula for jump process reads

$$
\begin{aligned}
f(X(t)) = {} & f(X(0)) + \int_0^t u_s f'(X(s)) dW(s) + \frac{1}{2} \int_0^t |u_s|^2 f''(X(s))\, dW(s) \\
& + \int_0^t v_s f'(X(s))\, ds + \int_0^t \eta_s (f(X(s)) - f(X(s-)))\, dN(s) \\
= {} & f(X(0) + \int_0^t u_s f'(X(s)) dW(s) + \frac{1}{2} \int_0^t |u_s|^2 f''(X(s))\, dW(s) \\
& + \int_0^t v_s f'(X(s))\, ds + \int_0^t \eta_s (f(X(s)) - f(X(s-)))\, dN(s).
\end{aligned}
$$

Given two Itô processes $\{(X(t)), t \geq 0\}$ and $\{(Y(t)), t \geq 0\}$written in differential notation as

$$dX(t) = u_t dW(t) + v_t dt + \eta_t dN(t),$$

and

$$dY(t) = a_t dW(t) + b_t dt + c_t dN(t),$$

the Itô's formula for jump processes can also be written as

$$d(X(t)Y(t)) = X(t)dY(t) + Y(t)dX(t) + dX(t)dY(t).$$

12.5 Itô's formula for Multivariate Processes

It is straightforward to extend our definitions to Brownian motions taking values in $\mathbb{R}^m$.

Definition 12.12 (Multi-variate Brownian motion) An $\mathbb{R}^m$-valued stochastic process $\mathbf{W}(t) = (W_1(t), W_2(t), \ldots, W_m(t))$ is an m-dimensional Wiener process (or Brownian motion) provided

(a) each $W_i(t)$ is a one-dimensional Brownian motion
(b) for $i \neq j$, the processes $W_i(t)$ and $W_j(t)$ are independent.

For $i \neq j$, define the sample cross variation of W_i and W_j on $[0, T]$, denoted by $\left[W_i, W_j\right](T)$, as

$$\begin{aligned}\left[W_i, W_j\right](T) &= \lim_{\|\Pi\|\to 0} \sum_{k=0}^{n} \left(W_i\left(t_{k+1}\right) - W_i\left(t_k\right)\right)\left(W_j\left(t_{k+1}\right) - W_j\left(t_k\right)\right) \\ &= \lim_{\|\Pi\|\to 0} C_{\Pi_{ij}},\end{aligned}$$

where $C_{\Pi_{ij}} = \left(W_i\left(t_{k+1}\right) - W_i\left(t_k\right)\right)\left(W_j\left(t_{k+1}\right) - W_j\left(t_k\right)\right)$. Observe that, since increments of W_i and W_j are independents and both have mean zero, thus we get that $E\left[C_{\Pi_{ij}}\right] = 0$. Again, by using independent increments of $W_i(t)$ and $W_j(t)$, we obtain

$$\begin{aligned}\operatorname{var}\left(C_{\Pi_{ij}}\right) = E\left[C_{\Pi_{ij}}^2\right] &= E\left[\sum_{k=0}^{n-1} \left(W_i\left(t_{k+1}\right) - W_i\left(t_k\right)\right)^2 \left(W_j\left(t_{k+1}\right) - W_j\left(t_k\right)\right)^2\right] \\ &= \sum_{k=0}^{n-1} E\left[\left(W_i\left(t_{k+1}\right) - W_i\left(t_k\right)\right)^2\right] E\left[\left(W_j\left(t_{k+1}\right) - W_j\left(t_k\right)\right)^2\right] \\ &= \sum_{k=0}^{n-1} \left(t_{k+1} - t_k\right)^2 \\ &\leq \| \Pi \| \, T \to 0 \text{ as } \| \Pi \| \to 0.\end{aligned}$$

This implies that the sample cross variation of $W_i(t)$ and $W_j(t)$ is zero, i.e., $\left[W_i, W_j\right](T) = 0$ for $i \neq j$.

Definition 12.13 Define the following:
a) An $M^{n\times m}$-valued stochastic process $\boldsymbol{G}(t) = \left[G_{ij}(t)\right]$ belongs to $\mathcal{Y}_{n\times m}(0, T)$ if

$$G_{ij}(t) \in \mathcal{Y}(0, T) \quad \forall 1 \leq i \leq n, 1 \leq j \leq m.$$

b) An $\mathbb{R}^n$-valued stochastic process $\mathbf{F}(t) = (F_1(t), F_2(t), \ldots, F_n(t))$ belongs to $L_n^1(0, T)$ if

$$F_i(t) \in L^1(0, T), \quad 1 \leq i \leq n.$$

Recall $L^1(0, T)$ is the space of $\mathcal{F}_t$-adapted, jointly measurable real-valued stochastic process $F(t)$ such that $E\left[\int_0^t |F|dt\right] < \infty$.

Definition 12.14 (Stochastic integral of multidimensional simple process) Let $\mathbf{W}(t) = (W_1(t), W_2(t), \ldots, W_m(t))$ be an m-dimensional Brownian motion. Then for any $\boldsymbol{G}(t) \in \mathcal{Y}_{n\times m}(0, T)$, define the stochastic integral $\int_0^T \boldsymbol{G}(t)d\mathbf{W}(t)$ as an $\mathbb{R}^n$-valued RV whose i-th component is given by

$$\sum_{j=1}^{m} \int_0^T G_{ij}(t)dW_j(t), \quad 1 \le i \le n.$$

Approximation by simple processes, one can arrive at the following lemma.

Lemma 12.1 *Let* $\mathbf{W}(t) = (W_1(t), W_2(t), \ldots, W_m(t))$ *be an* m*-dimensional Brownian motion and* $\mathbf{G}(t) \in \mathcal{Y}_{n\times m}(0, T)$*. Then*
i)

$$E\left[\int_0^T \mathbf{G}(t)d\mathbf{W}(t)\right] = 0$$

ii)

$$E\left[\left|\int_0^T \mathbf{G}(t)d\mathbf{W}(t)\right|^2\right] = E\left[\int_0^T \|\boldsymbol{G}(t)\|^2 dt\right]$$

where $\|\boldsymbol{G}(t)\|^2 := \sum_{1\le i\le n, 1\le j\le m} \left|G_{ij}(t)\right|^2$.

Definition 12.15 (Multidimensional Itô process) Let $\mathbf{W}(t)$ be an m-dimensional Brownian motion and $\{\mathcal{F}_t\}_{t\ge 0}$ be its associated filtration. An $\mathbb{R}^n$-valued Itô process is a stochastic process $\mathbf{X}(t) = (X_1(t), X_2(t), \ldots, X_n(t))$ of the form

$$\mathbf{X}(r) = \mathbf{X}(s) + \int_0^t \mathbf{F}(s)ds + \int_0^t \boldsymbol{G}(s)d\mathbf{W}(s),$$

for some

$$\mathbf{F}(t) = (F_1(t), F_2(t), \ldots, F_n(t)) \in \mathbb{L}_n^1(0, T)$$

and

$$\boldsymbol{G}(t) = \left[G_{ij}(t)\right] \in \mathcal{Y}_{n\times m}(0, T)$$

and for all $0 \le s \le r \le T$. $\mathbf{X}(\cdot)$ has the stochastic differential

$$d\mathbf{X}(t) = \mathbf{F}(t)dt + \boldsymbol{G}(t)d\mathbf{W}(t).$$

This means that,

$$dX_i(t) = F_i(t)dt + \sum_{j=1}^{m} G_{ij}(t)dW_j(t), \quad 1 \le i \le n.$$

Like in one-dimensional case, the family of Itô processes are stable under smooth maps.

Approximation by step/elementary processes, one can arrive at the following lemma.

We provide the Itô's formula for two-variable functions. Consider the following two-dimensional process:

$$dX(t) = \mu_t dt + \sigma_t dW_1(t) \tag{12.11}$$

$$dY(t) = \alpha_t dt + \beta_t dW_2(t) \tag{12.12}$$

where $\{W_1(t), t \ge 0\}$ and $\{W_2(t), t \ge 0\}$ are two Brownian motions with their covariance given by

$$cov\,(W_1(t), W_2(t)) = E\,(W_1(t) \cdot W_2(t)) = \rho t \tag{12.13}$$

where ρ is the correlation coefficient between the two Brownian motions. Assume that $Z_t = g(t, X(t), Y(t))$ and that $g(t, x, y)$ is a twice-differentiable function. Then $\{Z_t, t \ge 0\}$ is an Itô process as well, and satisfies:

$$\begin{aligned} dZ_t = {} & \frac{\partial g}{\partial t}(t, X(t), Y(t))dt + \frac{\partial g}{\partial x}(t, X(t), Y(t))dX(t) + \frac{\partial g}{\partial y}(t, X(t), Y(t))dY(t) \\ & + \frac{1}{2}\frac{\partial^2 g}{\partial x^2}(t, X(t), Y(t))(dX(t))^2 + \frac{1}{2}\frac{\partial^2 g}{\partial y^2}(t, X(t), Y(t))(dY(t))^2 \\ & + \frac{\partial^2 g}{\partial x \partial y}(t, X(t), Y(t))(dX(t))(dY(t)). \end{aligned}$$

The reader can find a proof in Karatzas and Shreve [4].

For more description, readers may refer to [1, 3, 5, 6]

12.6 Exercise Problems

1. Show that p-th variation of Brownian motion equals zero for all $3 \le p < \infty$ and almost everywhere.
2. Using Itô's formula, prove that the following stochastic processes are $\mathcal{F}_t$-martingales

   (a) $X(t) = e^{\frac{1}{2}t} \cos W(t)$
   (b) $X(t) = e^{\frac{1}{2}t} \sin W(t)$

(c) $X(t) = (W(t) + t)\exp\left(-W(t) - \frac{1}{2}t\right)$.

3. For which value of K the following stochastic process $\{X(t), t \geq 0\}$ with

$$X(t) = (W(t))^3 - K\int_0^t W(u)du$$

is a martingale w.r.t. natural filtration.

4. Suppose $\alpha \in \mathbb{R}$. Show that the amount of time Brownian motion spends at the point α is zero, i.e.,

$$\int_0^t 1_{\{\alpha\}}(W(s))\,ds = 0, \quad \text{a.s.}$$

for all $t > 0$.

5. Let

$$X(t) = \int_0^t e^{s-t} dW(s), t > 0$$

$$Y(t) = \sqrt{2(t+1)}X(\log((t+1)/2)), t > 0$$

Find (a) $E(X(t))$, (b) $var(X(t))$, (c) distribution of $Y(t)$.

6. Which of the following integrals are stochastic integrals.

(a) $\int_0^1 \left((W(u))^2 + 2W\left(\frac{u}{2}\right) + 1\right) dW(u)$.
(b) $\int_0^1 \frac{1}{(W(u))^\alpha} dW(u)$ for $\alpha < \frac{1}{2}$.
(c) $\int_0^1 W\left(\frac{1}{u}\right) dW(u)$.
(d) $\int_0^t \left(\int_0^u dW(v)\right) dW(u)$.

7. Consider

$$X(T) = \int_0^T \mu_t dt + \int_0^T \sigma_t dW(t).$$

Prove that, $\{X(t), t \geq 0\}$ is not a martingale unless $\mu_t = 0$, for all t.

8. Find the values of constants α and β such that $X(t) = \int_0^t (\alpha + \beta u)\, dW(u)$ is normal distributed RV with mean 0 and variance 1.

9. Using Itô-Doeblin formula, find $E[(W(t))^4]$.

10. Prove

$$\int_0^t s dW(s) = tW(t) - \int_0^t W(s)ds.$$

11. Prove

$$\int_0^t (W(s))^2 dW(s) = \frac{1}{3}(W(t))^3 - \int_0^t W(s)ds.$$

12. Prove that a multi-dimensional version of Lévy's theorem. Suppose that a $d-$ dimensional stochastic process is denoted as $\{(W_1(t), W_2(t), \ldots, W_d(t)), t \geq 0\}$

such that each component $\{W_i(t), t > 0\}$ is a continuous martingale w.r.t. $\{\mathcal{F}_t\}_{t\geq 0}$ and $dW_i(t)dW_j(t) = dt$ for $i = j$ and $dW_i(t)dW_j(t) = 0$ for $i \neq j$. Prove that $\{(W_1(t), W_2(t), \ldots, W_d(t)), t \geq 0\}$ is a d - dimensional Brownian motion.

13. Suppose that $dW_i(t)dW_j(t) = 0$ for $i \neq j$. Prove that $\{(W_1(t), W_2(t), \ldots, W_d(t)), t \geq 0\}$ is a d - dimensional Brownian motion.

References

1. Castaneda, L.B., Arunachalam, V., Dharmaraja, S.: Introduction to Probability and Stochastic Processes with Applications. Wiley, New Jersey (2012)
2. Dudley, R.M., Norvaiša, R.: *Differentiability of Six Operators on Non–smooth Functions and p-Variation*, 1st edn, vol. I. (Springer, Berlin/Heidelberg, 2006). ISBN: 978-3-540-48814-9
3. Gianin E.R., and Sgarra, C.: Mathematical Finance Theory Review and Exercises, Springer Cham, Second Edition (2023) https://doi.org/10.1007/978-3-031-28378-9
4. Karatzas, I., Shreve, S.E.: Brownian Motion and Stochastic Calculus. Springer, New York (1991)
5. Shreve, S.E.: Stochastic calculus for finance II: Continuous-time models. Vol. 11. New York: springer (2004)
6. Mikosch, T.: Elementary stochastic calculus with finance in view. World scientific (1998)

Chapter 13
Stochastic Differential Equations

Stochastic differential equation (SDEs) is a differential equation with respect to a stochastic process. Stochastic differential equations are a class of mathematical equations that involve both deterministic and stochastic (random) components. They are used to model systems where random fluctuations play a significant role in the evolution of a quantity over time. Stochastic processes are characterized by uncertainty and randomness, and SDEs provide a framework for describing how these uncertainties affect the behavior of a system. It can have three components, a diffusion part with respect to Brownian motion or its function, a deterministic component (dt) and a jump process. This chapter discusses various methods of solving stochastic differential equations.

13.1 Definition

When German mathematician Karl Weierstrass (31 October 1815–19 February 1897) constructed a continuous but non-differentiable real-valued function in the 19th century, it was regarded as nothing more than a mathematical curiosity. According to high-frequency data at any given time, the prices of currency rates, interest rates, and liquid assets are nearly continuous but are of unbounded variation. In particular, they are nowhere differentiable. Classical calculus is no longer appropriate for real-valued functions encountered in mathematical finance. As a result, classical calculus must be extended to include functions with unbounded variation.
To deal with functions of unbounded variation, stochastic calculus is a necessary extension of real analysis.
We introduce uncertainties by introducing as additive white noise term, i.e.,

$$dX(t) = b(t, X(t))dt + dW(t),$$

D. Selvamuthu, *Introduction to Stochastic Processes*, University Texts in the Mathematical Sciences, https://doi.org/10.1007/978-981-97-6152-4_13

where $b : [0, T] \times \mathbb{R} \to \mathbb{R}$ is a continuous function. Time derivative of a Brownian motion is called a *white noise process*. It is used in models where uncertainties arise from various sources.
In general, if $b : [0, T] \times \mathbb{R} \to \mathbb{R}$, and $\sigma : [0, T] \times \mathbb{R} \to \mathbb{R}$ are two suitable functions, then an integral equation of the form:

$$X(t) = X(0) + \int_0^t b(s, X(s))ds + \int_0^t \sigma(s, X(s))dW(s) \tag{13.1}$$

is called a stochastic integral equation and is defined by an integration of a stochastic process with respect to Brownian motion. Equation (13.1) can be written as

$$dX(t) = b(t, X(t))dt + \sigma(t, X(t))dW(t), \quad (0 \le t \le T). \tag{13.2}$$

The equation (13.2) is referred to as a stochastic differential equation. The interpretation of (13.2) tells us that the change $dX(t) = X(t + \Delta t) - X(t)$ is caused by a change Δt of time, with factor $b(t, X(t))$ in combination with a change $dW(t) = W(t + \Delta t) - W(t)$ of Brownian motion with factor $\sigma(t, X(t))$.

SDEs can be classified into two large groups: linear SDEs, and non-linear SDEs. Furthermore, scalar linear and vector-valued linear SDEs can be distinguished. Let us start with the simple case, the scalar linear SDEs. The extension of scalar SDEs is to allow $X(t)$ to be a vector in $\mathbb{R}^n$.

Definition 13.1 (Linear SDE) A stochastic differential equation

$$dX(t) = b(t, X(t))dt + \sigma(t, X(t))dW(t), \quad (0 \le t \le T), \tag{13.3}$$

with the initial condition $X(0)$ for a stochastic process $\{X(t), t \ge 0\}$ is called a linear SDE if the functions $b(t, X(t))$ and $\sigma(t, X(t))$ are affine functions of $X(t)$ and $W(t)$ is a Brownian motion.

In comparison with linear SDEs, non-linear SDEs are less understood. Most SDEs, especially non-linear SDEs, do not have analytical solutions so that one has to resort to numerical approximation schemes in order to simulate sample paths of solutions for the given equation.

13.1.1 Strong Solution of SDEs

Let $(\Omega, \mathcal{F}, P)$ be a probability space, and $\{W(t), t \ge 0\}$ be a Brownian motion defined on it. An almost surely continuous process $\{X(t), t \ge 0\}$ is called a strong solution of Equation (13.1) it is adapted to the completion of the filtration generated by the Brownian motion and satisfies the Equation (13.1) P-almost surely for $t \in [0, T)$.

Remark 13.1 Let $\{X(t), t \geq 0\}$ and $\{Y(t), t \geq 0\}$ defined on the same probability space are any two strong solutions of Equation (13.1) then they are almost surely unique , i.e.,

$$P\{X(t) = Y(t) \,\forall\, t\} = 1.$$

Remark 13.2 A strong solution is an explicit function f such that $X(t) = f(t, W(s)),\ s < t$.

13.1.2 Existence and Uniqueness of Solution

Now, let's discuss the existence of a strong solution.

Theorem 13.1 *Let $T > 0$, $b : [0, T] \times \mathbb{R} \to \mathbb{R}$ and $\sigma : [0, T] \times \mathbb{R} \to \mathbb{R}$ are the drift and diffusion coefficient of Equation (13.1). Suppose that the coefficients satisfy the following properties*

$$|b(t, x)| + |\sigma(t, x)| \leq C(1 + |x|), \quad t \geq 0, x \in \mathbb{R}.$$

for some constant $C > 0$,

$$|b(t, x) - b(t, y)| + |\sigma(t, x) - \sigma(t, y)| \leq K|x - y|, \quad t \geq 0, x, y \in \mathbb{R}$$

for some constant $K > 0$. Suppose that $X(0)$ is a random variable independent of the Brownian motion $\{W(t), t \geq 0\}$ and such that $E[X(0)^2] < \infty$. Then there exists a unique and adapted stochastic process $\{X(t), t \in [0, T)\}$ such that

$$E\left(\int_0^T |X(s)|^2 ds\right) < \infty$$

which satisfies Equation (13.1).

The first condition says that the drift and volatility increases no faster than a linear function in x. This condition ensures that the solution $\{X(t), t \in [0, T]\}$ does not explode in finite time t, i.e., it does not tend to ∞ for a finite time t. The second condition states that the functions are Lipschitz in the second argument. Further, this condition guarantees the solution's uniqueness.

13.1.3 Itô-Picard Iteration

Note that, the existence and uniqueness follow the standard Picard's method given by French mathematician Charles Emile Picard (24 July 1856–11 December 1941) for constructing solutions of ODE. The existence of a solution is proved using the

Picard method of successive approximations. For fixed intial value $X(0)$, and for $n = 1, 2, \ldots$, define

$$X^{n+1}(t) = X(0) + \int_0^t b(s, X^n(s))ds + \int_0^t \sigma(s, X^n(s))dW(s).$$

The iterations are well-defined. By the convergence of the iteration scheme, it is obtained

$$X(t) = \lim_{n\to\infty} X^n(t).$$

Example 13.1 Check whether the strong solution for the given stochastic differential equation exists or not

$$dX(t) = X(t)dW(t), \text{ with } X(0) = 1.$$

Solution: Here, $b(t, x) = 0$ and $\sigma(t, x) = x$. Hence, Lipschitz condition is satisfied and a strong solution exists.
Example 13.2 Consider the stochastic differential equation

$$dX(t) = \mu X(t)dt + X(t)dW(t), \text{ with } X(0) = X_0.$$

Find the strong solution, if it exists.
Solution: Here, $b(t, x) = \mu x$ and $\sigma(t, x) = x$.

$$|\mu x| + |x| = (1 + |\mu|)|x| \leq (1 + |\mu|)(1 + |x|)$$

$$|\mu x - \mu y| + |x - y| \leq (1 + |\mu|)(|x - y|).$$

Let $K = (1 + |\mu|)$. Hence, a strong solution exists.
Now consider,

$$\begin{aligned} d(e^{-\mu t}X(t)) &= -\mu e^{-\mu t}X(t)dt + e^{-\mu t}dX(t) \\ &= -\mu e^{-\mu t}X(t)dt + e^{-\mu t}(\mu X(t)dt + X(t)dW(t)) \\ &= e^{-\mu t}X(t)dW(t). \end{aligned}$$

Therefore

$$e^{-\mu t}X(t) - X_0 = \int_0^t e^{-\mu s}X(s)dW(s).$$

This implies

$$X(t) = e^{\mu t}X_0 + e^{\mu t}\int_0^t e^{-\mu s}X(s)dW(s).$$

Example 13.3 Let $S(t)$ be the stock price at time t. Let μ be the constant growth rate of the stock and σ be the volatility. Consider the SDE

$$dS(t) = \mu S(t)dt + \sigma S(t)dW(t), \quad S(0) = S_0$$

where constant growth rate of the stock $\mu \in (-\infty, \infty)$ and volatility $\sigma > 0$ are constants.
Solution: Assume that $S(t) = f(t, W(t))$. Using Itô's formula (version 2), we get

$$dS(t) = f_t dt + f_x dW(t) + \frac{1}{2} f_{xx} dt,$$

where

$$f_x = \sigma f \text{ and } f_t + \frac{1}{2} f_{xx} = \mu f.$$

After solving these pde, we obtain

$$f = e^{\sigma x} k(t),\ f_{xx} = \sigma^2 f,\ f_t = k'(t) e^{\sigma x}$$

which implies

$$k'(t)e^{\sigma x} = \left(\mu - \frac{\sigma^2}{2}\right) e^{\sigma x} k(t) \implies k(t) = S_0 e^{(\mu - \frac{\sigma^2}{2}))t}.$$

Hence

$$S(t) = S_0 e^{(\mu - \frac{\sigma^2}{2})t + \sigma W(t)}.$$

One can prove that $S(t)$ has lognormal distribution. Also,

1. $E(S(t)) = E(S_0)e^{\mu t}$.
2. $E[(S(t))^2] = E(S_0^2) \exp[(2\mu + \sigma^2)t]$.

Asymptotic behavior of the solution.

1. It $\mu > \frac{\sigma^2}{2}$, then $S(t) \to \infty$ as $t \to \infty$ almost surely.
2. If$\mu < \frac{\sigma^2}{2}$, then $S(t) \to 0$ as $t \to \infty$ almost surely.
3. It $\mu = \frac{\sigma^2}{2}$, then $S(t)$ will fluctuate between an arbitrary large and arbitrary small value as $t \to \infty$.

Remark 13.3 Here $S(t) = S_0 e^{(\mu - \frac{\sigma^2}{2})t + \sigma W(t)}$ is called an exponential martingale. Note that the intriguing additional factor $e^{-\frac{\sigma^2}{2}t}$ is due to the non-differentiabile paths of the Wiener process.

13.1.4 Weak Solution

The existence of the process $\{X(t), t \geq 0\}$ that solves the integral equation form of the SDE is required for both strong and weak solutions. The underlying probability space $(\Omega, \mathcal{F}, P)$ is the difference between the two. A strong solution is a process that satisfies the stochastic differential equation and is defined on a given probability space whereas a weak solution is a process that satisfies the stochastic differential equation in a probability space other than the given space. The main difference between weak and strong solutions in terms of the Brownian motion is that for a strong solutions Brownian motion is given on a given probability space. For the weak solutions there are no restrictions to choose the Brownian motion and the probability space.

Definition 13.2 (Weak Solution) Consider a stochastic differential equation

$$dX(t) = \mu(X(t))dt + \sigma(X(t))dW(t), \tag{13.4}$$

with initial condition $X_0 = x \in \mathbb{R}$.
Define, a probability space $(\Omega, \mathcal{F}, P)$, and an increasing family of sub sigma-fields of $\mathcal{F}$ as $\{\mathcal{F}_t\}_{t\geq 0}$. Consider the continuous $\mathcal{F}_t$-adapted Wiener processes $\{W(t), t \geq 0\}$ and $\{X(t), t \geq 0\}$ which satisfy the stochastic integral equation

$$X(t) = x + \int_0^t \mu(X(s))ds + \int_0^t \sigma(X(s))dW(s) \ \ a.s.. \tag{13.5}$$

Then, the family $(\Omega, \mathcal{F}, \{\mathcal{F}_t\}_{t\geq 0}, \{W(t), t \geq 0\}, \{X(t), t \geq 0\}, P)$ is called a weak solution of the stochastic differential equation (13.4).

Remark 13.4 A weak solution of the Equation (13.4) with initial condition x is a continuous stochastic process $\{X(t), t \geq 0\}$ defined on some probability space that for some Wiener process and some admissible filtration, is adapted and satisfies (13.4).

Definition 13.3 (Unique in Law) A weak solution of the stochastic differential equation (13.4) is said to be unique in law if two weak solutions $X_1(t)$ and $X_2(t)$ defined on probability spaces $(\Omega, \mathcal{F}, (\mathcal{F}_t, P_1, W_1(t))$ and $(\Omega, \mathcal{F}, (\mathcal{F}_t, P_2, W_2(t))$ have the same probability distribution, i.e.,

$$P_1\{X_1(\cdot), \omega \in B\} = P_2\{X_2(\cdot), \omega \in B\}$$

for any Borel set B in $\mathbb{R}$.

A strong solution distinguishes from a weak solution in the requirement that it should be adapted to the completion of minimal filtration. This makes each RV $X(t)$, a measurable function of the path $\{W(s), s \leq t\}$ of the driving Brownian motion. It turns out that for certain coefficient functions μ and σ, solutions to the stochastic integral equation (13.5) may exist for some Wiener processes and some admissible filtrations but not for others.

As a consequence of these definitions, consider the different notions of uniqueness. For strong solutions, pathwise unique solutions are searched, i.e., if $\{X_1(t), t \geq 0\}$ and $\{X_2(t), t \geq 0\}$ are strong solutions to (13.4) with the same initial condition, then pathwise uniqueness holds, i.e.,

$$P\{X_1(t) = X_2(t) \; \forall \, t\} = 1.$$

But for the weak solution, it doesn't make any sense to talk about the pathwise uniqueness. For instance, consider the following examples

Example 13.4 Consider an SDE

$$dX(t) = dW(t), \, X(0) = 0$$

Show that, $X_1(t) = W(t)$ and $X_2(t) = -W(t)$ are the weak solutions of this SDE.
Solution: Let $\{W(t), t \geq 0\}$ be a Brownian motion. It is known that $\{-W(t), t \geq 0\}$ is also a Brownian motion. Therefore, it can be chosen $X_1(t) = W(t)$ and $X_2(t) = -W(t)$ and both of them are Brownian motions. Then, $X_1(t) = W(t)$ and $X_2(t) = -W(t)$ are weak solutions to SDE $dX(t) = dW(t), X(0) = 0$. Further, it can be observed that although $W(t)$, and $-W(t)$ are defined on the same probability space, still path wise uniqueness need not be satisfied, i.e.,

$$P(X_1(t) = X_2(t)) = P(W(t) = -W(t)) = P(W(t) = 0).$$

Clearly, $P(X_1(t) = X_2(t)) < 1$.

The notion of uniqueness for weak solution is weak uniqueness, i.e., they have the same finite dimensional probability distribution (although they have different sample paths).

Example 13.5 Consider an SDE

$$dX(t) = -sign(X(t))dW(t), 0 \leq t \leq T$$

where $sign(x) = 1$ for $x > 0$ and $sign(x) = -1$ for $x \leq 0$ with initial condition $X_0 = 0$. Show that, it has weak solution, but no strong solution.
Solution: Firstly, it will be proved that the SDE has a weak solution. Let $\{(X(t), \mathfrak{F}_t), \, t \geq 0\}$ be some Brownian motion and define:

$$W(t) := -\int_0^t sign(X(s))dX(s).$$

It follows from Lévy's characterization that $(W(t), \mathfrak{F}_t)$ is also a Brownian motion. Since,

$$dW(t) = -sign(X(t))dX(t)$$

implies,

$$dX(t) = -sign(X(t))dW(t)$$

this means that $\{X(t), t \geq 0\}$ is a weak solution.

We claim that the given function $sign(x)$ doesn't satisfy the Lipschitz condition.

Take $x_1 = 0$ and $x_2 = \frac{1}{n}$ for any positive integer n. Note that $|sign(x_1) - sign(x_2)| = 2$, but $|x_1 - x_2| = \frac{1}{n}$.

As n approaches infinity, then $\frac{1}{n}$ is arbitrarily small. This shows that there is no constant K such that the Lipschitz condition holds for all x_1 and x_2, and thus, the sign function is not Lipschitz function which implies strong solution doesn't exist.

When there is no known explicit solution for a given SDE, we can approximate it with a numerical solution, substituting differences for differentials. As a result, the approximate solution method resembles numerical integration.

13.2 Time-Homogeneous Property

The time-homogeneous property of SDEs plays a crucial role in understanding and analyzing the behavior of stochastic processes.

Theorem 13.2 *A Itô diffusion is a stochastic process $X(t)(\omega) = X(t, \omega) : [0, \infty) \times \Omega \to \mathbb{R}^n$ that satisfies a stochastic differential equation of the form:*

$$dX(t) = b\left(X(t)\right) dt + \sigma\left(X(t)\right) dW(t), \quad t \geq s; \quad X(s) \text{is known}, \tag{13.6}$$

where $\mathbf{W(t)}$ is m-dimensional Brownian motion and $b : \mathbb{R}^n \to \mathbb{R}^n, \sigma : \mathbb{R}^n \to \mathbb{R}^{n\times m}$ that satisfies the following condition:

$$|b(x) - b(y)| + |\sigma(x) - \sigma(y)| \leq D|x - y|; \quad x, y \in \mathbb{R}^n, \tag{13.7}$$

where $|\sigma|^2 = \sum \left|\sigma_{ij}\right|^2$. Then $\{X(t), t \geq 0\}$ is time-homogeneous Itô process.

Proof Since, $X(t)$ written as $X(t)^{s,X(s)}$; $t \geq s$, can be written as the (unique) solution of (13.6). When $s = 0$, write $X(t)^{X(0)}$ instead of $X(t)^{0,X(0)}$. Note that in (13.6), it is assumed that b and σ depend solely on $X(s)$, and not on t. In the following sense, the process $X(t)$ will have the property of being time-homogeneous. Note that:

$$\begin{aligned} X(s+h)^{s,X(s)} &= X(s) + \int_s^{s+h} b\left(X(u)^{s,X(s)}\right) du + \int_s^{s+h} \sigma\left(X(u)^{s,X(s)}\right) dW(u) \\ &= X(s) + \int_0^{h} b\left(X(s+v)^{s,X(s)}\right) dv + \int_0^{h} \sigma\left(X(s+v)^{s,X(s)}\right) d\widetilde{W}(v), \end{aligned}$$

where $u = s + v$ and $\widetilde{W}(v) = W(s+v) - W(s)$; $v \geq 0$. On the other hand,

$$X(h)^{0,X(0)} = X(0) + \int_0^h b\left(X(v)^{0,X(0)}\right) dv + \int_0^h \sigma\left(X(v)^{0,X(0)}\right) dW(v). \quad (13.8)$$

Because $\left\{\widetilde{W}(v),\, v \geq 0\right\}$ and $\{W(v),\, v \geq 0\}$ have the same distributions, the solution of the stochastic differential equation

$$dX(t) = b\,(X(t))\,dt + \sigma\,(X(t))\,dW(t); \quad X(0) = X(0), \quad (13.9)$$

given by,

$$\left\{X(s+h)^{s,X(s)},\, h \geq 0\right\} \quad \text{and} \quad \left\{X(h)^{0,X(0)},\, h \geq 0\right\}, \quad (13.10)$$

has the same distributions, i.e., $\{X(t), t \geq 0\}$ is time-homogeneous Itô process (by uniqueness). □

13.3 Exact Solution for the SDE

Stochastic differential equations (SDEs) are used in standard stochastic models for instruments, securities or products such as asset prices, interest rates, and their derivatives. Unlike deterministic models such as ordinary differential equations, which have a unique solution for each appropriate initial condition, SDEs have solutions that are continuous-time stochastic processes. Similar to the case of ordinary differential equations, relatively few SDEs have closed form solutions. In this book, closed form solution are presented for the SDEs by solving them analytically. It is often necessary to use numerical approximation techniques to solve the SDEs which are not dealt with in this book.

The stochastic differential equation

$$dX(t) = a(t, W(t))dt + b(t, W(t))dW(t) \quad (13.11)$$

is called exact, if there is a differentiable function $f(t, x)$ such that

$$a(t, x) = \partial_t f(t, x) + \frac{1}{2}\partial_{xx}^2 f(t, x) \quad (13.12)$$

$$b(t, x) = \partial_x f(t, x). \quad (13.13)$$

Assume that the Equation (13.11) is exact. Then, substituting Equations (13.12) and (13.13) in Equation (13.11) yields:

$$dX(t) = \left(\partial_t f(t, W(t)) + \frac{1}{2}\partial_{xx}^2 f(t, W(t))\right) + \partial_x f(t, W(t))dW(t).$$

Applying Itô's formula, the previous equation become

$$dX(t) = df(t, W(t))$$

which implies $X(t) = f(t, W(t)) + c$, where c is a constant. Thus, finding a solution to the SDE (13.11) is equivalent to finding the solution to the partial differential equations (13.12) and (13.13). Next, some examples of exact SDEs are given with their solutions.

Example 13.6 Consider the exact stochastic differential equation

$$dX(t) = e^t(1 + (W(t))^2)dt + (1 + 2e^t W(t))dW(t), \ \ X(0) = 0.$$

Find the solution using Itô-Doeblin formula.
Solution: Here, comparing with the standard form in Equation (13.11) we get

$$a(t, x) = e^t(1 + x^2) \text{ and } b(t, x) = 1 + 2e^t x.$$

From Equations (13.12) and (13.13), the following Itô-Doeblin formula is obtained

$$\frac{\partial}{\partial t} f(t, x) + \frac{1}{2}\frac{\partial^2}{\partial x^2} f(t, x) = e^t(1 + x^2) \tag{13.14}$$

$$\frac{\partial}{\partial x} f(t, x) = 1 + 2e^t x. \tag{13.15}$$

Integrating Equation (13.15) partially with respect to x,

$$f(t, x) = x + e^t x^2 + H(t)$$

where, $H(t)$ is a function of t only. Substituting this value of $f(t, x)$ in Equation (13.14),

$$e^t x^2 + H'(t) + e^t = e^t(1 + x^2)$$

which implies $H(t) = c_1$, a constant. Therefore, $f(t, x) = x + e^t x^2 + c_1$ which is same as $X(t)$. Using the initial conditions $X(0) = 0$, $c_1 = 0$, the required solution is

$$X(t) = W(t) + e^t(W(t))^2.$$

Definition 13.4 (Ornstein-Uhlenbeck Process) Consider the SDE

$$dX(t) = -\mu X(t)dt + \sigma dW(t), \ \ X(0) = x_0, \tag{13.16}$$

where μ and σ are constants. This equation is often referred to as Langevin equation.This equation can be used to model interest rate. The stochastic process $\{X(t), t \geq 0\}$ is called Ornstein-Uhlenbeck process.

The sample path for the SDE given in Equation (13.16) with $x_0 = 1$, $\mu = 6$, and $\sigma = 5.5$ is shown in Fig. 13.1.

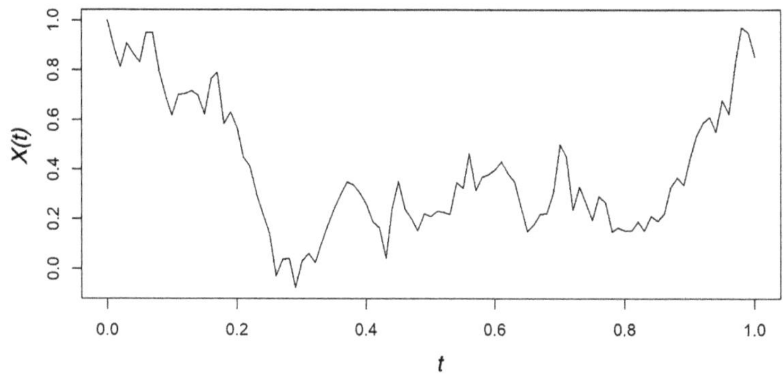

Fig. 13.1 Simulated path of the Ornstein-Uhlenbeck process

Example 13.7 Solve Equation (13.16) using Itô's formula (Version 2).
Solution: Set $dt = 1$ then

$$X(t+1) - X(t) = -\mu X(t) + \sigma(W(t+1) - W(t))$$
$$X(t+1) = -\mu' X(t) + \sigma(W(t+1) - W(t))$$

where $\mu' = \mu - 1$. This time series model can be considered as a discrete analogue of the solution of the Langevin equation (13.16). The stochastic differential equation (13.16) is interpreted as the following stochastic integral equation

$$X(t) = x_0 + \sigma W(t) - \mu \int_0^t X(s)ds.$$

Using the integration method and integration by parts, derive the solution $X(t)$ for each sample path of $W(t)$. The computation is somewhat complicated. Here, the solution is derived using Itô's formula which is easy. Let $f(t, x) = e^{\mu t}x$. Then,

$$\frac{\partial f}{\partial t} = \mu e^{\mu t}x, \ \frac{\partial f}{\partial x} = e^{\mu t}, \ \frac{\partial^2 f}{\partial x^2} = 0.$$

Use Itô's formula (Version 2) to provide the solution

$$X(t) = e^{-\mu t}x_0 + \sigma \int_0^t e^{-\mu(t-s)}dW(s). \tag{13.17}$$

This stochastic process $\{X(t), t \geq 0\}$ is called Ornstein-Uhlenbeck process. This is the strong solution (immaterial of the path of Brownian motion, the solution (13.17)) holds. If x_0 is constant or if x_0 follows standard normal distribution, then $X(t)$ is normal distributed.
If $x_0 = 0$, then

$$E(X(t)) = 0;\ var(X(t)) = \frac{\sigma^2}{2\mu}\left(1 - e^{-2\mu t}\right)$$

and

$$cov(X(t), X(s)) = \frac{\sigma^2}{2\mu}\left(e^{-\mu(t-s)} - e^{-\mu(t+s)}\right), s < t.$$

Since $\{X(t), t \geq 0\}$ is a zero-mean Gaussian process, the covariance function is characteristic for the Ornstein-Uhlenbeck process.

Remark 13.5 $\{X(t), t \geq 0\}$ is a Markov process with transition probability densities

$$p(t; x, y) = \frac{1}{\sqrt{2\pi}\sigma(t)} exp\left\{\frac{-(y - e^{-\mu t}x)^2}{2\sigma^2(t)}\right\}.$$

When $\mu > 0$, which corresponds to the stable case,

$$\lim_{t\to\infty} \sigma^2(t) = \frac{\sigma^2}{2\mu} = \theta$$

and

$$\lim_{t\to\infty} p(t; x, y) = \frac{1}{\sqrt{2\pi\theta}} exp\left\{\frac{-y^2}{2\theta}\right\}.$$

13.3.1 Integration by Inspection

When solving a stochastic differential equation by inspection, one tries to select some appropriate function and apply the product or the quotient formulas. For instance, if a stochastic differential equation can be written as:

$$dX(t) = f'(t)W(t)dt + f(t)dW(t),$$

where $f(t)$ is continuously differentiable function of t and the product rule brings the equation into the exact form

$$dX(t) = d\left(f(t)W(t)\right),$$

which after integration leads to the solution

$$X(t) = X_0 + f(t)W(t).$$

Example 13.8 Consider the Langevin equation

$$dX(t) = -\beta X(t)dt + \alpha dW(t)$$

where $\alpha \in \mathbb{R}$ and $\beta > 0$. The process $\{X(t); t \geq 0\}$ with $X(0) = X_0$ can be written as:

$$X(t) = X_0 + \alpha W(t) - \beta \int_0^t X(s)ds.$$

Solution: Let $f(t, x) = e^{\beta t}x$. Applying Itô-Doeblin formula,

$$\begin{aligned} d\left(e^{\beta t}X(t)\right) &= \beta e^{\beta t}X(t)dt + e^{\beta t}dX(t) \\ &= \beta e^{\beta t}X(t)dt + e^{\beta t}\left(-\beta X(t)dt + \alpha dW(t)\right) \\ &= \alpha e^{\beta t}dW(t). \end{aligned}$$

Integration of the above equation gives

$$X(t) = X(0)e^{-\beta t} + \alpha \int_0^t e^{-\beta(t-s)}dW(s).$$

Example 13.9 Consider an SDE

$$dX(t) = (W(t) + 3t^2)dt + tdW(t), \quad X(0) = c,$$

where c is a constant. Using integration by inspection method, find the solution of the SDE.
Solution: Since this equation can be rewritten as:

$$dX(t) = d(t^3) + d(tW(t))$$

which yields,

$$dX(t) = d(t^3 + tW(t))$$

which implies the solution is given by $X(t) = t^3 + tW(t) + c$, where c is a constant.

Example 13.10 Solve the following SDE,

$$dX(t) = e^{2t}(1 + 2(W(t))^2)dt + 2e^{2t}W(t)dW(t).$$

Solution: Note that this equation can further be written as:

$$dX(t) = (2e^{2t}dt)(W(t))^2 + e^{2t}(2W(t)dW(t) + dt).$$

which yields

$$dX(t) = d(e^{2t})(W(t))^2 + e^{2t}d((W(t))^2).$$

By the product rule, the right side becomes

$$dX(t) = d(e^{2t}(W(t))^2),$$

and hence the solution is $X(t) = e^{2t}(W(t))^2 + c$, where c is a constant.

Example 13.11 Consider an SDE,

$$dX(t) = (t + (W(t))^2)dt + 2tW(t)dW(t), \quad X(0) = c,$$

where c is a constant. Find the solution of this SDE using integration by inspection method.

Solution: Rewrite the given SDE as

$$dX(t) = (W(t))^2 dt + t(2W(t)dW(t) + dt).$$

This can be written as

$$dX(t) = (W(t))^2 dt + td((W(t))^2).$$

Using the product rule,

$$dX(t) = dt(W(t))^2$$

with the solution $X(t) = t(W(t))^2 + c$.

13.3.2 Linear SDEs and Integrating Factors

Consider the stochastic differential equation, with linear drift term in $X(t)$,

$$dX(t) = (\alpha(t)X(t) + \beta(t))\,dt + b(t, W(t))dW(t), \quad t \geq 0.$$

The SDE of the above form is known as linear SDE. This can also be written as

$$dX(t) - \alpha(t)X(t)dt = \beta(t)dt + b(t, W(t))dW(t), \quad t \geq 0.$$

Methodology to Solve Linear SDEs

Let the integrating factor be $A(t) = \int_0^t \alpha(s)ds$. Multiplying by the integrating factor $e^{-A(t)}$, the previous equation becomes an exact expression:

$$e^{-A(t)}\left(dX(t)-\alpha(t)X(t)dt\right)=e^{-A(t)}\beta(t)dt+e^{-A(t)}b(t,W(t))dW(t)$$
$$d\left(e^{-A(t)}X(t)\right)=e^{-A(t)}\beta(t)dt+e^{-A(t)}b(t,W(t))dW(t).$$

Integrating both sides yield,

$$X(t)=X_0e^{A(t)}+e^{A(t)}\left(\int_0^t e^{-A(s)}\beta(s)ds+\int_0^t e^{-A(s)}b(s,W(s))dW(s)\right).$$

The first integral within the previous parentheses is a Riemann integral, and the latter one is an Itô stochastic integral.

Example 13.12 Solve the linear stochastic differential equation

$$dX(t)=(2X(t)+1)dt+e^{2t}dW(t).$$

It can be written as

$$dX(t)-2X(t)dt=dt+e^{2t}dW(t).$$

and multiply by the integrating factor e^{-2t} to get

$$d\left(e^{-2t}X(t)\right)=e^{-2t}dt+dW(t),$$

which yields,

$$X(t)=X_0e^{2t}+e^{2t}\int_0^t e^{-2s}ds+e^{2t}\int_0^t dW(s),$$
$$=X_0e^{2t}+\frac{1}{2}(e^{2t}-1)+e^{2t}W(t).$$

Example 13.13 Consider the SDE:

$$dX(t)=\alpha(\mu-X(t))dt+\sigma dW(t),t\in[0,T].$$

Find the strong solution for the above SDE with given constant $X(0)$. Also, find the distribution of $X(t)$ at any t and $t\to\infty$.
Solution: The equation can be written as

$$dX(t)+\alpha X(t)dt=\alpha\mu dt+\sigma dW(t).$$

Multiplying by the integrating factor, $e^{\alpha t}$

$$d(e^{\alpha t}X(t))=\alpha\mu e^{\alpha t}dt+\sigma e^{\alpha t}dW(t),$$

which yields

$$e^{\alpha t}X(t) - X(0) = \mu\left(e^{\alpha t} - 1\right) + \sigma \int_0^t e^{\alpha s} dW(s)$$

$$\Longrightarrow X(t) = e^{-\alpha t}X(0) + \mu\left(1 - e^{-\alpha t}\right) + \sigma e^{-\alpha t} \int_0^t e^{\alpha s} dW(s).$$

Hence, we get for $\alpha > 0$, $X(t) \to \mu$ as $t \to \infty$.

Example 13.14 Consider the stochastic differential equation,

$$dX(t) = \frac{b - X(t)}{1 - t} dt + dW(t),\ 0 \leq t < 1,\ X(0) = a,$$

where a, b are constants. This is known as Brownian Bridge. Prove that, the solution is given by

$$X(t) = a(1 - t) + bt + (1 - t) \int_0^t \frac{1}{1 - s} dW(s),\quad 0 \leq t < 1.$$

Also show that

(a) $E(X(t)) = a(1 - t) + bt.$
(b) $var(X(t)) = t(1 - t).$
(c) $P\left(\omega : \lim_{t\to 1} X(t, \omega) = b\right) = 1.$

Solution: Take $Y(t) = b - X(t)$. Then, the given SDE becomes linear in $Y(t)$

$$dY(t) + \frac{1}{1 - t} Y(t)dt = -dW(t).$$

Multiplying the integrating factor $\frac{1}{1-t}$ yields

$$d\left(\frac{Y(t)}{1 - t}\right) = -\frac{1}{1 - t} dW(t).$$

By integration,

$$\frac{Y(t)}{1 - t} = c - \int_0^t \frac{1}{1 - s} dW(s).$$

Putting $t = 0$, $c = a - b$. Hence,

$$\frac{b - X(t)}{1 - t} = a - b - \int_0^t \frac{1}{1 - s} dW(s).$$

Now, solve for $X(t)$ yields

$$X(t) = a(1-t) + bt + (1-t)\int_0^t \frac{1}{1-s} dW(s), \quad 0 \le t < 1.$$

Consider,

$$\begin{aligned} E(X(t)) &= E\left(a(1-t) + bt + (1-t)\int_0^t \frac{1}{1-s} dW(s)\right) \\ &= a(1-t) + bt, \end{aligned}$$

and

$$\begin{aligned} E[(X(t))^2] &= a^2(1-t)^2 + b^2t^2 + (1-t)^2 E\left(\int_0^t \frac{1}{1-s} dW(s)\right)^2 \\ &= a^2(1-t)^2 + b^2t^2 + (1-t)^2 \int_0^t \frac{1}{(1-s)^2} ds \\ &= a^2(1-t)^2 + b^2t^2 + t(1-t) \end{aligned}$$

which implies

$$var(X(t)) = E[(X(t))^2] - (E[X(t)])^2 = t(1-t).$$

From the solution of $X(t)$ it can easily seen that $P\left(\omega : \lim_{t\to 1} X(t,\omega) = b\right) = 1$.

Example 13.15 Consider an SDE

$$dX(t) = 3X(t)dt + e^{3t} dW(t), \quad X(0) = 1.$$

Show that $X(t) = e^{3t}(1 + W(t))$.
Solution: Take the integrating factor e^{-3t}. Hence, the given SDE becomes

$$d(e^{-3t}X(t)) = dW(t).$$

Integration gives

$$e^{-3t}X(t) = W(t) + 1$$

Hence, the solution is $X(t) = e^{3t}(1 + W(t))$.

Now, some more examples are presented.

Example 13.16 Solve the following SDEs.

$$\begin{aligned}
&(a)\ dX(t) = (4X(t) - 1)dt + 2dW(t)\\
&(b)\ dX(t) = (3X(t) - 2)dt + e^{3t}dW(t)\\
&(c)\ dX(t) = (1 + X(t))dt + e^{t}W(t)dW(t)
\end{aligned}$$

Solution:

(a) Consider,

$$dX(t) = (4X(t) - 1)dt + 2dW(t).$$

Multiplying the above SDE by the integrating factor e^{-4t} gives

$$\begin{aligned}
d(e^{-4t}X(t)) &= -4e^{-4t}X(t)dt + e^{-4t}dX(t)\\
&= -4e^{-4t}X(t)dt + e^{-4t}[4X(t)dt - dt + 2dW(t)]\\
&= -e^{-4t}dt + 2e^{-4t}dW(t)
\end{aligned}$$

which yields

$$X(t) = X(0)e^{4t} + \frac{1}{4}(1 - e^{4t}) + 2\int_0^t e^{4(t-s)}dW(s).$$

(b) Consider

$$dX(t) = (3X(t) - 2)dt + e^{3t}dW(t).$$

Multiplying the above SDE by the integrating factor e^{-3t} gives

$$\begin{aligned}
d(e^{-3t}X(t)) &= -3e^{-3t}X(t)dt + e^{-3t}dX(t)\\
&= -3e^{-3t}X(t)dt + e^{-3t}[3X(t)dt - 2dt + e^{3t}dW(t)]\\
&= -2e^{-3t}dt + dW(t)
\end{aligned}$$

which yields

$$X(t) = X(0)e^{3t} + \frac{2}{3}(1 - e^{3t}) + 2\int_0^t e^{3t}dW(s).$$

(c) Consider

$$dX(t) = (X(t) + 1)dt + e^{t}W(t)dW(t).$$

Multiplying the above SDE by the integrating factor e^{-4t} gives

$$\begin{aligned} d(e^{-t}X(t)) &= -e^{-t}X(t)dt + e^{-t}dX(t) \\ &= -e^{-t}X(t)dt + e^{-t}[X(t)dt + dt + e^{t}W(t)dW(t)] \\ &= e^{-t}dt + W(t)dW(t). \end{aligned}$$

This yields

$$X(t) = X(0)e^{t} + (1 - e^{t}) + e^{t}\int_{0}^{t} W(s)dW(s)$$

and

$$\int_{0}^{t} W(s)dW(s) = \frac{1}{2}\left((W(t))^2 - t\right), \quad t \geq 0.$$

This implies

$$X(t) = X(0)e^{t} + (1 - e^{t}) + \frac{1}{2}e^{t}\left((W(t))^2 - t\right).$$

13.3.3 *The Method of Variation of Parameters*

The variation of constant parameters approach is one of the most often used methods for studying non-linear dynamical systems in the field of science and engineering. Knowing the existence of a solution process, the technique of variation of parameters is a very effective way of finding the solution of a system of differential equations.

13.3.3.1 Integrating Factor Method

Non-linear SDEs of the type given below are solved using the integrating factor approach.

$$dX(t) = f(t, X(t))dt + \sigma(t)X(t)dW(t), \quad X(0) = x. \tag{13.18}$$

Here, f is a continuous and deterministic function defined from $\mathbb{R}^{+} \times \mathbb{R} \to \mathbb{R}$.
Step 1. First solve:

$$dG(t) = \sigma(t)G(t)dW(t).$$

We get

$$G(t) = \exp\left(\int_{0}^{t} \sigma(s)dW(s) - \frac{1}{2}\int_{0}^{t} \sigma^2(s)ds\right).$$

We can write $G(t) = e^{Y(t)}$ where $Y(t) = \left(\int_0^t \sigma(s)dW(s) - \frac{1}{2}\int_0^t \sigma^2(s)ds\right)$. Now using Itô Doeblin formula,

$$\begin{aligned} dG(t) &= \left(-\frac{G(t)}{2}\sigma^2(t) + \frac{G(t)}{2}\sigma^2(t)\right)dt + G(t)\sigma(t)W(t) \\ &= \sigma(t)G(t)W(t). \end{aligned}$$

Let $F(t) = G^{-1}(t)$ denotes the integrating factor function. It is simple to verify that $F(t)$ satisfies.

$$dF(t) = -\sigma(t)F(t)dW(t) + \sigma^2(t)F(t)dt.$$

Step 2. Take $X(t) = G(t)C(t)$ and hence $C(t) = F(t)X(t)$, product rule implies

$$d(F(t)X(t)) = F(t)f(t, X(t))dt.$$

Therefore, the following "deterministic" ODE is satisfied by C_t

$$dC(t) = F(t)f(t, G(t)C(t)). \tag{13.19}$$

Step 3. Upon obtaining $C(t)$, use $X(t) = G(t)C(t)$ to obtain $X(t)$.

Remark 13.6 If (13.19) cannot be solved explicitly, we may still calculate $C(t)$ using various numerical techniques given in [6].

Example 13.17 Use the integrating factor method to solve the SDE

$$dX(t) = (X(t))^{-1}dt + \alpha X(t)dW(t), \quad X(0) = x > 0,$$

where α is a constant.
Solution: Here $f(t, x) = x^{-1}$ and $F(t) = \exp\left(-\alpha W(t) + \frac{\alpha^2}{2}t\right)$. All that is left to do is solve,

$$dC(t) = F(t)\,[G(t)C(t)]^{-1} = F^2(t)/C(t).$$

This provides us with $d(C(t))^2 = 2F^2(t)dt$. Therefore

$$(C(t))^2 = 2\int_0^t \exp\left(-2\alpha W(s) + \alpha^2 s\right)ds + x^2.$$

Because $x > 0$ is the starting condition, choose $Y(t) > 0$ such that

$$X(t) = G(t)C(t) = \exp\left(\alpha W(t) - \frac{\alpha^2}{2}t\right)\sqrt{2\int_0^t \exp\left(-2\alpha W(s) + \alpha^2 s\right)ds + x^2}.$$

Example 13.18 Show that the solution $X(t)$ of

$$dX(t) = (X(t))^2 dt + X(t)dW(t) \tag{13.20}$$

where, $X(0) = 1$ and $\{W(t), t \geq 0\}$ is a standard Brownian motion and has the representation

$$X(t) = \exp\left(\int_0^t X(s)ds - \frac{1}{2}t + W(t)\right).$$

Solution: Here $f(t, x) = x^2$ and $F(t) = \exp\left(-W(t) + \frac{t}{2}\right)$. Using above integrating factor method, we obtain

$$dC(t) = X(t)C(t)dt.$$

After solving this SDE, we obtain $C(t) = e^{\int_0^t X(s)ds}$. Hence, we get

$$X(t) = G(t)C(t) = \exp\left(\int_0^t X(s)ds - \frac{t}{2} + W(t)\right).$$

13.3.3.2 Moment Method

We can not get a solution that can be described clearly in terms of $W(t)$ for many SDEs, for example,

$$dX(t) = \kappa(\theta - X(t))dt + \sigma\sqrt{X(t)}dW(t), \quad X(0) = x. \tag{13.21}$$

The above SDE is called modified Cox-Ingersoll-Ross model and lacks an explicit solution. The process is often used to model short-term interest rates or stochastic volatility processes for stock prices. The moments of the process $X(t)$ has been discussed in more detail in [8]. Write the integral of (13.21) as

$$X(t) = x + \kappa \int_0^t (\theta - X(s))ds + \sigma \int_0^t \sqrt{X(s)}dW(s). \tag{13.22}$$

and use Itô's formula to obtain

$$(X(t))^2 = x^2 + \left(2\kappa\theta + \sigma^2\right) \int_0^t X(s)ds - 2\kappa \int_0^t (X(s))^2 ds + 2\sigma \int_0^t (X(s))^{3/2} dW(s). \tag{13.23}$$

Moments of the solution are determined using this equation and the properties of Itô integral. Taking expectation over both sides of (13.22) to obtain the first moment

$$M(t) := E[X(t)] = x + \kappa \left(\theta t - \int_0^t E[X(s)]ds \right),$$

because the stochastic integral part has a zero expectation. Then the following ODE can be solved:

$$dM(t) = \kappa \, (\theta - M(t)) \, dt$$

to obtain:

$$M(t) = \theta + (x - \theta)e^{-\kappa t}.$$

For the second moment

$$E\left[(X(t))^2\right] = x^2 + \left(2\kappa\theta + \sigma^2\right) \int_0^t E[X(s)]ds - 2\kappa \int_0^t E\left[(X(s))^2\right] ds.$$

This ODE is identical to the one obtained for $M(t)$. Then

$$E\left[(X(t))^2\right] = x^2 + \left(2\kappa\theta + \sigma^2\right) \left(\theta t + (x - \theta) \frac{\left(1 - e^{-\kappa t}\right)}{\kappa} \right) - 2\kappa \int_0^t E\left[(X(s))^2\right] ds.$$

Here, assume that $\int_0^t E\left[|X(s)|^3\right] ds < \infty$ such that $\int_0^t (X(s))^{3/2} dW(s)$ is an Itô integral with a square-integrable integrand and thus

$$E\left[\int_0^t (X(s))^{3/2} dW(s) \right] = 0.$$

Now, we present an example of a linear vector-valued SDE.

Example 13.19 Consider a two-dimensional geometric Brownian motion.

$$dS_1(t) = \mu_1 S_1(t)dt + S_1(t)(\sigma_{11} dW_1(t) + \sigma_{12} dW_2(t)), \tag{13.24}$$

$$dS_2(t) = \mu_2 S_2(t)dt + S_2(t)(\sigma_{21} dW_1(t) + \sigma_{22} dW_2(t)). \tag{13.25}$$

Find the correlation coefficient of $\ln\left(\frac{S_1(t)}{S_1(t-1)}\right)$ and $\ln\left(\frac{S_2(t)}{S_2(t-1)}\right)$.

Solution: Since, $S_1(t)$ and $S_2(t)$ are geometric Brownian motions which implies

$$S_1(t) = S_1(0) \exp\left\{ \left(\mu_1 - \frac{1}{2}\sigma_{11}^2 - \frac{1}{2}\sigma_{12}^2 \right) t + \sigma_{11} W_1(t) + \sigma_{12} W_2(t) \right\}$$

$$S_2(t) = S_2(0) \exp\left\{ \left(\mu_2 - \frac{1}{2}\sigma_{21}^2 - \frac{1}{2}\sigma_{22}^2 \right) t + \sigma_{21} W_1(t) + \sigma_{22} W_2(t) \right\}.$$

Now consider,

$$\ln \frac{S_1(t)}{S_1(t-1)} = \left(\mu_1 - \frac{1}{2}\sigma_{11}^2 - \frac{1}{2}\sigma_{12}^2\right) + \sigma_{11}\left(W_1(t) - W_1(t-1)\right) + \sigma_{12}\left(W_2(t) - W_2(t-1)\right)$$

$$\ln \frac{S_2(t)}{S_2(t-1)} = \left(\mu_2 - \frac{1}{2}\sigma_{21}^2 - \frac{1}{2}\sigma_{22}^2\right) + \sigma_{21}\left(W_1(t) - W_1(t-1)\right) + \sigma_{22}\left(W_2(t) - W_2(t-1)\right).$$

To find the correlation coefficient let us first find the covariance between them and variance. Consider,

$$cov\left(\ln \frac{S_1(t)}{S_1(t-1)}, \ln \frac{S_2(t)}{S_2(t-1)}\right) = \sigma_{11}\sigma_{21} + \sigma_{12}\sigma_{22}.$$

Now consider,

$$var\left(\ln \frac{S_1(t)}{S_1(t-1)}\right) = \sigma_{11}^2 + \sigma_{12}^2,$$

$$var\left(\ln \frac{S_2(t)}{S_2(t-1)}\right) = \sigma_{21}^2 + \sigma_{22}^2.$$

Hence, correlation coefficient of $\ln\left(\frac{S_1(t)}{S_1(t-1)}\right)$ and $\ln\left(\frac{S_2(t)}{S_2(t-1)}\right)$ is

$$\frac{\sigma_{11}\sigma_{21} + \sigma_{12}\sigma_{22}}{\left(\sqrt{\sigma_{11}^2 + \sigma_{12}^2}\right)\left(\sqrt{\sigma_{21}^2 + \sigma_{22}^2}\right)}.$$

This model can be easily extended to n-dimensional stochastic processes.

Example 13.20 Consider a non-linear SDE

$$dS(t) = dt + 2\sqrt{S(t)}dW(t),$$

with the initial condition $S(0) = S_0$. Prove that the solution is $S(t) = (W(t) + S_0)^2$.

Solution: Verify the solution using Itô's formula. Take $f(t, x) = S(t) = (X(t) + S_0)^2$ and $dX(t) = dW(t)$. The partial derivatives are $f_t = 0$, $f_x = 2(X(t) + S_0)$, and $f_{xx} = 2$. Thus

$$df(t, x) = [f_t + f_x \times 0 + 1 \times 2 f_{xx} \times 1]dt + f_x \times 1 dW(t), \tag{13.26}$$

$$df(t, x) = dt + 2(X(t) + S_0)dW(t), \tag{13.27}$$

implies

$$dS(t) = dt + 2\sqrt{S(t)}dW(t)$$

since $\sqrt{S(t)} = X(t) + S_0$.

For more description, readers may refer to [1–5, 7, 9–12]

13.4 Exercise Problems

1. Using Itô's formula, write the following stochastic processes in the differential form

   (a) $Y(t) = (W(t))^2$.
   (b) $Y(t) = 2 + t + e^{W(t)}$.
   (c) $Y(t) = (W_1)^2(t) + (W_2)^2(t)$ where $(W_1(t), W_2(t))$ is 2-dimensional Brownian motion.
   (d) $Y(t) = (t_0 + t, W(t))$.
   (e) Let $(W_1(t), W_2(t), W_3(t))$ be a 3-dimensional Brownian motion and $Y(t) = \big(W_1(t) + W_2(t) + W_3(t), W_2^2(t) - W_1(t)W_3(t)\big)$.

2. Find the SDE of

$$X(t) = \left(X(0)^3 + \frac{1}{3}W(t)\right)^{\frac{1}{3}}$$

   where $X(0) > 0$.
3. Prove that the limiting distribution of the SDE,

$$dX(t) = -\beta\left(X(t) - \frac{1}{2}\right)dt + \sqrt{\beta X(t)(1 - X(t))}dW(t)$$

   is uniform distribution on $(0, 1)$ for all $\beta > 0$.
4. Prove that the limiting distribution for the SDE

$$dX(t) = \beta(X(t) - (m + \gamma z))dt + \sigma\sqrt{z^2 - (X(t) - m)^2}dW(t)$$

   is beta distribution on $(m - z, m + z)$ with parameters $\frac{\beta(1-\gamma)}{\sigma^2}$ and $\frac{\beta(1+\gamma)}{\sigma^2}$.
5. Solve the SDE

$$dX(t) = -\theta\tan(X(t))dt + dW(t)$$

   for $\theta \geq \frac{1}{2}$.
6. Solve

$$dX(t) = -\beta(X(t) - \alpha)dt + \sqrt{2\beta(a(X(t))^2 + bX(t) + c)}dW(t),$$

   where α is the expectation of the invariant distribution and a, b, c are such that the square root is well defined.

7. Solve the system of SDEs

(a)

$$dX_1(t) = X_2(t)dt + \alpha_1 dW_1(t),$$

$$dX_2(t) = X_1(t)dt + \beta_1 dW_2(t),$$

where $(W_1(t), W_2(t))$ is a two-dimensional Wiener process and $\alpha_1, \beta_1 \in \mathbb{R}$.

(b)

$$dX(t) = Y(t)dt,$$

$$dY(t) = -\beta_2 X(t)dt - \alpha_2 Y(t)dt + \sigma dW(t),$$

where α_2, β_2 and $\sigma > 0$.

8. Consider the SDE of the form

$$dX(t) = \mu X(t)dt + \sigma dW(t),$$

where $X(0)$, μ and $\sigma > 0$ are constants. Find the strong solution of the above SDE. Also, find the distribution of $X(t)$.

9. Let $S(t) = e^{(\alpha-\alpha^2/2)t+\alpha W(t)}$. Derive the SDE satisfied by $[S(t)]^{-1}$.

10. Consider the stochastic differential equation

$$dX(t) = |X(t)|^r dW(t), \text{ with } X(0) = 0.$$

Find r such that it has a unique solution.

11. Let $X(t) = \int_0^t \sigma(s)dW(s)$. Prove that $\{Y(t), t \geq 0\}$ is a martingale where $Y(t) = (X(t))^2 - \int_0^t \sigma^2(s)ds$.

12. Prove that the process $\{X(t), t \geq 0\}$, $X(t) = \frac{1}{1+t}W(t)$ satisfies

$$dX(t) = -\frac{X(t)}{1+t}dt + \frac{1}{1+t}dW(t),$$

where $X(0) = 0$.

13. Show that $\{tW(t), t \geq 0\}$ is a Itô process. Find the SDE $d(tW(t))$.

14. Solve the SDE

$$dX(t) = (-\alpha X(t) + \beta)\,dt + \sigma dW(t),$$

where $X(0) = x_0$ and $\alpha > 0$. Verify that the solution can be written as

$$X(t) = e^{-\alpha t}\left(x_0 + \frac{\beta}{\alpha}\left(e^{\alpha t} - 1\right) + \sigma \int_0^t e^{\alpha s} dW(s)\right).$$

15. Solve the SDE

$$dX(t) = tX(t)dt + e^{t^2/2}dW(t),$$

with initial condition $X(0) = 1$.

16. Let

$$X(t) = X(0) + \int_0^t W(s)dW(s).$$

Show that $X(t)$ can also be written as

$$X(t) = X(0) + \frac{1}{2}\left(W(t)^2 - t\right).$$

17. Use proper coefficient matching to solve the following SDE:

$$dX(t) = -2\frac{X(t)}{1-t}dt + \sqrt{2t(1-t)}dW(t) \quad 0 \le t < 1 \text{ with } \quad X(0) = X_0.$$

Prove that the solution $\{X(t), t \ge 0\}$ follows a Gaussian process and find the covariance function $\text{cov}(X(s), X(t))$. Contrast this covariance function with the Brownian bridge covariance function.

18. Let $\{X(t), t \ge 0\}$ be a process satisfying $X(0) = 0$ and

$$dX(t) = a(X(t))\,dt + \sigma(X(t))\,dW(t),$$

where $a(\cdot)$ and $\sigma(\cdot)$ are smooth functions and $\sigma(\cdot) \ge \epsilon > 0$. Show that there exist a smooth function $b(\cdot)$ and a monotonically increasing function $f(\cdot)$ such that $Y(t) = f(X(t))$ and satisfies

$$Y(t) = Y(0) + \int_0^t b(Y(s))\,ds + W(t).$$

19. Prove that the solution of the SDE

$$dX(t) = \left((X(t))^3\right)dt + \left((X(t))^2\right)dW(t)$$

is $X(t) = \frac{1}{1-W(t)}$. Furthermore, the solution exists upto the first hitting time, $\tau = \inf\{t \ge 0 : W(t) = 1\}$. (Hint: Use the Itô's formula with $f(t, x) = \frac{1}{1-x}$.)

20. Consider a SDE

$$dX(t) = (m - X(t))dt + \alpha dN(t),$$

where $\{N(t); t \ge 0\}$ is a Poisson process with parameter $\lambda t > 0$. Prove that

(a) $X(t) = m + (X(0) - m)e^{-t} + \alpha e^{-t} \int_0^t e^s dN(s)$.
(b) $E(X(t)) = m + (X(0) - m)e^{-t} + \alpha\lambda(1 - e^{-t})$.
(c) $var(X(t)) = \frac{\lambda\alpha^2}{2}\left(1 - e^{-2t}\right)$.

21. Consider a SDE

$$t^3\,dX(t) = (3t^2X(t) + t)dt + t^6 dW(t), \quad X(1) = 0,$$

Using integration by inspection method, show that

$$X(t) = \frac{1}{4}\left(t^3 - \frac{1}{t}\right) + t^3(W(t) - W(1)), \quad t \geq 1.$$

22. Let $0 \leq s < t$. Find
   (a) $E[W(t)\int_0^s W(u)e^{au}du]$.
   (b) $E[\int_0^t W(u)e^{au}du \int_0^s W(v)e^{av}dv]$.

23. Let $u(s)$ be a bounded function. Let $X(t)$ be a solution of SDE $dX(t) = u(t)dt + dW(t)$. Consider the exponential process $M(t) = e^{-\int_0^t u(s)dW(s) - \frac{1}{2}\int_0^t u^2(s)ds}$. Prove that $Y(t) = M(t)X(t)$ is a $\mathcal{F}_t$ martingale.
24. Let $\alpha \in (0, \frac{1}{2})$ and $\sigma(x) = \min(1, |x|^\alpha)$. Find two solutions of

$$dX(t) = \sigma(X(t))dW(t)$$

and show that the two solutions are not equal in distribution.
Hint: One solution is identically zero. The other can be constructed by time changing a Brownian motion by the inverse of the increasing process

$$2\int_0^t (\min(1, |x|^{2\alpha}))^{-1}ds.$$

25. Consider the following model governing the dynamics of the positive price $\{S(t), 0 \leq t \leq T\}$ of an asset

$$dS(t) = \kappa(\theta - \ln(S(t)))\,S(t)dt + \sigma S(t)\,dW(t),$$

where κ, θ and σ are model parameters
   (a) Consider the log-price process $X(t) = \ln(S(t))$. Using Itô's formula, deduce that

$$dX(t) = \left(\kappa(\theta - X(t)) - \frac{1}{2}\sigma^2\right)dt + \sigma\,dW(t). \tag{13.28}$$

   (b) Solve SDE (13.28).

(c) Compute $E[X(T)]$, $E[(X(T))^2]$ by using the solution found in (b). **Hint**: If a function f is square-integrable on $[0, t]$, then $\int_0^t f(s)dW(s)$ is normal distributed with mean 0 and variance $\int_0^t f^2(s)ds$.

26. Consider a Poisson process $\{N(t), t \geq 0\}$ with mean $\lambda t > 0$. Assume $\{(S(t), t \geq 0\}$ represents the mean-reverting process specified by the following SDE:

$$dS(t) = -\alpha S(t)dt + \sigma\left(dN(t) - \beta dt\right),$$

where $S(0) > 0$ and $\alpha, \beta > 0$.

(a) Solve for $S(t)$.
(b) Compute $g(t) := E\left[S(t)\right]$ for all $t \geq 0$.
(c) Find the values of α, β, σ and λ such that the process $S(t)$ becomes a sub-martingale?
(d) Suggest a method for computing expectations of the form $E\left[\phi\left(S(T)\right)\right]$ where ϕ denotes a payoff function.

References

1. Castaneda, L.B., Arunachalam, V., Dharmaraja, S.: Introduction to Probability and Stochastic Processes with Applications. Wiley, New Jersey (2012)
2. Gianin, E.R. and Sgarra, C.: Mathematical Finance Theory Review and Exercises, Springer Cham, Second Edition, (2023). https://link.springer.com/book/10.1007/978-3-031-28378-9
3. Hull, J. and White, A.: Numerical Procedures for Implementing Term Structure Models I. Journal of Derivatives, **2**, 7–16 (1994)
4. Heston, S.L.: A Closed Solution for Options with Stochastic Volatility, with Application to Bond and Currency Options. Review of Financial Studies, **6**, 327–343 (1993)
5. Glosten, L.R., Jagannathan, R., and Runkle, D.E.: On the relation between the expected value and the volatility of the nominal excess return on stocks. The Journal of Finance **8**(5), 1779–1801 (1993)
6. Lakshmikantham, V., Deo, S.G.: Method of Variation of Parameters for Dynamic Systems. Routledge (2019)
7. Mathieu, K., Lindner, A., and Sørensen, M.: Statistical methods for stochastic differential equations. Monographs on Statistics and Applied Probability **124**, 7–12 (2012)
8. McKenna, J., Morrison, J.A.: Moments and correlation functions of solutions of a stochastic differential equation. J. Math. Phys. **11**(8), 2348–2360 (1970). https://doi.org/10.1063/1.1665402
9. Skiadas, C.H.: Exact solutions of stochastic differential equations: Gompertz, generalized logistic and revised exponential. Methodology and Computing in Applied Probability. **12**(2), 261–70 (2010)
10. Ross, S.M.: An Elementary Introduction to Mathematical Finance: Options and Other Topics. Cambridge, UK: Cambridge University Press (2003)
11. Shreve, S.E.: Stochastic calculus for finance II: Continuous-time models. Vol. 11. New York: springer (2004)
12. Mikosch, T.: Elementary stochastic calculus with finance in view. World scientific (1998)

Chapter 14
Applications to Finance - Derivatives Pricing

In this chapter, the problem of valuation of financial derivatives is discussed. Various stochastic models, including the diffusion models and jump-diffusion models, to model the dynamics of underlying securities are studied. This chapter begins with various bond pricing models and then the Black-Scholes option pricing formula is derived. Finally, greeks for options are presented.

14.1 Introduction

Derivatives are financial instruments whose values are dependent on some underlying financial asset or security, such as stocks, currency. Some of the common derivatives include forward contract, put option and call option. A forward contract is an agreement to buy or sell a security at a certain price at a future date. Such contracts can be used to reduce foreign exchange risk for future settlements. A buyer can lock in the current exchange rate for future settlement by using a forward contract.

Options are contracts which provide the buyer the right, but no obligation, to buy or sell, a risky asset at some future date at a predefined price to avoid any adverse future movements in stock price. Based on the rights of buying or selling a stock, options are classified as: call option and put option. A call option provides the holder of the option the right, but no obligation, to buy the underlying asset by a certain date at a certain price, whereas a put option provides the buyer of the contract the right, but no obligation, to sell the underlying asset at by a certain date at a certain price. Based on exercise date of the option contract, there are two types of options: American options and European options. American options can be exercised at any time before the expiration date of the contract, whereas European options can be exercised exactly at the expiration date of the contract. For example, a stock option is a derivative whose underlying security is a stock and whose value is dependent upon the stock value. This chapter discusses the problem of valuation of these derivatives.

D. Selvamuthu, *Introduction to Stochastic Processes*, University Texts in the Mathematical Sciences, https://doi.org/10.1007/978-981-97-6152-4_14

In this chapter, the underlying securities will be modeled with various stochastic models such as diffusion models, jump-diffusion models. The most fundamental forms of securities are bonds and stocks. Stocks, also known as equities, are securities that provide stockholders with a portion of the company's ownership. They carry a very high risk with the possibility of high returns. Bonds are fixed income traded securities with a fixed maturity date. They are usually the debts issued by companies to finance projects or operations. They are less risky compared to stocks. The bond price is dependent upon the prevailing interest rate in the market. The interest rate varies with time. Therefore, in this chapter, the interest rate will also be modeled and then an equation to obtain the bond price will be derived. Some of the examples of interest rate models include Vasicek model, Cox-Ingersoll-Ross model, etc.

One problem while pricing derivatives is the problem of arbitrage. Arbitrage refers to getting a positive return from zero investment with probability 1. This problem arises because of non-uniform pricing over markets. It occurs by simultaneously purchasing and selling same securities in two different markets with different price.

To avoid the problem of arbitrage and have uniform pricing, the approach of risk-neutral valuation to obtain derivative prices is used. A risk neutral probability measure (RNPM) is a probability measure such that each security price is exactly equal to the discounted expected value of future value of this security under this measure.

14.2 Derivative Security Pricing

Derivative securities are financial instruments whose values are derived from the value of an underlying asset, index, or rate. The pricing of derivative securities is a crucial aspect of financial markets and plays a central role in risk management, investment strategies, and financial decision-making.

There are three ways of computing a derivative security price.

1) Solving expectation using RNPM
2) Use Feynman-Kac theorem to solve the partial differential equation (PDE)
3) Monte Carlo simulations

According to the risk-neutral pricing concept, the price of a derivative on an asset $S(t)$ is unaffected by the risk preferences of market players. Hence, it may be assumed that they share the same risk aversion. In this situation, the valuation of the derivative price $f(t)$ at time t is performed as follows:

1. Assume that the expected return on the asset $S(t)$ is the risk-free rate, i.e., $\mu = r$.
2. Verify that the expected return on the derivative as of time t, given condition 1 is satisfied.
3. From time T to time t, discount is at the risk-free rate.

The first two steps involve analyzing the expectation at time t in a risk-free environment.

14.2.1 Change of Measure

Change of measure provides a method for transforming the probability measure associated with a stochastic process, allowing for a change in the source of randomness. This concept is particularly important in the pricing and risk management of financial derivatives.

Let us think about having two different problems on the same random experiment. For example, consider the SDE,

$$dX(t) = \sigma_t dW(t); \quad X(0) = x_0,$$

where $\{W(t), t \geq 0\}$ is a Brownian motion under the probability measure P. On the same random experiment under the probability measure Q, consider the SDE,

$$dX(t) = \mu_t dt + \sigma_t d\widetilde{W}(t); \quad X(0) = x_0,$$

where $\{\widetilde{W}(t), t \geq 0\}$ is a Brownian motion under the probability measure Q. Note that the process $\{X(t), t \geq 0\}$ is being modeled, and this process is an observable quantity whose distribution has to be obtained. Hence, the process $\{X(t), t \geq 0\}$ does not change, but its distribution from P to Q changes. Suppose we equate the above equations we get,

$$\mu_t dt + \sigma_t d\widetilde{W}(t) = \sigma_t dW(t)$$

$$\implies \frac{\mu_t}{\sigma_t} dt + d\widetilde{W}(t) = dW(t)$$

The above concept is captured in the following definition.

Definition 14.1 (Equivalent measure) We say that the probability measures P and Q are equivalent and we denote $P \sim Q$ if for all $A \in \mathfrak{F}$, $P(A) = 0 \iff Q(A) = 0$.

Definition 14.2 (Absolutely continuous) Q is absolutely continuous with respect to P, $Q << P$, if for any measurable set A, $P(A) = 0$ gives $Q(A) = 0$. That is, Q and P agree on null sets.

Definition 14.3 (Radon-Nikodym derivative) Let $(\Omega, \mathfrak{F}, P)$ be a probability space and Z be a non-negative RV with $E[Z] = 1$. Then the probability measure Q can be defined on $(\Omega, \mathfrak{F})$ such that $Q << P$ by

$$Q(A) = \int_A Z dP, \; A \in \mathfrak{F}. \tag{14.1}$$

In this case, Z is the Radon-Nikodym derivative of Q with respect to P and denoted by

$$Z = \frac{dQ}{dP}.$$

For any RV X, two expectation exist; one with respect to the original probability measure P denoted by $E_P(\cdot)$ and another with respect to a new probability measure Q denoted by $E_Q(\cdot)$.

Definition 14.4 (Radon-Nikodym derivative process) Eq. (14.1) can be written as

$$Q(A) = E_P\left[\frac{dQ}{dP} 1_A\right].$$

This quantity $\frac{dQ}{dP}$ is usually called the Radon-Nikodym derivative process or the likelihood ratio. It is easy to see that $\frac{dQ}{dP}$ is unique "almost surely".

Properties of Radon-Nikodym derivatives are as follows:

(a) $P\left(\frac{dQ}{dP} \geq 0\right) = 1$. In addition, if Q is equivalent to P, $P\left(\frac{dQ}{dP} > 0\right) = 1$.
(b) $E_P\left[\frac{dQ}{dP}\right] = 1$.
(c) For any $\mathfrak{F}$ measurable RV Y, $E_Q[Y] = E_P\left[Y\frac{dQ}{dP}\right]$.
(d) If Q is equivalent to P, $\frac{dP}{dQ} = \left(\frac{dQ}{dP}\right)^{-1}$.

Remark 14.1 For any RV X, we have $E_Q[X] = E_P[ZX]$. In addition, if $P(Z > 0) = 1$, then P and Q are equivalents, i.e., $Q << P$ and $P << Q$.

Example 14.1 Let P and Q be two probability measures on a space $(\Omega, \mathfrak{F})$. Suppose $\Omega = \{HH, HT, TH, TT\}$, the set of coin toss sequences of length 2. Let P correspond to probability $\frac{1}{3}$ for H and $\frac{2}{3}$ for T, let Q correspond to probability $\frac{1}{2}$ for H and $\frac{1}{2}$ for T. Find $Z(\omega)$, $\omega \in \Omega$, where Z is called the Radon-Nikodym derivative of Q with respect to P.

Solution: It is given that $P(H) = \frac{1}{3}$, $P(T) = \frac{2}{3}$ and $Q(H) = \frac{1}{2}$, $Q(T) = \frac{1}{2}$. Now in this question we need to find Z such that $Q(\omega) = Z(\omega)P(\omega)$ for all $\omega \in \Omega$.

Step 1: For $\omega = HH$, we have

$$Q(HH) = Z(HH)P(HH) \implies Z(HH) = \frac{9}{4}$$

for $\omega = HT$, we have

$$Q(HT) = Z(HT)P(HT) \implies Z(HT) = \frac{9}{8}$$

for $\omega = TH$, we have

$$Q(TH) = Z(TH)P(TH) \implies Z(TH) = \frac{9}{8}$$

and for $\omega = TT$, we have

$$Q(TT) = Z(TT)P(TT) \implies Z(TT) = \frac{9}{16}.$$

Hence, we get

$$Z(HH) = \frac{9}{4},\ Z(HT) = \frac{9}{8},\ Z(TH) = \frac{9}{8},\ Z(TT) = \frac{9}{16}.$$

Step 2: Now we need to show that $E_P[Z(\omega)] = 1$. Consider,

$$\begin{aligned}
E_P[Z(\omega)] &= Z(HH)P(HH) + Z(HT)P(HT) + Z(TH)P(TH) + Z(TT)P(TT) \\
&= \left(\frac{9}{4}\cdot\frac{1}{9}\right) + \left(\frac{9}{8}\cdot\frac{2}{9}\right) + \left(\frac{9}{8}\cdot\frac{2}{9}\right) + \left(\frac{9}{16}\cdot\frac{4}{9}\right) \\
&= 1.
\end{aligned}$$

Example 14.2 Let X be a random variable defined on $(\Omega, \mathfrak{F}, P)$ which follows $N(0, 1)$. Define $Z = \exp\{\theta X - \frac{1}{2}\theta^2\}$ where θ is a constant. Show that $X \sim N(\theta, 1)$ on $(\Omega, \mathfrak{F}, Q)$.

Solution: Let $f^P(x)$ and $f^Q(x)$ be the probability density function of random variable X w.r.t to probability measure P and Q respectively. Then,

$$\begin{aligned}
f^Q(x) &= Z.f^P(x) \\
&= \exp\left\{\theta x - \frac{1}{2}\theta^2\right\}\frac{1}{\sqrt{2\pi}}\exp\left\{\frac{-1}{2}x^2\right\} \\
&= \frac{1}{\sqrt{2\pi}}\exp\left\{\frac{-1}{2}x^2 + \theta x - \frac{1}{2}\theta^2\right\} \\
&= \frac{1}{\sqrt{2\pi}}\exp\left\{\frac{-1}{2}(x-\theta)^2\right\}.
\end{aligned}$$

Hence, $X \sim N(\theta, 1)$ on $(\Omega, \mathfrak{F}, Q)$.

14.2.2 Girsanov's Theorem for Diffusion Processes

This section includes the results which provide the way to change the probability measure under which a diffusion process evolves.

Lemma 14.1 *Let Y be a $\mathfrak{F}_t$-measurable RV and $Z(t)$ is the Radon-Nikodym derivative process which is defined as $Z(t) = E_P[Z|\mathfrak{F}_t]$. Then*

1. $E_Q[Y] = E_P[YZ(t)]$.
2. *For* $0 \le s \le t$, $E_Q[Y \mid \mathfrak{F}_s] = \frac{1}{Z(s)}E_P[YZ(t) \mid \mathfrak{F}_s]$.

Proof In view of the properties of conditional expectation, and definition of $Z(t)$,

$$E_Q[Y] = E_P[YZ] = E_P[E_P[YZ \mid \mathfrak{F}_t]] = E_P[YE_P[Z \mid \mathfrak{F}_t]] = E_P[YZ(t)]$$

To prove next part, it can be shown

$$\int_A \frac{1}{Z(s)} E_P\left[YZ(t) \mid \mathcal{F}_s\right] dQ = \int_A Y dQ, \quad \forall A \in \mathcal{F}_s.$$

$$\begin{aligned}\int_A \frac{1}{Z(s)} E_P\left[YZ(t)|\mathcal{F}_s\right] dQ &= E_Q\left[1_A \frac{1}{Z(s)} E_P\left[YZ(t)|\mathcal{F}_s\right]\right] \\ &= E_P\left[1_A E_P\left[YZ(t)|\mathcal{F}_s\right]\right] \\ &= E_P\left[E_P\left[1_A YZ(t)|\mathcal{F}_s\right]\right] = E_P\left[1_A YZ(t)\right] \\ &= E_Q\left[1_A Y\right] = \int_A Y dQ.\end{aligned}$$

Now, we present Girsanov's theorem, which is the main tool used in practice to change from physical or real-world probability measure to risk-neutral probability measure.

Theorem 14.1 (Girsanov's Theorem) *Let $\{W(t), t \geq 0\}$ be a Brownian motion on a filtered probability space $\left(\Omega, \mathcal{F}, \{\mathcal{F}_t\}_{t\geq 0}, P\right)$. Let $\{\theta(t), t \geq 0\}$ be an adapted process such that*

$$Z(t) = \exp\left\{-\int_0^t \theta(s) dW(s) - \frac{1}{2}\int_0^t \theta^2(s) ds\right\}$$

becomes a martingale. Define

$$\widetilde{W}(t) = \int_0^t \theta(s) ds + W(t), \quad 0 \leq t \leq T.$$

Then the process $\{\widetilde{W}(t), t \geq 0\}$ is a Brownian motion with respect to the new probability measure Q where

$$dQ(\omega) = Z(T) dP(\omega).$$

Proof Note that since $\{Z(t), 0 \leq t \leq T\}$ is a martingale $E_P[Z(T)] = Z(0) = 1$ and $Z(T)$ is $\mathcal{F}_T$-adapted RV such that, it is positive a.s.. Thus, Q is a probability measure. Lévy's theorem will be used to show that $\widetilde{W}(t)$ is a Brownian motion. Observe that $\widetilde{W}(0) = 0$ and the quadratic variation of $\widetilde{W}$ is the same as the quadratic variation of Brownian motion. Hence it remains to show that $\widetilde{W}$ is a martingale under Q. Since $Z(t)$ is a martingale, it can be seen that $Z(t) = E_P\left[Z(T) \mid \mathcal{F}_t\right]$ is the Radon-Nikodym derivative process. Next observe that $M(t) = \widetilde{W}(t) Z(t)$ is a martingale under P (proof is left to the readers as an exercise). Let $0 \leq s \leq t$. Then, by Lemma 14.1,

$$E_Q\left[\widetilde{W}(t) \mid \mathcal{F}_s\right] = \frac{1}{Z(s)} E_P\left[\widetilde{W}(t) Z(t) \mid \mathcal{F}_s\right] = \frac{1}{Z(s)} E_P\left[M(t) \mid \mathcal{F}_s\right] = \frac{M(s)}{Z(s)} = \widetilde{W}(s).$$

In other words, $\{\widetilde{W}(t), t \geq 0\}$ is martingale under Q. This completes the proof. □

Example 14.3 Let $Y(t) = t + W(t)$. For each $T > 0$, find a probability measure Q_T on $\mathfrak{F}_T$ such that $\{Y(t), 0 \leq t \leq T\}$ is a Brownian motion under Q_T. Show that there exists a probability measure Q on $\mathfrak{F}_\infty$ such that

$$Q|_{\mathfrak{F}_T} = Q_T \; \forall \, T > 0.$$

Solution: Taking $\theta(t) = 1$ in the Girsanov's theorem, it is seen that $Y(t) = t + W(t)$ is a Brownian motion under the probability measure Q_T on $\mathfrak{F}_T$, where Q_T is given by

$$dQ_T(\omega) = Z(T)dP(\omega); \quad Z(t) = e^{-W(t)-\frac{t}{2}}.$$

Note here that $Z(t)$ is a martingale under P. To prove the second part, first it can be shown that

$$Z(T)dP = Z(t)dP \text{ on } \mathfrak{F}_t, t \leq T.$$

Indeed, for any bounded $\mathfrak{F}_t$-measurable function, one has

$$\begin{aligned}\int_\Omega fZ(T)dP &= E[fZ(T)] = E\left[E\left[fZ(T) \mid \mathfrak{F}_t\right]\right] \\ &= E\left[fE\left[Z(T) \mid \mathfrak{F}_t\right]\right] = E[fZ(t)] = \int_\Omega fZ(t)dP\end{aligned}$$

Thus, $Q_T = Q_S$ on $\mathfrak{F}_t$ for all $t \leq \min\{T, S\}$. Hence there exists Q on $\mathfrak{F}_\infty$ such that $Q = Q_T$ on $\mathfrak{F}_T$ for all $T < \infty$ and the result follows.

Example 14.4 Find a probability measure Q on $\mathfrak{F}_T$ such that the process $\{Y(t), 0 \leq t \leq T\}$ given by

$$dY(t) = tdt + (2t + 1)dW(t)$$

becomes a martingale under the probability measure Q.

Solution: Observe that $s \mapsto \frac{s^2}{(2s+1)^2}$ is continuous and therefore $\int_0^T \frac{s^2}{(2s+1)^2}ds < +\infty$. Hence, the stochastic process

$$Z(t) = \exp\left\{\int_0^t \frac{s}{2s+1}dW(s) - \frac{1}{2}\int_0^t \frac{s^2}{(2s+1)^2}ds\right\}$$

is a martingale under P and $E[Z(T)] = 1$. Define

$$dQ(\omega) = Z(T)dP(\omega).$$

Then, Q is a probability measure and by Girsanov's theorem, the process

$$\widetilde{W}(t) = W(t) + \int_0^t \frac{s}{2s+1}ds : 0 \leq t \leq T$$

is a Brownian motion under Q. The process $\{Y(t), 0 \leq t \leq T\}$ can be expressed in terms of $\widetilde{W}(t)$ as

$$\begin{aligned} dY(t) &= tdt + (2t+1)dW(t) \\ &= tdt + (2t+1)\left\{d\widetilde{W}(t) - \frac{t}{2t+1}dt\right\} \\ &= (2t+1)d\widetilde{W}(t). \end{aligned}$$

Hence,

$$Y(t) = Y(0) + \int_0^t (2s+1)d\widetilde{W}(s).$$

Thus $\{Y(t), 0 \leq t \leq T\}$ is a martingale under the probability measure Q.

Changes in probability measures are essential in mathematical finance because they enable one to represent derivative pricing as discounted value of payoffs in "risk-neutral" form. As a result, many derivative pricing (for example, options) may be determined in closed-form.

Example 14.5 Prove that a measure Q is equivalent to measure P under which the discounted stock price process $\{D(t)S(t), t \geq 0\}$ is martingale where $dD(t) = -r(t)D(t)dt$ and $dS(t) = \alpha(t)S(t)dt + \sigma(t)S(t)dW(t)$.
Solution: The measure Q is computed directly using the Girsanov's theorem.

$$\begin{aligned} d(D(t)S(t)) &= -r(t)D(t)S(t)dt + D(t)dS(t) \\ &= (\alpha(t) - r(t))D(t)S(t)dt + D(t)S(t)\sigma(t)dW(t) \\ &= \sigma(t)D(t)S(t)(\theta(t)dt + dW(t)), \end{aligned}$$

where

$$\theta(t) = \frac{\alpha(t) - r(t)}{\sigma(t)}.$$

Define a new process $\{\widetilde{W}(t), t \geq 0\}$ by

$$d\widetilde{W}(t) = \theta(t)dt + dW(t),$$

and observe

$$d(D(t)S(t))dt = \sigma(t)D(t)S(t)d\widetilde{W}(t),$$

which shows that $\{D(t)S(t), t \geq 0\}$ is a martingale.

Definition 14.5 (Exponential process inducted by a function) Let $u : [0, T] \to \mathbb{R}$ be a measurable function such that $\int_0^t \mid u(s) \mid^2 ds < \infty$. Suppose u is a deterministic function. Then the stochastic process $\{M(t), t \geq 0\}$ where $M(t) = e^{\int_0^t u(s)dW(s) - \frac{1}{2}\int_0^t u^2(s)ds}$ is called the exponential process inducted by u.

Suppose $u(s) = \sigma$, a constant, then $M(t) = e^{\sigma W(t) = \frac{\sigma^2}{2}t}$ is a $\mathcal{F}_t$ martingale. Suppose $u(s) = s$. By integrating in $d(tW(t)) = tdW(t) - W(t)dt$ yields

$$\int_0^t s dW(s) = tW(t) - \int_0^t W(s)ds.$$

Then, $M(t) = e^{tW(t) - \frac{t^3}{6} - \int_0^t W(s)ds}$ is a $\mathcal{F}_t$ martingale.

14.3 Forward Contracts and Bond Pricing

A forward contract is an over-the-counter contract between two parties to buy or sell an asset at a predetermined price and at a predetermined date in the future that can be customised to fit specific needs. While forward contracts can be utilized for both hedging and trading, their non-standardized nature makes them especially well-suited for hedging purposes. Bond forwards are subject to cash settlement on expiry, this rule also applies on the transactions of a contract based on bond futures. A forward contract's value is equal to the bond price minus the present value of coupon payments minus the present value of the price at expiration. Forward contracts are used to reduce the risk of price volatility between now and a future date. Futures contracts are similar to forward contracts, but they are standardized. Let $S(t)$ be the price of unerlying asset at time t,

S_0 – the current spot price of the underlying asset,

r – the risk-free rate that applies to the life of the forward contract, and

T – the delivery date in years.

The forward price formula (which assumes zero dividends) denoted by F is given by

$$F = S_0 e^{rT}. \tag{14.2}$$

Example 14.6 A person is looking to buy a forward contract for an investment asset currently trading at Rs 1000. The risk-free rate in his country is 4%. The one year forward price for this asset can be calculated as:

$$F = 1000e^{((0.04)(1))} = Rs1,040.81.$$

14.3.1 Introduction to Bond Pricing

Bond pricing is a fundamental concept in finance that involves determining the present value of future cash flows associated with a bond. Bonds are debt securities that represent loans made by investors to entities such as governments, municipalities, or corporations. The price of a bond is crucial for both investors and issuers, as it influences investment decisions, yields, and overall financial planning. A zero-coupon bond is a type of bond that does not pay periodic interest (known as coupons) during its life. It is issued at a discount to its face value. If we consider the interest rate r to be constant, then the price of zero-coupon bond maturing at T denoted by P is given by

$$P = \frac{1}{(1+r)^T}.$$

If the rate r is continuously compounded, then

$$P = e^{-rT}.$$

Short Rate Models (zero coupon bond pricing only)
Short rate models are mathematical models used to evaluate interest rate derivatives that depict the variation in interest rates over time by identifying the evolution of the short rate at time t denoted r_t over time. Interest rate derivatives are priced via short rate modelling. Some of the important short rate models are

1. **Vasicek Model** The SDE for the spot rate r_t under a Vasicek model is given as:

$$dr_t = a\,(b - r_t)\,dt + \sigma dW(t).$$

2. **Cox-Ingersoll Ross (CIR) Model** The SDE for the spot rate r_t under a CIR model is given as:

$$dr_t = (\alpha - \beta r_t)dt + \sigma\sqrt{r_t}dW(t).$$

3. **Hull and White Model** The SDE for the spot rate r_t under a Hull and White model is given as:

$$dr_t = \kappa_t(\theta_t - r_t)dt + \sigma_t dW(t).$$

The pricing of a zero-coupon bond under these interest rate models is discussed in the following section.

14.3.2 Bond Pricing Under Vasicek Model

The Vasicek model is a mathematical model that describes the evolution of interest rates in finance. The model has been developed for credit markets and can be used to value interest rate derivatives. Czech mathematician Oldich Vasicek introduced it in 1977, and it can also be thought of as a stochastic investment model.

According to the model, the short rate is determined by the stochastic differential equation

$$dr_t = a\,(b - r_t)\,dt + \sigma dW(t). \tag{14.3}$$

The standard deviation parameter, σ, governs interest rate volatility and, in some ways, characterizes the magnitude of the instantaneous randomness inflow. The usual parameters b, a, and σ, along with the starting condition r_0, thoroughly define the dynamics and may be summarized as follows, provided a is non-negative:

- b: **long-term mean level**. In the long term, all future r trajectories will evolve around the mean level b.
- a: **speed of reversion**. It represents the velocity at which such trajectories will gather in time around b.
- σ: **instantaneous volatility**. It quantifies the magnitude of unpredictability/ randomness entering the system moment by moment. A higher σ indicates higher randomness.

14.3.2.1 Vasicek Interest Rate Model

Assume that the Lipschitz condition is satisfied for Equation (14.3). Now,

$$\begin{aligned} d(e^{at}r_t) &= e^{at}dr_t + ae^{at}r_t dt \\ &= e^{at}(ab - ar_t)dt + e^{at}\sigma dW(t) + ae^{at}r_t dt \\ e^{at}r_t &= r_0 + ab\int_0^t e^{as}ds + \sigma\int_0^t e^{as}dW(s). \end{aligned}$$

The strong solution is

$$r_t = r_0 e^{-at} + b(1 - e^{-at}) + \sigma e^{-at}\int_0^t e^{as}dW(s).$$

If r_0 is constant, then r_t is Gaussian distributed with

$$E(r_t) = r_0 e^{-at} + b(1 - e^{-at})$$
$$var(r_t) = \frac{\sigma^2}{2a}(1 - e^{-2at}).$$

As $t \to \infty$

$$r_t \sim N\left(b, \frac{\sigma^2}{2a}\right).$$

If $r_t >> b$ then the process will bring it back to b and if $r_t << b$ then the process will bring it up to b. Hence, this model is also known as mean reverting Ornstein-Uhlenbeck process.

1. If $r_t = b$, then the drift is zero.
2. If $r_t < b$, then the drift is greater than zero, and this implies that the process will try to pull back.
3. If $r_t > b$, then the drift is less than zero, and this implies that the process will try to pull down.

Thus, the interest rate does not go too high or too low. This is applicable from economic viewpoint too. Note that, the values of interest rates in this model assume negative values with positive probability which is an undesirable property.

Proposition 14.1 *Consider the spot rate r_t satisfying Vasicek's mean-reverting model*

$$dr_t = a\,(b - r_t)\,dt + \sigma dW(t).$$

The price of the zero-coupon bond that pays off \$1 *at time T is given by*

$$P(t,T) = A(t,T)e^{-B(t,T)r_t},$$

where

$$B(t,T) = \frac{1 - e^{-a(T-t)}}{a},$$
$$A(t,T) = e^{\frac{(B(t,T)-T+t)\left(a^2 b - \sigma^2/2\right)}{a^2} - \frac{\sigma^2 B(t,T)^2}{4a}}.$$

Proof The deduction of the previous formula by integrating the stochastic differential equation between t and T yields

$$-a\int_t^T r_s ds = r_T - r_t - \sigma\,(W(T) - W(t)) - ab(T-t).$$

Taking the exponential

$$e^{-\int_t^T r_s ds} = e^{\frac{1}{a}(r_T - r_t) - \frac{\sigma}{a}(W(T)-W(t)) - b(T-t)}. \tag{14.4}$$

Multiplying in the stochastic differential equation by e^{as} yields the exact equation

$$d\left(e^{as} r_s\right) = abe^{as} ds + \sigma e^{as} dW(s).$$

Integrating between t and T to get

$$e^{aT} r_T - e^{at} r_t = b\left(e^{aT} - e^{at}\right) + \sigma \int_t^T e^{as} dW(s).$$

Solving for r_T and subtracting r_t yields

$$r_T - r_t = -r_t\left(1 - e^{-a(T-t)}\right) + b\left(1 - e^{-a(T-t)}\right) + \sigma e^{-aT} \int_t^T e^{as} dW(s).$$

Dividing by a and using the notation for $B(t, T)$ yields

$$\frac{1}{a}\left(r_T - r_t\right) = -r_t B(t, T) + bB(t, T) + \frac{\sigma}{a} e^{-aT} \int_t^T e^{as} dW(s).$$

Substituting into (14.4) and using that $W(T) - W(t) = \int_t^T dW(s)$,

$$e^{-\int_t^T r_s ds} = e^{-r_t B(t,T)} e^{bB(t,T)-b(T-t)} e^{\frac{\sigma}{a}\int_t^T (e^{as-aT}-1) dW(s)}.$$

Taking the measurable part out and dropping the independence condition, the price of the zero-coupon bond at time t becomes

$$\begin{aligned} P(t, T) &= E\left[e^{-\int_t^T r_s ds} \mid \mathfrak{F}_t\right] \\ &= e^{-r_t B(t,T)} e^{bB(t,T)-b(T-t)} E\left[e^{\frac{\sigma}{a}\int_t^T (e^{as-aT}-1) dW(s)} | \mathfrak{F}_t\right] \\ &= e^{-r_t B(t,T)} e^{bB(t,T)-b(T-t)} E\left[e^{\frac{\sigma}{a}\int_t^T (e^{as-aT}-1) dW(s)}\right] \\ &= e^{-r_t B(t,T)} A(t, T), \end{aligned}$$

where

$$A(t, T) = e^{bB(t,T)-b(T-t)} E\left[e^{\frac{\sigma}{a}\int_t^T (e^{as-aT}-1) dW(s)}\right]. \tag{14.5}$$

As a stochastic integral, $\int_t^T \left(e^{as-aT} - 1\right) dW(s)$ is normal distributed with mean zero and variance

$$\int_t^T \left(e^{as-aT}-1\right)^2 ds = e^{-2aT}\frac{e^{2aT}-e^{2at}}{2a} - 2e^{-aT}\frac{e^{aT}-e^{at}}{a} + (T-t)$$
$$= B(t,T)\frac{2-aB(t,T)}{2} - 2B(t,T) + (T-t)$$
$$= -\frac{a}{2}B(t,T)^2 - B(t,T) + (T-t).$$

Then, the log-normal variable $e^{\frac{\sigma}{a}\int_t^T (e^{as-aT}-1)dW(s)}$ has the mean

$$E\left[e^{\frac{\sigma}{a}\int_t^T (e^{as-aT}-1)dW(s)}\right] = e^{\frac{1}{2}\frac{\sigma^2}{a^2}\left[-\frac{a}{2}B(t,T)^2 - B(t,T)+(T-t)\right]}.$$

Substituting into (14.5) and using that,

$$bB(t,T) - b(T-t) + \frac{1}{2}\frac{\sigma^2}{a^2}\left[-\frac{a}{2}B(t,T)^2 - B(t,T) + (T-t)\right]$$
$$= -\frac{\sigma^2}{4a}B(t,T)^2 + (T-t)\left(\frac{\sigma^2}{2a^2} - b\right) + \left(b - \frac{\sigma^2}{2a^2}\right)B(t,T)$$
$$= -\frac{\sigma^2}{4a}B(t,T)^2 + \frac{1}{a^2}\left(a^2 b - \sigma^2/2\right)\left(B(t,T) - T + t\right).$$

We obtain,

$$A(t,T) = e^{\frac{(B(t,T)-T+t)\left(a^2b-\sigma^2/2\right)}{a^2} - \frac{\sigma^2 B(t,T)^2}{4a}}.$$

Note that $B(t,T)$ depends on the time to maturity, $T-t$, and the spot rate r_t.

14.3.3 Bond Pricing Under CIR Model

The Cox-Ingersoll-Ross model is a popular mathematical framework used to model interest rates and, by extension, bond prices. It is an extension of the Vasicek model and is widely employed in finance for interest rate risk management and derivative pricing. The CIR model incorporates mean-reverting properties, reflecting the observation that interest rates tend to revert to a long-term mean over time. In addition, this model is non-negative which make it more useful to model the interest rates.

14.3.3.1 Cox-Ingersoll-Ross Model

The Cox-Ingersoll-Ross model describes the evolution of interest rates. The model can be used in the valuation of interest rate derivatives. It was introduced in 1985 by John C. Cox, Jonathan E. Ingersoll, and Stephen A. Ross as an extension of the Vasicek model. The interest rate process $\{r_t, t \geq 0\}$ is defined by the SDE

$$dr_t = (\alpha - \beta r_t)dt + \sigma\sqrt{r_t}dW(t)$$

where $\alpha, \beta, \sigma > 0$, are positive constants. In this model, volatility is also dependent on the interest rate unlike in Vasicek interest rate model where it was constant. This model is also a mean-reverting model. As soon as r_t tends to zero, the model becomes

$$dr_t = \alpha dt.$$

This implies

$$r_t = r_0 + \alpha t.$$

This stochastic process will make sure that r_t linearly increases as $r_0 > 0$ and $\alpha t > 0$. The model does not have a closed form solution. Only numerical schemes can be used to solve the CIR model. $E(r_t)$ can be obtained as follows:

$$d(e^{\beta t}r_t) = \beta e^{\beta t}r_t dt + e^{\beta t}dr_t + 0(dr_t dr_t).$$

The above equation in the integral form is

$$\begin{aligned} e^{\beta t}r_t &= r_0 + \int_0^t \beta e^{\beta s}r_s ds + \int_0^t e^{\beta s}((\alpha - \beta r_s)ds + \sigma\sqrt{r_s}dW(s)), \\ &= r_0 + \int_0^t \alpha e^{\beta s}ds + \int_0^t e^{\beta s}\sigma\sqrt{r_s}dW(s), \\ &= r_0 + \frac{\alpha}{\beta}(e^{\beta t} - 1) + \int_0^t e^{\beta s}\sigma\sqrt{r_s}dW(s). \end{aligned}$$

It is known that $\int_0^t \sigma\sqrt{r_s}dW(s)$ is an Itô integral and hence a martingale. Consequently, its expectation is given by

$$E(e^{\beta t}r_t) = r_0 + \frac{\alpha}{\beta}\left(e^{\beta t} - 1\right).$$

Thus,

$$E(r_t) = r_0 e^{-\beta t} + \frac{\alpha}{\beta}\left(1 - e^{-\beta t}\right).$$

Variance of r_t is given by

$$var(r_t) = \frac{\sigma^2}{\beta} r_0 \left(e^{-\beta t} - e^{-2\beta t}\right) + \frac{\alpha\sigma^2}{2\beta^2}\left(1 - 2e^{-\beta t} + e^{-2\beta t}\right).$$

As $t \to \infty$,

$$E(r_t) = \frac{\alpha}{\beta} \text{ and } var(r_t) = \frac{\alpha\sigma^2}{2\beta^2}.$$

If α and β are chosen appropriately, variance can be brought down and the model will be more stable. The bond pricing equation can be solved explicitly. Now, the bond pricing formula for CIR model is presented without the proof.

Theorem 14.2 *Let r_t follow the CIR model, i.e., $dr_t = a(b - r_t)dt + \sigma\sqrt{r_t}dW(t)$, then the zero-coupon bond price for this model can be derived similarly to Vasicek's model and is given by*

$$P(t, T) = A(t, T)e^{-B(t,T)r_t}.$$

The function $A(t, T)$ and $B(t, T)$ are given by

$$A(t, T) = \left(\frac{2\psi e^{(a+\psi)(T-t)/2}}{(\psi + a)(e^{\psi(T-t)} - 1) + 2\psi}\right)^{2ab/\sigma^2},$$

$$B(t, T) = \frac{2(e^{\psi(T-t)} - 1)}{(\psi + a)(e^{\psi(T-t)} - 1) + 2\psi}$$

where $\psi = (a^2 + 2\sigma^2)^{1/2}$.

14.3.4 Bond Pricing Under Hull–White Model

The Hull-White model is a model of future interest rates in financial mathematics. It belongs to the family of no-arbitrage models that can match today's term structure of interest rates in their most basic form. Interest rate derivatives such as Bermuda swaptions can be valued in the model since it is relatively simple to transpose the mathematical description of the evolution of future interest rates onto a tree or lattice. The first Hull–White model was described by John C. Hull and Alan White in 1990. The model is still popular in the market today. Short rate no-arbitrage models include the Hull–White model. It is applied to the pricing of derivatives involving interest rates, such as caps and floors. It extends Vasicek's pioneering equilibrium model (1977). The model is defined as

$$dr_t = \kappa_t(\theta_t - r_t)dt + \sigma_t dW(t).$$

Two of the major model characteristics are

1. The short rate r_t is mean-reverting (if rates deviate significantly from the long-term mean θ, they will probably converge back to this level and the speed of mean reversion is κ_t).
2. The short rate r_t follows a normal distribution (negative interest rates are therefore feasible, which was formerly regarded as a disadvantage but is now beneficial).

The model is simple and provides the closed form pricing formulas for zero-coupon bonds, bond options (therefore caps and floors), and swaptions. Calibration is thus quite simple.

The model belongs to the family of affine term-structure models, which allow the price of a zero-coupon bond to be written as

$$P(t, T) = e^{(A(t,T)+r_t B(t,T))}.$$

This class also includes the Cox-Ingersoll-Ross (1985) model. Monte Carlo simulations are also fairly straightforward to implement because the model assumes a normal distribution. To simulate several trajectories, this model utilize an easy Euler approximation:

$$r_{t+\Delta t} = r_t + \kappa_t(\theta_t - r_t)\Delta t + \sigma_t\sqrt{\Delta t}Z,$$

where $Z \sim N(0, 1)$.

14.4 Option Pricing

Options are a type of derivative financial instrument in which two parties agree to trade an asset at a predetermined price before a future date. An option grants its holder the right to buy or sell an asset at the exercise price, but the holder is not required to exercise (buy or sell) the option.

European option: A European option is a type of option contract in which execution is possible only at expiration date. In other words, if the underlying security, such as a stock, has changed in value, an investor will be unable to exercise the option and take delivery or sell the shares before the option's maturity date. Instead, the call or put action will take place only on the option's maturity date.

American option: The American option is another type of option contract that can be exercised at any time up to and including the expiration date. The names of these two versions should not be confused with the geographical location because the name only represents the right of execution.

14.4.1 Asset Price Models

Let $S(t)$ denotes the price of stock at time t. The basic model for stock price process follows the following SDE

$$dS(t) = \mu S(t)dt + \sigma S(t)dW(t)$$

where σ and μ represent the standard deviation (or volatility) and expected return of the short-term returns, respectively. If $\sigma = 0$, the model simplifies to $dS(t) = \mu S(t)dt$, indicating $S(t) = S(0)e^{\mu t}$, which is just the value of an interest-bearing, risk-free asset.

14.4.2 Option Pricing Under Black–Scholes Model

The Black–Scholes model, sometimes known as the Black–Scholes–Merton model, is a mathematical model that attempts to describe the behavior of financial derivatives, most notably options. American economist Fischer Sheffey Black (January 11, 1938–August 30, 1995) and Myron Samuel Scholes (born July 1, 1941) proposed it in 1973. It provided theoretical backing for trading options to hedge positions, which had previously been practiced but lacked strong backing. The model allows us to compute the price of an option depending on a variety of different parameters. There are several variants of the Black–Scholes model available today, each of which aims to enhance the model based on certain criteria, generally at the expense of a large increase in complexity. The initial model will be the main focus of attention which serves as the foundation for all subsequent models.

Notation Before getting to the equations, there are certain terms that will be stated out, which will be spelt.

C = Price of Call option
K = Strike price of the option
r = risk-free interest rate
σ = standard deviation of the stock returns
T = time to option maturity (in years)

A European call option with strike price K at time T gives its holder an opportunity (i.e., the right but not the obligation) to buy from the writer one share of the prescribed stock at time T for price K. Note that if, at time T the value of the stock is less than K, then the option is worthless and will not be exercised, but if the value of the stock is greater than K, then the option is valuable and therefore it will be exercised. Thus the value of option at time T is $(S(T) - K)^+ = \max\{0, S(T) - K\}$. Our goal is to determine the value of this option at time 0. We will write C to denote the value of the option. Since C depends on both time and on the underlying stock, we see that $C(t, S(t))$ denotes the value of option at time t, $0 \leq t \leq T$.

Black–Scholes Equation

The Black–Scholes equation illustrates how an option's price changes over time. We derive the Black–Scholes formula for the European call option using two different approaches. We apply the results of the risk-neutral measure for a European call option and in second approach we will use Feynman-Kac theorem.

Risk-Neutral Measure Approach: Let us consider a European call option on a stock with strike price K. The claim has a payoff function $H = \max(S(t) - K, 0)$ at expiry time T. The stock price follows the geometric Brownian motion

$$dS(t) = \mu S(t)dt + \sigma S(t)dW(t)$$

where $\mu \in \mathbb{R}$ and $\sigma > 0$.

Assume that

- $S(t)$ follows a geometric Brownian motion with constant drift μ and volatility σ.
- There are no arbitrage opportunities, i.e., there is no way to make riskless profit.
- There are no transaction costs or taxes.
- The underlying security does not pay a dividend.
- It is possible to borrow and lend cash at a known constant risk-free interest rate.
- It is possible to buy and sell any asset, even fraction of stock.

It is known that $S(t)$ satisfies the equation

$$dS(t) = \mu S(t)dt + \sigma S(t)dW(t)$$

where $\{W(t), t \geq 0\}$ is a Brownian motion. Stock price $S(t)$ under RNPM Q with risk-free interest rate r, is of the form

$$dS(t) = rS(t)dt + \sigma S(t)d\widetilde{W}(t)$$

where $\{\widetilde{W}(t), t \geq 0\}$ is the Brownian motion under Q, and after simplification above SDE become

$$S(t) = S(0)e^{\left(r-\frac{1}{2}\sigma^2\right)t+\sigma\widetilde{W}(t)} \tag{14.6}$$

where $S(0)$ is the price of the stock at time $t = 0$. Under a RNPM the fair price of the call option is:

$$C(t) = E_Q\left(e^{-r(T-t)}\max(S(T) - K, 0) \mid \mathcal{F}_t\right). \tag{14.7}$$

For any $0 \leq t \leq T$, Equation (14.6) can be rewrite as

$$\begin{aligned} S(T) &= S(t)\exp\left\{\left(r - \frac{1}{2}\sigma^2\right)(T-t) + \sigma\left(\widetilde{W}(T) - \widetilde{W}(t)\right)\right\} \\ &= S(t)\exp\left\{\left(r - \frac{1}{2}\sigma^2\right)(T-t) + \sigma\sqrt{T-t}Z\right\} \end{aligned}$$

where

$$Z = \frac{\widetilde{W}(T) - \widetilde{W}(t)}{\sqrt{T-t}} \sim N(0, 1).$$

Substitute this in Equation (14.7),

$$\begin{aligned} C(t) &= E_Q\left(e^{-r(T-t)} \max\left(S(T) - K, 0\right)\right) \\ &= E_Q\left(e^{-r(T-t)} S(T) 1_{\{S(T)>K\}}\right) - K e^{-r(T-t)} E_Q\left(1_{\{S(T)>K\}}\right) \\ &= (I) - (II). \end{aligned}$$

First, the term (II) is evaluated.

$$\begin{aligned} E_Q\left(1_{\{S(T)>K\}}\right) &= Q\left(S(T) > K\right) \\ &= Q\left(\ln S(T) > \ln K\right) \\ &= Q\left(Z > -\left(\frac{-\ln K + \ln S(t) + \left(r - \frac{1}{2}\sigma^2\right)(T-t)}{\sigma\sqrt{T-t}}\right)\right). \end{aligned}$$

But, it is known that, under measure Q, $Z \sim N(0, 1)$. Hence,

$$\begin{aligned} Q\left(S(T) > K\right) &= Q\left(Z > -\frac{\ln\left(\frac{S(t)}{K}\right) + \left(r - \frac{1}{2}\sigma^2\right)(T-t)}{\sigma\sqrt{T-t}}\right) \\ &= Q\left(Z < \frac{\ln\left(\frac{S(t)}{K}\right) + \left(r - \frac{1}{2}\sigma^2\right)(T-t)}{\sigma\sqrt{T-t}}\right) \\ &= \Phi\left(d_2\right) \end{aligned}$$

where

$$d_2 = \frac{\ln\left(\frac{S(t)}{K}\right) + \left(r - \frac{1}{2}\sigma^2\right)(T-t)}{\sigma\sqrt{T-t}}.$$

Now, compute the term (I):

$$
\begin{aligned}
E_Q & \left(e^{-r(T-t)} S(T) 1_{\{S(T)>K\}}\right) \\
& = e^{-r(T-t)} E_Q\left(S(t) e^{(r-\frac{1}{2}\sigma^2)(T-t)-\sigma\sqrt{T-t}\,Z} 1_{\{S(T)>K\}}\right) \\
& = E_Q\left(S(t) e^{-\frac{1}{2}\sigma^2(T-t)-\sigma\sqrt{T-t}\,Z} 1_{\{S(T)>K\}}\right) \\
& = \frac{1}{\sqrt{2\pi}} \int_{-\infty}^{d_2} \left(S(t) e^{-\sigma\sqrt{T-t}z-\frac{1}{2}\sigma^2(T-t)}\right) e^{-\frac{z^2}{2}} dz \\
& = S(t) \frac{1}{\sqrt{2\pi}} \int_{-\infty}^{d_2} e^{-\frac{1}{2}(z+\sigma\sqrt{T-t})^2} dz \\
& = S(t) \frac{1}{\sqrt{2\pi}} \int_{-\infty}^{+d_2+\sigma\sqrt{T-t}} e^{-\frac{x^2}{2}} dx \quad (\text{as } x = z + \sigma\sqrt{T-t}) \\
& = S(t)\Phi\left(d_2 + \sigma\sqrt{T-t}\right) \\
& = S(t)\Phi\left(d_1\right)
\end{aligned}
$$

where

$$d_1 = d_2 + \sigma\sqrt{T-t}.$$

Thus, for any $0 \le t < T$

$$C(t) = S(t)\Phi\left(d_1\right) - Ke^{-r(T-t)}\Phi\left(d_2\right)$$

where d_1 and d_2 are given by

$$
\begin{aligned}
d_1 &= \frac{\ln\left(\frac{S(t)}{K}\right) + \left(r + \frac{1}{2}\sigma^2\right)(T-t)}{\sigma\sqrt{T-t}}, \\
d_2 &= \frac{\ln\left(\frac{S(t)}{K}\right) + \left(r - \frac{1}{2}\sigma^2\right)(T-t)}{\sigma\sqrt{T-t}} = d_1 - \sigma\sqrt{T-t}.
\end{aligned}
$$

PDE Approach: Let us now turn our attention to present another famous and significant theorem in stochastic of finance. The Feynman-Kac Theorem also known as Feynman-Kac Formula, named after R. Feynman and M. Kac, establish a link between parabolic partial differential equation and stochastic processes.

Theorem 14.3 *Let the stochastic process $\{X(t), 0 \le t \le T\}$ satisfy the following SDE*

$$dX(t) = \mu(t, X(t))dt + \sigma(t, X(t))dW(t), \tag{14.8}$$

where $\mu(t, X(t))$ and $\sigma(t, X(t))$ are smooth functions on $[0, T] \times \mathbb{R}$, called the drift and diffusion functions respectively. In addition, let the initial condition on $X(t)$ be $X(0) = x$, for some $x \in \mathbb{R}$. Then the solution of the following PDE

$$g_t(t,x)+\mu(t,x)g_x(t,x)+\frac{1}{2}\sigma^2(t,x)g_{xx}(t,X)-rg(t,x)=0, \tag{14.9}$$

subject to the boundary condition $g(T, X(T)=x)=h(x)$, $x\in\mathbb{R}$, *is a function* $g:[0,T]\times\mathbb{R}\to\mathbb{R}$ *given by*

$$g(t,x)=E\left(e^{-r(T-t)}h(X(T))|X(t)=x\right). \tag{14.10}$$

The Feynman–Kac theorem can be used in both directions. That is, if we know that $X(t)$ follows the SDE (14.8) and we are given the PDE with boundary condition, then we can always obtain the solution $g(t,x)$ as described in (14.10). On the other hand if we know that the solution $g(t,x)$ is given by (14.10) and that $X(t)$ follows the process in (14.8), then we are assured that $g(t,x)$ satisfies the PDE(14.9) The generator of the process in (14.8) is defined as the operator

$$\mathcal{A}=\mu(t,x(t))\frac{\partial}{\partial x}+\frac{1}{2}\sigma^2(t,x(t))\frac{\partial^2}{\partial x^2}.$$

The PDE (14.9) can then be rewritten as

$$\frac{\partial g}{\partial t}+\mathcal{A}g-rg=0.$$

Let us now examine how we can apply the above theorem to derive the BS formula for a derivative security. Let the stock price $S(t)$ be driven by the process

$$dS(t)=\mu S(t)dt+\sigma S(t)dW(t). \tag{14.11}$$

The risk-neutral process is given by

$$dS(t)=rS(t)dt+\sigma S(t)d\widetilde{W}(t), \tag{14.12}$$

where $\widetilde{W}(t)=W(t)+\frac{\mu-r}{\sigma}t$. Suppose a derivative is written on this stock. Let $V(t,S(t))$ be the price of this security at any time $t\in[0,T)$, and $V(T,S(T))$ be its payoff on maturity. Note $V:[0,T]\times\mathbb{R}^+\to\mathbb{R}^+$, $\mathbb{R}^+$is the set of non-negative reals. By Itô's Lemma, we have

$$dV(t)=dV(t,S(t))=\left(\frac{\partial V}{\partial t}+rS(t)\frac{\partial V}{\partial x}+\frac{1}{2}\sigma^2 S^2(t)\frac{\partial^2 V}{\partial x^2}\right)dt \tag{14.13}$$
$$+\sigma S(t)\frac{\partial V}{\partial x}d\widetilde{W}(t).$$

Suppose the derivative security can be hedged. We replicate the derivative security by taking a portfolio in which we take $a(t)$ shares of underlying stock and $b(t)$ shares of a risk-free asset whose price is governed by an ODE. Then, we will have,

$$dV(t) = a(t)dS(t) + b(t)r\beta(t)dt \qquad (14.14)$$
$$= a(t)(rS(t)dt + \sigma S(t)d\widetilde{W}(t)) + rb(t)\beta(t)dt.$$

From (14.13) and (14.14), it follows that

$$\frac{\partial V}{\partial x} = a(t) \quad \text{and} \quad \frac{\partial V}{\partial t} + \frac{1}{2}\sigma^2 S^2(t)\frac{\partial^2 V}{\partial x^2} = rb(t)\beta(t).$$

Now using $b(t)\beta(t) = V(t) - a(t)S(t) = V(t) - \frac{\partial V}{\partial x}S(t)$, we finally get that the derivative price satisfies the following Black- Scholes PDE

$$\frac{\partial V}{\partial t} + rS\frac{\partial V}{\partial S} + \frac{1}{2}\sigma^2\frac{\partial^2 V}{\partial S^2} - rV = 0.$$

The generator of the process is given by $\mathcal{A} = rS\frac{\partial}{\partial S} + \frac{1}{2}\sigma^2\frac{\partial^2}{\partial S^2}$. By the Feynman–Kac theorem, the time t value of the derivative is the solution of

$$V(t, S(t)) = e^{-r(T-t)}E_Q\left(h(S(T))|\mathcal{F}_t\right),$$

where Q is a RNPM, and $h(S(T))$ is payoff of the derivative security at maturity. For instance, $h(T) = (S(T) - K)^+$ for European call option maturing at T with strike price K.

One allows continuous trading, and if as the number of shares held a time u, the net gain through the trading the stock is $\int_0^t a_u dS(u)$. One has a similar net gain of $\int_0^t b_u dW(u)$, when trading bonds, if b_u is the number of bonds held at time s. The pair (a, b) is called a trading strategy. Let $V(t) = a_t S(t) + b_t W(t)$ the amount of wealth one has a time t. The strategy is self-financing if , for all $t \geq 0$,

$$V(t) = V(0) + \int_0^t a_u dS(u) + \int_0^t b_u dW(u).$$

The first integral represents the net gain from trading in the stock. The second integral is the net gain from the trading in the bond. The above equation says that one's wealth at time t is equal to what one starts with plus what one has realized through trading in the stock and bond. Assume that there are no transaction costs, i.e., no brokerage fees.

We will prove the previous theorem by the partial differential equation approach proposed by Black and Scholes in 1973. We consider the stock price process that follows geometric Brownian motion

$$dS(t) = \mu S(t)dt + \sigma S(t)dW(t)$$

where $\mu \in \mathbb{R}$ and $\sigma > 0$ and $\{W(t), t \geq 0\}$ is a Brownian motion. Consider a portfolio at time t which consists of $\alpha(t)$ shares on stock and $W(t)$ units of bonds with value $R(t)$. The bond is assumed to be risk-free with deterministic equation

$$dR(t) = rR(t)dt,$$

where $r > 0$ is the interest rate. At time t, the value of the portfolio is:

$$V_t = a(t)S(t) + b(t)R(t).$$

The portfolio is supposed to hedge a European call option with value $C = C(t) = C(t, S(t))$. The European call option with strike price K and expiry date T has payoff value $C(t) = \max(S(t) - K, 0)$. From Itô's formula for the call option C, we have

$$\begin{aligned} dC &= \frac{\partial C}{\partial t}dt + \frac{\partial C}{\partial S}dS + \frac{\partial^2 C}{\partial S^2}(dS)^2 \\ &= \left(\frac{\partial C}{\partial t} + \mu S\frac{\partial C}{\partial S} + \frac{1}{2}\sigma^2 S^2\frac{\partial^2 C}{\partial S^2}\right)dt + \sigma S\frac{\partial C}{\partial S}dW(t). \end{aligned} \tag{14.15}$$

The value of the portfolio satisfies

$$\begin{aligned} dV_t &= a(t)dS(t) + b(t)dR(t) \\ &= a(t)\left(\mu S(t)dt + \sigma S(t)dW(t)\right) + W(t)rR(t)dt \\ &= \left(\alpha\mu S(t) + \beta rR(t)\right)dt + \alpha(t)\sigma S(t)dW(t). \end{aligned} \tag{14.16}$$

Under a self-financing strategy, (14.16) must coincide with (14.15). Since terms dt and $dW(t)$ are independent, the respective coefficients must be equal. Otherwise there shall be an arbitrage opportunity. Therefore:

$$a(t)S(t) + b(t)R_t = C_t \tag{14.17}$$

$$\alpha\mu S + \beta rR = \frac{\partial C}{\partial t} + \mu S\frac{\partial C}{\partial S} + \frac{1}{2}\sigma^2 S^2\frac{\partial^2 C}{\partial S^2} \tag{14.18}$$

$$\alpha\sigma S = \sigma S\frac{\partial C}{\partial S}. \tag{14.19}$$

From equation (14.19), we have

$$\alpha = \frac{\partial C}{\partial S}. \tag{14.20}$$

Substituting (14.20) in (14.17), we have:

$$\beta = \frac{1}{R}\left(C - S\frac{\partial C}{\partial S}\right). \tag{14.21}$$

Again substituting equations (14.20) and (14.21) in (14.18), we get

$$\frac{\partial C}{\partial t} + rS\frac{\partial C}{\partial S} + \frac{1}{2}\sigma^2 S^2 \frac{\partial^2 C}{\partial S^2} = rC. \tag{14.22}$$

Thus we have shown that the value of a call option satisfies the above partial differential equation. This equation is also known as the Black–Scholes equation, which can be solved under appropriate conditions. Equation (14.22) can be solved for the final condition

$$C(T) = C(S(t), T) = \max(S(t) - K, 0)$$

along with the boundary conditions

$$C(0, t) = 0 \text{ and } C(S, t) \to S \text{ as } t \to \infty.$$

Finding the solution of the above equation is a long computational procedure and the reader may refer to Atkinson et al. [1]. The solution for the value of a European call option at time T with strike price K, underlying stock price S, risk-free interest rate r and volatility σ starting from time t is given as

$$C(S(t), t) = S\Phi(d_1) - Ke^{-r(T-t)}\Phi(d_2) \tag{14.23}$$

where

$$d_1 = \frac{\ln(S/K) + (r + \sigma^2/2)(T - t)}{\sigma\sqrt{T - t}},$$

$$d_2 = \frac{\ln(S/K) + (r - \sigma^2/2)(T - t)}{\sigma\sqrt{T - t}}.$$

Example 14.7 Suppose you wish to purchase a Rs. 210 strike European call option. The stock is presently selling at Rs. 208.99 per share. You estimate the stock's volatility to be 17%. You may borrow and lend money at a 5% interest rate. The option has a time to maturity of 77 days. How much should you pay for the option contract (per share)?

Solution: Using the given information, first find d_1 and d_2.

$$\mathrm{d}_1 = \frac{1}{0.17\sqrt{0.21095}}\left[\ln\left(\frac{208.99}{210}\right) + 0.21095\left(0.05 + \frac{0.17^2}{2}\right)\right] = 0.1123799.$$

$$\mathrm{d}_2 = \frac{1}{0.17\sqrt{0.21095}}\left[\ln\left(\frac{208.99}{210}\right) + 0.21095\left(0.05 - \frac{0.17^2}{2}\right)\right] = 0.0343001.$$

Using the table values of CDF of standard normal distribution (refer Appendix A.4), find $\Phi(d_1)$ and $\Phi(d_2)$.

$$\Phi(d_1) = \Phi(0.1123799) = \frac{1}{\sqrt{2\pi}}\int_{-\infty}^{0.1123799} \mathrm{e}^{-\frac{1}{2}z^2}\,dz = 0.3516077.$$

$$\Phi(d_2) = \Phi(0.0343001) = \frac{1}{\sqrt{2\pi}} \int_{-\infty}^{0.0343001} e^{-\frac{1}{2}z^2} dz = 0.3135993.$$

Now, substitute these values in (14.23) to get the call option price (per share).

$$\begin{aligned} C(208.99, 0.21095) &= (0.3516077)(208.99) - (0.3135993)(210)e^{-(0.05)(0.21095)} \\ &\approx 2.3454. \end{aligned}$$

The options contract should cost roughly \$2.35 (per share). The fact that options contracts are issued in lots of 100 shares means that this specific contract would sell for around \$235.

14.4.3 Option Greeks

Option Greeks are used to calculate an option's risk and sensitivity to the risk factors. The variables are symbolized by the Greek letters Delta, Gamma, Theta, Vega, and Rho. There are other additional Greeks that may be derived from those listed above.

The Greeks reflect the various levels of risk involved in options trading. The Greeks are the collective name for these dimensions. The Greeks are important risk management tools that can assist option traders in making informed judgments about what to trade and when to trade. They assist in determining how price changes, interest rate fluctuations, volatility, and time that affect the price of an option contract.

Definition 14.6 (Option Greek Delta) A measure of an option's price sensitivity to changes in the value of the underlying asset is called the Delta Δ of the option.

$$\Delta = \frac{\partial V}{\partial S},$$

where ∂ is first derivative. V is the option price and S is the underlying asset price.

For example, if the price of the underlying asset rises by \$1, the option's price will rise by Δ. Since the derivative is a measure of the variable's rate change over a specific period, it is important to recall the first derivative of the option and the prices of the underlying assets. The delta is typically computed as a decimal value between -1 and 1. Call options have a delta ranging from 0 to 1, whereas puts have a delta ranging from -1 to 0. The option is more in-the-money =if delta closes to 1. The weighted average of all the options' deltas inside a portfolio makes up the option's portfolio delta.

Delta is sometimes referred to as a hedge ratio. If a trader is aware of the option's delta, he can hedge his position by purchasing or selling the amount of underlying assets multiplied by delta.

Several facts relating to delta are provided below

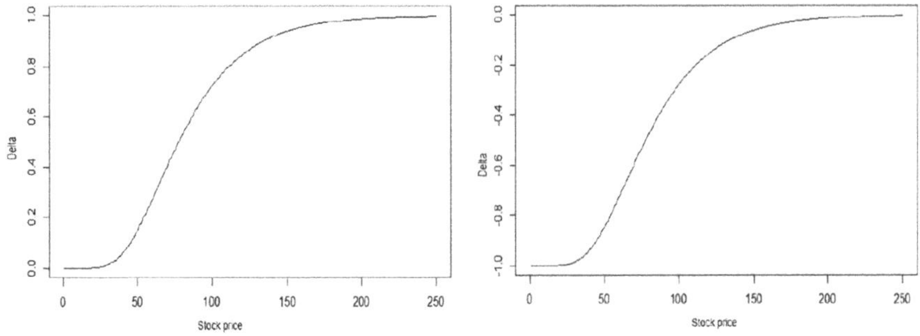

(a) call delta (b) put delta

Fig. 14.1 Option Greek Delta

1. The value of a call increases as the stock price increases. However, the increase in the value of the call is always less than or equal to the increase in the stock price. Thus, $0 \leq \Delta_{call} \leq 1$.
2. The value of a put decreases as the stock price increases. However, the decrease in the value of the put is always less than or equal to the increase in the stock price. Thus, $-1 \leq \Delta_{put} \leq 0$.
3. Under the Black-Scholes framework, the delta for a call is given by $\Delta_{call} = e^{-\delta T}\Phi(d_1)$ and the delta for a put is given by $\Delta_{put} = -e^{-\delta T}\Phi(-d_1)$. The derivations of these formulas are omitted.
4. Graphs of Δ as a function of S are shown in Fig. 14.1. The other parameters are held constant at $K = 100$, $\sigma = 0.3$, $r = 0.05$, $\delta = 0.02$, and $T = 1$.

Definition 14.7 (Option Greek Gamma) Gamma Γ is a metric for comparing changes in the delta to those in the underlying asset's price. The option's delta will vary by the gamma amount if the price of the underlying asset rises by \$1. The principal use of gamma is to calculate the delta of an option.

$$\Gamma = \frac{\partial \Delta}{\partial S} = \frac{\partial^2 V}{\partial S^2}.$$

The gamma of long options is positive. When an option is at-the-money, (option strike price equals the price of the underlying asset) it has the highest gamma. Gamma, on the other hand, falls when an option is deep-in-the-money or out-the-money. Several facts relating to Gamma are provided below

1. Gamma for a call is always equal to gamma for a put with the same parameters. That is, $\Gamma_{call} = \Gamma_{put}$.
2. Gamma is always greater than or equal to zero.
3. A graph of Γ as a function of S is shown in Fig. 14.2. The other parameters are held constant at $K = 100$, $\sigma = 0.3$, $r = 0.05$, $\delta = 0.02$, and $T = 1$.

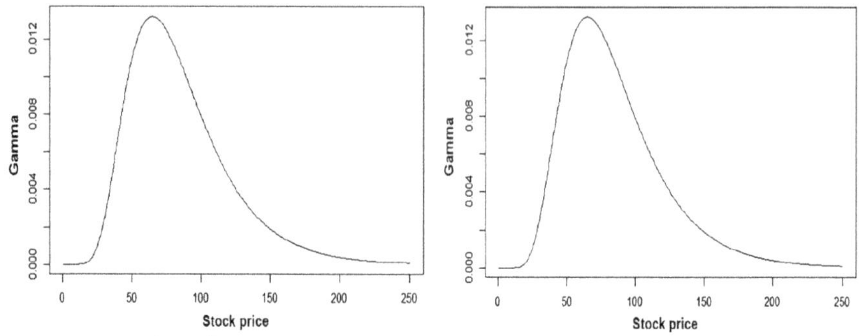

(a) call gamma (b) put gamma

Fig. 14.2 Option Greek Gamma

Definition 14.8 (Option Greek Vega) The option Greek ν quantifies the sensitivity of an option price to the volatility of the underlying asset. If the underlying asset's volatility rises by 1%, the option price will rise by the vega amount.

$$\nu = \frac{\partial V}{\partial \sigma},$$

where σ is the volatility of the underlying asset.

Rather than a decimal number, the vega is expressed as a monetary value. In general, a rise in vega correlates to an increase in option value (both calls and puts).

1. Vega for a call is always equal to vega for a put with the same parameters. That is, $\nu_{call} = \nu_{put}$.
2. Vega is always greater than or equal to zero.
3. A graph of vega as a function of S is shown in Fig. 14.3. The other parameters are held constant at $K = 100, \sigma = 0.3, r = 0.05, \delta = 0.02$, and $T = 1$.

Definition 14.9 (Option Greek Theta) Theta Θ is a measurement of the sensitivity of the option price in relation to the time to maturity of the option. The price of the option will vary by the theta amount if the time to maturity of the option is shortened by one day. Theta option Greek is also known as time decay.

$$\Theta = -\frac{\partial V}{\partial \tau},$$

where τ denotes the time to maturity of the option.

Theta is usually negative for options. For some European options, it could, nevertheless, be positive. When the option is at-the-money, theta is the most negative.

1. Theta is occasionally defined to be $\theta = V_t/365$. This definition of theta results from measuring t in days rather than years.

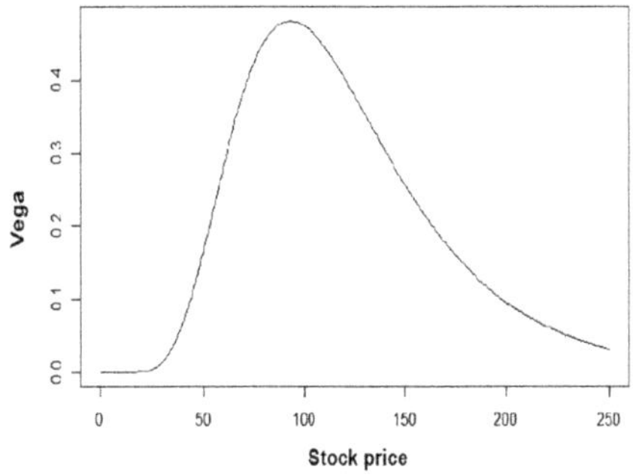

(a) call vega

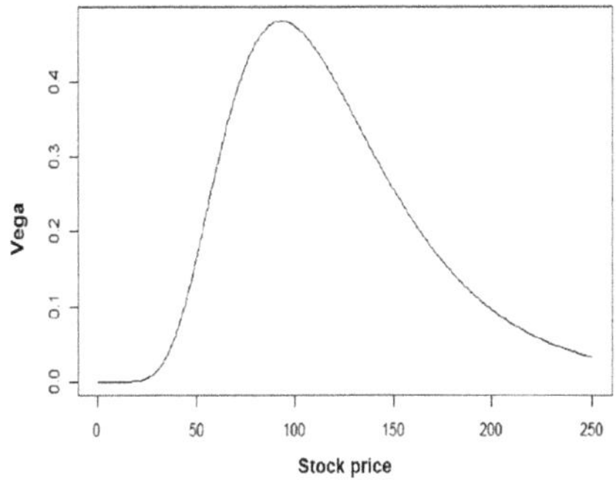

(b) put vega

Fig. 14.3 Option Greek Vega

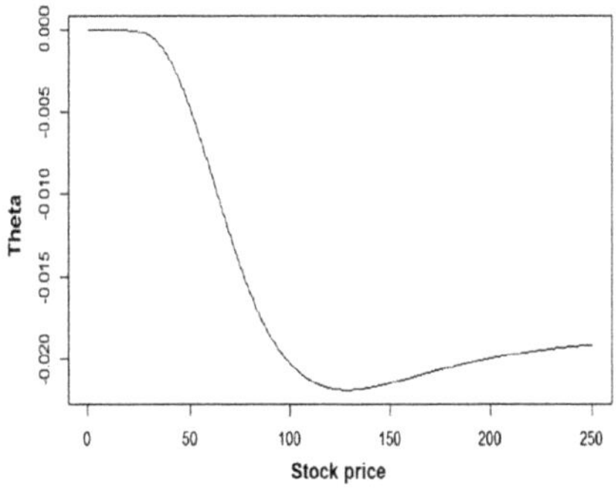

(a) call theta

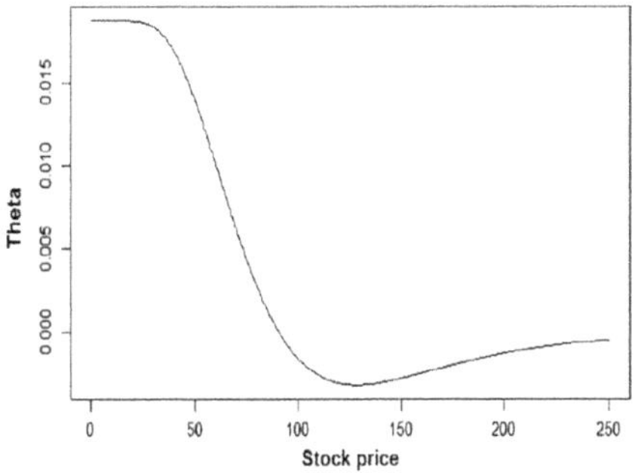

(b) put theta

Fig. 14.4 Option Greek Theta

2. Graphs of θ as a function of S are shown in Fig. 14.4. The other parameters are held constant at $K = 100, \sigma = 0.3, r = 0.05, \delta = 0.02$, and $T = 1$.

Definition 14.10 (Option Greek Rho) The sensitivity of the option price to interest rates is measured by Rho ρ. The option prices vary by the amount rho if a benchmark interest rate rises by 1%. The rho is regarded as the least significant option Greek because option prices are often less sensitive to changes in interest rates than to changes in other factors.

$$\rho = \frac{\partial V}{\partial r},$$

where r represents the interest rate.

In general, call options have a positive rho, whereas put options have a negative rho.

1. Rho for a call is always greater than or equal to zero. Rho for a put is always less than or equal to zero. These facts can be easily recalled by noting to role of r in

the expressions

$$Call = S(0)e^{-rT}\Phi(d_1) - Ke^{-rT}\Phi(d_2)$$

and

$$Put = Ke^{-rT}\Phi(-d_2) - x_0 e^{-rT}\Phi(-d_1).$$

2. Graphs of ρ as a function of S are shown in Fig. 14.5. The other parameters are held constant at $K = 100, \sigma = 0.3, r = 0.05, \delta = 0.02$, and $T = 1$.

Table 14.1 shows the summary of Greeks.

The famous Black and Scholes theory of option pricing has two remarkable features: The hedging strategy eliminates the risk entirely, and the option price does not depend at all on the average return of the underlying asset. Although everyone agrees that the geometric Brownian motion model of stock prices considered by Black and Scholes (1973) is only a rough first approximation to the real world, there is no consensus on how to price an option in a more general, hopefully more realistic model. It is important to realize that in most models of stock price movements, except for a few special cases (of which the geometric Brownian model is one), risk in option trading cannot be eliminated. Even in the geometric Brow-

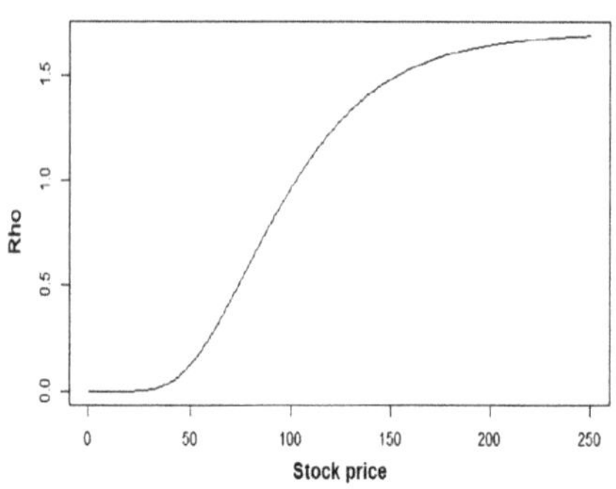

(a) call rho

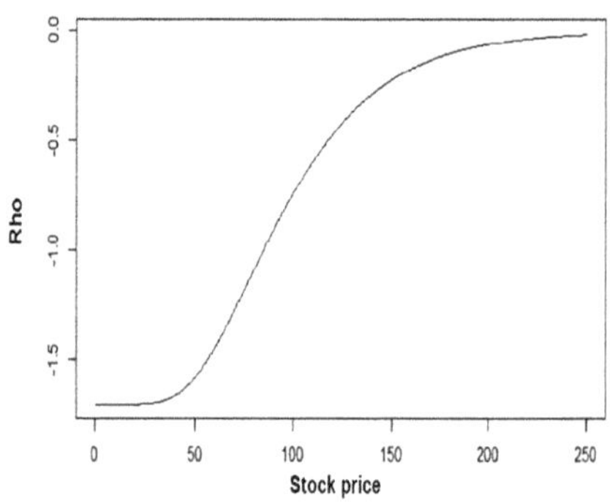

(b) put rho

Fig. 14.5 Option Greek Rho

Table 14.1 Summary table of the Greeks

Greek	Symbol	Measures	Definition
Delta	$\Delta = \frac{\partial v}{\partial S}$	Equity exposures	Change in option price due to spot price
Gamma	$\Gamma = \frac{\partial \Delta}{\partial S} = \frac{\partial^2 v}{\partial S^2}$	Payout convexity	Change in delta due to spot price
Vega	$\nu = \frac{\partial V}{\partial \sigma}$	Volatility exposure	Change in option price due to volatility
Theta	$\Theta = -\frac{\partial V}{\partial \tau}$	Time decay	Change in option price due to time
Rho	$\rho = \frac{\partial V}{\partial r}$	Interest rate exposure	Change in option price due to interest rate

nian model, risk cannot be eliminated if trading is only allowed at discrete times, as noted already by Black and Scholes themselves ([7]). Arbitrage arguments hence cannot be used to fix the option price in a general price movement model. The problem of pricing inherently risky options has been addressed quite frequently in the finance literature, and obviously also by practitioners, for whom it is a fact of life. Broadly speaking, one can find two main approaches in the published literature (see, e.g. Hull 1994, [4]). The first consists of various more or less ad hoc modifications of the basic Black-Scholes formula, where, in particular, the volatility of the stock is adjusted to account in some way for risk, and also for trading costs.

The Heston-Hull-White model

Consider a risk-neutral measure Q which can be determined by choosing as numerate the T-bond and by selecting the market price of one volatility-dependent asset and Heston [5]. The risk-neutral dynamics of these stochastic processes are as follows:

$$dS(t) = r(t)S(t)dt + \sqrt{V(t)}S(t)dW_1(t), \tag{14.24}$$

$$dV(t) = k_v(\theta_v - V(t))dt + \omega_v\sqrt{V(t)}dW_2(t), \tag{14.25}$$

$$dr(t) = k_r(\theta_r - r(t))dt + \omega_r dW_3(t), \tag{14.26}$$

where $\{W_1(t), t \geq 0\}$, $\{W_2(t), t \geq 0\}$ and $\{W_3(t), t \geq 0\}$ are Wiener processes such that $\text{cov}(dW_1(t), dW_2(t)) = \rho_v dt$, $cov(dW_1(t), dW_3(t)) = \rho_r dt$ and $cov(dW_2(t), dW_3(t)) = 0$.

Equation (14.24) describes the evolution with respect to time t of the price of an equity $S(t)$. Its volatility is given by $\sqrt{v(t)}$ where $v(t)$ is described by Equation (14.25). This is exactly the dynamic from the well known *Heston stochastic volatility model*. Equation (14.26) describes the dynamic of short-term interest rate with time-dependent but deterministic mean-reversion level θ_t. This is known as the *Hull-White model*. That is why the model (14.24)–(14.26) is called the *hybrid Heston-Hull-White model*.

Example 14.8 Compute the Greeks for a European put option, $\Delta = \frac{\partial P}{\partial S}$, $\Gamma = \frac{\partial^2 P}{\partial S^2}$, $\nu = \frac{\partial P}{\partial \sigma}$, $\Theta = -\frac{\partial P}{\partial \tau}$, $\rho = \frac{\partial P}{\partial r}$. Note that Γ and Vega for a European put option with strike price K are the same as Γ and Vega for a European call option with strike price K.

Solution: The Black-Scholes model provides a formula for the price of a European put option:

$$P = Ke^{-r\tau}\Phi(-d_2) - S_0\Phi(-d_1)$$

where P is the price of the put option, S_0 is the current price of the underlying asset, K is the strike price of the option, r is the risk-free interest rate, τ (tau) is the time to expiration, σ (sigma) is the volatility of the underlying asset, $N(\cdot)$ is the cumulative distribution function of the standard normal distribution, and d_1 and d_2 are given by:

$$d_1 = \frac{\ln\left(\frac{S_0}{K}\right) + (r + \frac{\sigma^2}{2})\tau}{\sigma\sqrt{\tau}} \quad \text{and} \quad d_2 = d_1 - \sigma\sqrt{\tau}.$$

Now, let's calculate the Greeks:

1. Delta (Δ):

$$\Delta = \frac{\partial P}{\partial S} = -\Phi(-d_1).$$

2. Gamma (Γ):

$$\Gamma = \frac{\partial^2 P}{\partial S^2} = \frac{e^{-d_1^2/2}}{S_0\sigma\sqrt{\tau}}.$$

3. Vega (ν):

$$\nu = \frac{\partial P}{\partial \sigma} = Ke^{-r\tau}\sqrt{\tau}\Phi(d_2)$$

where $\Phi(\cdot)$ is the probability density function of the standard normal distribution.

4. Theta (Θ):

$$\Theta = -\frac{\partial P}{\partial \tau} = \frac{Ke^{-r\tau}\sigma}{2\sqrt{\tau}}\Phi(d_2) - rKe^{-r\tau}\Phi(-d_2).$$

5. Rho (ρ):

$$\rho = \frac{\partial P}{\partial r} = -K\tau e^{-r\tau}\Phi(-d_2).$$

These formulas give the sensitivity of the option price to changes in the underlying asset price (Δ), changes in volatility (ν), changes in time to expiration (Θ), and changes in the risk-free interest rate (ρ).

For more description, readers may refer to [1–11].

14.5 Exercise Problems

1. Assume the spot rate r_t satisfies the mean reverting model with jumps.

$$dr_t = (\alpha - \beta r_t)dt + \sigma dW(t) + JdN(t)$$

where jump sizes J follow a normal distribution $J \sim N(\mu_J, \sigma_J^2)$, and $\{N(t), t \geq 0\}$ represent the Poisson process with intensity λ. Find the price of the zero-coupon bond that pays off Re 1 at time T.

2. Consider the Black, Derman and Toy model

$$d(\ln r_t) = \left[\theta_t + \frac{\sigma_t'}{\sigma_t} lnr_t\right] dt + \sigma_t dW(t). \tag{14.27}$$

(a) Solve the above SDE in the case when σ is a constant.
(b) Show that in this case the spot rate r_t is log-normal distributed.
(c) Find the mean and the variance of r_t.

3. Consider the spot rate model by Black and Karasinski

$$d(\ln r_t) = [\theta_t - a_t \ln r_t] dt + \sigma_t dW(t). \tag{14.28}$$

Find the explicit formula for r_t in this case.
4. Find the price of an infinitely lived bond also known as a perpetual bond, this type of bond has no maturity date and provides interest payments indefinitely. As a result, the issuer is not obligated to repay the principal amount in the case when spot rates satisfy Vasicek's model.
5. Let

$$dr_t = \alpha - \beta r_t dt + \sigma\sqrt{r_t} dW(t),$$

where α, β and σ are positive constants. Show that r_t doesn't have closed form solution.
6. Let $\{W(t), t \geq 0\}$ be a Brownian motion under P. Define Q on $\{\mathcal{F}\}_{t\geq 0}$ by

$$\frac{dQ}{dP} = e^{-bW(t) - b^2 t/2}.$$

By the Girsanov theorem, prove that, $\widetilde{W}(t) = W(t) + bt$ is a Brownian Motion under Q.
7. Let $dS(t) = rS(t)dt + \sigma S(t)dW(t)$. Find the probability that $S(3t) > 3S(t)$.
8. A European put option that gives one the right to sell a stock at price K at time t has value $(K - S_t)^+$ at time t. Find the present-day value of the European put at time 0.
9. Let the interest rate r be equal to 0 and at time t_e an option V has payoff $\sup_{s\leq t_e} S(s)$. What is the price of V at time 0?
10. Consider $S(t) = S(0) + rt + \sigma W(t)$. Then find the European call option price at the strike price of K.
11. Find the prices for a call and a put options under BS model if $S(0) = 100$, $K = 110$, $T = 0.5$, $r = 5\%$, $\mu = 8\%$, and $\sigma = 35\%$.
12. Let $a, b > 0$. Let Y_i, $i = 1, 2, \ldots$ be a sequence of IID RVs that take only values b and $-a$ and let $S_n = \sum_{i=1}^n Y_i$. Show that if $P(Y_i = b) > 0$ and $P(Y_i = -a) > 0$ there exist a probability measure Q equivalent to P under which $\{S_n, n = 1, 2, \ldots\}$ is a martingale. Describe the Radon Nikodym derivative of Q w.r.t P.
13. Consider a European option in Black-Scholes model with payoff of $\max\{(S(T))^2 - K, 0\}$ at time T. The price of this option will be given by $V(0) = e^{-rT}E[(S_T^2 - K)^+]$. Prove that

$$V(0) = (S(0))^2 e^{(\sigma^2+r)T} \Phi\left(\frac{\ln\left(\frac{(S(0))^2}{K}\right) + T\left(2r + 3\sigma^2\right)}{2\sigma\sqrt{T}}\right)$$
$$- Ke^{-rT} \Phi\left(\frac{\ln\left(\frac{(S(0))^2}{K}\right) + T\left(2r - \sigma^2\right)}{2\sigma\sqrt{T}}\right).$$

14. **Black-Scholes-Merton Model:** Assume that the total value of the firm $V = \{V(t), 0 \leq t \leq T\}$ follows a geometric Brownian motion

$$dV(t) = rV(t)dt + \sigma V(t)dW(t), \quad V_0 > 0,$$

where r is the risk-free interest rate, $\sigma > 0$ is the asset volatility. Assume that default can only occur at maturity and is said to occur if the value of the firm is below the threshold L at time T. Show that the probability of default is given by

$$P_{def}(T) = P(V(T) \leq L) = \Phi(d_2),$$

where

$$d_2 = \frac{\ln(V(t)/L) + (r + \frac{\sigma^2}{2})(T - t)}{\sigma\sqrt{T - t}} - \sigma\sqrt{T - t}.$$

15. Prove that if the firm has survived up to time t, the conditional survival probability up to maturity in the risk-neutral setting is given by

$$P_{Surv}(T \mid \mathcal{F}_t) = P(\tau > T \mid \tau > t) = \Phi(d_3) - \left(\frac{L}{V_t}\right)^{(2r/\sigma^2)-1} \Phi(d_4)$$

where Φ is the distribution function of a standard normal RV and

$$d_3 = \frac{\ln(V_t/L) + (r + \frac{\sigma^2}{2})(T - t)}{\sigma\sqrt{T - t}}$$
$$d_4 = d_3 - \sigma\sqrt{T - t}.$$

16. Prove that $\Theta_P = \Theta_C + rKe^{-r(T-t)}$, where Θ_C and Θ_P are theta on call and put respectively, with the same strike K and expiry T.

References

1. Atkinson, C., Wilmott, P.: Portfolio management with transaction costs: an asymptotic analysis of the Morton and Pliska model. Math. Finance **5**(4), 357–367 (1995)
2. Castaneda, L.B., Arunachalam, V., Dharmaraja, S.: Introduction to Probability and Stochastic Processes with Applications. Wiley, New Jersey, (2012)
3. Gianin, E.R. and Sgarra, C. Mathematical Finance Theory Review and Exercises, Springer Cham, Second Edition, (2023). https://link.springer.com/book/10.1007/978-3-031-28378-9
4. Hull, J., White, A.: Numerical procedures for implementing term structure models I. J. Deriv. **2**, 7–16 (1994)
5. Black, F., Scholes, M.: The pricing of options and corporate liabilities. J. Polit. Econ. **8**, 637–654 (1973)
6. Glosten, L.R., Jagannathan, R. and Runkle, D.E.: On the relation between the expected value and the volatility of the nominal excess return on stocks. The Journal of Finance **48**(5), 1779–1801 (1993)
7. Heston, S.L.: A closed solution for options with stochastic volatility, with application to bond and currency options. Rev. Financ. Stud. **6**, 327–343 (1993)
8. Ross, S.M.: An Elementary Introduction to Mathematical Finance: Options and Other Topics. Cambridge, UK: Cambridge University Press (2003)
9. Karatzas, I. and Shreve, S.E.: Brownian Motion and Stochastic Calculus. New York: Springer (1991)
10. Shreve, S.E.: Stochastic calculus for finance II: Continuous-time models. Vol. 11. New York: springer (2004)
11. Mikosch, T.: Elementary stochastic calculus with finance in view. World scientific (1998)

Chapter 15
Applications to Finance—Credit Risk

The derivatives considered in Chap. 14 are mainly concerned with market risk. This chapter deals with another important category of risk for financial institutions: credit risk, also known as default risk. In this chapter, we present various stochastic models of credit risk. The chapter begins with an introduction to credit risk in Sect. 15.1. Then, we introduce the idea of modeling credit risk using stochastic models in Sect. 15.2, in particular, the reduced-form models and structural models in Sects. 15.2.1 and 15.2.2, respectively. Section 15.3 presents the pricing of European call option when the option writer can default on obligations. Finally, the models for the dynamics of credit ratings are presented in Sect. 15.4.

15.1 Introduction

Credit risk, also known as default risk, is the risk that an unexpected change in the credit quality of counterparty (e.g., issuer of a contract or any trading partner) brings to the value of the portfolio. This includes losses due to default of the counterparty as well as the losses that arise due to a decline in the credit quality, usually measured by credit ratings issued by internal systems or external agencies, of the counterparty. Credit risk is prevalent in the portfolios of most financial institutions and is closely tied to the fundamental activities of banks. For instance, credit risk affects the lending portfolios of a bank as a borrower won't be able to pay back its debts. In fact, any derivative traded in over-the-counter (OTC) markets involves such risk. Further, regulatory authorities also require banks to maintain capital to account for the default risk they are bearing. Therefore, credit risk analysis is a central topic in modern finance, and hence, financial institutions allocate substantial resources to assess and mitigate credit risk. Given the importance, several practitioners and researchers have developed sophisticated models toward modeling the credit risk.

D. Selvamuthu, *Introduction to Stochastic Processes*, University Texts in the Mathematical Sciences, https://doi.org/10.1007/978-981-97-6152-4_15

As evident, one of the main objectives of credit risk modeling is to model the probability of default of a reference entity or the joint probability of default for a group of reference entities. The reference entity could be a borrower in case of a bank loan, or a counterparty in a derivative contract. The goal is to develop stochastic models to obtain the probability that the default event will occur. Credit risk models are broadly categorized into two clusters: structural models (also known as firm-value models) and intensity-based models (also known as reduced-form models or hazard rate models). These categories differ in the way they define the credit default event. The main difficulties when modeling credit risk arise from the fact that default events are quite rare and that they occur unexpectedly. When default events take place, they often lead to significant losses, the size of which is unknown before default. Although default events occur very seldom, credit risk is, by definition, inherent in any payment obligation. This is due to the fact that the probability that a certain obligor might default can be close to zero but it is seldom 0.

Credit rating is one of the important parameters. Credit rating of a credit risky bond, issued by a company, is an evaluation of its likelihood of default and ability to pay back the loan. Better the credit rating of a bond, safer it is. International organizations like Standard & Poor's, CRISIL, ICRA AND CARE issues ratings to the companies that issues bond in order to evaluate the credit risk. More specifically, the credit ratings issued to a firm by a rating agency give the creditworthiness of the firm, i.e., its capacity to repay the debt. In practice, there are the following ratings given by Standard and Poor's rating agency:

$$S = \{AAA, AA, A, BBB, BB, B, CCC, D\}.$$

The bonds having rating above BB are investment grade bonds whereas those having BB or below BB are speculative bonds and state D corresponds to default. Since credit ratings also serve as an important input in several applications in risk management, such as portfolio optimization, pricing of credit derivatives, modeling the term structure of credit risk premium, hence their accurate estimation is critical. Since the creditworthiness of the firm changes over time, the level of rating changes too. Therefore, it needs to be modeled by using an appropriate stochastic process.

Financial institutions also have an alternative to hedge, transfer, or mitigate their credit risk (partially or wholly) using credit derivatives. Credit derivatives can be considered as an insurance against the default. Banks and other financial institutions utilize credit derivatives to lower the risk concentrations on their balance sheets and to mitigate the strict capital requirements enforced by the Basel II. The basic idea behind these products is that the credit risk can be transferred to the other party without reallocating the ownership of the underlying reference entity. These are bilateral contracts in which one party, called the protection seller, provides default protection against the default of a single reference entity or the defaults in a reference portfolio. The other party, known as the protection buyer, pays periodic premiums to the protection seller in return for the protection received. The credit derivatives can be classified mainly into two categories: the single-name credit derivatives, whose payoffs are determined by the default of a single reference name, e.g., Credit Default

Swap (CDS) and the multi-name credit derivatives, whose payoffs are hinged with the default of a number of reference names in a portfolio, e.g., Collateralized Debt Obligations (CDOs).

15.2 Credit Risk Modeling

As mentioned earlier, the primary objective of credit risk analysis is to model the probability of default. Let us denote the time of occurence of the default event by τ and the probability that a default event will not occur between 0 and time t, i.e., the survival probability, by $P_{surv}(t)$ which is given by $P(\tau > t)$. Equivalently, one can define the default probability, which is denoted by $P_{def}(t)$, as follows:

$$P_{def}(t) = 1 - P_{surv}(t), \quad 0 \le t \le T$$

where T is the time horizon under study, for instance, the maturity of the derivative contract. We can similarly define the joint survival $P^{ij}_{surv}(t)$ and joint default probabilities $P^{ij}_{def}(t)$ as follows:

$$P^{ij}_{surv}(t) = P(\tau_i > t, \tau_j > t), \quad 0 \le t \le T \tag{15.1}$$

$$P^{ij}_{def}(t) = P(\tau_i < t, \tau_j < t), \quad 0 \le t \le T. \tag{15.2}$$

where τ_i and τ_j denote the default times of the two reference entities (say i and j). If both the reference entities are independent,

$$P^{ij}_{surv}(t) = P^{i}_{surv}(t)P^{j}_{surv}(t), \tag{15.3}$$

$$P^{ij}_{def}(t) = P^{i}_{def}(t)P^{j}_{def}(t). \tag{15.4}$$

This is practically very unlikely because different reference entities either have common business relationships or they are dependent because they operate in the same financial market and thus are impacted by the common macro-economic variables.

In the next section, we discuss the reduced form of models to find the probability of default.

15.2.1 *Intensity-Based Models*

These categories of models are built upon the assumption that at any instant of time, there is always a probability, even if it is close to zero, that the reference entity will default, and this probability depends on the overall health of the reference entity. The underlying idea of these models is that default can be defined as first jump time of a counting process. More specifically, if $N = \{N(t),\ t \geq 0\}$ is a counting process with the intensity process $\lambda = \{\lambda(t),\ 0 \leq t \leq T\}$, then the occurrence of the first jump in the process N denotes the occurrence of the default event. The intensity process λ is known as the default rate of the reference entity.

15.2.1.1 Poisson Model

A classic example of a counting process is the (homogeneous) Poisson process with constant intensity $\lambda > 0$ (refer Sect. 5.5). It is known that for a Poisson process, the probability that the counting process $N = \{N(t), 0 \leq t \leq T\}$ equals integer n is given by

$$P(N(t) = n) = \frac{e^{-\lambda t}(\lambda t)^n}{n!}, \quad n = 0, 1, \ldots.$$

Therefore, the default event by a Poisson process is modeled (refer Jarrow and Turnbull (1995) [5]), the probability of survival between 0 to t can be obtained as

$$P_{surv}(t) = P(N(t) = 0) = e^{-\lambda t}.$$

In general, under the assumption that the reference entity has survived till time t, the intensity of default is defined as

$$\lambda(t) = \lim_{h \to 0} \frac{P(t < \tau \leq t + h \mid \tau > t)}{h},$$

which is equivalent to

$$P(\tau \leq t + \Delta t \mid \tau > t) \approx \lambda(t)\Delta t,$$

which means that given that the reference entity has survived till time t, the probability of its default in a small interval Δt is approximately given by $\lambda(t)\Delta t$.

More generally, the intensity could be defined as a stochastic process, in which case, the model is known as doubly stochastic model or Cox model.

15.2.1.2 CIR Model

Consider the CIR model for default interests (refer Sect. 14.3.3.1) as

$$d\lambda(t) = \kappa(\theta - \lambda(t))dt + \xi\sqrt{\lambda(t)}dW(t),$$

where $\{W(t), t \geq 0\}$ is a Brownian motion, κ, θ, ξ are positive constants denoting the mean reverting rate, mean-reversion level, and the volatility of the intensity process, respectively. The survival probability in this case is given by

$$P_{surv}(t) = P(N(t) = 0) = E\left(e^{-\int_0^t \lambda(s)ds}\right),$$

where the expectation can be calculated similar to Proposition 14.2 and is given by

$$P_{surv}(t) = A(t)e^{-B(t)\lambda(0)},$$

with the functions $A(t)$ and $B(t)$ given by

$$B(t) = \frac{2(e^{\psi t} - 1)}{(\psi + \kappa)(e^{\psi t} - 1) + 2\psi}, \quad A(t) = \left(\frac{2\psi e^{(\kappa+\psi)(t)/2}}{(\psi + \kappa)(e^{\psi t} - 1) + 2\psi}\right)^{2\kappa\theta/\xi^2}$$

where, $\psi = \sqrt{\kappa^2 + 2\xi^2}$.

15.2.1.3 Hawkes Model

In the previous examples that were based on the traditional approach in intensity based models default is defined as the first jump time of a point process. Here, it is assumed that any firm has some resistance, measured by d, to avoid default. Therefore, the default is no longer defined as the first jump time of a point process. The arrival of adverse events that cause default of the firm (such as credit rating downgrade or unfavorable news in the market) is modeled by a point process $\{N(t), t \geq 0\}$ such that the default probability is given by

$$P_{def}(T) = 1 - E[(1-d)^{N(T)}], \quad d \in (0, 1]. \tag{15.5}$$

Therefore, to find the default probability, it is sufficient to find the PGF of the point process. Suppose that the stochastic process $\{N(t), t \geq 0\}$ is a self-exciting point process with intensity process $\{\lambda(t), t \geq 0\}$ given by

$$d\lambda(t) = \delta(\eta - \lambda(t))dt + \sigma dW(t) + d\left(\sum_{j=1}^{N(t)} Z_j\right), \tag{15.6}$$

where

- $\lambda(0)$ is the initial intensity at time $t = 0$.
- η is the equilibrium level of the intensity process.
- δ is the constant mean-reversion speed of the intensity process.
- σ represents the volatility of the diffusion component.
- $\{Z_j\}_{j=1,2,\ldots}$ is a sequence of IID positive RV's representing the size of self-excited jumps. It is assumed that these RV's have a common cumulative distribution function $G(z)$, $z > 0$ at the corresponding random times $\{T_j\}_{j=1,2,\ldots}$. Further, it is assumed that Z_j follows exponential distribution with parameter β.

- $\{T_j\}_{j=1,2,\ldots}$ are the arrival times of the process $\{N(t), t \geq 0\}$.
- $W := \{W(t)|t \geq 0\}$ is a standard Brownian motion with respect to its natural filtration.
- $\{Z_j\}_{j=1,2,\ldots}$, $\{T_j\}_{j=1,2,\ldots}$ and $\{W(t), t \geq 0\}$ are independent of each other.

Under suitable conditions on the structure of $\lambda(t)$, the process $\{(\lambda(t), N(t)), t \geq 0\}$ is a Markov process, which implies that the infinitesimal generator $\mathcal{A}$ of the process $\{(\lambda(t), N(t)), t \geq 0\}$ applied to a function $f(\lambda, n, t)$ in $S(\mathcal{A})$, the domain of the generator, is given by

$$\mathcal{A} f(\lambda, n, t) = \frac{\partial f}{\partial t} - \delta(\lambda - \eta)\frac{\partial f}{\partial \lambda} + \frac{1}{2}\sigma^2 \frac{\partial^2 f}{\partial \lambda^2} + \lambda \int_0^\infty [f(\lambda + z, n+1, t) - f(\lambda, n, t)] dG(z). \tag{15.7}$$

Here, $\mathcal{A} \in S(\mathcal{A})$ is such that the function $f(\lambda, n, t)$ is differentiable with respect to λ, t for all λ, n and t, and

$$\left| \int_0^\infty [f(\lambda + z, n+1, t) - f(\lambda, n, t)] dG(z) \right| < \infty.$$

Theorem 15.1 *Assume that jumps size Z_j follows exponential distribution with parameter β such that $\beta\delta > 1$. Then,*

$$E(\theta^{N(T)} e^{-v\lambda(T)} \mid \lambda(0)) = e^{-B(0)\lambda(0)} e^{-(c(T)-c(0))}, \tag{15.8}$$

where $0 \leq \theta \leq 1$, $v \geq 0$, are constants and for any $0 \leq t \leq T$, the function $c(t)$ is given by

$$c(t) - c(0) = \eta\delta \int_0^t B(s) ds + \frac{1}{2}\sigma^2 \int_0^t B^2(s) ds, \tag{15.9}$$

and $B(t)$ is given in parametric form as follows:

$$B(t(u)) = \frac{(1 + \delta\beta)}{\delta} u - \beta, \tag{15.10}$$

where the time t is a function of the parameter u in the following form

$$t(u) = T - \frac{1}{\delta}\left(\frac{1}{2}\ln\left(\frac{u_0^2 - u_0 - \alpha_1}{u^2 - u - \alpha_1}\right) + \frac{\tan^{-1}(\frac{2u_0-1}{\sqrt{-4\alpha_1-1}}) - \tan^{-1}(\frac{2u-1}{\sqrt{-4\alpha_1-1}})}{\sqrt{-4\alpha_1 - 1}}\right), \tag{15.11}$$

and $u_0 = \frac{(\beta+v)\alpha_1}{\alpha_2}$ *with* $\alpha_1 = \frac{-\beta\theta\delta}{(1+\beta\delta)^2}$ *and* $\alpha_2 = \frac{-\beta\theta}{1+\beta\delta}$.

Proof Assume that the function $f(\lambda, n, t)$ takes the following affine form

$$f(\lambda, n, t) = A^n(t)e^{-B(t)\lambda}e^{c(t)}, \tag{15.12}$$

with the terminal condition $B(T) = v$. Substituting $f(\lambda, n, t)$ from (15.7) to (15.12) yields

$$(-B'(t)\lambda + n\frac{A'(t)}{A(t)} + c'(t))f \; + B(t)\delta(\lambda - \eta)f + \frac{1}{2}\sigma^2 B^2(t)f$$
$$+\lambda f\int_0^\infty (A(t)e^{-B(t)z} - 1)dG(z) = 0.$$

Since this equation is true for any possible choice of λ and n, the coefficients of λ and n must be zero. Hence, it follows that

$$\frac{A'(t)}{A(t)} = 0, \quad A(T) = \theta, \tag{15.13}$$

$$B'(t) - \delta B(t) - A(t)\frac{\beta}{\beta + B(t)} + 1 = 0, \quad B(T) = v, \tag{15.14}$$

$$c'(t) - \eta\delta + \frac{1}{2}\sigma^2 B^2(t) = 0. \tag{15.15}$$

From (15.13), $A(t) = \theta$ and (15.15) is obtained as

$$c(t) - c(0) = \eta\delta\int_0^t B(s)ds + \frac{1}{2}\sigma^2\int_0^t B^2(s)ds,$$

which is same as (15.9). In order to get a closed-form expression, it is needed to solve explicitly for $B(t)$, the solution of which is given in the theorem statement.

Given the parametric solution of $B(t)$, since the process $(\lambda(t), N(t))$ is a Markov process,

$$M_t^f := f((\lambda(t), N(t))) - f((\lambda(0), N(t))) - \int_0^t \mathcal{A}f((\lambda(s), N(s)))ds, \tag{15.16}$$

is a martingale. Taking $f((\lambda(t), N(t)))$ of the form $\theta^{N(T)}e^{-B(T)\lambda(T)}e^{c(T)}$ and noting that $\mathcal{A}f((\lambda(t), N(t))) = 0$, (15.16) yields that the process $\theta^{N(T)}e^{-B(T)\lambda(T)}e^{c(T)}$ −

$e^{-B(0)\lambda(0)}e^{c(0)}$ is a martingale with zero mean, hence

$$E(\theta^{N(T)}e^{-v\lambda(T)} \mid \lambda(0)) = e^{-B(0)\lambda(0)}e^{-(c(T)-c(0))}.$$

Corollary 15.1 *The PGF of $N(T)$ can be obtained from Theorem 15.1 upon substituting $v = 0$ and is given by*

$$E(\theta^{N(T)} \mid \lambda(0)) = e^{-B(0)\lambda(0)}e^{-(c(T)-c(0))}, \tag{15.17}$$

where $B(0)$ and $c(T) - c(0)$ are given in Theorem 15.1 with $v = 0$.

15.2.2 Structural Models

The structural models, also known as firm-value models, link the occurrence of default to the value of the financial assets of the reference entity unlike reduced-form approach that directly models the default event by a stochastic process. More specifically, the event of default is defined in terms of some boundary conditions on the value process of the firm, denoted by $V = \{V_t,\ 0 \le t \le T\}$, a stochastic process governing the dynamics of the total value of the firm. For example, the simplest example of the boundary condition is that default has occurred at time t if the value of the firm at time t is less than some pre-specified threshold L, i.e.,

$$P_{def}(t) = P(V_t < L),\ 0 \le t \le T.$$

Similar to what we did in the previous chapter while pricing bonds under different dynamics of interest rates, we now need to model the dynamics of the value process V in order to find $P(V_t < L)$. The simplest dynamics one can consider is the geometric Brownian motion, which is presented in the next section. Some other possible dynamics include geometric Brownian motion with stochastic volatility, a Poisson jump-diffusion process, or a Hawkes jump-diffusion process.

15.2.2.1 Black–Scholes–Merton Model

Observe that the above idea can be seen as a call option on the firm's underlying value. More specifically, the default occurs if $V_t < L$ with probability $P(V_t < L)$, the bondholder will receive the recovery value V_t. Otherwise, if $V_t > L$, the bond holder receives L and the equity holder is entitled to receive $V_t - L$, which can be represented as $(V_t - L)^+$ in general.

Assume that the total value of the firm $V = \{V_t, 0 \le t \le T\}$ follows a geometric Brownian motion

$$dV_t = \mu V_t dt + \sigma_V V_t dW(t),\ \ V_0 > 0, \tag{15.18}$$

where μ is the expected rate of return, $\sigma_V > 0$ is the asset volatility, and $W = \{W(t), 0 \leq t \leq T\}$ is a standard Brownian motion.

Proposition 15.1 *Assume that default can only occur at maturity and is said to occur if the value of the firm is below the threshold L at time T. Show that under the dynamics (15.18), the probability of default is given by*

$$P_{def}(T) = 1 - \Phi(d_2),$$

where Φ is the cumulative distribution function of a standard normal RV and

$$d_2 = \frac{\log(V_t/L) + (\mu + \frac{\sigma_V^2}{2})(T-t)}{\sigma_V\sqrt{T-t}} - \sigma_V\sqrt{T-t}.$$

Proof Follow Sect. 14.4.2 and observe that

$$P_{def}(T) = 1 - P_{surv}(T) = 1 - P(V_T > L) = 1 - \Phi(d_2).$$

15.2.2.2 First Passage Time Model (Black and Cox (1976))

First passage models include the possibility of an early default for the reference entity. More specifically, default takes place the first time the asset value $V = \{V_t, 0 \leq t \leq T\}$ goes below a default barrier L, which can be given either exogenously or endogenously with respect to the model, i.e.,

$$\tau = \inf\{t > 0 \mid V_t \leq L\}.$$

In other words, the firm has to maintain an account balance above the level L throughout time, and therefore default occurs at the first time the process S_t hits the level L.

Proposition 15.2 *Assume that the firm value follows the dynamics in equation (15.18), prove that the probability of survival when L is a constant is given by*

$$P_{Surv}(T \mid \mathcal{F}_t) = P(\tau > T \mid \tau > t) = \Phi(d_3) - \left(\frac{L}{V_t}\right)^{(2\mu/\sigma_V^2)-1}\Phi(d_4),$$

where Φ is the cumulative distribution function of a standard normal RV and

$$d_3 = \frac{\log(V_t/L) + (\mu - \frac{\sigma_V^2}{2})(T-t)}{\sigma_V\sqrt{T-t}},$$
$$d_4 = d_3 - \sigma_V\sqrt{T-t}.$$

Proof From Sect. 11.2.5, the reflection principle allows us to find the distribution of the first hitting time of a Brownian motion.

In practice, the default or a change in credit quality of a reference entity is often sudden; i.e., it is often shock-driven. Therefore, it is reasonable to assume jumps in the firm-value process of underlying reference entities. We now present a Poisson jump-diffusion model to obtain the probability of default using a different approach based on the characteristic function.

15.2.2.3 Poisson Jump-Diffusion Model

Assume that the dynamics of the values of the firm, denoted by V_t follows Poisson jump-diffusion processes (refer Sect. 11.5 for more details)

$$\frac{dV_t}{V_{t-}} = (\mu - \beta\lambda)dt + \sigma dW_t + d\left[\sum_{i=1}^{N_t} Z_i\right], \tag{15.19}$$

for some real parameter μ. The process $W(t)$ is the standard Brownian motion and σ denotes the corresponding volatility. The sequences $\{Z_i\}_{i=1,2,\ldots}$ representing the magnitudes of jumps are assumed to be IID log-normal RV's with $E[Z_i] = \beta$. Precisely, $\log(1 + Z_i) \sim N(a, b^2), i = 1, 2, \ldots$ such that $\beta = e^{a+\frac{1}{2}b^2} - 1$. Further, $N(t)$ is a Poisson process with intensity λ. Let T denote the maturity time of a debt contract and L denote the default boundary of the firm.

Proposition 15.3 *Assuming the above Poisson jump-diffusion dynamics for the firm value V_t, find the probability of default.*

Proof The probability of default is given by

$$P(V_T \le L) = P(\log(V_T) \le \log(L)) = P(X_T \le \log(L)) = F_{X_T}(\log(L))$$

where $X_T = \log(V_T)$ and F_{X_T} denotes the CDF of X_T. The CDF of a RV can be obtained using the characteristic function as follows,

$$F_{X_T}(x) = \frac{1}{2} - \frac{1}{2\pi}\int_0^{\infty} \Delta_u\left[\frac{\phi_{X_T}(u)e^{-iux}}{iu}\right]du. \tag{15.20}$$

Therefore, in order to find the default probability, we need to find the characteristic function $\phi_{X_T}(u)$ of X_T. Using Itô's lemma, the dynamics of X_T can be obtained as

$$X_T = \log(V_0) + (\mu - \frac{1}{2}\sigma^2 - \beta\lambda)T + \sigma W(T) + \left[\sum_{i=1}^{N(T)} \log(1 + Z_i)\right].$$

The characteristic function of X_T is given by

$$\begin{aligned}\phi_{X_T}(u) &= E\left[\exp(iuX_T)\right] \\ &= E\left[\exp\left(iu(\gamma T + \sigma W(T) + \sum_{i=1}^{N(T)} \log(1+Z_i))\right)\right].\end{aligned}$$

where $\gamma = \mu - \frac{1}{2}\sigma^2 - \beta\lambda$. By independence of $W(T)$ and $N(T)$ we get

$$\phi_{X_T}(u) = \exp(iu\gamma T)E\left[\exp(iu\sigma W(T))\right]E\left[\exp\left(iu\sum_{i=1}^{N(T)} \log(1+Z_i))\right)\right]. \tag{15.21}$$

Since $\sigma W(T) \sim N(0, \sigma^2 T)$, we can write

$$E\left[\exp(iu\sigma W(T)\right] = \exp(-\frac{1}{2}\sigma^2 u^2 T),$$

The second term, i.e., $E\left[\exp\left(iu\sum_{i=1}^{N(T)} \log(1+Z_i)\right)\right]$ can be computed as follows

$$\begin{aligned}E\left[\exp\left(iu\sum_{i=1}^{N(T)} \log(1+Z_i)\right)\right] =& E\left[E\left[\exp\left(iu\sum_{i=1}^{N} \log(1+Z_i)\right)|N(T)=N\right]\right] \\ =& E\left[E\left[\prod_{i=1}^{N} \exp\left(iu\log(1+Z_i)\right)|N(T)=N\right]\right] \\ \overset{Z_i \text{i.i.d}}{=}& E\left[\prod_{i=1}^{N} \underbrace{E\left[\exp\left(iu\log(1+Z_1)\right)\right]}_{=:M}\right] \\ =& E\left[M^{N(T)}\right] = \exp(\lambda T(M-1))\end{aligned}$$

where $M = E\left[\exp\left(iu\log(1+Z_1)\right)\right]$ is characteristic function of $\log(1+Z_1)$, which follows a normal distribution. Hence,

$$M = E\left[\exp\left(iu\log(1+Z_1)\right)\right] = (e^{iua - \frac{1}{2}b^2u^2} - 1).$$

Substituting in (15.21), we obtain the characteristic function Φ_{X_T},

$$\Phi_{X_T}(u) = \exp\left(T\left(iu\gamma - \frac{1}{2}\sigma^2 u^2 + \lambda(e^{iua - \frac{1}{2}b^2u^2} - 1)\right)\right),$$

which when substituted in (15.20) gives the probability of default.

To model correlated defaults and account for potential "domino effects" we consider two given default times, τ_1 and τ_2, as correlated RVs. Specifically, given τ_1 and τ_2 as two default times, we define the correlation coefficient ρ as

$$\rho = \frac{\text{cov}(\tau_1, \tau_2)}{\sqrt{\text{var}[\tau_1] \cdot \text{var}[\tau_2]}} \in [-1, 1].$$

Next, we present how to find joint default/survival probability for multiple reference entities when the underlying entities follow a Hawkes jump-diffusion model (refer Sect. 11.6). It is discussed here the case for two entities, but one can easily extend it to n reference entities. The Hawkes jump-diffusion model can easily be relaxed to Poisson jump-diffusion model. Therefore, the approach presented in next example can also be used to obtain the default probability in addition to the characteristic function approach presented in the previous example.

15.2.2.4 Hawkes Jump-Diffusion Model

Now, we consider a Hawkes jump-diffusion model for the dynamics of the value of the firm. Hawkes jump-diffusion process is an extension to Poisson jump-diffusion model and that several advantages including it allows to model the clustering of jumps.

Example 15.1 Assume that the dynamics of the values of two reference firms, denoted by $V_{1,t}$ and $V_{2,t}$, follow the two-dimensional Hawkes jump-diffusion processes, for $n = 1, 2$,

$$\frac{dV_{n,t}}{V_{n,t-}} = (\mu_n - \beta_n\lambda(t) - \beta_n\lambda_n)(t)dt + \sigma_n dW_n(t) + d\left[\sum_{i=1}^{N(t)+N_n(t)} Z_{n,i}\right] \tag{15.22}$$

where the stochastic processes $\{W_n(t),\ t \geq 0\}$ are correlated standard Brownian motions with $dW_1(t)dW_2(t) = \rho dt$ where ρ is the correlation coefficient and σ_n denotes the corresponding volatility. The sequences $\{Z_{n,i}\}_{i=1,2,...}$ representing the magnitudes of jumps are assumed to be IID log-normal RVs with $E[Z_{n,i}] = \beta_n$. Precisely, $\ln(1 + Z_{n,i}) \sim N(a_n, b_n^2), i = 1, 2, \ldots$ such that $\beta_n = e^{a_n + \frac{1}{2}b_n^2} - 1$ for $n = 1, 2$. Further, $N(t)$, $N_1(t)$ and $N_2(t)$ are three independent Hawkes processes with $\{T_i\} > 0$, $\{T_{1,i}\} > 0$ and $\{T_{2,i}\} > 0$ representing the corresponding jump times and the corresponding intensity processes are $\lambda(t)$, $\lambda_1(t)$ and $\lambda_2(t)$, given by

$$d\lambda(t) = \delta(\lambda_\infty - \lambda(t))dt + \theta dN(t) \tag{15.23}$$
$$d\lambda_n(t) = \delta_n(\lambda_{n,\infty} - \lambda_n(t))dt + \theta_n dN_n(t),\ n = 1, 2 \tag{15.24}$$

where $\lambda_\infty, \theta, \delta, \lambda_{n,\infty}, \theta_n, \delta_n$ are positive constants.

Proposition 15.4 *Assume that the dynamics of the values of two reference firms, denote by $V_{1,t}$ and $V_{2,t}$, follow the two-dimensional Hawkes jump-diffusion processes, refer Example 15.1. Let T denote the maturity time of a debt contract and $L_n,\ n = 1, 2$ denote the default boundaries of the firm. Find the joint probability of default and the default correlation.*

Proof We need to find the marginal and joint default probabilities of $(V_{1,T}, V_{2,T})$, defined by $P(V_{1,T} \le L_1)$, $P(V_{2,T} \le L_2)$, and $P(V_{1,T} \le L_1, V_{2,T} \le L_2)$, respectively. If we define the log-value of the firm, i.e., $Y_{n,t} = \ln(V_{n,t})$ for $n = 1, 2$, then the required default probabilities are $P(Y_{1,T} \le \log(L_1))$, $P(Y_{2,T} \le \log(L_2))$ and $P(Y_{1,T} \le \log(L_1), Y_{2,T} \le \log(L_2))$. Clearly, we need the joint distribution of the RVs $(Y_{1,T}, Y_{2,T})$.

Using Itô's lemma, one can show that, for $n = 1, 2$,

$$V_{n,t} = X_{n,t} J_{n,t}, \tag{15.25}$$

where

$$X_{n,t} = V_{n,0} \exp\Bigg\{\sigma_n W_{n,t} + \widetilde{\eta}_n t + \sum_{k=1}^{N_t} \frac{\beta_n \theta}{\delta}\left(e^{-\delta(t-T_k)} - 1\right) + \sum_{k=1}^{N_{n,t}} \frac{\beta_n \theta_n}{\delta_n}\left(e^{-\delta_n(t-T_{n,k})} - 1\right)\Bigg\},$$

and $J_{n,t} = \prod_{k=1}^{N(t)+N_n(t)} (1 + Z_{n,k})$ with $\widetilde{\eta}_n = \mu_n - \frac{1}{2}\sigma_n^2\left(1+\gamma_n^2\right) - \beta_n\lambda(0) - \beta_n\lambda_n(0)$.

Define few notations for the simplification. For $n = 1, 2$, set

$$x_n = \eta_n - \beta_n\lambda(0) - \beta_n\lambda_n(0), \qquad \eta_n = \mu_n - \frac{1}{2}\sigma_n^2\left(1+\gamma_n^2\right)$$

$$Y_n = \sum_{j=1}^{\tilde{n}} \frac{\beta_n\theta}{\delta}(e^{-\delta(t-T_j)} - 1) + \sum_{j=1}^{n_n} \frac{\beta_n\theta_n}{\delta_n}(e^{-\delta_n(t-T_{n,j})} - 1).$$

Define the natural filtration of the stochastic processes. Let $\mathfrak{F}_{n,t} = \sigma\{N(s), N_n(s) : 0 \le s \le t\}$, and define $\mathcal{G}_t = \mathfrak{F}_{1,t} \bigcup \mathfrak{F}_{2,t}$. Similarly, let $\mathfrak{F}_{V_{n},t} = \sigma\{V_{n,s} : 0 \le s \le T\}$, $\mathfrak{F}_{\lambda_n}(t) = \sigma\{\lambda_n(s) : 0 \le s \le T\}$, $n = 1, 2$ and $\mathfrak{F}_{\lambda,t} = \sigma\ \{\lambda(s) : 0 \le s \le T\}$. Define $\mathfrak{F}_t = \mathfrak{F}_{V_1,t} \bigcup \mathfrak{F}_{V_2,t} \bigcup \mathfrak{F}_{\lambda,t} \bigcup \mathfrak{F}_{\lambda_1(t)} \bigcup \mathfrak{F}_{\lambda_2(t)}$.

Assume that the information is contained in $\mathfrak{F}_{i,T}$, since $N(t) = \tilde{n}$, $N_i(T) = n_i$. That is, we assume that the number of jumps and corresponding jump times of the Hawkes process are known. By taking logarithm of Equations (15.25),

$$\ln(V_{i,T}) = \ln(V_{i,0}) + x_i T + \sigma_i W_i(T) + Y_i + \sum_{k=1}^{\tilde{n}+n_i} \ln(1 + Z_{i,k}). \tag{15.26}$$

Clearly, for $n = 1, 2$, the process $Y_{n,t} = \ln(V_{n,t})$ follows a bivariate normal distribution with the following means, variances, and correlation,

$$M_i = E[\xi_{i,T}] = \log(V_{i,0}) + x_i T + Y_i + (\tilde{n} + n_i)a_i,$$
$$V_i^2 = Var(\xi_{i,T}) = \sigma_i^2 T + (\tilde{n} + n_i)b_i,$$

and

$$\rho_{1,2} = \frac{cov\left(Y_{1,T}, Y_{2,T}\right)}{V_1 V_2} = \frac{\sigma_1 \sigma_2 \rho T}{V_1 V_2}. \tag{15.27}$$

Therefore, the conditional probability of default is

$$\begin{aligned} P(V_{i,T} \le L_i \mid \mathfrak{F}_{i,T}) &= P\left(Y_{i,T} \le \ln(L_i) \mid \mathfrak{F}_{i,T}\right) \\ &= P\left(\frac{Y_{i,T} - E[Y_{i,T}]}{\sqrt{Var(Y_{i,T})}} \le \frac{\ln(L_i) - E[Y_{i,T}]}{\sqrt{Var(Y_{i,T})}} \mid \mathfrak{F}_{i,T}\right) \\ &= P\left(\frac{Y_{i,T} - M_i}{V_i} \le \frac{\ln(L_i) - M_i}{V_i} \mid \mathfrak{F}_{i,T}\right) \\ &= \Phi\left(\frac{\ln(L_i) - M_i}{V_i}\right) \end{aligned} \tag{15.28}$$

where $\Phi(\cdot)$ is the cumulative distribution function (CDF) of the standard normal distribution. Now, we need to find the unconditional default probability, i.e., take expectation with respect to the paths of Hawkes processes. Therefore, taking the mean value of the conditional probability of default (15.28), get the unconditional probability of default for ith firm as

$$p_i(T) = E_i\left[\Phi\left(\frac{\ln(L_i) - M_i}{V_i}\right)\right]. \tag{15.29}$$

Similarly, given $\mathfrak{F}_T$, the conditional joint probability of default is

$$\begin{aligned} &P(V_{1,T} \le L_1, V_{2,T} \le L_2 \mid \mathfrak{F}_T) \\ &\quad = P\left(Y_{1,T} \le \ln(L_1), Y_{2,T} \le \ln(L_2) \mid \mathfrak{F}_T\right) \\ &\quad = P\left(\frac{Y_{1,T} - M_1}{V_1} \le \frac{\ln(L_1) - M_1}{V_1}, \frac{Y_{2,T} - M_2}{V_2} \le \frac{\ln(L_2) - M_2}{V_2} \mid \mathfrak{F}_T\right) \\ &\quad = \Phi_2\left(\frac{\ln(L_1) - M_1}{V_1}, \frac{\ln(L_2) - M_2}{V_2}, \rho_{1,2}\right) \end{aligned} \tag{15.30}$$

where $\Phi_2(\cdot, \cdot, \rho)$ is the CDF of the bivariate normal distribution with zero mean and covariance matrix given by $\begin{pmatrix} 1 & \rho \\ \rho & 1 \end{pmatrix}$. As before, taking the mean value of the conditional joint probability of default (15.30), get the unconditional joint probability of default as

$$p_{1,2}(T) = E_{1,2}\left[\Phi_2\left(\frac{\ln(L_1) - M_1}{V_1}, \frac{\ln(L_2) - M_2}{V_2}, \rho_{1,2}\right)\right]. \quad (15.31)$$

Consequently, the default correlation is obtained as given by

$$\rho_{1,2}(T) = \frac{p_{1,2}(T) - p_1(T)p_2(T)}{\sqrt{p_1(T)(1 - p_1(T))}\sqrt{p_2(T)(1 - p_2(T))}}. \quad (15.32)$$

To find the default probabilities and default correlation, one needs to simulate the paths of Hawkes processes, evaluate the terms inside the expectation for each path, and finally take the mean to obtain the required probability or correlation.

Remark 15.1 In previous example, if $\theta = \theta_n = 0$ are set, the dynamics of firm value degenerates to jump-diffusion process where jumps are described by Poisson processes.

15.3 Pricing Vulnerable Option Under Black–Scholes Model

In the previous chapter, we discussed pricing of a European call option under the Black–Scholes framework. In this section, we extend that model and consider the risk of default by the option writer. The options with credit risk are also known as vulnerable options. We consider a structural approach to model the credit risk and assume that the option written defaults when it's total value V_t is below a default barrier D^* at the time of maturity T of the option contract. The default boundary is assumed to be independent of the option value. In the case of a default at T, the recovery rate of the outstanding nominal claims is set to $\frac{(1-\alpha)V_T}{D^*}$, where α is the percentage loss of the asset value of the option writer, representing the dead weight costs of bankruptcy. More specifically, the value of a vulnerable call C^* is given by

$$C^* = e^{-r(T-t)}E^*\left[\max(S_T - K, 0)\left(1_{\{V_T > D^*\}} + \frac{(1-\alpha)V_T}{D^*}1_{\{V_T \le D^*\}}\right)\right]. \quad (15.33)$$

where E^* denotes risk-neutral expectation. The price of these options under Black–Scholes framework could be obtained in a similar manner as in the previous chapter. We assume that the dynamics of the underlying asset S_t is given by

$$\frac{dS}{S} = \mu_s dt + \sigma_s dW_1(t).$$

The dynamics of the total value of the option writer is assumed to follow the following dynamics

$$\frac{dV}{V} = \mu_v dt + \sigma_v dW_2(t).$$

We assume that the diffusion processes $W_1(t)$ and $W_2(t)$ follow standard Wiener processes with instantaneous correlation ρ. We further assume that both S_t and V_t are traded in the market.

Based on these assumptions, it can easily be shown following the previous chapter, and the appropriate risk-neutral processes are:

$$\frac{dS}{S} = rdt + \sigma_s dW_1(t), \tag{15.34}$$

$$\frac{dV}{V} = rdt + \sigma_v dW_2(t). \tag{15.35}$$

Applying Itô's lemma results in the following processes for $\ln(S)$ and $\ln(V)$:

$$\frac{d\ln S}{dt} = (r - \frac{\sigma_s^2}{2})dt + \sigma_s dW_1(t), \tag{15.36}$$

$$\frac{d\ln V}{dt} = (r - \frac{\sigma_v^2}{2})dt + \sigma_v dW_2(t). \tag{15.37}$$

From Eqs. (15.36) and (15.37), it is clear that $\ln S$ and $\ln V$ are bivariate normal distributed, i.e.,

$$\begin{bmatrix} \ln S_T \\ \ln V_T \end{bmatrix} \sim N_2\left(\begin{bmatrix} \ln S_t + (r - \frac{\sigma_s^2}{2})(T-t) \\ \ln V_t + (r - \frac{\sigma_v^2}{2})(T-t) \end{bmatrix}, \begin{bmatrix} \sigma_s^2(T-t) & \rho\sigma_s\sigma_v(T-t) \\ \rho\sigma_s\sigma_v(T-t) & \sigma_v^2(T-t) \end{bmatrix} \right).$$

Equation (15.33) can be restated as follows:

$$\begin{aligned} C^* = e^{-r(T-t)} \Big(& E^*\left[S_T 1_{\{S_T > K, V_T > D^*\}}\right] - E^*\left[K 1_{\{S_T > K, V_T > D^*\}}\right] \\ & + E^*\left[S_T (1-\alpha)\frac{V_T}{D^*} 1_{\{S_T > K, V_T \le D^*\}}\right] - E^*\left[K(1-\alpha)\frac{V_T}{D^*} 1_{\{S_T > K, V_T \le D^*\}}\right]. \end{aligned} \tag{15.38}$$

Proposition 15.5 *The price of the vulnerable European call option under the Black–Scholes framework is given by*

$$\begin{aligned} C^* = {} & S_t\left(\Phi_2(a_1, a_2, \rho) + e^{(r+\rho\sigma_S\sigma_V)(T-t)} \frac{(1-\alpha)V_t}{D^*} \Phi_2(c_1, c_2, -\rho) \right) \\ & - e^{-r(T-t)} K \left(\Phi_2(b_1, b_2, \rho) + e^{r(T-t)} \frac{(1-\alpha)V_t}{D} \Phi_2(d_1, d_2, -\rho) \right), \end{aligned}$$

where Φ_2 denotes the bivariate standard normal cumulative distribution function and

Fig. 15.1 State transition diagram for a credit rating model

$$a_1 = \frac{\ln\left(\frac{S_t}{K}\right) + \left(r + \frac{\sigma_s^2}{2}\right)(T-t)}{\sigma_s\sqrt{T-t}} = b_1 + \sigma_s\sqrt{T-t},$$

$$a_2 = \frac{\ln\left(\frac{V_t}{D^*}\right) + \left(r - \frac{\sigma_v^2}{2} + \rho\sigma_s\sigma_v\right)(T-t)}{\sigma_v\sqrt{T-t}} = b_2 + \rho\sigma_s\sqrt{T-t},$$

$$b_1 = \frac{\ln\left(\frac{S_t}{K}\right) + \left(r - \frac{\sigma_s^2}{2}\right)(T-t)}{\sigma_s\sqrt{T-t}},$$

$$b_2 = \frac{\ln\left(\frac{V_t}{D^*}\right) + \left(r - \frac{\sigma_v^2}{2}\right)(T-t)}{\sigma_v\sqrt{T-t}},$$

$$c_1 = \frac{\ln\left(\frac{S_t}{K}\right) + \left(r + \frac{\sigma_s^2}{2} + \rho\sigma_s\sigma_v\right)(T-t)}{\sigma_s\sqrt{T-t}} = b_1 + (\sigma_s + \rho\sigma_v)\sqrt{T-t},$$

$$c_2 = -\frac{\ln\left(\frac{V_t}{D^*}\right) + \left(r + \frac{\sigma_v^2}{2} + \rho\sigma_s\sigma_v\right)(T-t)}{\sigma_v\sqrt{T-t}} = -\left(b_2 + (\sigma_v + \rho\sigma_s)\sqrt{T-t}\right),$$

$$d_1 = \frac{\ln\left(\frac{S_t}{K}\right) + \left(r - \frac{\sigma_s^2}{2} + \rho\sigma_s\sigma_v\right)(T-t)}{\sigma_s\sqrt{T-t}} = b_1 + \rho\sigma_v\sqrt{T-t},$$

$$d_2 = -\frac{\ln\left(\frac{V_t}{D^*}\right) + \left(r + \frac{\sigma_v^2}{2}\right)(T-t)}{\sigma_v\sqrt{T-t}} = -\left(b_2 + \sigma_v\sqrt{T-t}\right).$$

Proof Left as an exercise.

15.4 Credit Ratings

Let $\{Z(t), t \geq 0\}$ be the stochastic process governing the dynamics of credit ratings of a firm. More specifically, $Z(t)$ denotes the rating of the firm at time t. Clearly, it is a discrete-state continuous-time stochastic process and hence can be represented by a state transition diagram, as shown in Fig. 15.1. The state space is $\{AAA, AA, A, BBB, BB, B, C, D\}$ where D denotes the default, an absorbing state.

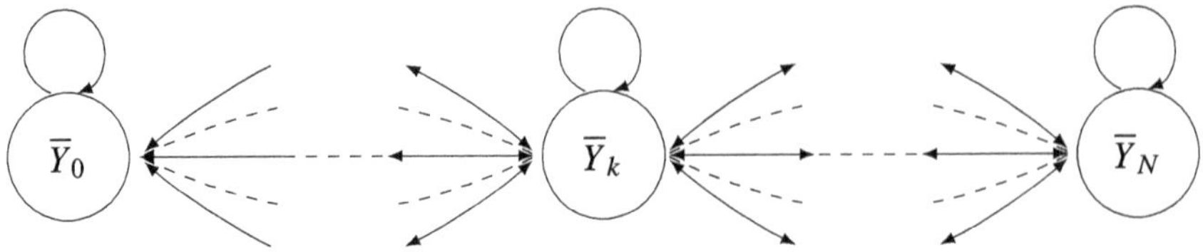

Fig. 15.2 General state transition diagram for discrete-time Markov chain

15.4.1 Credit Rating Model Based On Markov Chain

In this section, we relate the structural credit risk model with a credit rating model. More specifically, consider a financial market with one risky asset and one risk-free asset. The price of the risk-free asset is taken equal to 1 (that is the riskless interest rate is supposed to be equal zero). We assume that the price S of the stock, discounted with respect to the price of the bond, is a stochastic process evolving according to the representation

$$S(t) = S_0 \exp\{Y(t)\}$$

with $t > 0$ and $S_0 > 0$. The stochastic process Y, called the log-return price, is modeled by a discrete-time Markov chain model having an absorbing state, which represents the default.

In the framework of credit risk, to model a defaultable asset, it is sensible to assume a discrete-time Markov chain having a finite number of states. Hence, given a positive integer constant N, the log-return price Y is assumed to be a Markov chain taking values in a finite set $\mathcal{Y} := \{\overline{Y}_0, \overline{Y}_1, \ldots, \overline{Y}_N\}$, such that $\overline{Y}_i \in \mathbb{R}$, for each $i = 0, 1, \ldots, N$, and $\overline{Y}_0 < \ldots < \overline{Y}_N$. The corresponding state transition diagram for a credit rating model such that transition is possible from every state i to every other state j, $i, j \in \{y_1, \ldots, y_N\}$ is shown in Fig. 15.2. The state y_0 is an absorbing state. A transition is possible from each state to the state y_0. Hence, for $p_{0,0} = 1$ and, as usual, $p_{k,0} + p_{k,1} + \cdots + p_{k,N} = 1$, for $k = 1, 2, \ldots, N$.

The default time is the first hitting time to the state labeled by 0, which is the catastrophe of the Markov chain.

Definition 15.1 The mean time to the default is defined as

$$\tau_{k,0} := \min\left\{n \geq 0 : Y_n = \overline{Y}_0 \text{ given } Y_0 = \overline{Y}_k\right\}.$$

Proposition 15.6 *Show that the mean time to the default from each nodes is such that*

$$E[\tau_{k0}] = 1 + \sum_{j=1}^{N} E[\tau_{j0}] p_{kj}, \qquad k = 1, 2, \ldots, N.$$

Proof Since, for $k, j = 1, \ldots, N$,

$$E[\tau_{k0} | Y_1 = \overline{Y}_0] = 1, \quad \text{and} \quad E[\tau_{k0} | Y_1 = \overline{Y}_j] = 1 + E[\tau_{j0}],$$

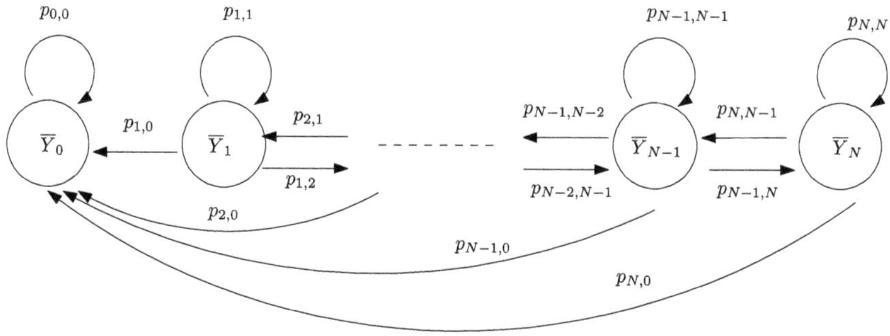

Fig. 15.3 State transition diagram for discrete-time Markov chain

then,

$$E[\tau_{k0}] = \sum_{j=0}^{N} E[\tau_{k0}|Y_1 = \overline{Y}_j]P(Y_1 = \overline{Y}_j|Y_0 = \overline{Y}_k) = 1 + \sum_{j=1}^{N} E[\tau_{j0}]p_{kj}.$$

Summing up, in the general case, for $k, j = 1, \ldots, N$, setting $c_{kk} = p_{kk} - 1$ and $c_{kj} = p_{kj}$, for the following system of N linear equations in N variables

$$\sum_{j=1}^{N} E[\tau_{j0}]c_{kj} = -1, k \in \{1, 2, \ldots, N\}.$$

As well known, this system admits a unique solution, which can be achieved by classical methods.

However, in the framework of credit ratings, it is sensible to assume that the evolution of the system is possible only to neighboring nodes or to default. Therefore, consider such scenario as shown in Fig. 15.3. Let's find the mean time to the default for this scenario.

Proposition 15.7 *Assuming that $p_{kk} \neq 1$, $k = 1, 2, \ldots, N$, the mean time to the default from each state is*

$$E[\tau_{10}] = \frac{1}{1 - p_{11}} + E[\tau_{20}]\frac{p_{12}}{1 - p_{11}}, \tag{15.39}$$

$$E[\tau_{k0}] = \frac{1}{1 - p_{kk}} + E[\tau_{k-1,0}]\frac{p_{k,k-1}}{1 - p_{kk}} + E[\tau_{k+1,0}]\frac{p_{k,k+1}}{1 - p_{kk}}, \quad k = 2, 3, \ldots, N-1, \tag{15.40}$$

$$E[\tau_{N0}] = \frac{1}{1 - p_{NN}} + E[\tau_{N-1,0}]\frac{p_{N,N-1}}{1 - p_{NN}}. \tag{15.41}$$

Proof The conditional mean value is such that

$$E[\tau_{k0}|Y_1 = \overline{Y}_0] = 1 \qquad k = 1, 2, \ldots, N, \tag{15.42}$$

$$E[\tau_{k0}|Y_1 = \overline{Y}_k] = 1 + E[\tau_{k,0}] \qquad k = 0, 1, \ldots, N, \tag{15.43}$$

$$E[\tau_{k0}|Y_1 = \overline{Y}_{k-1}] = 1 + E[\tau_{k-1,0}] \qquad k = 2, 3, \ldots, N, \tag{15.44}$$

$$E[\tau_{k0}|Y_1 = \overline{Y}_{k+1}] = 1 + E[\tau_{k+1,0}] \qquad k = 1, 2, \ldots, N-1. \tag{15.45}$$

Hence,

$$\begin{aligned} E[\tau_{10}] &= E[\tau_{10}|Y_1 = \overline{Y}_0]P(Y_1 = \overline{Y}_0|Y_0 = \overline{Y}_1) \\ &\quad + E[\tau_{10}|Y_1 = \overline{Y}_1]P(Y_1 = \overline{Y}_1|Y_0 = \overline{Y}_1) \\ &\quad + E[\tau_{10}|Y_1 = \overline{Y}_2]P(Y_1 = \overline{Y}_2|Y_0 = \overline{Y}_1) \\ &= p_{10} + (1 + E[\tau_{10}])p_{11} + (1 + E[\tau_{20}])p_{12} = 1 + E[\tau_{10}]p_{11} + E[\tau_{20}]p_{12}. \end{aligned}$$

For $k = 2, 3, \ldots, N-1$, taking into account (15.42) (15.43) (15.44), and (15.45),

$$\begin{aligned} E[\tau_{k0}] &= E[\tau_{k0}|Y_1 = \overline{Y}_0]P(Y_1 = \overline{Y}_0|Y_0 = \overline{Y}_k) \\ &\quad + E[\tau_{k0}|Y_1 = \overline{Y}_k]P(Y_1 = \overline{Y}_k|Y_0 = \overline{Y}_k) \\ &\quad + E[\tau_{k0}|Y_1 = \overline{Y}_{k-1}]P(Y_1 = \overline{Y}_{k-1}|Y_0 = \overline{Y}_k) \\ &\quad + E[\tau_{k0}|Y_1 = \overline{Y}_{k+1}]P(Y_1 = \overline{Y}_{k+1}|Y_0 = \overline{Y}_k) \\ &= 1 + E[\tau_{k,0}]p_{k,k} + E[\tau_{k-1,0}]p_{k,k-1} + E[\tau_{k+1,0}]p_{k,k+1}, \end{aligned}$$

and, analogously,

$$E[\tau_{N0}] = 1 + E[\tau_{N,0}]p_{N,N} + E[\tau_{N-1,0}]p_{N,N-1}.$$

Proposition 15.8 *For the mean time to the default from each states, the following recursive relations holds*

$$E[\tau_{N0}] = \frac{1}{1 - p_{NN}} + E[\tau_{N-1,0}]\frac{p_{N,N-1}}{1 - p_{NN}} \tag{15.46}$$

and, for $k = 1, 2, \ldots, N-1$,

$$E[\tau_{k0}] = a_k + b_k E[\tau_{k+1,0}] \tag{15.47}$$

where $a_0 = 0$, $b_0 = 0$, *and*

$$a_k = \frac{1 + a_{k-1}p_{k,k-1}}{1 - p_{kk} - b_{k-1}p_{k,k-1}} \quad \text{and} \quad b_k = \frac{p_{k,k+1}}{1 - p_{kk} - b_{k-1}p_{k,k-1}}. \tag{15.48}$$

Proof By (15.39), (15.40), and (15.41), setting $a_1 = \frac{1}{1-p_{11}}$ and $b_1 = \frac{p_{12}}{1-p_{11}}$, then

$$E[\tau_{10}] = a_1 + E[\tau_{20}]b_1.$$

For $k = 2$, successively,

$$\begin{aligned}E[\tau_{20}](1 - p_{22}) &= 1 + E[\tau_{10}]p_{21} + E[\tau_{30}]p_{23}\\ &= 1 + (a_1 + E[\tau_{20}]b_1)p_{21} + E[\tau_{30}]p_{23}\\ &= 1 + a_1 p_{21} + E[\tau_{20}]b_1 p_{21} + E[\tau_{30}]p_{23}.\end{aligned}$$

This in turn implies that $E[\tau_{20}] = a_2 + E[\tau_{30}]b_2$, setting

$$a_2 = \frac{1 + a_1 p_{21}}{1 - p_{22} - b_1 p_{21}} \quad \text{and} \quad b_2 = \frac{p_{23}}{1 - p_{22} - b_1 p_{21}}.$$

Again, for $k = 3$, successively

$$\begin{aligned}E[\tau_{30}](1 - p_{33}) &= 1 + E[\tau_{20}]p_{32} + E[\tau_{40}]p_{34}\\ &= 1 + (a_2 + E[\tau_{30}]b_2)p_{32} + E[\tau_{40}]p_{34}.\end{aligned}$$

Hence, $E[\tau_{30}] = a_3 + E[\tau_{40}]b_3$, where

$$a_3 = \frac{1 + a_2 p_{32}}{1 - p_{33} - b_2 p_{32}} \qquad \text{and} \qquad b_3 = \frac{p_{34}}{1 - p_{33} - b_2 p_{32}}.$$

By induction, for $k = 2, 3, \ldots, N - 1$, assuming that

$$E[\tau_{k-1,0}] = a_{k-1} + b_{k-1}E[\tau_{k,0}],$$

then

$$E[\tau_{k0}](1 - p_{kk} - b_{k-1}p_{k,k-1}) = 1 + a_{k-1}p_{k,k-1} + E[\tau_{k+1,0}]p_{k,k+1}.$$

Proposition 15.9 *For the mean time to the default of each state, by all the previous results,*

$$\begin{aligned}E[\tau_{N,0}] &= \frac{1 + a_{N-1}p_{N,N-1}}{1 - p_{NN} - b_{N-1}p_{N,N-1}},\\ E[\tau_{N-1,0}] &= \frac{a_{N-1}(1 - p_{NN}) + b_{N-1}}{1 - p_{NN} - b_{N-1}p_{N,N-1}}\end{aligned}$$

and, for $k = 1, 2, \ldots, N - 2$,

$$E[\tau_{k,0}] = \sum_{j=2}^{N-k} a_{N-j} \prod_{i=j+1}^{N-k} b_{N-i} + \frac{a_{N-1}(1 - p_{NN}) + b_{N-1}}{1 - p_{NN} - b_{N-1}p_{N,N-1}} \prod_{j=2}^{N-k} b_{N-j}.$$

Proof Recalling (15.46), (15.47), and (15.48), for $k = N - 1$,

$$E[\tau_{N-1,0}] = a_{N-1} + b_{N-1}E[\tau_{N,0}],$$
$$E[\tau_{N,0}] = \frac{1}{1-p_{NN}} + E[\tau_{N-1,0}]\frac{p_{N,N-1}}{1-p_{NN}}.$$

The above two equations form a system of two equations in two variables with unknowns $E[\tau_{N-1,0}]$ and $E[\tau_{N,0}]$. Solving this system of equation, $E[\tau_{N0}]$ and $E[\tau_{N-1,0}]$.

Therefore, by (15.47), successively, substituting,

$$\begin{aligned} E[\tau_{N-2,0}] &= a_{N-2} + b_{N-2}E[\tau_{N-1,0}] \\ &= a_{N-2} + b_{N-2}\frac{a_{N-1}(1-p_{NN}) + b_{N-1}}{1-p_{NN} - b_{N-1}p_{N,N-1}}, \end{aligned}$$

and analogously,

$$E[\tau_{N-3,0}] = a_{N-3} + b_{N-3}a_{N-2} + b_{N-3}b_{N-2}\frac{a_{N-1}(1-p_{NN}) + b_{N-1}}{1-p_{NN} - b_{N-1}p_{N,N-1}}.$$

Thus, in general, for $h = 2, 3, \ldots, N-1$,

$$E[\tau_{N-h,0}] = \sum_{j=2}^{h} a_{N-j} \prod_{i=j+1}^{h} b_{N-i} + \frac{a_{N-1}(1-p_{NN}) + b_{N-1}}{1-p_{NN} - b_{N-1}p_{N,N-1}} \prod_{j=2}^{h} b_{N-j},$$

which, setting $k = N - h$, allows us to achieve the required result. For $k = 1, 2, \ldots, N-2$, by induction, recalling that $E[\tau_{k0}] = a_k + b_k E[\tau_{k+1,0}]$, the result follows.

15.4.2 Credit Rating Model Based on Markov Regenerative Process

Markov regenerative process is discussed in Sect. 9.1.2.2 in details. A Markov regenerative process is a non-Markov process which allows a) a non-exponential inter-arrival distribution; b) a transition from one state to another state between two regeneration epochs. Since there is a rating momentum, as described above, in the dynamics of credit ratings, one can view it as that the underlying process $\{Z(t), t \geq 0\}$ satisfies the Markov property only at the time instants when it gets a better rating than the previous one whereas the time epoch of a downgrade cannot be considered as a regeneration time epoch. As a result, the regeneration instances exactly correspond to times of entering state i from a state j such that $j < i$. Equivalently, the rating of a firm can change between two regenerations (i.e., upgrades) because of downgrades. Consider the sequence of epochs $\{T_n, n = 0, 1, \ldots\}$ at which the process $\{Z(t), t \geq 0\}$ gets a better rating than the previous one. The state of the process can change between T_n and T_{n+1} because of movement downwards. Hence, observe that $\{Z(t), t \geq 0\}$ is a

Markov regenerative process (MRGP) and the sequence $(Z(T_n^+), T_n)$ is a Markov renewal sequence (MRS). In order to determine the transition probability matrix, determine the global and local kernel, which can be obtained as follows:

Global Kernel: Let p_{ij} be the one-step transition probabilities for the embedded Markov chain. Then, for $i, j \in S'$

$$p_{ij} = \lim_{t\to\infty} k_{ij}(t).$$

Let $F_{ij}(t); i \in \{1, 2, \ldots, 7\}$ and $j \in S$ represent the distribution of migrating from state i to state j. It basically represents the distribution function of the waiting time in each state i, given that next state j is known.

$$F_{ij}(t) = P\{T_{n+1} - T_n \le t | X(T_n) = i, X(T_{n+1}) = j\},\ t \ge 0.$$

The global kernel $K(t) = [k_{ij}(t)]_{i,j\in S'}$ can be obtained as follows:

$$k_{ij}(t) = P(Z(T_1) = j; T_1 \le t \mid Z(0) = i) = \begin{cases} F_{ij}(t)p_{ij} & if\ \ p_{ij} \neq 0 \\ 0 & if\ \ p_{ij} = 0 \end{cases}.$$

Local Kernel: The local kernel $E(t) = [e_{ij}(t)]_{i\in S', j\in S}$. Now, determine $e_{ij}(t),\ i \in S'$ and $j \in S$, the probability of migrating from i to j without any regeneration (upward movement). Clearly,

$$e_{ij}(t) = 0\ \ if\ \ i > j,\ t \ge 0,\ \ i \in S', j \in S$$

since a transition from i to j when $i > j$, i.e., from a lower rating category to a higher rating category, is not possible without an upgrade (regeneration). For a transition from i to j when $i \le j$, let $S^{(i)}$ for each $i \in S'$ be the set of all states reachable from state i by downward movement that is $S^{(i)} = \{i, i+1, ..., 8\}$. Let $\{M^{(i)}(t); t \ge 0\}, i \in S'$ be the subordinated semi-Markov process (since sojourn time in each state has general distribution in which there are only downward movements from state i but no upward movement. Here, by choosing parameters such that the probability of downward movement is high, the problem of rating momentum can be addressed. The transition probabilities of this semi-Markov process for each $i \in S'$ will give us the local kernel of MRGP.

Given that the system is in state i at time 0, the transition probabilities of subordinated SMP's can be obtained by solving SMP say $\{M^{(i)}(t), t \ge 0\}$ with state space $S^{(i)}$ and with only downward movement possible from state j to state k $(k > j), j, k \in S^{(i)} = \{i, i+1, \ldots, 8\}$. Let $(Y_n^{(i)}, S_n^{(i)})$ be the Markov renewal sequence where $Y_n^{(i)}$ represents the state at the nth transition and $S_n^{(i)}, n \ge 1$ with state space equal to $\mathbb{R}^+$ represents the time of nth transition. The kernel $Q^{(i)}(t) = [Q_{jk}^{(i)}(t)],\ \ j, k \in S^{(i)}$ of the subordinated SMP is defined by

$$Q_{jk}^{(i)}(t) = P(Y_{n+1}^{(i)} = k; S_{n+1}^{(i)} - S_n^{(i)} \le t \mid Y_n^{(i)} = j),\ \ j, k \in S^{(i)}$$

and it follows that

$$q_{jk}^{(i)} = \lim_{t\to\infty} Q_{jk}^{(i)}(t),\ \ j,k \in S^{(i)}$$

are the transition probabilities of the embedded Markov chain in the process. Therefore, for $j, k \in S^{(i)}$

$$G_{jk}^{(i)}(t) = P(S_{n+1}^{(i)} - S_n^{(i)} \le t \mid Y_n^{(i)} = j, Y_{n+1}^{(i)} = k) = \begin{cases} \frac{Q_{jk}^{(i)}(t)}{q_{jk}^{(i)}} & \text{if } q_{jk}^{(i)} \neq 0 \\ 1 & \text{if } q_{jk}^{(i)} = 0 \end{cases}.$$

Clearly, $G_{jk}^{(i)}(t)$ is same as $F_{jk}(t)$ defined in Sect. 8.4. Furthermore, the probability that the process will leave state j before time t is given by

$$H_j^{(i)}(t) = P(S_{n+1}^{(i)} - S_n^{(i)} \le t \mid Y_n^{(i)} = j),\ \ j \in S^{(i)}.$$

Since

$$H_j^{(i)}(t) = \sum_{k=j}^{8} Q_{jk}^{(i)}(t).$$

Now, the transition probabilities of the SMP $\{M^{(i)}(t), t \ge 0\}$ are defined by

$$\phi_{jk}^{(i)}(t) = P(M^{(i)}(t) = k \mid M^{(i)}(0) = j),\ \ j,k \in S^{(i)}$$

and are obtained by solving the renewal equation

$$\phi_{jk}^{(i)}(t) = \delta_{jk}(1 - H_j^{(i)}(t)) + \sum_{\beta=j}^{k} \int_0^t \phi_{\beta k}^{(i)}(t-y)dQ_{j\beta}^{(i)}(y).$$

It is observed that

$$\phi_{jk}^{(i)}(t) = 0 \ \text{ if } j > k.$$

In the matrix form, it can be written as

$$\phi^{(i)}(t) = H^{(i)}(t) + Q^{(i)}(t) * \phi^{(i)}(t)$$

where $*$ denotes the convolution, $H^{(i)}(t)$ is the diagonal matrix with jth diagonal entry $1 - H_j^{(i)}(t)$. $\phi^{(i)}(t)$ will be an upper triangular matrix.

Therefore, for given initial state $i \in S'$, the local kernel of MRGP $e_{ij}(t)\ \forall\ j \ge i$ can be obtained as follows:

$$\begin{aligned}e_{ij}(t) &= P\{Z(t) = j, T_1 > t | Z_0 = i\} \\ &= P\{Z(t) = j | T_1 > t, Z_0 = i\} P\{T_1 > t | Z_0 = i\} \\ &= \phi_{ij}^{(i)}(t) \times (1 - P(T_1 \leq t | Z_0 = i)) \\ &= \phi_{ij}^{(i)}(t) \times (1 - \sum_{k \in S'} k_{ik}(t)).\end{aligned}$$

For each $i \in S^{'}, e_{ij}(t), j \in S$ of the MRGP describes the behavior of rating evolution between two regeneration epochs as to how the rating moves to a lower rating before going to upper rating and is given by

$$e_{ij}(t) = \begin{cases} 0 & if\ i > j \\ \phi_{ij}^{(i)}(t) \times (1 - \sum_{k \in S'} k_{ik}(t)) & if\ i \leq j \end{cases}.$$

Given $E(t)$ and $K(t)$, it can be obtained the transition probability matrix for $Z(t)$ as

$$V_{ij}(t) = e_{ij}(t) + \sum_{\gamma \in S'} \int_0^t V_{\gamma j}(t - y) dk_{i\gamma}(y); i \in S^{'}, j \in S.$$

For more description, readers may refer to [1–4].

15.5 Exercise Problems

1. Find the probability of default of a firm at time $T = 1$, assuming that the default is modeled as a first jump time of a Poisson process with intensity $\lambda = 2$.
2. Explain how the default of a firm is modeled in Black–Scholes–Merton's model. Further, discuss the relation between total value of the firm V, the default barrier (debt of the firm) L and European option on the total value of the firm.
3. Consider a firm whose total value follows a geometric Brownian motion with drift $\mu = 0.12$ and volatility $\sigma_V = 0.25$. Assume that the current value of the firm is $V_0 = 220$ and the debt that firm needs to pay at maturity time $T = 1$ year is 110. The risk-free rate is 5% per annum. Find the probability of survival of the firm at time T when it can default only at the maturity.
4. Repeat the above exercise assuming that firm can default at any time between $t = 0$ and T.
5. The current value of the firm is \$4 billion and the volatility is 60%. The debt that firm has to pay in 2 years is \$15 billion. The risk-free interest rate is 5.5% per annum. Find the probability of default using Merton's model.
6. Assume that $V_0 > L$ and $\mu > 0$ in Black–Scholes–Merton model. Show that the default probability at the maturity date T is an increasing function of the volatility σ_V.

7. In Theorem 15.1, solve the differential (15.14) corresponding to $B(t)$ and verify that the solution is given by (15.10).
8. Fix $n = 1$. Assume that the dynamics of the values of a firm $V_{n,t}$ is given in SDE (15.22). Prove that the solution to this SDE is given by (15.25).
9. Prove the Proposition 15.5.
10. Consider Example 15.4.2 and derive the transition probability matrix for a semi-Markov process based credit rating model.

References

1. Dharmaraja, S., Pasricha, P., and Tardelli, P.: Hedging and utility valuation of a defaultable claim driven by Hawkes processes, Applied Stochastic Models in Business and Industry **38**(2), 334–352 (2022). https://onlinelibrary.wiley.com/doi/abs/10.1002/asmb.2663
2. Durrett, R.: Essentials of Stochastic Processes. 2nd edition, Springer, New York (2012)
3. Glosten, L.R., Jagannathan, R., and Runkle, D.E.: On the relation between the expected value and the volatility of the nominal excess return on stocks. The Journal of Finance **48**(5), 1779–1801 (1993)
4. Hawkes, A.G.: Hawkes Processes and Their Applications to Finance: a Review. Quantitative Finance **18**(2), 193–198 (2018)
5. Jarrow, R.A., Turnbull, S.M.: Pricing derivatives on financial securities subject to credit risk. J. Financ. **1**, 53–85 (1995)

Appendix A

A.1 Modes of Convergence

1. **Convergence in Probability**
   Let $X_1, X_2, \ldots$ be a sequence of RVs and X be another RV. Then X_n is said to converge in probability to the RV X (denoted as $X_n \xrightarrow{p} X$) iff, for every $\epsilon > 0$,

$$P\left(|X_n - X| > \epsilon\right) \to 0 \quad \text{as } n \to \infty.$$

2. **Convergence in Distribution**
   Let us define $C(F_X)$ as the continuity set of F_X where F_X is the distribution function of RV X. Let $X_1, X_2, \ldots$ be a sequence of RVs and X be another RV. Then X_n is said to converge in distribution to the RV X (denoted as $X_n \xrightarrow{d} X$) IFF

$$F_{X_n}(x) \to F_X(x) \quad \text{as } n \to \infty, \ \forall \ x \in C(F_X).$$

   Convergence in distribution (weak convergence, convergence in Law) is defined as pointwise convergence of the CDF at all values of x except those at which $F(x)$ is discontinuous. Of course if the limiting distribution is absolutely continuous (for example the normal distribution as in the Central Limit Theorem), then $F_n(x) \to F(x)$ does hold for all values of x.
3. **Convergence in r-order moments**
   Let $X_1, X_2, \ldots$ be a sequence of RVs and X be another RV. Then X_n is said to converge in r-order moments (convergence in L^r) to the RV X (denoted as $X_n \xrightarrow{r} X$) IFF

$$E\left(|X_n - X|^r\right) \to 0 \quad \text{as } n \to \infty.$$

4. **Almost Sure Convergence**
   Let $X_1, X_2, \ldots$ be a sequence of RVs and X be another RV. Then X_n is said to converge almost surely to the RV X (denoted as $X_n \xrightarrow{a.s.} X$) IFF

$$P\left(\{\omega : X_n(\omega) \to X(\omega) \text{ as } n \to \infty\}\right) = 1.$$

D. Selvamuthu, *Introduction to Stochastic Processes*, University Texts in the Mathematical Sciences, https://doi.org/10.1007/978-981-97-6152-4

A.2 Some Important Results on Probability

1. **Monotone Convergence Theorem** Suppose that $0 \leq f_1 \leq f_2 \leq \dots$ is a monotonically increasing sequence of non-negative measurable functions on R^n, and let $f(x) = \lim_{n\to\infty} f_n(x)$ (which may $= \infty$ for some x). Then

$$\int f = \lim_{n\to\infty} \int f_n.$$

2. **Fatou's Lemma** Suppose $\{f_n, n = 1, 2, \dots\}$ is a sequence of non-negative measurable functions. Let $f(x) = \lim_{n\to\infty} \inf f_n(x)$. Then

$$\int f \leq \lim_{n\to\infty} \inf \int f_n.$$

3. **The Dominated Convergence Theorem** Let $\{f_n \in L^1 : n \in N\}$ be a sequence of functions such that (a) $f_n \to f$ almost everywhere and
(b) there exists a non-negative $g \in L^1$ such that $|f_n| \leq g$ almost everywhere for all $n \in N$. Then,

$$f \in L^1 \text{ and } \int f = \lim_{n\to\infty} \int f_n.$$

4. **Continuity Theorem** Let $\{X_n, n = 1, 2, \dots\}$ and X have MGF $\{M_{X_n}(t), n = 1, 2, \dots\}$ and $M_X(t)$ with CDF's $F_{X_n}(x)$ and $F(x)$ respectively. Then a necessary and sufficient condition for $\lim_{n\to\infty} F_{X_n}(X) = F(X)$ is that for every t,

$$\lim_{n\to\infty} M_{X_n}(t) = M_X(t), \text{ where } M_X(t) \text{ is continuous at } t = 0.$$

5. **The Weak Law of Large Numbers (WLLN)** Let $X_1, X_2, \dots$ be a sequence of i.i.d. r.v.s with mean μ and finite variance σ^2. Then, for any $\epsilon > 0$, we have that

$$\lim_{n\to\infty} P\left(\left|\frac{X_1 + \dots + X_n}{n} - \mu\right| > \epsilon\right) = 0.$$

6. **The Strong Law of Large Numbers (SLLN)** Let $X_1, X_2, \dots$ be a sequence of i.i.d. r.v.s with finite mean μ and finite variance σ^2. Then:

$$\overline{X_n} = \frac{X_1 + X_2 + \dots + X_n}{n} \xrightarrow[n\to\infty]{a.s.} \mu.$$

7. **Central Limit Theorem** Let $X_1, X_2, \ldots$ be a sequence of i.i.d. r.v.s with mean μ and finite variance σ^2. Let $S_n := \sum_{j=1}^{n} X_j$ and $Y_n := \frac{S_n - n\mu}{\sigma\sqrt{n}}$. Then, the sequence of r.v.s $Y_1, Y_2, \ldots$ converges in distribution to a r.v. $Z \sim N(0, 1)$. In other words:

$$\lim_{n\to\infty} P\left(\frac{S_n - n\mu}{\sigma\sqrt{n}} \leq x\right) = \Phi(x) \quad \text{for all } x \in \mathbb{R}$$

where $\Phi(.)$ denotes the CDF of $N(0, 1)$.

A.3 Standard Distributions with Their Moments

See Tables A.1 and A.2.

Table A.1 Discrete probability distributions with their moments

Distribution	Parameters	Probability mass function	Mean	Variance
Bernoulli	$0 < p < 1$, $q = 1 - p$	$\begin{cases} p, & x = 1 \\ q, & x = 0 \end{cases}$	p	pq
Binomial	$n = 1, 2, \ldots$, $0 < p < 1$, $q = 1 - p$	$\binom{n}{x} p^x q^{n-x}$, $x = 0, 1, \ldots, n$	np	npq
Poisson	$\lambda > 0$	$e^{-\lambda} \frac{\lambda^x}{x!}, x = 0, 1, \ldots$	λ	λ
Geometric	$0 < p < 1$, $q = 1 - p$	$pq^{x-1}, x = 1, 2, \ldots$	$\frac{1}{p}$	$\frac{q}{p^2}$
Modified geometric	$0 < p < 1$, $q = 1 - p$	$pq^x, x = 0, 1, \ldots$	$\frac{1-p}{p}$	$\frac{q}{p^2}$
Negative binomial	$r > 0$, $0 < p < 1$, $q = 1 - p$	$\binom{x+r-1}{x} p^r q^x, x = 0, 1, \ldots$	$\frac{rq}{p}$	$\frac{rq}{p^2}$
Hypergeometric	$N = 1, 2, \ldots$, $n = 1, 2, \ldots, N$ $p = 0, \frac{1}{N}, \ldots, 1$	$\frac{\binom{Np}{x}\binom{Nq}{n-x}}{\binom{N}{n}}, x = 0, 1, 2, \ldots$ $q = 1 - p$	np	$npq \frac{N-n}{N-1}$
Discrete Uniform	$a, b \in \mathbb{Z}$, $a \leq b$ $N = b - a + 1$	$\frac{1}{N}$, $x = a, a+1, \ldots, b-1, b$	$\frac{a+b}{2}$	$\frac{N^2-1}{12}$

Table A.2 Continuous distributions with their moments

Distribution	Parameters	Probability density function	Mean	Variance
Uniform over interval a to b	$-\infty < a < b < \infty$	$\frac{1}{b-a}, a < x < b$	$\frac{a+b}{2}$	$\frac{(b-a)^2}{12}$
Normal	$-\infty < \mu < \infty$, $\sigma > 0$	$\frac{1}{\sigma\sqrt{2\pi}} e^{-\frac{1}{2}\left(\frac{x-\mu}{\sigma}\right)^2}$	μ	σ^2
Exponential	$\lambda > 0$	$\lambda e^{-\lambda x}, x > 0$	$\frac{1}{\lambda}$	$\frac{1}{\lambda^2}$
Gamma	$r > 0, \lambda > 0$	$\frac{\lambda}{\Gamma(r)} (\lambda x)^{r-1} e^{-\lambda x}, x > 0$	$\frac{r}{\lambda}$	$\frac{r}{\lambda^2}$

A.4 CDF of Standard Normal Distribution

$$\Phi(x) = \int_{-\infty}^{x} \frac{1}{\sqrt{2\pi}} e^{-t^2/2}\, dt, \quad \Phi(-x) = 1 - \Phi(x).$$

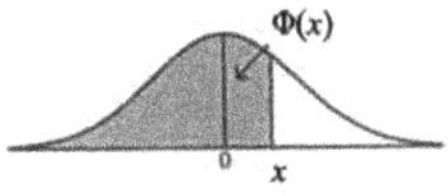

x	0.00	0.01	0.02	0.03	0.04	0.05	0.06	0.07	0.08	0.09
0.0	0.5000	0.5040	0.5080	0.5120	0.5160	0.5199	0.5239	0.5279	0.5319	0.5359
0.1	0.5398	0.5438	0.5478	0.5517	0.5557	0.5596	0.5636	0.5675	0.5714	0.5753
0.2	0.5793	0.5832	0.5871	0.5910	0.5948	0.5987	0.6026	0.6064	0.6103	0.6141
0.3	0.6179	0.6217	0.6255	0.6293	0.6331	0.6368	0.6406	0.6443	0.6480	0.6517
0.4	0.6554	0.6591	0.6628	0.6664	0.6700	0.6736	0.6772	0.6808	0.6844	0.6879
0.5	0.6915	0.6950	0.6985	0.7019	0.7054	0.7088	0.7123	0.7157	0.7190	0.7224
0.6	0.7257	0.7291	0.7324	0.7357	0.7389	0.7422	0.7454	0.7486	0.7517	0.7549
0.7	0.7580	0.7611	0.7642	0.7673	0.7704	0.7734	0.7764	0.7794	0.7823	0.7852
0.8	0.7881	0.7910	0.7939	0.7967	0.7995	0.8023	0.8051	0.8078	0.8106	0.8133
0.9	0.8159	0.8186	0.8212	0.8238	0.8264	0.8289	0.8315	0.8340	0.8365	0.8389
1.0	0.8413	0.8438	0.8461	0.8485	0.8508	0.8531	0.8554	0.8577	0.8599	0.8621
1.1	0.8643	0.8665	0.8686	0.8708	0.8729	0.8749	0.8770	0.8790	0.8810	0.8830
1.2	0.8849	0.8869	0.8888	0.8907	0.8925	0.8944	0.8962	0.8980	0.8997	0.9015
1.3	0.9032	0.9049	0.9066	0.9082	0.9099	0.9115	0.9131	0.9147	0.9162	0.9177
1.4	0.9192	0.9207	0.9222	0.9236	0.9251	0.9265	0.9279	0.9292	0.9306	0.9319
1.5	0.9332	0.9345	0.9357	0.9370	0.9382	0.9394	0.9406	0.9418	0.9429	0.9441
1.6	0.9452	0.9463	0.9474	0.9484	0.9495	0.9505	0.9515	0.9525	0.9535	0.9545
1.7	0.9554	0.9564	0.9573	0.9582	0.9591	0.9599	0.9608	0.9616	0.9625	0.9633
1.8	0.9641	0.9649	0.9656	0.9664	0.9671	0.9678	0.9686	0.9693	0.9699	0.9706
1.9	0.9713	0.9719	0.9726	0.9732	0.9738	0.9744	0.9750	0.9756	0.9761	0.9767
2.0	0.9772	0.9778	0.9783	0.9788	0.9793	0.9798	0.9803	0.9808	0.9812	0.9817
2.1	0.9821	0.9826	0.9830	0.9834	0.9838	0.9842	0.9846	0.9850	0.9854	0.9857
2.2	0.9861	0.9864	0.9868	0.9871	0.9875	0.9878	0.9881	0.9884	0.9887	0.9890
2.3	0.9893	0.9896	0.9898	0.9901	0.9904	0.9906	0.9909	0.9911	0.9913	0.9916
2.4	0.9918	0.9920	0.9922	0.9925	0.9927	0.9929	0.9931	0.9932	0.9934	0.9936
2.5	0.9938	0.9940	0.9941	0.9943	0.9945	0.9946	0.9948	0.9949	0.9951	0.9952
2.6	0.9953	0.9955	0.9956	0.9957	0.9959	0.9960	0.9961	0.9962	0.9963	0.9964
2.7	0.9965	0.9966	0.9967	0.9968	0.9969	0.9970	0.9971	0.9972	0.9973	0.9974
2.8	0.9974	0.9975	0.9976	0.9977	0.9977	0.9978	0.9979	0.9979	0.9980	0.9981
2.9	0.9981	0.9982	0.9982	0.9983	0.9984	0.9984	0.9985	0.9985	0.9986	0.9986
3.0	0.9987	0.9987	0.9987	0.9988	0.9988	0.9989	0.9989	0.9989	0.9990	0.9990
3.1	0.9990	0.9991	0.9991	0.9991	0.9992	0.9992	0.9992	0.9992	0.9993	0.9993
3.2	0.9993	0.9993	0.9994	0.9994	0.9994	0.9994	0.9994	0.9995	0.9995	0.9995
3.3	0.9995	0.9995	0.9995	0.9996	0.9996	0.9996	0.9996	0.9996	0.9996	0.9997
3.4	0.9997	0.9997	0.9997	0.9997	0.9997	0.9997	0.9997	0.9997	0.9997	0.9998

A.5 Chebyshev Polynomials

Chebyshev polynomials form an important class of orthogonal polynomials as they have zeros in closed form. Some important properties of Chebyshev polynomials which occur in the transient solutions of a finite BDP are given below for future reference. Chebyshev polynomials of the first kind are defined by

$$T_n(x) = \cos(n\theta) = \cos(n\cos^{-1} x), \quad n = 0, 1, \ldots. \tag{A.1}$$

Chebyshev polynomials of the second kind $U_n(x)$ and *modified* Chebyshev polynomials of the second kind $V_n(x)$ and $W_n(x)$ are defined as follows:
For $n = 0, 1, \ldots$

$$U_n(\cos(\theta)) = \frac{\sin((n+1)\theta)}{\sin(\theta)}, \tag{A.2}$$

$$V_n(\cos(\theta)) = \frac{\cos\left((n+\frac{1}{2})\theta\right)}{\cos\left(\frac{\theta}{2}\right)}, \tag{A.3}$$

$$W_n(\cos(\theta)) = \frac{\sin\left((n+\frac{1}{2})\theta\right)}{\sin(\frac{\theta}{2})}. \tag{A.4}$$

The significance of these observations is that the properties of sines and cosines can be used to establish many of the properties of Chebyshev polynomials. Chebyshev polynomials have acquired great practical importance in polynomial approximation methods. These polynomials satisfy the recurrence relation

$$U_{n+1}(x) + U_{n-1}(x) = 2xU_n(x), \quad n = 1, 2, \ldots$$

with $U_0(x) = 1$ and $U_1(x) = 2x$. We observe that

$$V_n(x) = U_n(x) - U_{n-1}(x) \text{ and } W_n(x) = U_n(x) + U_{n-1}(x), \quad n = 1, 2, \ldots$$

where $V_0(x) = W_0(x) = 1$. It satisfies the generating function

$$\frac{1}{1-2xt+t^2} = \sum_{n=0}^{\infty} U_n(x)t^n.$$

We can write (A.2) as

$$U_n(x) = \sum_{k=0}^{\lfloor n/2 \rfloor} \binom{n-k}{k} (-1)^k (2x)^{n-2k}.$$

These polynomials $U_n(x)$, $V_n(x)$ and $W_n(x)$ are orthogonal in $(-1, 1)$ with weight functions $\sqrt{1-x^2}$, $\sqrt{\frac{1+x}{1-x}}$ and $\sqrt{\frac{1-x}{1+x}}$ respectively (Szegö [1]).

A.6 Laplace Transforms

One particular transform that is useful in stochastic models is the Laplace transform (LT). It is often easier to solve the equations in terms of a Laplace transform of the function. The resulting solution gives the transform of the function, which must then be inverted back into the original space to give the function of interest.

Let $f(t)$ be a function of the variable t which is defined for all positive values of t. Let s be a complex variable. If the integral $\int_0^\infty e^{-st} f(t)\mathrm{d}t$ exist and is equal to $\widehat{F(s)}$, then $\widehat{F(s)}$ is called the Laplace transform of $f(t)$ and is denoted by the symbol $L[f(t)]$. i.e., $L[f(t)] = \int_0^\infty e^{-st} f(t)\mathrm{d}t = \widehat{F(s)}$. The Laplace transform of $f(t)$ is said to exist if the integral converges for some values of s, otherwise it does not exist.

The sufficient conditions for the existence of the Laplace transforms is as follows. Let $f(t)$ be defined and continuous for all positive values of t. The Laplace transform of $f(t)$ exists if the following conditions are satisfied.

1. $f(t)$ is piecewise continuous or sectionally continuous.
2. $f(t)$ should be of exponential order.

Note that, a function $f(t)$ is said to be of exponential order if $\lim_{t\to\infty} e^{-st} f(t) = 0$.

Here the operator L is called the Laplace transform operator which transforms the functions $f(t)$ into $\widehat{F(s)}$ and $\mathrm{Lim}_{s\to\infty} \widehat{F(s)} = 0$.

Remark A.1 1. A related transform is the Laplace–Stieltjes transform (LST). Let $F(t)$ be a real-valued function define on the interval $0 \le t < \infty$. The LST of $F(t)$ is

$$L^*[f(t)] = F^*(s) = \int_0^\infty e^{-st} \mathrm{d}F(t).$$

The above integral is the Lebesgue–Stieltjes integral (Tables A.3 and A.4).

2. Let X be a continuous type RV with a PDF f, then the Laplace transform of f is given as the expectation of e^{-sX}, which is also referred to as the Laplace transform of the RV X itself.
3. Let $\{F_n(t), n = 1, 2, \ldots\}$ be a sequence of CDF's such that $F_n \to F$. Define $\{F_n(s)\}$ as the sequence of LT such that $L(F_n(t)) = \widehat{F_n(s)}$. Then $\widehat{F_n(s)} \to \widehat{F(s)}$ and conversely if $\widehat{F_n(s)} \to \widehat{F(s)}$, then $F_n(t) \to F(t)$.

The Laplace transformation has several properties, such as:

The following Laplace transform table can help solve differential equations for different functions:

Example A.1 Find the PDF of $X = X_1 + X_2$ where $X_1 \sim Exp(\lambda_1)$ and $X_2 \sim Exp(\lambda_2)$ are independent RVs and $\lambda_1 \neq \lambda_2$.

Solution: Since $X_1 \sim Exp(\lambda_1)$ and $X_2 \sim Exp(\lambda_2)$, the LT of PDF of X_1 and X_2

$$L(f_i(t)) = \frac{\lambda_i}{\lambda_i + s}, \quad i = 1, 2.$$

Now, the LT of PDF of X, denoted by $L(f(t))$ is given by

$$L(f(t)) = L(f_1(t))L(f_2(t)) = \frac{\lambda_1\lambda_2}{(\lambda_1 + s)(\lambda_2 + s)}.$$

Using the partial fraction,

$$L(f(t)) = L(f_1(t))L(f_2(t)) = \frac{\lambda_1\lambda_2}{\lambda_2 - \lambda_1}\left[\frac{1}{s + \lambda_1} - \frac{1}{s + \lambda_2}\right].$$

Taking inverse transforms, the PDF of X is given by

$$f(t) = \frac{\lambda_1\lambda_2}{\lambda_2 - \lambda_1}\left[e^{-\lambda_1 t} - e^{-\lambda_2 t}\right], \quad t \geq 0.$$

Table A.3 Properties in Laplace transform

S. No.	Property	Function	Laplace transform
1	Linearity property	$Af_1(t) + Bf_2(t)$	$A\widehat{F_1(s)} + B\widehat{F_2(s)}$
2	Frequency shifting property	$e^{s_0 t} f(t))$	$\widehat{F(s - s_0)}$
3	Integration	$\int_0^t f(t)\mathrm{d}t$	$1/s\widehat{F(s)}$
4	Multiplication by time	$Tf(t)$	$-\mathrm{d}\widehat{F(s)}\mathrm{d}s$
5	Complex shift property	$f(t)e^{-at}$	$\widehat{F(s + a)}$
6	Time reversal property	$f(-t)$	$\widehat{F(-s)}$
7	Time scaling property	$f(t/a)$	$a\widehat{F(as)}$

Table A.4 Laplace transform of different functions

S. No.	$f(t)$	$L(f(t)) = \widehat{F(s)}$	S. No.	$f(t)$	$L(f(t)) = \widehat{F(s)}$
1	1	$\frac{1}{s}$	11	$e^{(at)}$	$\frac{1}{(s-a)}$
2	t^n at $t = 1, 2, 3, \ldots$	$\frac{n!}{s^{(n+1)}}$	12	t^p, at $p > -1$	$\frac{\Gamma(p+1)}{s^{(p+1)}}$
3	$\sqrt{(t)}$	$\sqrt{\frac{\pi}{2s^{(3/2)}}}$	13	$t^{(\frac{n-1}{2})}$ at $n = 1, 2, \ldots$	$\frac{(1.3.5...(2n-1)\sqrt{\pi})}{(2n_s(n+1/2)}$
4	$\sin(at)$	$\frac{a}{(s^2+a^2)}$	14	$\cos(at)$	$\frac{s}{(s^2+a^2)}$
5	$t\sin(at)$	$\frac{2as}{(s^2+a^2)^2}$	15	$t\cos(at)$	$\frac{(s^2-a^2)}{(s^2+a^2)^2}$
6	$\sin(at + b)$	$\frac{(s\sin(b)+a\cos(b)}{(s^2+a^2)}$	16	$\cos(at + b)$	$\frac{(s\cos(b)-a\sin(b))}{(s^2+a^2)}.$
7	$\sinh(at)$	$\frac{a}{(s^2-a^2)}$	17	$\cosh(at)$	$\frac{s}{(s^2-a^2)}$
8	$e^{(\mathrm{at})}\ \sin(bt)$	$\frac{b}{((s-a)^2+b^2)}$	18	$e^{(at)}\cos(bt)$	$\frac{(s-a)}{((s-a)^2+b^2)}$
9	$e^{(ct)}f(t)$	$F(s - c)$	19	$t^n f(t)$ at $n = 1, 2, 3.$	$(-1)^n F^n s$
10	$f'(t)$	$s\widehat{F(s)} - f(0)$	20	$f''(t)$	$s^2\widehat{F(s)} - sf(0) - f'(0)$

A.7 Modified Bessel Function of the First Kind

Consider the second-order ordinary differential equation

$$x^2 \frac{d^2 y}{dx^2} + x \frac{dy}{dx} - \left(x^2 + n^2\right) y = 0. \tag{A.5}$$

The solutions of the above equation are the modified Bessel functions of the first and second kinds and can be written

$$y = a_1 J_n(-ix) + a_2 Y_n(-ix) \tag{A.6}$$

$$= c_1 I_n(x) + c_2 K_n(x), \tag{A.7}$$

where $J_n(x)$ is a Bessel function of the first kind, $Y_n(x)$ is a Bessel function of the second kind, $I_n(x)$ is a modified Bessel function of the first kind, and $K_n(x)$ is modified Bessel function of the second kind.

The modified Bessel function of the first kind $I_n(x)$ can be defined by the contour integral

$$I_n(x) = \frac{1}{2\pi i} \oint e^{(x/2)(t+1/t)} t^{-n-1} dt,$$

where the contour encloses the origin and is traversed in a counterclockwise direction. In terms of $J_n(x)$,

$$I_n(x) \equiv i^{-n} J_n(ix) = e^{-n\pi i/2} J_n\left(x e^{i\pi/2}\right).$$

For a real number v, the function can be computed using

$$I_v(x) = \left(\frac{1}{2}x\right)^v \sum_{k=0}^{\infty} \frac{\left(\frac{1}{4}x^2\right)^k}{k!\Gamma(v+k+1)},$$

where $\Gamma(z)$ is the gamma function. An integral formula is

$$I_v(x) = \frac{1}{\pi} \int_0^{\pi} e^{x\cos\theta} \cos(v\theta) d\theta - \frac{\sin(v\pi)}{\pi} \int_0^{\infty} e^{-x\cosh t - vt} dt,$$

which simplifies for v an integer n to

$$I_n(x) = \frac{1}{\pi} \int_0^{\pi} e^{x\cos\theta} \cos(n\theta) d\theta.$$

A derivative identity for expressing higher-order modified Bessel functions in terms of $I_0(x)$ is

$$I_n(x) = T_n\left(\frac{\mathrm{d}}{\mathrm{d}x}\right) I_0(x),$$

where $T_n(x)$ is a Chebyshev polynomial of the first kind.

A.8 Continued Fractions

Continued fraction (CF) approximations often provide good representations for transcendental functions that are much more generally valid than the classical representation by power series. A CF is denoted by

$$\cfrac{a_1}{b_1 + \cfrac{a_2}{b_2 + \cfrac{a_3}{b_3 + \dots}}} \quad \text{or} \quad \frac{a_1}{b_1+}\,\frac{a_2}{b_2+}\,\frac{a_3}{b_3+}\cdots \tag{A.8}$$

where a_n's and b_n's are real or complex numbers. This fraction can be terminated by retaining the terms $a_1,\ b_1,\ a_2,\ b_2,\ldots,a_n,\ b_n$ and dropping all the remaining terms $a_{n+1},\ b_{n+1},\ldots$. The number obtained by this operation is called the nth convergent or n^{th} approximate and is denoted by A_n/B_n. Both A_n and B_n satisfy the recurrence relation

$$U_n = a_n U_{n-2} + b_n U_{n-1} \tag{A.9}$$

with initial values $A_0 = 0,\ A_1 = a_1$ and $B_0 = 1,\ B_1 = b_1$.

The CF (A.8) is said to be convergent if at most a finite number of its denominators B_n vanish and the limit of its sequence of approximates, $\lim_{n\to\infty}\frac{A_n}{B_n}$, exists and is finite. Otherwise, the CF is said to be divergent. The value of a CF is defined to be the limit (A.8) of its sequence of approximates. For any CF, the exact value of the fraction lies between any two neighboring convergent. All even numbered convergent lie to the left of the exact value, that is, they give a lower bound approximation to the exact value. All odd numbered convergent lie to the right of the exact value, that is they give a upper bound approximation to the exact value. For a terminating CF, that is, a fraction with finite number of terms, the last convergent coincides with the exact value. If a CF is non-terminating, the sequence of convergent is infinite.

Different types of CFs exist and they are C-fractions, regular C-fractions, S-fractions, g-fractions, associated CFs, J-fractions, T-fractions, general T-fraction and M-fractions.

The CF we have used in this book is the J-fraction, and it is given by

$$\frac{1}{z+b_1-}\,\frac{a_1}{z+b_2-}\,\frac{a_2}{z+b_3-}\cdots.$$

The nth convergent of this fraction is denoted by $\frac{A_n}{B_n}$.

By using the recurrence relation given in (A.9), the nth numerator and denominator polynomials can be represented in a determinant form. The zeros of these polynomials are the eigenvalues of a tri-diagonal matrix and are obtained by using the properties of tri-diagonal determinants. A detailed working procedure of this type of CFs in the study of BDPs is given in Chap. 5.

A.9 Some Identities of Tridiagonal Determinants

Here we give some identities for tridiagonal determinants which are useful in finding the time-dependent solution of the simple Markov models under consideration.

Identity 1

$$\begin{vmatrix} A+a_0 & a_0 & \cdot & \cdots & \cdot & \cdot \\ b_1 & A+a_1+b_1 & a_1 & \cdots & \cdot & \cdot \\ \cdot & b_2 & A+a_2+b_2 & a_2 \cdot & \cdot & \cdot \\ \cdot & \cdot & \cdot & \cdots & \cdot & \cdot \\ \cdot & \cdot & \cdot & \cdot\ \cdot\ b_{n-1} & A+a_{n-1}+b_{n-1} & a_{n-1} \\ \cdot & \cdot & \cdot & \cdots & b_n & A+b_n \end{vmatrix}_{n+1}$$

$$= A \times \begin{vmatrix} g_0 & b_1 & \cdot & \cdot & \cdot & \cdot & \cdot & \cdot \\ a_1 & g_1 & b_2 & \cdot & \cdot & \cdot & \cdot & \cdot \\ \cdot & a_2 & g_2 & b_3 & \cdot & \cdot & \cdot & \cdot \\ \cdot & \cdot & \cdot & \cdot & \cdot & \cdot & \cdot & \cdot \\ \cdot & \cdot & \cdot & \cdot & \cdot & a_{n-2} & g_{n-2} & b_{n-1} \\ \cdot & \cdot & \cdot & \cdot & \cdot & \cdot & a_{n-1} & g_{n-1} \end{vmatrix}_n$$

where $g_j = A + a_j + b_{j+1}, \quad j = 0, 1, \ldots, n-1.$

Identity 2

$$\begin{vmatrix} \theta_1 & d_1 & \cdot & \cdot & \cdot & \cdot & \cdot & \cdot \\ -1 & \phi & d_2 & \cdot & \cdot & \cdot & \cdot & \cdot \\ \cdot & -1 & \theta_2 & d_3 & \cdot & \cdot & \cdot & \cdot \\ \cdot & \cdot & \cdot & \cdot & \cdot & \cdot & \cdot & \cdot \\ \cdot & \cdot & \cdot & \cdot & \cdot & -1 & \theta_m & d_{2m-1} \\ \cdot & \cdot & \cdot & \cdot & \cdot & \cdot & -1 & \phi \end{vmatrix}_{2m} = \begin{vmatrix} \gamma_1 & d_1 & \cdot & \cdot & \cdot & \cdot & \cdot & \cdot \\ d_2 & \gamma_2 & d_3 & \cdot & \cdot & \cdot & \cdot & \cdot \\ \cdot & d_4 & \gamma_3 & d_5 & \cdot & \cdot & \cdot & \cdot \\ \cdot & \cdot & \cdot & \cdot & \cdot & \cdot & \cdot & \cdot \\ \cdot & \cdot & \cdot & \cdot & \cdot & d_{2m-4} & \gamma_{m-1} & d_{2m-3} \\ \cdot & \cdot & \cdot & \cdot & \cdot & \cdot & d_{2m-2} & \gamma_m \end{vmatrix}_m$$

where $\gamma_j = d_{2j-2} + \theta_j \phi + d_{2j-1}, \quad j = 1, 2, \ldots, m$ with $d_0 = 0$.

Identity 3

$$\begin{vmatrix} A & a & \cdot & \cdot & \cdot & \cdot & \cdot & \cdot \\ b & A & a & \cdot & \cdot & \cdot & \cdot & \cdot \\ \cdot & b & A & a & \cdot & \cdot & \cdot & \cdot \\ \cdot & \cdot & \cdot & \cdot & \cdot & \cdot & \cdot & \cdot \\ \cdot & \cdot & \cdot & \cdot & \cdot & b & A & a \\ \cdot & \cdot & \cdot & \cdot & \cdot & \cdot & b & A \end{vmatrix}_N = \prod_{j=1}^{N} \left(A - 2\sqrt{ab}\, \cos\left(\frac{j\pi}{N+1} \right) \right).$$

Reference

1. G. Szegö, in *Orthogonal Polynomials*, vol. 23, 4th edn. (American Mathematical Society Colloquium Publications, Providence, RI, 1975)

Index

D. Selvamuthu, *Introduction to Stochastic Processes*, University Texts in the Mathematical Sciences, https://doi.org/10.1007/978-981-97-6152-4

www.ingramcontent.com/pod-product-compliance
Ingram Content Group UK Ltd.
Pitfield, Milton Keynes, MK11 3LW, UK
UKHW022218051026
15624UKWH00001B/213
* 9 7 8 9 8 1 9 7 6 1 5 1 7 *